Nuclear and Electron Resonance Spectroscopies Applied to Materials Science

MATERIALS RESEARCH SOCIETY SYMPOSIA PROCEEDINGS

Nuclear and Electron Resonance Spectroscopies Applied to Materials Science

Proceedings of the Materials Research Society Annual Meeting,
November 1980, Copley Plaza Hotel, Boston, Massachusetts, U.S.A.

EDITORS:

E.N. Kaufmann
Bell Laboratories, Murray Hill, New Jersey, U.S.A.

and

G.K. Shenoy
Argonne National Laboratories, Argonne, Illinois, U.S.A.

NORTH HOLLAND
NEW YORK · OXFORD

6388 -2437

PHYSICS

Published by:

Elsevier North Holland, Inc.
52 Vanderbilt Avenue, New York, New York 10017

Sole distributors outside USA and Canada:

North-Holland Publishing Company
P.O. Box 103, Amsterdam, The Netherlands

Library of Congress Cataloging in Publication Data

Main entry under title:

Nuclear and electron resonance spectroscopies applied to materials science.
  (Materials Research Society symposia proceedings; v.3 ISSN 0272-9172)

    Bibliography: p.
    Includes index.
    1. Nuclear magnetic resonance spectroscopy—Congresses. 2. Electron
    paramagnetic resonance spectroscopy—Congresses. 3. Materials—Congresses.
    I. Kaufmann, E. N. II. Shenoy, G.K. III. Materials Research Society.
    IV. Series: Materials Research Society. Materials Research Society symposia
    proceedings; v. 3.
QC762.N85   538′.362   80-28496
ISBN 0-444-00597-8

Manufactured in the United States of America

PREVIOUSLY PUBLISHED MRS SYMPOSIA

Phase Transitions 1973
H.K. Henisch, R. Roy, L.E. Cross, editors
Pergamon, New York 1973

Scientific Basis for Nuclear Waste Management-I
G.J. McCarthy, editor
Plenum Press, New York 1979

Conference on In-Situ Composites-III
J.L. Walter, M.F. Gigliotti, B.F. Oliver, H. Bibring, editors
Ginn & Co., Lexington, Massachusetts 1979

Laser Solid Interactions and Laser Processing 1978
S.D. Ferris, H.J. Leamy, J.M. Poate, editors
American Institute of Physics, New York 1979

Scientific Basis for Nuclear Waste Management-II
C.J. Northrup, editor
Plenum Press, New York 1980

Ion Implantation Metallurgy
C.H. Preece and J.K. Hirvonen, editors
The Metallurgical Society--A.I.M.E., New York 1980

Laser and Electron Beam Processing of Materials
C.W. White and P.S. Peercy, editors
Academic Press, New York 1980

Scientific Basis for Nuclear Waste Management-III
J.G. Moore, editor
Plenum Press, New York 1981

CONTENTS

PREFACE

The symposium on Nuclear and Electron Resonance Spectroscopies Applied to Materials Science was held at the annual meeting of the Materials Research Society in Boston the week of November 17, 1980. The four technique groups covered by the Symposium were Nuclear Resonance, Electron Resonance, the Mössbauer Effect, and Spin Precession as they relate to materials research. All static and dynamic properties of a material are reflected in the local environment of its constituent electrons and nuclei. Resonance methods which measure the interaction of electrons and/or nuclei with this environment are, therefore, looking at the material from this internal microscopic viewpoint. The "microview" is valuable and complementary to other methods of characterizing or diagnosing materials and is naturally included in interdisciplinary materials research programs.

As reflected in these proceedings, a wide variety of *both* technique variations *and* materials types were treated in the Symposium. This "two-dimensional" breadth lent a unique quality to the meeting and was particularly apropos to inclusion among the concurrent topical symposia of the Materials Research Society forum.

In the past, single-disciplines as physics, chemistry, biology, etc., have relied heavily on resonance methods which often bear on fundamental aspects of the science itself. In this interdisciplinary Symposium, the contributions ranged from those of a rather fundamental nature to those which are quite pragmatic in their approach to materials diagnosis. The Symposium was the first materials-oriented meeting to bring together specialists in the four (heretofore considered distinct) resonance techniques. The awareness of the common and complementary contributions of the techniques to the broad spectrum of materials topics, which developed during the sessions, added to the vitality of scientific exchange and bolstered the conceptual unity of resonance methods. An animated evening discussion session further confirmed these thoughts.

About 150 participants, including strong international representation, attended the two and one-half day Symposium which consisted of two poster sessions and five plenary sessions comprising 18 invited lectures and 51 contributed posters. This volume contains the corresponding 18 invited papers and 48 of the contributions. An overview of the application of the four resonance methods to materials science can be gleaned from the invited papers which are collected at the beginning of the volume. These are followed by the contributed reports of recent work on specific materials.

The Symposium was financially supported by the Office of Naval Research (L. Cooper), the Defense Advanced Research Projects Agency (R. Reynolds), the United States Bureau of Mines (E. Amey), and also sponsored by the Argonne National Laboratory and Bell Telephone Laboratories. Overall logistical support was provided by the Materials Research Society. A Symposium reception was supported by the New England Nuclear Corp., the North-Holland Publishing Co. (Amsterdam), Unipub. Inc., and the Harwell Mössbauer Group. We are grateful to the scientists who served as our program advisors and to those who were kind enough to act as session chairpersons during the Symposium. Useful discussions on several aspects of the Symposium planning with P. W. Anderson, R. L. Cohen, J. E. Fischer, P. K. Gallagher, A. C. Gossard, A. Gottlieb, A. Heeger, J. K. Hirvonen, C. Kurkjian, G. E. Peterson, J. M. Poate, C. M. Preece, D. A. Shirley, J. G. Stevens, P. Vashishta, W. M. Walsh, R. E. Walstedt and J. Weiss are greatfully acknowledged. Special thanks are due to R. Littauer and to L. Stanion for their help with the manuscripts and to S. Koscielniak of the Elsevier North-Holland Publishing Company for overseeing the publication process. We are particularly grateful to our many diligent and enthusiastic authors and excellent reviewers whose participation has insured the high quality of these proceedings.

E. N. Kaufmann
G. K. Shenoy

SYMPOSIUM ON

NUCLEAR AND ELECTRON RESONANCE SPECTROSCOPIES APPLIED TO MATERIALS SCIENCE

NUCLEAR RESONANCE ELECTRON RESONANCE

MOSSBAUER EFFECT SPIN PRECESSION

Chairpersons

E. N. Kaufmann
G. K. Shenoy

Program Advisors

J. I. Budnick
A. T. Fiory
R. A. Levy
A. G. Redfield
C. P. Slichter
G. D. Watkins

Session Chairpersons

J. H. Brewer
J. I. Budnick
J. W. Corbett
M. L. Good
P. M. Richards

Partially supported by

Office of Naval Research (L. Cooper)
Defense Advanced Research Projects Agency (R. Reynolds)

Special supplementary support from

United States Bureau of Mines (E. Amey)

Nuclear Resonance

Published 1981 by Elsevier North Holland, Inc.
Kaufmann and Shenoy, editors
Nuclear and Electron Resonance Spectroscopies Applied to Materials Science

NUCLEAR MAGNETIC RESONANCE IN ALLOYS

L. H. Bennett
National Bureau of Standards, Metallurgy Division, Washington, D.C. 20234

ABSTRACT

Many papers on NMR in alloys are addressed to NMR
specialists rather than to metal or alloy specialists, and
talk about the "potential" of NMR for application to
alloys. This presentation emphasizes a review of some
useful results of NMR experiments in alloys, including
applications in diffusion, phase diagrams, magnetic
materials, ordering in intermetallic compounds, liquid
alloys and amorphous alloys.

INTRODUCTION

There are three distinct "branches" of nuclear magnetic resonance (NMR).
Roughly speaking, these are: molecular chemistry, nonmetallic solids, and
alloys. The most developed of these from the point of view of commercial,
well-engineered equipment is the first -- molecular chemistry, growing out of
the early days of high-resolution NMR. The application of NMR to chemical
molecules is advanced and sophisticated and its use is widespread, with
multipulse techniques routine, and the operation of the equipment
"black-box". NMR thus has joined infra-red and other spectrocopies as a tool
in chemical manufacturing as well as in research.

The second branch -- non-metallic solids, is rapidly catching up. In
this area, I include new techniques such as NMR imaging (or zeugmatography or
NMR "diffraction") and applications to relaxation processes in polymers and a
host of other such developments. In this branch, as in the first, ^{1}H and
^{13}C resonances predominate.

The third branch -- the application of NMR to the study of alloys, is the
theme of this talk. Here the situation is quite different. Although there
is available high-quality, commercial equipment, most laboratories doing NMR
in alloys have at least partially home-made apparatus, designed to perform
some specialized function. Most papers on NMR in alloys are written for the
NMR specialist, not for the user. Many papers still talk, after about thirty
years, of the "potential" of NMR for application to alloys.

In this talk, I will highlight some particular applications of NMR to
alloys which are important to the material scientist who has interests other
than NMR. But first I want to say a few words about what is, in some sense,
the most important application of NMR in alloys, namely, to the understanding
of the theory of alloys. NMR is most powerful when it tests detailed,
careful, first principle theories of alloys. In my opinion, for a theory to
be finally accepted, it must be able to reproduce the NMR results. For
example, consider an energy-band calculation for an intermetallic compound.
Many such calculations give controversial results for the charge transfer
from atom A to atom B, or vice-versa. Often, there is charge left over, in
the muffin-tin region. How can such a theory be tested? It should be
required to predict NMR measurables such as the Knight shifts, relaxation
times, and quadrupole interactions for both atoms A and B. Such calculations
are feasible but they require more care and cost to perform, including taking
proper account of relativistic effects, spin-polarization of core electrons,

4

and orbital contributions to the charge densities. The ability of NMR to give well-defined measurables, which, though predictable, require more care in the calculations, makes NMR ideal as the ultimate test of any alloy theory.

Turning now from the philosophical point of view to the practical, I will review some specific applications as listed in Table I. The other speakers in this symposium will spell out in greater detail specific applications that those resonators have themselves been so effective in achieving.

TABLE I

Applications of NMR in Alloys

Diffusion in Alloys
Alloy Phase Diagrams
Magnetic Materials
Ordering in Intermetallic Compounds
Liquid Alloys
Amorphous Alloys

DIFFUSION IN METALS AND ALLOYS

One area in which the potential of NMR has been realized, and in which NMR results must be taken seriously by the user, is in the measurement of diffusion. Spin-echo experiments in NMR provide the means for making direct measurements of the self-diffusion coefficient, D. By direct measurement, it is meant that there is some label for the diffusing atoms. In tracer diffusion, some radioactive decay provides such a label. In other diffusion experiments, the mass of the diffusing species is used. In NMR, the nuclear spin is the label. If the diffusion occurs in a magnetic field gradient, then the diffusion of the atoms will produce changes in the Larmor precession frequencies associated with the nuclear spin.

In measurements of self-diffusion by tracer techniques, it is necessary to introduce the concept of a geometrical correlation factor to account for the fact that the atomic jump probabilities depend on the direction of the preceeding jump [1]. For example, consider a tracer atom jumping from position A in Fig. 1 to a vacant position B. After the tracer jumps to position B, the vacancy is then on position A. On its next jump, the tracer atom is not equally likely to jump into any of the four near-neighbor positions, but is more likely to jump back to A, resulting in a correlated motion. The tracer self-diffusion coefficient, D^{tracer}, then equals a correlation factor, f, times the macroscopic self-diffusion coefficient, D^{macro}.

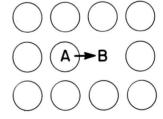

Fig. 1. Schematic two-dimensional square lattice with a vacancy diffusion mechanism to illustrate correlated random walk.

$$D^{tracer} = f \cdot D^{macro} \qquad (1)$$

For a bcc lattice, $f = 0.72$ in the case of a single vacancy jump mechanism.

With increasing theoretical understanding of the details of NMR diffusion experiments [2], the value of D^{macro} can be extracted from them. Most recently, the self-diffusion in Na was determined from the NMR spin-lattice relaxation times T_1 and $T_{1\rho}$ by Brunger, Kanert and Wolf [3] and was used by them in conjunction with tracer diffusion measurements to understand curvature in the Arrhenius plot. The NMR results showed that the correlation factor was temperature dependent, and that both mono- and divacancies contributed to the observed tracer diffusion. In an independent experiment using quasielastic neutron scattering [4], a similar conclusion was reached. The neutron experiments were carried out over a temperature range 323-371 K which did not overlap with the range of 70-280 K accessible to the NMR relaxation measurements.

For higher diffusion rates, the stimulated spin-echo pulsed magnetic-field-gradient technique [5], illustrated schematically for a simple pulse sequence in Fig. 2, gives a direct measurement of diffusion. The amplitude of the echo at 2t is approximately given by

$$M = M_0 \exp \left[-2t/T_2 - \gamma^2 G^2 D \delta^2 (t - \delta/3) \right] \qquad (2)$$

where T_2 is the spin-spin relaxation time, γ is the nuclear gyromagnetic ratio, and the other terms are defined in Fig. 2. By fitting data for M as a function of G, t, and temperature, the value of D and the activation energy for atomic jumps can be obtained. This method has been used by Murday and Cotts to determine the pre-exponential factor D_0, as well as the activation energy for liquid Li [6] and liquid Na [7].

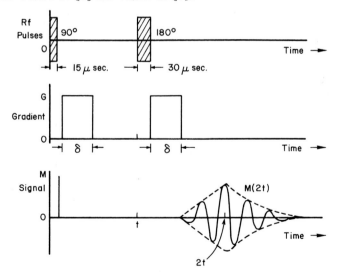

Fig. 2. A simplified representation of the stimulated spin-echo pulsed magnetic-field-gradient method for measuring diffusion coefficients.

In another example of the application of the measurement of the diffusion coefficient by the stimulated echo pulsed-magnetic-field-gradient technique, consider γ-TiH$_x$ (1.5 = x = 2.0) illustrated in Fig. 3. NMR lineshape experiments [8] established that the hydrogen nuclei randomly occupy the tetrahedral positions within the fcc titanium structure. These hydrogen sites form a simple cubic lattice. The conclusion of tetrahedral site occupancy was confirmed by NMR spin-lattice relaxation time (T_1) measurements [9]. For any hydrogen concentration less than TiH$_2$, there is thus a permanent vacancy concentration of $1 - x/2$. NMR results [8,10] have shown that the diffusion rate is directly proportional to the number of hydrogen-site vacancies.

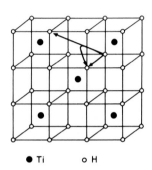

Fig. 3. The structure of the γ-phase of the titanium-hydrogen alloy system. The titanium atoms occupy a fcc lattice. The hydrogen atoms randomly occupy some of the sites shown, with the occupancy depending on the hydrogen concentration. The arrows illustrate an elementary jump to a third near-neighbor site, and two different jump paths to a near-neighbor site.

● Ti   o H

Bustard, Cotts and Seymour [11] provided an interesting example of the determination of the elementary jump step by nuclear magnetic resonance of the protons. The diffusion constant, D, is measured via Eq. 2, and related to the jump length, L, as

$$D = fL^2/6t_d ,\qquad\qquad (3)$$

where f is the tracer correlation factor of Eq. 1, and t_d is the mean time between atomic jumps. Using a detailed Monte Carlo calculation [12], Bustard found that it was possible to relate the time t_d to $(T_1)_d$, the diffusion contribution to the spin-lattice relaxation time, T_1. From this, it was concluded that the hydrogen motion in γ-TiH$_x$ is predominantly by the first nearest neighbor jump mechanism, in agreement with a theoretical calculation of Bisson and Wilson [13], but in contrast to an earlier suggestion that the elementary jump was to the third neighbor site. Fig. 3 shows such a third neighbor jump, as well as two different jump paths to a near neighbor site. The NMR experiments are unable to distinguish between a curved and a straight jump path.

ALLOY PHASE DIAGRAMS

Among the earliest applications of NMR in metals was the determination and the understanding of alloy phase diagrams [14,15]. NMR can be used in two different ways to determine phase boundaries and phase transitions in alloys. First, the NMR parameters, (e.g. linewidth, Knight shift, etc.) may change in a measurable way when the alloy changes phase. The NMR parameter then provides a signature by which to identify each phase. Second, in a two-phase mixture, two NMR signals may be observed. In such a case, quantitative information on the integrated intensity of the two signals can be used to give quantitative information on the positions of the phase boundaries.

In using a technique such as NMR, which depends on sensing atomic nuclei, there is an advantage over the usual lever rule of metallurgy. In the usual lever rule, concentrations at phase boundaries are calculated from the ratio of the phase concentrations in the two-phase region. NMR (or another nuclear technique) measures the proportion of the nuclear species in the two phases. This preferential sense leads to an "enhanced" lever rule [16], which for an $\alpha + \beta$ two-phase region in an A-B alloy can be expressed as

$$R = C_\alpha f_\alpha / C_\beta f_\beta \qquad (4)$$

Here R is the ratio of the intensity of the NMR (or Mossbauer, etc.) signal from the B atoms in the α phase to the intensity from those in the β phase. C_α and C_β are the concentrations of the B atoms in the α and β phases, respectively, and f_α and f_β are the fractions of α-phase and β-phase present, as given by the usual lever rule. When close to the α-phase boundary, the "enhancement" of the signal from the minority β-phase arises from the high concentration of B atoms in the β-phase. An example of the NMR signals obtained in the two-phase region of the pseudo-binary system AuAl$_2$-AuIn$_2$ is shown in Fig. 4. From this single measurement, the solubility of AuIn$_2$ in AuAl$_2$ at the homogenization temperature can be obtained very accurately. The ability of the enhanced lever rule to be used for high precision phase boundary determinations is greatest when the two phase region is widest.

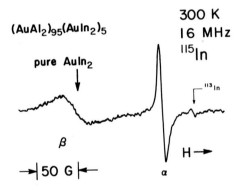

Fig. 4. The ^{115}In NMR absorption derivative spectrum for the two-phase alloy, (AuAl$_2$)$_{95}$(AuIn$_2$)$_5$, showing a distinct resonance for the indium nuclei in each of the two phases. From Bennett and Carter [16].

Measurement of the NMR in alloys at high temperatures is a useful technique which has not yet been fully exploited. Warren [17] has shown that such a measurement can be useful in phase diagram determinations. Although integrated intensities are weaker at high temperature due to the Boltzmann

8

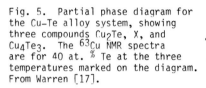

Fig. 5. Partial phase diagram for the Cu-Te alloy system, showing three compounds Cu_2Te, X, and Cu_4Te_3. The ^{63}Cu NMR spectra are for 40 at. % Te at the three temperatures marked on the diagram. From Warren [17].

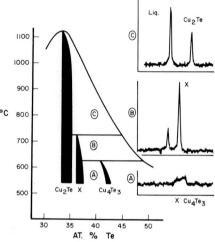

factor, substantially increased signal amplitudes are found due to motional narrrowing by atomic diffusion. Three intermediate phases ("compounds") are shown in the portion of the Cu-Te phase diagram of Fig. 5. The existence of the phase called "X" had been in doubt [18]. The NMR measurement not only confirmed the existence of "X" but, using Eq. (4), found that the composition of the phase at 700°C extended to 38 at. % Te instead of the 36 at. % Te given in Hansen [18].

Utilizing the large electric quadrupole moment of ^{93}Nb, an NMR study by Hwang, Torgeson, and Barnes [19] has revealed the occurrence of a low-temperature pseudo-cubic phase near $NbH_{0.75}$ in which the point symmetry at the Nb sites is axially symmetric. At high temperatures, the rapid hydrogen diffusion washes out the crystalline electric field gradient. At sufficiently low temperatures, the hydrogen motion is effectively frozen, and the hydrogen atoms order on their sublattice. The appearance of quadrupole satellites in the NMR spectra make it possible to measure the separate phases in two-phase regions, even though the two spectra overlap. This sort of NMR experiment is most valuable when part of a program that includes neutron diffraction determinations as well.

A good review of the structure and phase diagrams of V, Nb and Ta with hydrogen and deuterium has been given by Hauck [20]. Of particular interest is the data review for the symmetry of the hydride lattice and the H or D site preference as determined by six different methods: x-ray, neutron, or electron diffraction, inelastic neutron scattering, channelling, and NMR.

Other phase diagram measurements on metal-hydrogen systems have been performed using ^1H or ^2D resonances (e.g.[21]). Here it must be borne in mind that the two isotopes of hydrogen can give different phase diagrams. Since the ^1H nucleus provides the highest sensitivity of any nuclear isotope, many measurements have been and will be made with it. Future applications of proton resonances in metals, using spin-locked apparatus with fast Fourier transforms, are sure to be rewarding for phase diagram studies, as well as for diffusion, structural studies, and for understanding of hydrogen embrittlement and metal hydrides for hydrogen storage.

MAGNETIC MATERIALS

It is difficult to discuss applications of NMR to magnetic materials without some attention to experimental details. In a conventional NMR apparatus utilizing a magnet with a large homogeneous magnetic field, even a small per cent of ferromagnetic precipitate or impurity would severely distort and weaken the resonance. Nonetheless, strong resonances are observed in ferromagnetic materials such as Fe, Co, Ni, Ni_3Fe, Mn_4N, Fe_2B, $GdAl_2$, Heusler alloys, etc. due to the magnetic domain-wall-produced "enhancement" factor. It is traditional to call NMR in ferromagnetic materials ferromagnetic nuclear resonance (FNR). Generally, the experimental apparatus to observe FNR is different from the conventional NMR equipment. No large magnet is needed, the Zeeman splitting of the nuclear states being accomplished by the large effective internal magnetic field at the nuclear site. Applications of FNR to magnetism are wide ranging, and the annual conference on Magnetism and Magnetic Materials always has sessions devoted to it. Applications of FNR to order-disorder, measurements of sublattice magnetization, stacking faults, domain wall dynamics, and precipitation phenomena in magnetic materials have been presented. In disordered alloys, satellite lines due to neighboring solute atoms have been observed and interpreted on the basis of local environments. Combining FNR with EXAFS (extended x-ray absorption fine structure) may be a particularly fruitful means of studying local atomic arrangements in ferromagnetic alloys.

In itinerant ferromagnetic or antiferromagnetic materials, which are generally characterized by relatively small moments and low critical (Curie or Néel) temperatures, T_c, electron spin fluctuations near T_c can be studied by NMR. Most such measurements have been made by conventional NMR above T_c. For example, Borsa and Lecander [22] measured the [11]B nuclear spin-lattice relaxation time, T_1, as a function of temperature in the weak antiferromagnet CrB_2 from room temperature down to 81 K. The Néel temperature for CrB_2 is not accurately known, but thought to lie between 76 K and 85 K. In metals near magnetic phase transitions, the electron-electron interaction produces an enhancement of the relaxation rate T_1^{-1} over the Korringa value (i.e. T_1T = constant). Even in weakly magnetic systems, the effect of spin fluctuations is to produce a hyperbolic or parabolic divergence at T_c, and indeed, it was found [22] that T_1^{-1} had a very sharp maximum in the vicinity of the Néel temperature in CrB_2. The details of the shape of the critical behavior was compared with the predictions of two different theories.

In the itinerant ferromagnet $CrBe_{12}$, the [9]Be resonance has been observed [23] below the Curie temperature (≈ 50 K) in an applied field, using conventional NMR apparatus. The lineshapes and shifts of the [9]Be resonance are illustrated in Figs. 6a and 6b. The Be site is only weakly coupled to the moments, which reside on the Cr atoms. As a result, the resonant frequency of the [9]Be nucleus is determined primarily by the applied steady magnetic field, even in the ferromagnetic state. Well above T_c, the resonance position is determined by the Knight shift, which is proportional to the applied field. The solid line in Fig. 6b, which represents a Curie-Weiss law with a 50 K Curie temperature, is seen to fit the data above 65 K extremely well. Below this temperature, the shifts are not Knight shifts. At 4.2 K the Cr moments induce a small, field dependent, magnetic field at the Be site leading to a ferromagnetic shift which is not proportional to the applied field, but instead follows the magnetization. An extrapolation of the paramagnetic state Knight shift data versus temperature yields a T_c in good agreement with that found by bulk magnetization measurements.

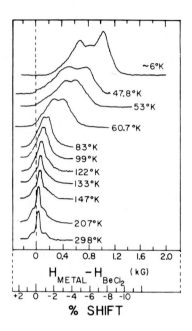

Fig. 6a. The ^{9}Be NMR dispersion derivatives in the itinerant ferromagnet CrBe$_{12}$. The dashed vertical line represents zero Knight or chemical shift. The shifts are true Knight shifts for the spectra from 83 K to 298 K. The shift at the lowest temperature is a magnetization shift. From Wolcott et al. [23].

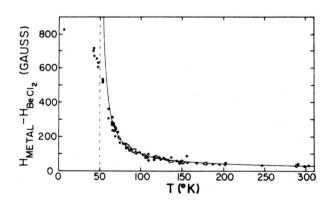

Fig. 6b. Magnetic field shifts for the itinerant ferromagnet CrBe$_{12}$ at 8 MHz vs temperature. These shifts produce negative Knight shifts for temperatures above 65 K. The vertical dashed line marks the Curie temperature. From Wolcott et al. [23].

ORDERING IN INTERMETALLIC COMPOUNDS

NMR is especially useful in investigating details of atomic order in intermetallic compounds. A classic example is NiAl. The intermediate phase based on the equiatomic "compound" NiAl exists over a relatively wide range of composition. On the Ni-rich side, Al atoms are replaced by Ni, but on the Al-rich side Ni vacancies form. From measurements of NMR line intensity versus the deviation from the equiatomic composition, West [24] was able to deduce that the vacancies have a tendency to order. This information is difficult to obtain from x-ray studies. More detailed measurements of the quadrupole interaction in NiAl is possible using the transient NMR quadrupolar echo technique [25]. In this technique, echoes which represent the individual satellite contributions uniquely may be seen at times other than the time 2t at which dipolar echoes form. The satellite echo width is inversely proportional to the satellite contributions to the NMR linewidth. The pulsed echo method is capable of measuring the distribution of electric field gradients directly, because the Fourier transform of the echo is the frequency spectrum of the NMR lineshape. Thus, the quadrupole echo technique is particularly useful in cases where inhomogeneous broadening due to quadrupole interactions dominate the linewidth, which often occurs in concentrated alloys.

Another example of vacancy ordering in an intermediate phase is vanadium monocarbide, which does not exist at 50 at. % C, but is only stable from 39 to 47 at. % C. Knight shift and quadrupole measurements [26] revealed that this "phase" is actually a series of "line compounds" such as V_8C_7 and V_6C_5. The wealth of structural information available from a single NMR experiment is seen in Fig. 7.

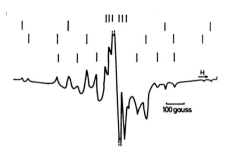

Fig. 7. The ^{51}V NMR absorption derivative in $V_{0.53}C_{0.47}$, with structure arising from vacancy ordering. Four different sets of quadrupole satellites indicate that there are at least four different local environments for the vanadium atoms. From Froidevaux and Rossier [26].

NaTl is an ordered, congruently melting, intermediate phase existing over a range of compositions from about 48 to 53 at. % Tl. Schone and Knight [27] showed by NMR that vacancies appear on the Na sublattice. The ^{205}Tl and ^{203}Tl resonances in NaTl [28] are shown in Fig 8a. Since NaTl is chemically very reactive with water, it was interesting to observe the changes in the NMR spectrum as water vapor was passed over the powdered NaTl. The sequence of Fig. 8 showed that the NaTl lines disappeared and a new phase, identified as $NaTl_2$ appeared. What is probably happening is the formation of NaOH and the depletion of Na from the NaTl structure. The NaTl$_2$ phase decomposes at a peritectic and is difficult to prepare as a pure phase. So far, x-ray and neutron difraction measurements have not been able to unambiquously identify the structure of NaTl$_2$. Other NaTl-type compounds have also been studied by NMR. A defect lattice, with vacancies on the alkali sublattices, has been found [27] for NaTl, LiAl, LiGa and LiIn, whereas the absence of a defect lattice was found [29] for the isostructural compounds LiCd and LiZn. The first group are alloys of alkali metals with group IIIB elements (Al, Ga, In, Tl), while the second are with group IIB (Cd, Zn). Thus a relation between the electronic structure and the defect structure has been suggested [29, 30].

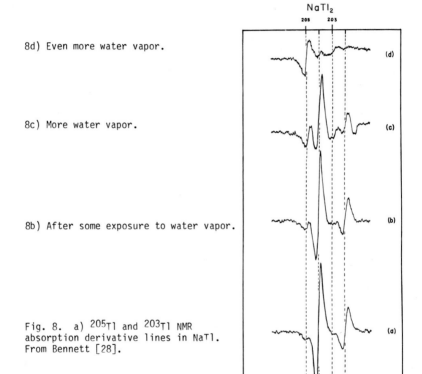

8d) Even more water vapor.

8c) More water vapor.

8b) After some exposure to water vapor.

Fig. 8.   a) ^{205}Tl and ^{203}Tl NMR absorption derivative lines in NaTl. From Bennett [28].

LaNi$_5$, LaPt$_5$, and other rare-earth-Ni$_5$ compounds have the hexagonal structure shown in Fig. 9. Compounds of this structure have attracted considerable attention for technological as well as for scientific reasons, with applications to hard magnets (e.g. SmCo$_5$) and to hydrogen storage. The LaNi$_5$-LaPt$_5$ pseudobinary alloy system was studied by NMR relaxation time and Knight shift measurements and by x-ray diffraction by Weisman et al. [31]. Note that there are two crystallographically inequivalent transition metal sites (Fig. 9). The A sites, holding two of the five transition metal atoms per molecular unit, lie in the basal plane with the lanthanum atoms. The B sites, holding the remaining three, lie between. Two distinct ^{195}Pt shifts are seen for Pt atoms on the A and B sites in $LaPt_5$. Monitoring the relative intensities in the two eesonance lines provides a measure of the relative populations of Pt at A and B sites in $LaNi_{5-x}Pt_x$, and thus it was found possible to measure quantitatively the sublattice ordering of the pseudobinary alloy. The NMR demonstrated that there is a high degree of ordering of Ni on the A sites and of Pt on the B sites. The ordering at $LaNi_2Pt_3$ is more than two-thirds complete, as can be seen in Fig. 10, where it is so strong that it approaches compound formation of a new structural type. Although the fact that there was ordering at this composition could also be inferred from the c/a ratio measured by x-ray diffraction [31], the NMR method was more sensitive to changes in A-site occupation than conventional x-ray methods.

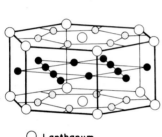

○ Lanthanum
◯ Nickel, Platinum A-site
● Nickel, Platinum B-site

Fig. 9. Atomic structure of $LaNi_5$

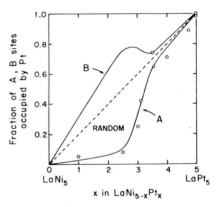

x in $LaNi_{5-x}Pt_x$

Fig. 10. Fraction of A and B sites (defined in Fig. 9) occupied by Pt in $LaNi_5$-$LaPt_5$ pseudobinary alloys, as determined by NMR. From Weisman et al. [31].

In addition to the structural information obtained by NMR on $LaNi_{5-x}Pt_x$, the NMR results were interesting as a model system for an understanding of hydride formation in this class of compounds. $LaNi_5$ is one of the best hydrogen storage alloys per unit volume. $LaPt_5$, on the other hand, stores little or no hydrogen. Since Ni and Pt, being in the same column of the periodic chart, have the same number of valence electrons, and since $LaPt_5$ has larger lattice parameters and larger available interstitial volumes than the isostructural $LaNi_5$, this result seems strange at first. The NMR measurements [31] showed, however, that $LaNi_5$ is more transition-metal-like than $LaPt_5$, with more d-band effects and more exchange enhancement. Thus these detailed electronic structure factors appear to be more important than simple volume effects in determining hydrogen absorption capabilities. It is also interesting to note that the hydrogen absorption capacities of these compounds do not obey the simple Miedema equations.

NMR quadrupole effects, measured as a function of temperature and pressure, have been used by Schirber et al. [32] to demonstrate that ReO_3 undergoes a cubic to a lower symmetry structure at high pressure. The temperature vs. pressure phase boundary was determined from the NMR results.

LIQUID ALLOYS

Quadrupole relaxation rate, R_Q, measurements in liquids, whether by conventional NMR or by measuring the perturbed angular distributions (PAD) of γ-radiation after pulsed excitation of an appropriate isomer, are complementing other structural investigations using x-rays or neutron scattering, to probe the microscopic spatial arrangements and the motion of the atoms [33]. It was found [33] that R_Q in alloys is essentially determined by the valence difference between the alloy components.

The electronic structure of metals such as sodium or aluminum is not very different in the crystalline solid from in the liquid. Thus, for metals such as these, the Knight shift does not change appreciably upon melting. On the other hand, the melting of a semiconductor can involve appreciable change in the electronic structure. For example, Allen and Seymour [34] showed that the ^{115}In resonance frequency in the semiconductor InSb abruptly changes from a zero to a substantial Knight shift on melting. From the behavior of the Knight shift, and of the electrical resistivity and magnetic susceptibility, it was concluded that there is a change to the metallic state in the liquid.

In contrast to liquid InSb which is metallic, liquid CsAu has been shown to be ionic. Dupree et al. [35] have demonstrated by NMR Knight shift and relaxation time measurements that liquid Cs-Au alloys change abruptly with increasing Au concentration from metallic for Cs-rich alloys to "ionic alloys" near the CsAu stoichiometry. Electrical transport properties and magnetic susceptibility measurements also suggest that the electrons introduced by small excess metal concentrations in CsAu are localized. The NMR results [35] give information on the nature of the localized states indicating that they are analogous to F-centers previously proposed for molten metal-halide solutions.

In a similar manner, NMR measurements have confirmed [36] the assumption of a partially salt-like mixture around the composition Li_4Pb in liquid Li-Pb alloys and Li_4Sn in liquid Li-Sn alloys. The behavior of the Knight shift for liquid Li-Pb is seen in Fig. 11. The theoretical understanding of the electronic structure revealed in these measurements are expected to be important in the understanding of amorphous ("glassy") alloys.

Fig. 11. The sharp reduction of the ^7Li Knight shift at Li_4Pb is evidence for a metal-non metal transition in liquid Li-Pb alloys. From van der Marel et al. [36].

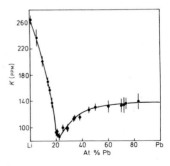

AMORPHOUS ALLOYS

NMR measurements of Ni-P glassy alloys prepared in two different ways revealed [37] that there are at least two different structures for these structureless (= a - morphous) alloys. Fig. 12 shows the Knight shift vs P concentration for Ni-P alloys prepared by electroplating and by a chemical deposition method known as "electroless". This figure illustrates that small composition deviations are not responsible for the different Knight shifts. Annealing experiments suggested [37] that the electroless, higher Knight shift, Ni-P alloy at least partially converted at elevated temperature to the structure of the electroplated, lower Knight shift, Ni-P alloy.

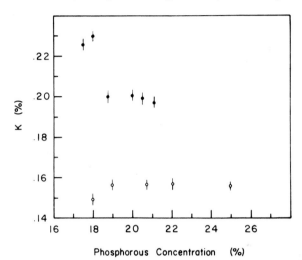

Fig. 12. Knight shift vs phosphorous concentration (in atomic per cent) for electroplated (o) and chemically deposited (•) samples. From Bennett et al. [36].

The NMR linewidths in these glassy Ni-P alloys are magnetic-field dependent, implying inhomogeneities as the origin of the additional widths. The extrapolated zero-field widths are very small, quantitatively in agreement with the model of the structure of these amorphous alloys wherein phosphorous atoms have only nickel neighbors.

NMR measurements have also been reported [38,39] on ^{31}P, ^{195}Pt, and $^{63,65}Cu$ in a number of Ni-Pd-P, Ni-Pt-P, Ni-Cu-P, and Ni-P-B amorphous alloys. The results have been discussed in terms of the usual narrow d-band and broad overlapping conduction band typical of crystalline alloys of transition metals with non-transition metals.

NMR quadrupole studies [40] in the amorphous alloys $La_{75}Ga_{25}$, $Mo_{70}B_{30}$, $Mo_{48}Ru_{32}B_{20}$ and $Ni_{78}P_{14}B_8$ showed that these amorphous structures retain to a significant degree the local site symmetry at the s-p elements of the appropriate crystalline counterpart, cubic La_3Ga, tetragonal Mo_2B and orthorhombic Ni_3B. The local symmetry was found to be independent of the fabrication technique for the amorphous alloy, e.g. by sputtering or by quenching from the liquid.

CONCLUDING REMARKS

In this review, I have tried to give some insight to those who are not expert in NMR for some of the ways in which NMR in alloys has been applied to problems of importance to the metallurgist and the alloy physicist. In almost every example given, the NMR results were crucial to a full understanding of the alloy problem. In many cases, the NMR results were unique. Details of the wide variety of experimental techniques used in NMR [41] have generally been omitted from this review. Also omitted are references to a number of useful reviews and textbooks on NMR in alloys by Rowland, Slichter, Narath, and many others. A list of these can be found on page 2 of [15].

It should be emphasized that the various areas of applications cited in Table I, and reviewed in this paper, are not orthogonal to each other, nor are they exhaustive. Other important applications of NMR in alloys are listed in Table II. The application of NMR to these and, in fact, almost all areas of modern alloy research and development could be the subject of extensive individual reviews. Some of these applications are discussed later in this symposium.

TABLE II

More Applications of NMR in Alloys

Hydrogen in Metals
Catalysis
Superconductivity
One and Two Dimensional Alloys
Alloy Age Hardening
Solid Solution Softening
Guinier–Preston Zones
Charge Density Waves
Kondo Effect

What are the obstacles to further exploitation of the great power of NMR techniques? The most important is the specialized nature of NMR. Unlike many metallurgical tools, NMR cannot be applied with nearly equal sensitivity to all alloy systems. The magnitudes of the nuclear dipole moments, the natural abundance of appropriate isotopes, the magnitudes of the nuclear quadrupole moments, the site symmetries -- these and other parameters play a role in determining the strength of the resonance. Hence, the NMR investigator must use great ingenuity in choosing model systems and problems to study, and in applying the most useful NMR technique. Despite these limitations, I expect NMR to be used as a probe for the study of alloys, in conjunction with other techniques, by an increasing number of leading research centers.

References

1. J. R. Manning, Diffusion Kinetics for Atoms in Crystals (D. Van Nostrand Company, 1968) p. 75 ff.

2. D. Wolf, Spin-Temperature and Nuclear-Spin Relaxation in Matter (Clarendon Press, 1979).

3. G. Brunger, O. Kanert and D. Wolf, Solid State Comm. 33, 569 (1980); Phys. Rev. B 22, 4247 (1980).

4. G. Goltz, A. Heidemann, H. Mehrer, A. Seeger and D. Wolf, Phil. Mag. A 41, 723 (1980).

5. E. O. Stejskal and J. E. Tanner, J. Chem. Phys. 42, 288 (1965). J. E. Tanner, J. Chem. Phys. 52, 2523 (1970).

6. J. S. Murday and R. M. Cotts, J. Chem. Phys. 48, 4938 (1968).

7. J. S. Murday and R. M. Cotts, J. Chem. Phys. 53, 4724 (1970).

8. B. Stalinski, C. G. Coogan and H. S. Gutowsky, J. Chem. Phys. 34, 1191 (1961).

9. H. T. Weaver and J. P. VanDyke, Phys. Rev. B 6, 694 (1972).

10. C. Korn and D. Zamir, J. Phys. Chem. Solids 31, 489 (1970).

11. L. D. Bustard, R. M. Cotts and E. F. W. Seymour, Phys. Rev. B 22, 12 (1980).

12. L. D. Bustard, Phys. Rev. B 22, 1 (1980).

13. C. L. Bisson and W. D. Wilson, in Effect of Hydrogen on Behavior of Materials, A. W. Thomson and I. M. Bernstein, Editor (A.I.M.E., 1976) p.416

14. T. J. Rowland, Progr. Mater. Sci. 9, 1 (1961).

15. G. C. Carter, L. H. Bennett, and D. J. Kahan, Progr. Mater. Sci. 20, 1 (1977).

16. L. H. Bennett and G. C. Carter, Metall. Transactions 2, 3079 (1971).

17. W. W. Warren, Jr., in Charge Transfer/Electronic Structure of Alloys, L. H. Bennett and R. H. Willens, Editors (A.I.M.E., 1974) p. 223

18. M. Hansen and K. Anderko, Constitution of Binary Alloys (McGraw-Hill, 1958).

19. Y. S. Hwang, D. R. Torgeson and R. G. Barnes, Scripta Metall., 12, 507 (1978).

20. J. Hauck, Acta Cryst. A34, 389 (1978).

21. R. R. Arons, H. G. Bohn and H. Lutgemeier, J. Phys. Chem. Solids 35, 207 (1974).

18

22. F.Borsa and R. G. Lecander, Solid State Comm. 20, 389 (1976).

23. N. M. Wolcott, R. L. Falge, Jr., L. H. Bennett and R. E. Watson, Phys. Rev. Letters 21, 546 (1968).

24. G. W. West, Phil. Mag. 9, 979 (1964).

25. I. D. Weisman and L. H. Bennett, Phys. Rev. 181, 1341 (1969).

26. C. Froidevaux and D. Rossier, J. Phys. Chem. Solids 28, 1197 (1967).

27. H. E. Schone and W. D. Knight, Acta Metall. 11, 179 (1963).

28. L. H. Bennett, Acta Metall. 14, 997 (1966).

29. L. H. Bennett, Phys. Rev. 150, 418 (1966).

30. K. Kishio and J. O. Brittain, J. Phys. Chem. Solids 40, 933 (1979).

31. I. D. Weisman, L. H. Bennett, A. J. McAlister and R. E. Watson, Phys. Rev. B 11, 82 (1975).

32. J. E. Schirber, L. J. Azevedo and A. Narath, Phys. Rev. B 20, 4746 (1979).

33. J. Rossbach, M. von Hartrott, D. Hohne, D. Ouitmann, E. Weihreter and F. Willeke, J. Phys. F: Metal Phys. 10, 729 (1980).

34. P. S. Allen and E. F. W. Seymour, Proc. Phys. Soc. 85, 509 (1965).

35. R. Dupree, D. J. Kirby, W. Freyland and W. W. Warren, Jr., Phys. Rev. Letters 45, 130 (1980).

36. C. van der Marel, W. Geertsma and W. van der Lugt, J. Phys. F: Metal Phys. 10, 2305 (1980).

37. L. H. Bennett, H. E. Schone and P. Gustafson, Phys. Rev. B 18 2027 (1978).

38. W. A. Hines, K. Glover, L. T. Kabacoff, C. U. Modzelewski, R. Hasegawa and P. Duwez, Phys. Rev. B 21, 3771 (1980).

39. I. Bakonyi, I. Kovacs, A. Lovas, L. Takacs, K. Tompa and L. Varga, Proc. Conf. on Metallic Glasses, Budapest, Hungary (1980).

40. P. Panissod, D. A. Guerra, A. Amamou, J. Durand, W. L. Johnson, W. L. Carter and S. J. Poon, Phys. Rev. Letters 44, 1465 (1980).

41. I. D. Weisman, L. S. Swartzendruber and L. H. Bennett, in Techniques of Metals Research, R. F. Bunshah, Editor (Interscience Publ. 1972) Vol. VI-2, p.1.

HYDRIDES EXAMINED BY NUCLEAR MAGNETIC RESONANCE

R. G. BARNES
Ames Laboratory-USDOE[*] and Department of Physics, Iowa State
University, Ames, Iowa 50011

ABSTRACT

Nuclear magnetic resonance (NMR) methods offer many
opportunities for studying both crystal and electronic
structure in hydrides, as well as the changes that occur in
these at phase transitions. NMR also affords an almost
uniquely powerful approach to the study of hydrogen diffusion in hydrides. The NMR of all three hydrogen isotopes
can be utilized, as well as of a range of technolgically
significant metals. Single crystals are not required,
samples need not be single-phase, and measurements can be
conveniently made over a wide range of temperatures.

INTRODUCTION

Nuclear magnetic resonance (NMR) methods offer many opportunities for obtaining
information on crystal structure and atom locations and on electronic structure
in hydrides, as well as on the changes that occur in these properties at structural and electronic phase transitions. In addition, NMR methods afford an
almost uniquely powerful approach to the study of hydrogen motion (diffusion)
in hydrides. Nuclei interact with the microscopic local environment via their
magnetic dipole and electric quadrupole moments, and since these moment properties are invariable for a given nuclear species, the interactions serve to
probe the magnitudes and symmetries of the electric and magnetic fields at the
nuclear sites. A variety of experimental techniques may be utilized, including
both steady-state (wide-line) and transient (pulsed measurements. These may be
applied to study the NMR of all three hydrogen isotopes (^{1}H, ^{2}D, ^{3}T) as well as
the NMR of a considerable number of transition metals. Important practical
features of NMR methods relevant to the case of metal hydrides include the fact
that single crystal specimens are not required, that samples need not be single
phase, that measurements can be conveniently made over a wide range of temperatures, and that information pertaining to both the metal and hydrogen sites in
the structure can frequently be obtained in a given hydride.

In this brief survey we illustrate with examples the capabilities of NMR in
studying metal hydride systems. Recent review of these topics should be con-
sulted for greater detail and thoroughness [1,2].

HYDROGEN LOCATIONS AND STRUCTURAL PHASE TRANSITIONS

A problem fundamental to most other metal-hydrogen system studies is that of
determining the locations of the hydrogen in the metal lattice. Whether one is
concerned with hydrogen solubility in the α-phase, with a simple binary hydride
phase, or with a more complex intermetallic (ternary) hydride, accurate knowledge of hydrogen locations is frequently essential to the interpretation of
other measurements (e.g., optical) and to formulating correct theoretical de-
scriptions (e.g., band structure). Since hydrogen is essentially undetected by
x-rays, the principal experimental methods available for its location are neutron diffraction and nuclear magnetic resonance. NMR provides information on

hydrogen locations and structure in two ways: (1) the magnetic dipole interaction between nuclei, and (2) the interaction of quadrupolar nuclei with the
gradient of the crystalline electric field (CEF). In the following we consider
some specific applications of these methods.

NUCLEAR DIPOLAR INTERACTION

The magnetic field produced by a nuclear dipole moment at the site of a
neighbor nucleus is on the order of 1-10 mT. For protons, and other spin 1/2
nuclei, barring the possible presence of paramagnetic ions in the lattice, this
dipolar field is responsible for the width ΔH and in particular the second
moment, M_2, of the resonance. Experimentally, the second moment has its conventional statistical meaning, i.e., $M_2 = \int (\nu - \nu_o)^2 g(\nu) d\nu$ where the integral is
taken over all frequencies. Here $g(\nu)$ is the shape function of the resonance,
frequently Gaussian or nearly so, and ν_o is the center frequency of the resonance. In a powder sample M_2 depends on the sum of r^{-6} taken over all nuclear
sites in the lattice, where r is the distance from the probe nucleus to the
contributing nucleus, and on the strength of the magnetic moment of the contributing nucleus [3]. Since M_2 is affected (reduced) by rapid motion of the
hydrogen (see below in connection with diffusion), for structural
determinations one measures M_2 at a temperature low enough that ΔH and M_2 have
become temperature-independent. This value of M_2 is referred to as the rigid-
lattice value.

In many important metals and hydride phases, hydrogen can occupy several
different sets of interstitial sites. For example, in the non-stoichiometric
dihydride phases of Sc, Y, La, Ti, Zr, etc., tetrahedral (T) and octahedral (O)
sites exist in the fcc metal lattice which may or may not be occupied by hydrogen. For this type of situation M_2 can be expressed in a general way by sums
of contributions from protons in the two types of sites and from the host metal
nuclei:

$$M_2 = \frac{2\beta}{2\beta + \alpha} (\beta M_{TT} + \alpha M_{OT} + M_{HT}) + \frac{\alpha}{2\beta + \alpha} (\alpha M_{OO} + \beta M_{TO} + M_{HO}) \tag{1}$$

where M_{ij} (i,j=T or O) denotes the contribution of i-site protons to the M_2 of
a j-site proton. The notation M_{HT} denotes the contribution of host metal
nuclei to the M_2 of T-site hydrogen, etc. The fractional occupancy of O and T
sites is represented by α and β, respectively, which are constrained by the
total hydrogen content, $x = 2\beta + \alpha$, in MH_x.

Second moment measurements are most discriminating as to hydrogen locations
when M_{HT} and M_{HO} are small. This happens when the metal nuclei have small magnetic moments (e.g., [89]Y) or are of low abundance, as is the case with the
isotopes of titanium and zirconium, for example.

Referring to the case of the fcc dihydride phases of the IIIB and IVB transition metals, the NMR measurements show clearly that in the case of TiH_2 negligible occupation of the O sites occurs [4]. The octahedral sites appear to be
inaccessible to hydrogen. The same may apply in the case of ScH_2, but the
results are not so clear-cut [5] since [45]Sc has a substantial magnetic moment
which makes Eq. (1) less discriminating.

The case of YH_2 is particularly interesting because certain low-energy,
composition-dependent features observed in the optical absorptivity [6] could
not be interpreted in terms of self-consistent band structure calculations [7]
based on the perfect dihydride structure (CaF_2 type) in which hydrogen occupies
only the T sites. NMR measurements made in support of this work showed unam-

biguously that at $YH_{1.98}$ 15 percent of the O sites were already occupied [8]. The O site occupancy factor decreases with decreasing hydrogen content, reaching 10 percent at $YH_{1.92}$ [8]. These results were subsequently confirmed by neutron diffraction measurements made on some of the same (powder) samples used in the NMR study [9]. Hydrogen occupation of O sites appears also to occur, but to a lesser degree, in LaH_2. This was determined in the pioneering NMR study of lanthanum hydrides by Schreiber and Cotts [10] who found approximately 5 percent O site occupancy at LaH_2.

When we deal with new and unusual metal hydrides for which single crystals have not been obtained, NMR proves to be especially useful. An excellent example is provided by the hydride phases of the zirconium monohalides, ZrCl and ZrBr [11], which are two-dimensional metals. A schematic of the structure of ZrBr is shown in Fig. 1; that of ZrCl differs only in the stacking sequence

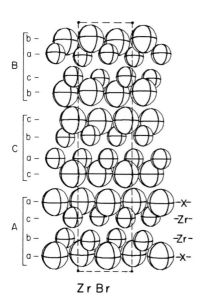

Zr Br

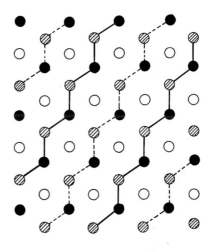

Fig. 1. Projection along [100] of the structure of ZrBr. The lower-case letters refer to relative positions of close-packed layers along [001]; the capital letters to packing of the four-layer sheets. [After Ref. 11.]

Fig. 2. Schematic representation of the close-packed Zr layers in ZrX; filled circles represent Zr in one Zr layer, shaded circles those in the adjacent layer above. A tetrahedral (T) site occurs above each Zr in the lower layer and below each Zr in the upper layer. The octahedral (O) sites are shown by open circles. The solid lines show two zig-zag chains of occupied T sites, and the dashed lines show the locations of the intervening chains of vacant T sites.

of the Cl-Zr-Zr-Cl "sandwiches". The two close-packed metal layers contain the same ratio of octahedral to tetrahedral interstitial sites (1:2) as in the f.c.c. CaF_2 structure. The NMR M_2 measurements show that in both $ZrBrH_{0.5}$ and

$ZrBrH_1$ hydrogen occupies predominantly the T sites with about 5% of the O-sites ones being occupied [12]. In $ZrBrH_{0.5}$ the measured M_2 is only consistent with a zig-zag chain superlattice arrangement of hydrogen in the T-sites, every other chain of sites being empty. Again, about 5% of the O-sites are also occupied. The hydrogen structure is shown schematically in Fig. 2. Long range ordering of these chains in successive layers is not determined by NMR—it is probably quite complex.

ELECTRIC QUADRUPOLE INTERACTION

The interaction of the nuclear electric quadrupole moment of nuclei having spin $\geqslant 1$ with the second derivative of the net electrostatic potential at the nuclear site—the electric field gradient or EFG—also furnishes information on atom locations and site symmetries in the lattice. Immediate application to hydrogen in hydrides occurs through the study of the NMR of the heavy isotope deuterium, ^2D. In addition, most metal nuclei of interest possess electric quadrupole moments so that their NMR also furnishes such information. Detailed treatment of the quadrupole interaction may be found in standard NMR texts [13]. Application to hydrides is given in Ref. 2, and the special case of ^2D NMR (DMR) spectra has also been described in detail [14].

An example is provided by recent work on lanthanum deuterides, LaD_x, in the composition range $2 \leqslant x \leqslant 3$ [15]. In Fig. 3 are shown deuteron spectra for x=2.28

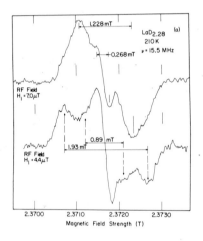

Fig. 3. Deuteron spectra in $LaD_{2.28}$ and $LaD_{2.48}$ showing the marked difference in features resulting from the different ordered deuterium superlattice structures at these two compositions. [After Ref. 18.]

and x=2.48. In both cases the narrow central line can be assigned to deuterons in O sites in the f.c.c. metal lattice. In the x=2.48 spectrum the outermost features arise from deuterons in T sites experiencing completely asymmetric EFG tensors, exactly consistent with the x=2.5 deuterium superlattice determined by both neutron diffraction [16] and electron spin resonance [17] measurements. On the other hand, the outermost features of the x=2.28 spectrum arise from deuterons experiencing axially symmetric EFG tensors and are not consistent with this superlattice. These features are due to tetrahedral site deuterons in a different ordered structure characteristic of x=2.25 which precedes the x=2.5 structure [18].

The fact that the EFG vanishes at a site of cubic symmetry, coupled with the integral spin of the deuteron, makes DMR a particularly sensitive probe of structural phase transitions, especially those between cubic and non-cubic phases. This property is illustrated in Fig. 4 which shows a sequence of DMR

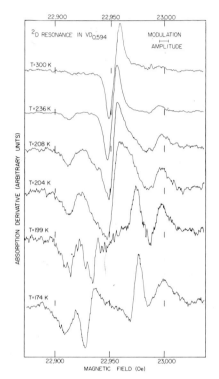

Fig. 4. Sequence of deuteron resonance spectra in $VD_{0.6}$ showing the transition from the cubic α' phase at 300K to the two-phase $\eta+\delta$ region at low temperatures. The narrow central line is characteristic of the cubic phase and the symmetric satellite pairs of the non-cubic phases. [After Ref. 19.]

derivative spectra in $VD_{0.59}$ over a range of temperatures spanning the transition from the cubic α' phase at high temperature (300K) down through the $\alpha'+\beta$ and $\alpha'+\eta$ regions at intermediate low temperatures to the $\eta+\delta$ region at low temperature (174K) [19]. The β and η phases are essentially tetragonal [20] and their DMR shows little difference. The δ phase is orthorhombic [20]. In the cubic α' phase the DMR spectrum is distinguished by a single narrow line, whereas in the non-cubic phases the spectrum consists of a symmetric pair of "satellites". Hence, at the lowest two temperatures shown, the spectrum consists entirely of two such satellite pairs, the central line characteristic of the α' phase having been extinguished.

The NMR of transition metal nuclei has also proved extremely useful in establishing site symmetries and hydrogen locations in hydrides. These results depend mainly on the quadrupole interaction, and because of the typically large spins and quadrupole moments of these nuclei, analysis of the spectra is relatively complex. Recent applications of these methods have dealt with hydrides of niobium and lanthanum [21-22].

ELECTRON STRUCTURE

Information concerning the electronic structure of metal hydrides may be obtained from both the Knight shift, K, and spin-lattice relaxation rate, $R=(T_1T)^{-1}$, where T_1 is the spin-lattice relaxation time and Ti is the temperature, of both the metal and hydrogen nuclei. Both of these quantities are related to the density-of-states (d.o.s.) at the Fermi level via the effective hyperfine fields. The magnetic susceptibility χ is also related to the density-of-states, and when all three quantities are measured, i.e., K, R, and χ, the d.o.s. can be partitioned into its s and d band contributions. For example, recent measurements of K and R of ^{93}Nb in mixed titanium-niobium dihydride [23] which has the cubic CaF$_2$ structure show that the conduction band in this system has predominantly d character, in agreement with the expectations of band structure calculations [24]. The latter calculations predict very small changes in d-like states and substantial modifications of s-like states when transition metals form hydrides. The reliance in the NMR of ^{93}Nb in this instance is motivated by the very poor NMR characteristics of the titanium nuclear species.

A somewhat more qualitative example is provided by measurements of the ^{93}Nb Knight shift in niobium-hydrogen system. The solid solution phase is cubic (b.b.c.), and extensive measurements of K have been made as a function of hydrogen concentration, showing behavior remarkably similar to alloying Nb with metals further to the right in the periodic table (e.g., Mo and Tc) [25]. Measurements of K(^{93}Nb) have also been made in the orthorhombic β-phase hydride in the composition range NbH$_{0.73}$-NbH$_{0.88}$ at temperatures just below the α'-β transition [26]. These measurements, which are severely complicated by the strong electric quadrupole interaction of ^{93}Nb, show that the isotropic component of the Knight shift tensor has the same value in the hydride phase as in the solid solution phase at the same composition, indicating that no significant change in band structure details occurs as a consequence of the structural transition.

HYDROGEN DIFFUSION

Fast absorption and desorption kinetics are desirable characteristics of a practical hydrogen-storing metal hydride. One factor determining these kinetics is the diffusion of hydrogen within the metallic lattice. NMR is sensitive to atomic diffusion via the effects which such motion has on the proton spin-lattice and spin-spin relaxation rates, $(T_1)^{-1}$ and $(T_2)^{-1}$, respectively. These rates are functions of power spectra of the randomly varying dipolar fields resulting from the atomic diffusive jumps, and depend on the strength of the dipolar interaction between protons and the diffusion jump frequency, ν_j [2,13].

The characteristic temperature dependence of T_1 resulting from diffusion is illustrated in Fig. 5 by measurements on ZrBrH$_{0.5}$ [27]. The measured rate $(T_1)^{-1}$ is the sum of the diffusion (or dipolar) controlled rate $(T_{1d})^{-1}$ and the conduction electron contribution $(T_{1e})^{-1}$ which dominates at low temperatures. $(T_{1d})^{-1}$ reaches its maximum (and T_1 its minimum) when $\nu_j \simeq \omega_o$, the nuclear Larmor precession frequency in the applied field. T_1 in the rotating frame, $T_{1\rho}$, has the same significance as T_{1d}, but the Larmor frequency is replaced by the precession frequency in the rotating radiofrequency field H_1. Since the latter is usually on the order of 10^{-3} x the Larmor frequency, this measurement extends the range of accessible ν_j values downward by this factor to the range of truly "slow" diffusive motion [28].

When hydrogen occupies only a single sublattice of interstitial sites, as in TiH$_2$ or in the hydride phases of V, Nb, and Ta, a single minimum occurs in T_{1d} and in $T_{1\rho}$. However, as seen in Fig. 5, a well-defined secondary minimum appears in both T_{1d} and $T_{1\rho}$ in the case of ZrBrH$_{0.5}$ where we know from the M_2 measurements that although hydrogen occupies predominantly T-sites, about 5% of the O-sites are also occupied. Diffusion of these O-site hydrogen occurs with a lower activation energy than that of the T-site hydrogen and causes the secondary, low-temperature minima in T_{1d} and $T_{1\rho}$.

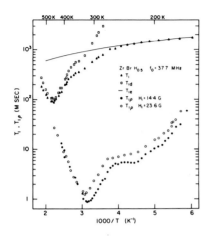

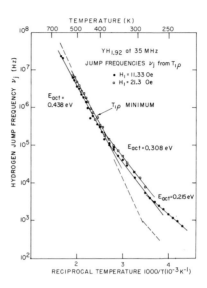

Fig. 5. Measured proton spin-lattice relaxation times T_1 and $T_{1\rho}$ in ZrBrH$_{0.05}$ as a function of reciprocal temperature. The $T_{1\rho}$ values are shown for two values of the rotating field H_1. Decomposition of the T_1 measurements into electronic (T_{1e}) and diffusion-controlled (T_{1d}) components is also shown. The low-temperature, secondary minima in T_1 and $T_{1\rho}$ are due to the diffusion of O-site hydrogen. [After Ref. 27.]

Fig. 6. Values of the hydrogen jump frequency ν_j for YH$_{1.92}$ derived from $T_{1\rho}$ measurements. The equations of the straight-line segments are listed in Table I. Also shown (dashed lines) are the corresponding measurements for TiH$_{1.90}$ [Ref. 29] in which O-sites in the f.c.c. CaF$_2$ structure are not occupied. [After Ref. 28.]

The same situation occurs in the dihydrides of Y, La, and Lu, where premature (at x<2 in MH$_x$) occupation of O-sites is coupled with the presence of vacant T-sites. Figure 6 shows the dependence of jump frequency ν_j, deduced from $T_{1\rho}$ data, on temperature in YH$_{1.92}$ [29]. Also shown in Fig. 6 for comparison are ν_j values for TiH$_{1.90}$, also from $T_{1\rho}$ measurements [30]. Whereas the TiH$_{1.90}$ data conform to a single activated process, $\nu_j = \nu_o \exp(-E_{act}/kT)$, over the entire range apart from a short region at low temperatures, the YH$_{1.92}$ data clearly show that two processes occur in addition to a third region at low temperatures. As suggested in Ref. 29, the low-temperature region may indicate non-classical diffusion, i.e., quantum-assisted tunneling. This three-stage character of hydrogen diffusion has been observed throughout the yttrium dihydride phase [29], and the results of measurements on three compositions are summarized in Table I.

The most striking feature of these results is the difference in behavior of YH$_{1.63}$ as contrasted with YH$_{1.92}$ and YH$_{1.98}$. The jump frequency is about three orders of magnitude smaller in YH$_{1.63}$ than in the higher compositions. Secondly the E_{act} values are consistently higher by a factor of at least two.

Table I. Summary of the best fits to the ν_j data for yttrium dihydrides. The transition temperatures were empirically determined before making the linear fits to the data, and the rotating field H_1 was not corrected for local field effects. [After Ref. 28.]

	$YH_{1.63}$	$YH_{1.92}$	$YH_{1.98}$
ν_j^{I} (sec^{-1})	$4.5 \times 10^8 \exp(-0.457 \text{ eV}/kT)$	$2.35 \times 10^7 \exp(-0.208 \text{ eV}/kT)$	$6.96 \times 10^7 \exp(-0.218 \text{ eV}/kT)$
Transition Temperature	465K	280K	270K
ν_j^{II} (sec^{-1})	$1.45 \times 10^{13} \exp(-0.877 \text{ eV}/kT)$	$1.80 \times 10^9 \exp(-0.308 \text{ eV}/kT)$	$1.052 \times 10^{10} \exp(-0.336 \text{ eV}/kT)$
Transition Temperature	600K	370K	330K
ν_j^{III} (sec^{-1})	$1.066 \times 10^{15} \exp(-1.1 \text{ eV}/kT)$	$1.58 \times 10^{11} \exp(-0.438 \text{ eV}/kT)$	$1.58 \times 10^{11} \exp(-0.417 \text{ eV}/kT)$

And thirdly, the transition temperatures between regions of different E_{act} are much higher than in the higher concentration samples.

The marked decrease in E_{act} coupled with the substantial hydrogen occupation of O-sites as hydrogen concentration increases supports the interpretation first advanced by Schreiber and Cotts [10] in the case of lanthanum dihydride that O-site occupation is responsible for the decrease in E_{act}. In that system O-site occupancy appears to be substantially less (~2.5%) than in YH_2, and the decrease in E_{act} occurs at a somewhat higher concentration, about $LaH_{1.95}$. More extensive measurements of T_{1d} and $T_{1\rho}$ show that E_{act} continues to decline in magnitude up to the limiting composition LaH_3 [10,31], reaching a value of about 0.2 eV/atom at that point. The exact nature of the mechanism by which O-site occupancy reduces E_{act} remains to be elucidated.

To gain complete understanding of the hydrogen diffusion process requires finally the determination of the actual jump path ℓ taken by the diffusing hydrogen. This can be accomplished by combining NMR measurements of the jump frequency (as above) with measurements of the diffusion coefficient, D, also by NMR methods. The latter quantity is related to the other two by $D=f_T\ell^2\nu j/6$, where f_T is the tracer correlation factor. The NMR measurement of D depends on a multiple-pulse spin-echo sequence combined with the application of magnetice field gradient pulses which serve to dephase the precessing spins as they diffuse [2].

An elegant application of these methods has recently been made to the case of titanium dihydride [32]. As remarked above, in this system the octahedral sites remain unoccupied. The T-sites in the CaF_2 structure form a simple-cubic lattice in which the metal atoms occupy the body-centers of alternate cells, as shown in Fig. 7 [32]. Theoretical calculations indicate that the two most likely diffusion jump paths of the hydrogen occur from one T-site to its nearest and third nearest neighbor (1NN and 3NN) T-sites directly or via the intervening vacant O-site, as shown. The presence of the Ti atoms in half the body-centered positions inhibits 3NN jumping from each T-site.

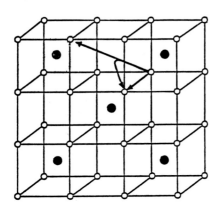

Fig. 7. The simple cubic lattice of T-site hydrogen in the f.c.c. CaF_2 structure (TiH_x). In this representation the metal atoms occupy the body-center positions in every-other s.c. cell. Theoretical calculations [Ref. 33] indicate that the two most likely hydrogen diffusion jump paths are from a T-site to the first and third nearest neighbor T-sites. [After Ref. 32.]

● Ti   o H .

The final results of the measurements of Bustard, Cotts, and Seymour [32] for TiH$_{1.71}$ are shown in Fig. 8. Agreement with the predicted values of D based on jumping between 1NN sites is seen to be excellent (however, the measurements do not distinguish the two 1NN paths shown in Fig. 7). This technique, although difficult, can be expected to contribute significantly to our understanding of hydrogen diffusion in the future.

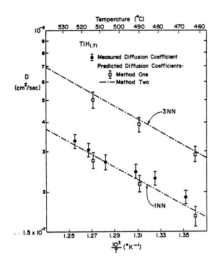

Fig. 8. Measured and predicted values of the diffusion coefficient for hydrogen in TiH$_{1.71}$. [After Ref. 32.]

Finally, knowledge of the diffusion jump rate also reflects on structural determinations. Thus, in the sequence of vanadium deuteride spectra shown in Fig. 4, ν_j is known to be on the order of 10^7Hz for T=250K in the β-phase, not greatly reduced from its value at 300K in the α' phase. The tetrahedral sites in the b.c.c. lattice have $\bar{4}$ symmetry (not cubic) and form three sublattices according to the orientation of the EFG tensor in each. In the cubic phase, the high value of ν_j coupled with motion among all the T-sites averages the quadrupole coupling, whose strength is only $\sim10^4$Hz, to zero, yielding a single line spectrum. In the β-phase, however, although ν_j is still much faster than the quadrupole coupling, a well-resolved quadrupole splitting of the spectrum appears. This reveals unambiguously that the diffusion, and hence site occupancy, is restricted to one of the three sublattices of T-sites, forming a superlattice. The same conclusion is reached in the cases of Nb-D [25] and Ta-D [34]. In the former case, measurement of the quadrupole coupling parameters of ^{93}Nb in the β-phase, coupled with the knowledge that the deuterium is restricted to a single T-site sublattice, enables the detailed ordered superlattice of the β-phase to be deduced [26].

SUMMARY

Examples have been presented to illustrate how NMR contributes to our knowledge of hydrogen locations, diffusion, and to a lesser extent electronic structure in metal hydrides. The fact that polycrystalline (powder) and mixed phase samples can be readily and profitably investigated constitutes an important plus for NMR methods in dealing with interesting hydrides which are seldom available in single-crystal form.

ACKNOWLEDGMENTS

The author expresses his appreciation to D. R. Torgeson for his helpful comments and criticism during the preparation of this review, to L. D. Bustard and R. M. Cotts for permission to use figures from their recent work, and to Janet Hartman for her careful typing of the manuscript.

REFERENCES

*Operated for the U.S. Department of Energy by Iowa State University under contract No. W-7405-Eng-82. This research was supported by the Director for Energy Research, Office of Basic Energy Sciences, WPAS-KC-02-02-02.

1. R. M. Cotts, Ber. Bunsenges. phys. Chem. 76, 760 (1972).
2. R. M. Cotts, "Nuclear Magnetic Resonance on Metal-Hydrogen Systems" in Hydrogen in Metals I. Basic Properties, ed. by G. Alefeld and J. Völkl, Springer Verlag (Berlin, 1978).
3. J. H. Van Vleck, Phys. Rev. 74, 1168 (1948).
4. B. Stalinski, C. K. Coogan, and H. S. Gutowsky, J. Chem. Phys. 34, 1191 (1961).
5. H. T. Weaver, Phys. Rev. B 5, 1663 (1972).
6. J. H. Weaver, R. Rosei, and D. T. Peterson, Phys. Rev. B 19, 4855 (1979).
7. D. J. Peterman, B. N. Harmon, J. Marchiando, and J. H. Weaver, Phys. Rev. B 19, 4867 (1979).
8. D. L. Anderson, R. G. Barnes, D. T. Peterson, and D. R. Torgeson, Phys. Rev. B 21, 2625 (1980).
9. D. Khatamian, W. A. Kamitakahara, R. G. Barnes, and D. T. Peterson, Phys. Rev. B 21, 2622 (1980).
10. D. L. Schreiber and R. M. Cotts, Phys. Rev. 131, 1118 (1963).
11. A. W. Struss and J. D. Corbett, Inorg. Chem. 16, 360 (1977).
12. T. Y. Hwang, D. R. Torgeson, and R. G. Barnes, Phys. Letters 66A, 137 (1978).
13. C. P. Slichter, "Principles of Magnetic Resonance," Springer Series in Solid-State Sciences, Vol. 1, 2nd ed. Springer Verlag (Berlin, 1978).
14. R. G. Barnes, in Advances in Nuclear Quadrupole Resonance, Vol. I., ed. by J. A. S. Smith, Heyden, and Sons, London (1974), p. 335.
15. D. G. de Groot, R. G. Barnes, B. J. Beaudry, and D. R. Torgeson, J. Less-Common Metals 73, 233 (1980).
16. C. G. Titcomb, A. K. Cheetham, and B. E. F. Fender, J. Phys. C 7, 2409 (1974).
17. K. Knorr, B. E. F. Fender, and W. Drexel, Z. Physik B 30, 265 (1978).
18. R. G. Barnes, D. G. de Groot, D. Misemer, and D. R. Torgeson, to be published.
19. K. P. Roenker, Ph.D. Thesis (unpublished), Iowa State University, 1973.
20. T. Schober and H. Wenzl, "The Systems NbH(D), TaH$_z$(D), VH(D): Structures, Phase Diagrams, Morphologies, Methods of Preparation," in Hydrogen in Metals I. Basic Properties, ed. by G. Alefeld and J. Völkl, Springer Verlag, Berlin (1978).
21. Y. S. Hwang, D. R. Torgeson, and R. G. Barnes, Sripta Met. 12, 507 (1978).
22. D. G. de Groot, R. G. Barnes, B. J. Beaudry, and D. R. Torgeson, Z. phys. Chem. N. F. 114, 83 (1979).
23. B. Nowak, O. J. Zogal, and M. Minier, J. Phys. C 12, 4591 (1979).
24. A. C. Switendick, "The Change in Electronic Properties on Hydrogen Alloying and Hydride Formation," in Hydrogen in Metals I. Basic Properties, ed. by G. Alefeld and J. Völkl, Springer Verlag, Berlin (1978).
25. R. G. Barnes, K. P. Roenker, and H. R. Brooker, Ber. Bunsenges. phys. Chem. 80, 876 (1976).

26. Y. S. Hwang, D. R. Torgeson, and R. G. Barnes, Solid State Commun. $\underline{24}$, 773 (1977).
27. T. Y. Hwang, D. R. Torgeson, and R. G. Barnes, to be published.
28. C. P. Slichter and D. C. Ailion, Phys. Rev. $\underline{135}$, A1099 (1964).
29. D. L. Anderson, R. G. Barnes, T. Y. Hwang, D. T. Peterson, and D. R. Torgeson, J. Less-Common Metals $\underline{73}$, 243 (1980).
30. D. Korn and S. D. Goren, J. Less-Common Metals (in press).
31. R. B. Creel, R. G. Barnes, B. J. Beaudry, D. R. Torgeson, and D. G. de Groot, Solid State Commun. $\underline{36}$, 105 (1980).
32. L. D. Bustard, R. M. Cotts, and E. F. W. Seymour, Phys. Rev. B $\underline{22}$, 12 (1980).
33. C. L. Bisson and W. D. Wilson, in Proceedings of an International Conference on the Effects of Hydrogen on Behavior of Materials, A. W. Thompson and I. M. Bernstein, eds. (Moran, WY, 1976), p. 416.
34. K. P. Roenker, R. G. Barnes, and H. R. Brooker, Ber. Bunsenges. phys. Chem. $\underline{80}$, 470 (1976).

Published 1981 by Elsevier North Holland, Inc.
Kaufmann and Shenoy, editors
Nuclear and Electron Resonance Spectroscopies Applied to Materials Science

NUCLEAR MAGNETIC RESONANCE STUDIES OF TYPE II SUPERCONDUCTORS*

F. Y. FRADIN
Materials Science Division, Argonne National Laboratory, Argonne, IL   60439

ABSTRACT

   The results of nuclear magnetic resonance (nmr)
experiments on simple type-I superconductors were among
the first and most important verifications of the BCS
theory of superconductivity. In this paper, the application of nmr techniques to the study of superconducting
properties in the more complex type-II superconductors
will be reviewed. The discussion will include the effect
of material parameters (e.g., degree of long range crystalline order, density of states at the Fermi level,
effects of magnetic dopants) on the superconducting
properties, including size of the superconducting gap,
vortex structure, upper-critical field H_{c2}, and variations in T_c. Emphasis will be placed on high T_c,
high H_{c2} materials, i.e., A15 compounds and the ternary
Chevrel phases.

INTRODUCTION

   This paper will focus on the materials properties important for the
understanding and the development of useful type-II superconductors. Nuclear
magnetic resonance (nmr) techniques have a historical coupling to the
phenomena of superconductivity starting with the elegant experiments of Hebel
and Slichter [1] that verified a crucial aspect of the microscopic theory of
superconductivity proposed by Bardeen et al. [2]. This article will deal only
peripherally with the important topic of nmr experiments in the superconducting state; an excellent review article dealing with nmr studies of the
physical phenomena associated with superconductivity has been written by
MacLaughlin [3]. The emphasis in this article will be on the relationship
between nmr and normal state parameters, such as the degree of crystalline
order, the density of states at the Fermi level, and the coupling of the
conduction electrons to magnetic dopants that determine the strength of the
Cooper pairing and thus the critical temperature T_c and critical field H_{c2}.
   Although much of the physics of superconductivity was first established on
the simple type-I, nontransition metal superconductors, in recent years a
great deal of research has dealt with the more complex type-II superconductors
(see for example, the two volume review edited by Parks [4]). These
transition metal alloys and intermetallic compounds are of technological
importance due to their high values of T_c, H_{c2}, and critical currents J_c.
Because of the complexity of the metallurgical behavior, the electronic
structure, and the physical properties of these transition metal systems, a
large number of problems must be solved by the materials scientist in order to
optimize their superconducting properties. Since this field is already very
broad, this review will not attempt to be comprehensive but will try to
illustrate the utility of nmr by selective examples mostly taken from the
authors own work.

*Work supported by the U.S. Department of Energy.

MIXED STATE

The principal feature that makes type-II superconductivity technologically important is the ability to sustain the superconducting state in the presence of large applied magnetic fields. Phenomenologically this is due to a low surface energy between normal and superconducting regions that allows for the formation of the so called mixed state consisting of magnetic flux tubes penetrating the sample for applied fields $H > H_{c1}$, the lower critical field. At the core of the flux tube the superconducting order parameter ψ^2 goes to zero resulting in normal regions. The magnetic field is a maximum at the vortex center and one flux quantum $\phi_0 = hc/2e$ is associated with each vortex. The order parameter ψ^2 recovers its full value over a characteristic distance ξ, the coherence length, and the magnetic field decays to zero over a characteristic distance λ, the penetration length. For a strong type-II superconductor $\lambda \gg \xi$. Because of attractive interactions between vortices, a regular 2-dimensional vortex lattice can be formed.

In a series of elegant low-field experiments Redfield and co-workers [5,6] determined the internal field distribution in the mixed state of vanadium (see Fig. 1), and compared their results to theoretical shape functions calculated for a triangular and a square vortex lattice (see Fig. 2). In Fig. 1 the absorption signal is measured as the size of the dip in signal relative to background. For applied magnetic fields $H_e > H_{c2}$, the upper critical field, superconductivity is quenched and there is a sharp absorption signal as shown in the upper right hand corner of Fig. 1. Redfield deduced that the results for $H \approx 1/2 \, H_{c2}$ are consistent with a triangular vortex lattice. At low fields $H \gtrsim H_{c1}$ the internal field distributions are consistent with a superposition model of independent vortices.

SUPERCONDUCTING ENERGY GAP

In a BCS [2] superconductor the size of the energy gap about the Fermi energy at $T = 0$, $2\Delta(o)$, measures the condensation energy of the Cooper pairs in the superconducting ground state. BCS predict the relation between $2\Delta(o)$ and the superconducting transition temperature T_c given by

$$2\Delta(o) = 3.5 \, k_B \, T_c \quad . \qquad (1)$$

This relation appears to be obeyed for weak coupled (low T_c superconductors). For the strong coupled superconductors of interest in this review, it is of interest to study the systematic dependence of $\Delta(o)/T_c$ on the normal state properties, i.e., the Fermi energy electronic density of states $N(E_F)$ and the phonon modes. The nuclear spin-lattice relaxation rate $1/T_1$ in a metal measures the density of available conduction electron states that can undergo mutual spin flips with the nuclei thereby exchanging energy between the "lattice", which in a metal is the bath of conduction electron spins σ at E_F, and the nuclear spins I. This interaction of the form $A I_\mp \sigma_\pm$ yields a normal state nuclear spin-lattice relaxation rate $1/T_1 \sim |A|^2 \, N(E_F)^2 \, kT$, since only electrons within $k_B T$ of E_F can exchange an energy of the size of the nuclear Zeeman energy. In the superconducting state electrons can spin flip with the nuclei only if they can be excited across the gap, i.e. $1/T_1 \sim \exp[-\Delta(o)/k_B T]$ for T sufficiently below T_c. In Fig. 3, the ^{51}V relaxation rate is plotted versus reciprocal temperature for a number of pseudobinary $V_3Ga_{1-x}Sn_x$ A15 compounds [7]. As can be seen in Table I both T_c and $\Delta(o)$ are strong functions of the B-sublattice occupation, i.e., the relative concentration of Ga and Sn. It should be noted that for T_1 values much longer than ~10 sec, the relaxation rate is short-circuited by other

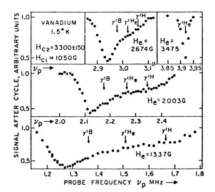

Fig. 1. Nuclear-resonance signal after a field cycle, as a function of probe frequency during the time the sample is superconducting. The sample is 25 pieces of 0.05-cm diameter wire 1 cm long, loosely packed into the receiver coil perpendicular to the applied dc field. The magnetism M is measured ballistically, and B and H calculated assuming $B = H_e + 2\pi M$ and $B = H + 4\pi M$, as appropriate for this geometry. γ' is $(1 + K)/2\pi$ times the nuclear gyromagnetic ratio, and K is the normal-state Knight shift. The probe rf-field amplitude is roughly 0.7-G peak for all the curves. [After Ref. 5.]

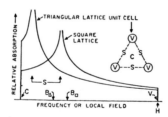

Fig. 2. Theoretical line shapes calculated by Lasher from the solutions to the Ginsburg-Landau equations appropriate for H near H_{c2}. Actual line shapes would be smeared by sample inhomogeneity and nuclear dipolar interaction. Also shown is the unit cell of the triangular lattice, showing the points in space corresponding to the maximum (V), minimum (C), and saddle-point (S) fields. S is twice as close to C in space as it is to a vortex V, whereas for the square lattice these two distances are the same. The square ends of the line are a consequence of the fact that the field distribution in real space has zero gradient at the points V and C. [After Ref. 5.]

processes involved with vortex lattice motion. Returning to Table I, we note that the values of $2\Delta(o)/k_B T_c$ span a range of values from $\lesssim 3.5$ for low T_c V_3Sn to ≈ 5 for the high T_c Ga-rich compounds. The large values of $2\Delta(o)/k_B T_c$ for the Ga-rich compounds are an indication of an important contribution to the attractive electron-electron interaction due to the virtual exchange of soft (low frequency) phonon modes. The presence of soft modes in V_3Ga has more recently been found in inelastic neutron scattering experiments [8].

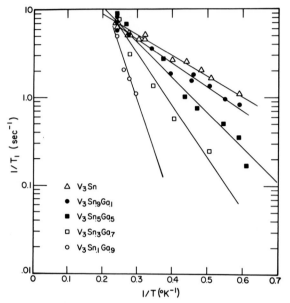

Fig. 3. Spin-lattice relaxation rate as a function of reciprocal temperature in the superconducting state. [After Ref. 7.]

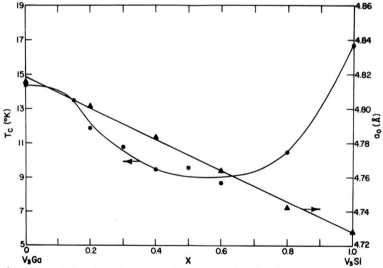

Fig. 4. Composition dependence of the superconducting transition temperature T_c and the lattice parameter a_o for the $V_3Ga_{1-x}Si_x$ compounds. [After Ref. 10.]

DENSITY OF STATES IN THE A15 COMPOUNDS

The question of the systematic relationship between the normal state electronic properties, in particular the value of $N(E_F)$, and the strength of the phonon mediated electron-electron interaction strength λ has been a matter of much discussion and research in transition metals since the work of McMillan [9]. McMillan derived an equation for T_c given by

$$T_c = \frac{\langle\omega^2\rangle^{1/2}}{1.20} \ \exp\left[-\frac{1+\lambda}{0.96\lambda - \mu^*(1+0.6\lambda)}\right]$$

with

$$\lambda = \frac{N(E_F) \langle I^2\rangle}{M\langle\omega\rangle/\langle\omega^{-1}\rangle} \ . \tag{2}$$

Here the $\langle\omega^n\rangle$ are moments of the phonon distribution, μ^* is the Coulomb pseudopotential, $\langle I^2\rangle$ is a Fermi surface average of the electron-ion interaction, and M is the ion mass.

Although there have been a number of band structure calculations of the A15 compounds there is some argument as to their accuracy and resolution. Therefore, we have had to rely on various probes of the electronic structure to obtain information about the Fermi surface electrons. Except for the very low T_c, low density of states $N(E_F)$, compound Nb_3Sb, no detailed deHaas-vanAlphen work exists on the A15 compounds. Similarly, the optical experiments have been unable to determine features of $N(\epsilon)$ for ϵ near ϵ_F. However, the electronic heat capacity coefficient γ, the spin susceptibility χ_s, the electrical resistivity ρ, and the nuclear spin-lattice relaxation rate $1/T_1$ yield information about $N(E_F)$. In addition, the temperature dependences of these quantities yield information about the shape of $N(\epsilon)$ near ϵ_F, which is quite sharp for V_3Ga, V_3Si, and Nb_3Sn.

In Fig. 4 we show the variation of both T_c and the lattice parameter a_o versus concentration x in the $V_3Ga_{1-x}Si_x$ system. By comparison in Fig. 5 the value of $1/T_1$ multiplied by reciprocal temperature for ^{51}V is shown [10]. There is a clear correlation between $(T_1T)^{-1}$ and T_c in these compounds.

TABLE I
Superconducting transition temperature and energy gap in $V_3Ga_{1-x}Sn_x$ [7]

x	0.0	0.1	0.3	0.5	0.7	0.9	1.0
T_c ($^\circ K$)	13.8	10.7	7.2	5.6	4.8	3.8	3.4
$2\Delta(0)$ (meV)	---	4.4(6)	2.3(2)	1.8(1)	---	1.03(5)	0.84(5)
$2\Delta(0)/k_BT_c$	---	4.7(7)	3.7(3)	3.6(2)	---	3.2(2)	2.9(2)

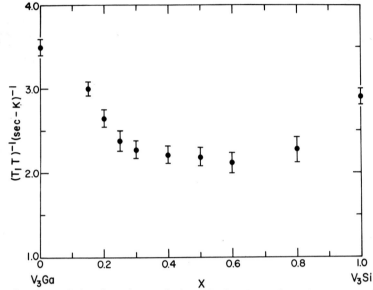

Fig. 5. Composition dependence of the spin-lattice relaxation-rate-reciprocal-temperature product at 77 K for ^{51}V in the $V_3Ga_{1-x}Si_x$ compounds. [After Ref. 10.]

TABLE II
Irreducible representation of $D_{2d}(1 = 2)$

Bases	Assignments	Atomic Functions	Spherical Harmonics[a]
A_1	σ_z	$(5/4\pi)^{1/2}(3z^2 - r^2)/2r^2$	Y_2^0
E	π_{1z}	$(15/4\pi)^{1/2}\, zx/r^2$	Y_2^1,c
E	π_{2z}	$(15/4\pi)^{1/2}\, zy/r^2$	Y_2^1,s
B_1	δ_{1z}	$(15/16\pi)^{1/2}(x^2 - y^2)/r^2$	Y_2^2,c
B_2	δ_{2z}	$(15/4\pi)^{1/2}\, xy/r^2$	Y_2^2,s

a $Y_1^{m,c} \equiv (Y_1^m + Y_1^{-m})/\sqrt{2}$; $Y_1^{m,s} = -i(Y_1^m - Y_1^{-m})/\sqrt{2}$; $Y_1^m(\theta,\phi)$ are normalized spherical harmonics. The polar angle θ is measured with respect to a z-axis parallel to the tetragonal or chain axis.

The A15 (Cr$_3$O) structure in which the A$_3$X (A = V or Nb and X = nontransition element) compounds crystallize in a cubic lattice with the X atoms on a body-centered-cubic sublattice. We focus attention on the A sites. These are the sites for which a LCAO description of the $\ell = 2$ component of the wave functions will yield some insights into the dominant character of the Fermi surface electrons. The A sites have tetragonal symmetry; the point group is D$_{2d}$($\overline{4}$2m), and the A atoms lie on chains arranged in three orthogonal families. In Table II we list the irreducible representation of the d wave functions ($\ell = 2$) for an A atom on a chain with axis parallel to the $\overline{z}$ direction.

In a tight-binding picture of the V(Nb) d-states in the A15 compounds, the nuclear spin-lattice relaxation rate $1/T_1$ of ^{51}V (^{93}Nb) is given in terms of bilinear products of the subband density of states N^m(ϵ) evaluated at the Fermi level [7] where

$$N^m(0) = \sum_\mu \sum_k |C_{\mu mk}|^2\, \delta(E_{\mu k} - E_F) \quad . \qquad (3)$$

Here the $C_{\mu mk}$ are the fractional admixture coefficients in the LCAO prescription for the wave function for band index μ and wave-vector k, and m represents the appropriate tight binding basis function, i.e., σ, π, δ_1, or δ_2. The results for the subband densities of states derived from the experimental values [7,10,11] of $1/T_1$ are shown in Fig. 6 together with the values of λ [12]. In general for the V$_3$X compounds, we find a correlation of $1/T_1$ with $[N_{\gamma 0}(0)]^2$, the bare density of states at the Fermi level, principally due to the dominance of N$^\pi$(0). No correlation between $1/T_1$ and $[N_{\gamma 0}(0)]^2$ is found in the Nb$_3$X compounds;

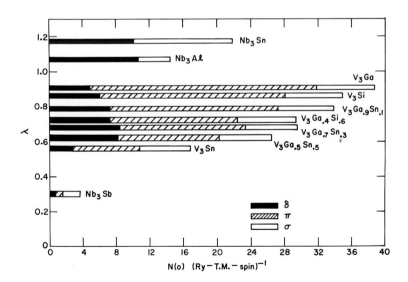

Fig. 6. Dependence of the strength of the electron-phonon interaction on the bare subband density of states for the A15 compounds. N$^\delta$(0) = N$^{\delta 1}$(0) + N$^{\delta 2}$(0). [After Ref. 12.]

also, $N^\pi(0)$ is quite small for the Nb_3X compounds. The large value of $N_{\gamma 0}(0)$ for Nb_3Sn relative to Nb_3Al, is due to a large value of $N^\sigma(0)$ for Nb_3Sn. The value of $N_{\gamma 0}(0)$ for the very low T_c compound Nb_3Sb is very small; the Fermi-level appears to lie in a valley in $N(\varepsilon)$.

DISORDER IN A15 COMPOUNDS

For both the V_3X and Nb_3X compounds where X is a nontransition element T_c is known [13] to increase with the degree of long range crystalline order. There has been considerable effort experimentally to characterize the types of disorder and the mechanism by which disorder affects T_c. Since both ^{51}V ($I = 5/2$) and ^{93}Nb ($I = 9/2$) have large nuclear electric quadrupole moments Q, there is high sensitivity to disorder. In Fig. 7, we show the experimental and theoretical ^{93}Nb nmr spectrum for well ordered Nb_3Sb. Both the 1st order quadrupole split satellites for the $\pm 1/2$ $\pm 3/2$, $\pm 3/2$ $\pm 5/2$, $\pm 5/2$ $\pm 7/2$, and $\pm 7/2$ $\pm 9/2$ transitions, as well as the 2nd order quadrupole split central $\pm 1/2$ $-1/2$ transition, is visible [11]. The effect of disorder between Ga and Si on the X sublattice is shown for $V_3Ga_{1-x}Si_x$ in Fig. 8 [10]; here the nuclear electric quadrupole interaction is fit by a Gaussian distribution of electric field gradients instead of a delta function at a mean value q_0 as was done in Fig. 7. Note the fractional width $\Delta q/q_0$ of the distribution varies from a low of <10% in V_3Si to a high of ~60% in $V_3Ga_{0.6}Si_{0.4}$. It has been shown [10] that the major influence on the distribution of q is from the compositional fluctuation broadening of the local density of states $N(\varepsilon)$ near E_F.

Disordering effects between the A and B sublattices has a very strong effect on T_c. This can be seen in Fig. 9 for off-stoichiometric $V_{0.75+x}Ga_{0.25-x}$ and in fast neutron damaged V_3Ga [14]. There is an obvious correlation of T_c to the value of ^{51}V $1/T_1$. Following an analysis similar to that in the previous section, we find the correlation is due to the variation in density of states at the Fermi level shown in Fig. 10. Again due to the statistical fluctuation in the environment in the disordered compound, $N(\varepsilon)$ is broadened resulting in a drop in $N(E_F)$ and a concomitant drop in T_c [14].

RARE-EARTHS IN TERNARY COMPOUNDS

Although the A15 compounds have the highest known T_c's (≈ 23 K for Nb_3Ge), the Chevrel phases have the highest known values of H_{c2} (≈ 600 kOe for $PbMo_6S_8$). The addition of paramagnetic ions to superconductors is known to have a depressing effect on T_c due to the pair breaking effect of exchange scattering of the conduction electrons off of the paramagnetic moment. Surprisingly in the Chevrel phases, Fisher et al. [15] found that the addition of rare-earth paramagnetic ions exhibited a very weak depression of T_c and an actual enhancement of H_{c2} in these extremely strong type-II superconductors. Since the Mo d-electrons are known to be principally responsible for the Cooper pairing in these compounds, the Mo nmr is an obvious probe to measure the spin polarization transferred from the rare-earth ions to Mo [16]. The Knight shift K of the ^{95}Mo resonance is shown together with the reduced magnetization for $Sn_{0.5}Eu_{0.5}Mo_6S_8$ as a function of H/T in Fig. 11. Both from the J = 7/2 Brillouin function behavior of the magnetization and the ^{151}Eu Mössbauer isomer shift it is found that Eu is in the divalent $^8S_{7/2}$ state. The interesting feature of the ^{95}Mo K-shift are the large negative values at low temperature that persist in the superconducting state. The small positive change in K below T_c is similar to that in pure $SnMo_6S_8$ and is due to the quenching of the d-band spin susceptibility in the superconducting state. The enhanced H_{c2} in $Sn_{0.5}Eu_{0.5}Mo_6S_8$ can be understood as follows. We consider the system to contain two bands which we label the "s band" and the "d

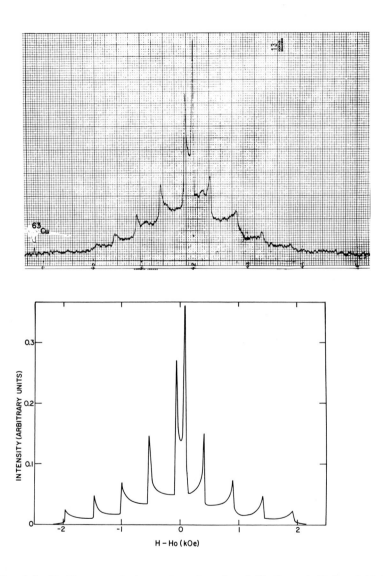

Fig. 7. (a) Field sweep of the quadrupole echo of Nb$_3$Sb at 4.33 K and 43.98
MHz. Field increases to the right at 460 Oe per major division. ^{63}Cu resonance
in the rf coil is the small peak at the left-hand side of the trace. (b)
Synthetic spectra for e^{2}Qq/h = 24.2 MHz and K$_{ax}$ = -0.08%. [After Ref. 11.]

40

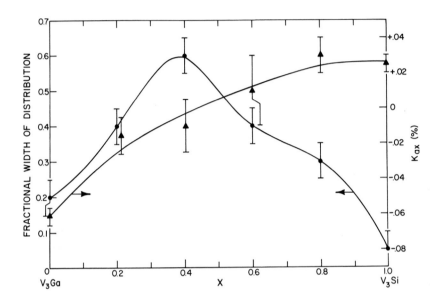

Fig. 8. Composition dependence of the mean axial Knight shift K_{ax} and the fractional width of the Gaussian distribution of the electric field gradient and K_{ax} in the $V_3Ga_{1-x}Si_x$ compounds. [After Ref. 10.]

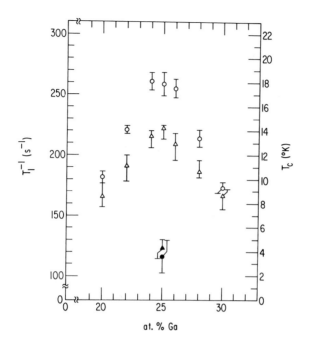

Fig. 9. Composition dependence of the spin-lattice relaxation
rate T_1^{-1} at ≈ 77 K ($\nu_o = 12$ MHz, $H_o \approx 10.6$ kG) (open circles) and the
superconducting-transition temperature T_c (open triangles) for annealed powders
of unirradiated $V_{0.75-x}Ga_{0.25+x}$ samples. The solid figures denote the
corresponding value for the neutron-irradiated sample of V_3Ga. The vertical bars
on the T_c data points represent the width ΔT_c of the superconducting
transition. [After Ref. 14.]

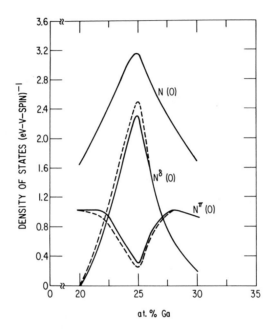

Fig. 10. Composition dependence of the density of states at the Fermi level. Total density of states N(0) interpolated from heat capacity data. The $N^\delta(0)$ and $N^\pi(0)$ subband densities of states for $N^\sigma(0) = 0.55$ $(eV-V-spin)^{-1}$ and either $N_s(0) = 0.073$ $(eV-V-spin)^{-1}$ (solid curves) or $N_s(0) = 0.147$ $(eV-V-spin)^{-1}$ (dashed curves). [After Ref. 14.]

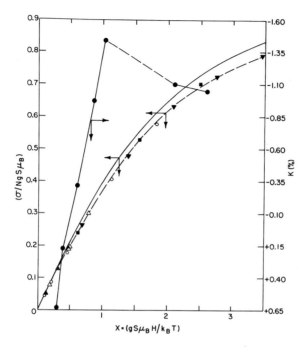

Fig. 11. Reduced magnetization ($\sigma/NgS\mu_B$) and ^{95}Mo Knight shift K as a function of $x = gS\mu_B H/k_B T$ for $Sn_{0.5}Eu_{0.5}Mo_6S_8$. Normal-state magnetization given by the dashed curve. Solid curve for $\sigma/NgS\mu_B$ is the Brillouin function $B_{7/2}(x)$. Knight-shift data for $x > 1.5$ was obtained in the superconducting state. The values of H are the applied field corrected for the demagnetizing field. Absolute shift can be obtained by multiplying 46, 863 G by $K/(1 + K)$. [After Ref. 16.]

band". The superconducting properties are dependent on the d band, presumably derived largely from the Mo ions. Below the superconducting transition temperature, a gap opens in the d band causing the observed change in the Knight shift. When paramagnetic ions are added to the material, a rather weak exchange interaction occurs, largely with the s band. Also, a shielding of the Mo d electrons from the Eu site, which is surrounded by sulfur atoms, inhibits exchange scattering of the d electrons off the Eu^{2+} moments, and so T_c is weakly dependent on the magnetic-ion concentration. Therefore, the primary effect of the magnetic ions is to cause a polarization of the s-band electrons when the Eu^{2+} moments are aligned in an external field. This polarization is parallel to the field at the Eu site, but shows a spatial dependence such that it is negative at the Mo site. Below T_c, no gap opens in the s band so this polarization remains. As a result, there is a partial compensation of the applied field at the Mo sites due to weak s-d exchange coupling, inhibiting pair breaking of the d-band electrons and causing an enhancement of the critical field over that in the pure $SnMo_6S_8$.

CONCLUSION

In this brief review, I have attempted to illustrate a number of important insights into the difficult materials and solid state physics aspects of type-II superconductivity that can be obtained using various nuclear magnetic resonance techniques. Since the detailed behavior of these transition metal superconducting compounds and alloys is strongly dependent on the character of the conduction electrons within a small energy shell of order the Debye energy around E_F, an extremely sensitive probe of this small number of conduction electrons is required. Because these materials generally have small mean free paths, the important tool of the fermiologists, the deHaas vanAlphen effect, usually can not be applied. Thus, nmr, which in metals is strongly affected by hyperfine interactions with the Fermi energy electrons, is an ideal probe of these interesting and important superconductors.

REFERENCES

1. L. C. Hebel and C. P. Slichter, Phys. Rev. 113, 1504 (1957).

2. J. Bardeen et al., Phys. Rev. 108, 1175 (1957).

3. D. E. MacLaughlin in Solid State Physics, Vol. 31, H. Ehrenreich, F. Seitz and D. Turnbull eds. (Academic Press 1976) p. 1.

4. R. D. Parks, ed. Superconductivity, Vols. 1 & 2 (Dekker 1969).

5. W. Fite and A. G. Redfield, Phys. Rev. Lett. 17, 381 (1966).

6. A. G. Redfield, Phys. Rev. 162, 367 (1967).

7. F. Y. Fradin and D. Zamir, Phys. Rev. B 7, 4861 (1973).

8. B. P. Schweiss et al., in Superconductivity in d- and f-band Metals, D. H. Douglass ed. (Plenum Press 1976) p. 189.

9. W. L. McMillan, Phys. Rev. 167, 331 (1968).

10. F. Y. Fradin and J. D. Williamson, Phys. Rev. B 10, 2803 (1974).

11. F. Y. Fradin and G. Cinader, Phys. Rev. B 16, 73 (1977).

12. F. Y. Fradin et al., in Superconductivity in d- band f-band Metals, D. H. douglass ed. (Plenum Press 1976) p. 297.

13. See for example, J. M. Poate et al., in Superconductivity in d- and f-band Metals, D. H. Douglass ed. (Plenum Press 1976) p. 489.

14. F. Y. Fradin et al., Solid State Commun. 30, 737 (1979).

15. Ø. Fischer et al., J. Phys. C 8, L474 (1975); Ø. Fischer et al. in Superconductivity in d- and f-band Metals, D. H. Douglass ed. (Plenum Press 1976) p. 175.

16. F. Y. Fradin et al., Phys. Rev. Lett. 38, 719 (1977).

MAGNETIC RESONANCE AS A PROBE OF ANISOTROPIC CONDUCTORS

W. G. CLARK
Physics Department, University of California at Los Angeles, Los Angeles,
California 90024

ABSTRACT

A review is given of the wide range of magnetic resonance
methods used to study the special properties of quasi one-
dimensional (1-d) conductors. Specific examples are presented which show how magnetic resonance methods have been exploited (a) to tell if the conduction is 1-d, (b) to locate
the conduction electrons on a molecular distance scale,
(c) to probe the unusual effects of disorder in 1-d conductors, (d) to study the charge density wave transition, and
(e) to verify exotic electrical transport mechanisms, such
as charged solitons in trans-$(CH)_x$.

INTRODUCTION

In the past few years, a large number of highly anisotropic, or quasi one-
dimensional (1-d) conductors have been discovered. They have evoked strong
interest on the part of researchers because many of them are organic metals and
because 1-d conductors exhibit special properties and exotic conduction mechanisms which are particular to their one-dimensionality. Some examples of these
are: (a) their instability against a Peierls, or charge density wave (CDW)
transition [1] from a high temperature metallic state to a low temperature
(modified) metallic, semiconducting, or insulating state, (b) nonlinear and
frequency dependent conduction by the CDW's [2], (c) conduction by charged solitons in lightly doped polyacetylene, $(CH)_x$ [3,4], and (d) the unusual importance of disorder on the properties of 1-d conductors [5-7].
    A wide range of magnetic resonance methods have been employed to investi-
gate some of the most basic questions posed by 1-d conductors. In this paper,
several examples are described which demonstrate how magnetic resonance experiments have been used to study the basic properties of 1-d conductors. Even
though this session of the Symposium is formally on NMR, I shall include as
well such topics as ESR and three types of double resonance, as all have played
a key role in research on 1-d conductors. The rest of this paper is organized
in terms of several key questions regarding 1-d conductors.
    In order to keep this paper from being unduly long, some topics which
could be reported have been left out, and the goal of a truly complete bibliography has not been met. Instead, the more recent or comprehensive references
are emphasized, as they are most useful at leading the interested reader to a
more complete bibliography.

IS THE CONDUCTION REALLY 1-D?

    The essential feature of a 1-d conductor is that the motion of the elec-
trons be along only one direction in the solid [8]. For any material proposed
to be a 1-d conductor, it is important that this point be established and, if
possible, the microscopic transport times be established. The traditional and
most widely used method to investigate this question has been to measure the
electric conductivity (σ) parallel and perpendicular to the high conductivity

axis. Such measurements are done over macroscopic distances and may be misleading because of drastic inhomogeneities which are often present in 1-d conductors. An alternative method which has been applied in several cases is to
measure the frequency (ω) dependence of the nuclear (usually proton) relaxation
rate $1/T_1$. As discussed below, there is a special signature for 1-d electronic
motion which helps to so identify it and, in favorable cases, determine the
rate for electron transfer along the high conductivity axis as well as the
much slower rate in perpendicular directions. The ratio of such rates is a
measure of how 1-d is electric conduction in the material. In practice, these
methods probe electron motion over microscopic distances; they are much less
subject to problems of crystal morphology and electrical contact formation than
the usual transport methods. On the other hand, they are rather complicated
to carry out, require a moderately large sample, and require care in the interpretation of the results.

The physical basis for the method is the following. The spins of the
nuclei (fixed in space) are relaxed by the moving conduction electron spins,
which generate a fluctuating magnetic field at the site of the nucleus. These
fluctuations have a power spectrum which reflects the dimensionality of the
electronic motion; it is $\omega^{-1/2}$ for purely 1-d motion, logarithmic for purely
2-d motion, and frequency independent for 3-d motion. For the non-ideal case
of strongly anisotropic diffusion, an ordered crossover going from 1-d to 3-d
behavior occurs at progressively longer times [9]. The formula which describes
all three cases is [10,11]:

$$1/T_1 = \Omega^z F_z(\omega_N) + \Omega^+ F_+(\omega_e) \quad , \tag{1}$$

where Ω^z and Ω^+ are related to the electron-nuclear couplings [12], ω_N and ω_e
are respectively the nuclear and electron Larmor frequencies, and
$F(\omega)$ is the frequency spectrum of the appropriate electron spin fluctuations at
the site of the nucleus. Two terms are important:

$$F_z(\omega) = \int_{-\infty}^{\infty} <S^z(t)S^z(0)> e^{i\omega t} dt \tag{2}$$

for those (dipolar) parts of the electron-nuclear interaction which flip a
nuclear spin but leave the electron spin unchanged in direction, such as I^+S^z,
and

$$F_+(\omega) = \int_{-\infty}^{\infty} <S^+(t)S^-(0)> e^{i\omega t} dt \tag{3}$$

for the parts (dipolar and hyperfine) of the electron-nuclear interaction which
flip both, such as I^-S^+. In both cases, I is the nuclear spin operator, $\vec{S}$
is the electron spin operator, and t is the time. For a diffusive model of the
electron motion, the electron spin correlation functions $<S^z(t) S^z(0)>$ and
$<S^+(t) S^-(0)>$ have a slow decay at long t which varies as $(Dt)^{-d/2}$, where d is
the dimensionality and D is the diffusion coefficient. For the purely 1-d case
(d=1), this leads to $1/T_1 \propto (D\omega)^{-1/2}$.

The time scales that are probed depend on the details of the electron-
nuclear coupling. From Eqs. (1)-(3) it is seen that fluctuations at ω_N are important for the correlation function $<S^z(t) S^z(0)>$ and fluctuations
at ω_e are important for $<S^+(t) S^-(0)>$. For fields up to 10T, this means that
the two terms probe time scales down to $1/\omega_N \simeq 0.4\times10^{-9}$s (protons) and $1/\omega_e \simeq$
0.7×10^{-12}s.

Most of the experimental studies of 1-d conductors via the ω dependence of $1/T_1$ have been carried out on highly conducting TCNQ (tetracyanoquinodi-methane) salts, such as TTF-TCNQ (tetrathifulvalene-TCNQ) [10,13], NMP-TCNQ (N-methylphenazinium-TCNQ) [10], and Qn(TCNQ)$_2$ (quinolinium-TCNQ) [10,14,15].

The analysis of the experiments is rather complicated, and direct refer-ence to the literature should be made for those details, especially with regard to other relaxation mechanisms not discussed in this paper. Here, we point out that clear evidence for 1-d electronic motion has been found in Qn(TCNQ)$_2$ [10, 15] and NMP-TCNQ [10]. Work on the former is especially noteworthy, as com-bined T_1 and σ measurements as a function of temperature (T) have shown that the observed $\sigma(T)$ is to be associated mainly with changes in the electrical mobility, and not in the number of carriers (n). This point is important for 1-d materials such as Qn(TCNQ)$_2$, as conventional Hall measurements of n have not been successful. It also argues against one of the models proposed for $\sigma(T)$ in such materials [16].

The nuclear relaxation work on the microscopic dimensionality of electron-ic conduction in TTF-TCNQ [10,13] shows that it is highly anisotropic, but there is some controversy as to whether it is 1-d or 2-d. Similar nuclear re-laxation work on (CH)$_x$ is discussed later in this paper along with other work on that material.

WHERE ARE THE CONDUCTION ELECTRONS?

An important topic of research in 1-d conductors is to identify on a molecular scale the location of the conduction electrons. This is especially true for many organic charge transfer salts, such as TTF-TCNQ, where more than one chain of molecules can be responsible for the conduction. For most of the work described in this section, locating the electrons also permits measuring their spin susceptibility, which is intimately related to their 1-d properties.

Since many organic 1-d conductors have a relatively narrow ESR line asso-ciated with the electrons responsible for conduction, ESR methods have been widely used to investigate their properties. The two main pieces of informa-tion obtained from such studies are the g-shift and the intensity of the ESR absorption. From the former, one is often able to obtain the molecular loca-tion of the spins, and from the latter, their magnetic susceptibility χ.

One important example of this approach has been the work of Tomkiewicz et al. [17] on the organic conductor TTF-TCNQ and related materials [18]. In TTF-TCNQ, electronic conduction occurs on both the TTF and TCNQ chains, each of which has its own g value (g_F and g_Q respectively). But because of a substan-tial interaction between electrons on the two chains, a single ESR sign is seen which lies between them and is given by

$$g = \frac{\chi_F}{\chi} g_F + \frac{\chi_Q}{\chi} g_Q \tag{4}$$

$$\chi(T) = \chi_F(T) + \chi_Q(T) \tag{5}$$

where g is the observed g-value, χ is the total spin susceptibility of the two chains, and χ_F and χ_Q are the individual chain susceptibilities. Values for g_F and g_Q are known from ESR work on the individual molecules. The quantities obtained directly in an experiment are g(T) and $\chi(T)$ from the ESR field and the area under the absorption curve respectively. Equations (4) and (5) are then used to obtain separately $\chi_F(T)$ and $\chi_Q(T)$. They reflect the individual filling and widths of both TTF and TCNQ conduction bands, as well as their separate Peierls transitions, which occur at different temperatures.

A different magnetic resonance experiment aimed at the same physical problems was done by Rybaczewski et al. [19]. They measured the Knight shift of enriched ^{13}C located on the CN groups of the TCNQ molecule. This shift is proportional to $\chi_Q(T)$. When combined with static measurements of $\chi(T)$ and proton T_1 measurements, an overall separation into $\chi_F(T)$ and $\chi_Q(T)$ was achieved which was in substantial agreement with the ESR experiments.

A different kind of experiment to locate the electrons responsible for conduction has been carried out by Clark et al. [6]. This was a double resonance experiment on the disordered organic metal Qn(TCNQ)$_2$ in which the hyperfine shift ("Day" shift) of the ESR signal was observed upon changing the proton polarization. The physical origin of the effect is the hyperfine field (H_{hf}) at a (delocalized) electron, which is given by

$$H_{hf} = a_H <I_z> \qquad (6)$$

where a_H is the hyperfine constant for the molecule of interest (typically ~ 1 G per proton) and $<I_z>$ is the average value of the z-component of the proton spin. At normal temperatures and magnetic fields, the thermal equilibrium $<I_z>$ is very small.

The experiment was carried out in several steps [20]. The ESR signal in fully protonated Qn(TCNQ)$_2$ was partially saturated by a 9.5 GHz microwave field at resonance. In this way an enhanced $<I_z>$ was obtained via the Overhauser effect. Then a second saturating rf field was swept through the proton NMR frequency (ω_p). When this frequency reached ω_p, the enhanced $<I_z>$ was set to zero, and the corresponding change [Eq. (6)] in the position of the ESR signal was observed. This measurement was repeated on a sample in which the TCNQ protons had been replaced by deuterons. No comparable shift was observed in the latter sample, thereby showing that the protons which interact directly with the electrons are on the TCNQ chains. From this it was concluded that the electrons are also located on the TCNQ chains.

Finally, there has been a series of studies in which the molecular location of the electron spin density in organic conductors has been investigated by measurement of the proton Knight shift, linewidth, and T_1 in selectively deuterated materials. An example of this is the work of Devreux et al. [12] in TTF and TCNQ salts. By selective deuteration of individual chains, they are able to change the electron-nuclear interaction on a local basis. Analysis of their results yields values of the individual chain susceptibilities in several materials, as well as numbers for both the isotropic and anisotropic parts of the electron-nuclear interaction. For further details of the work in this area, one should consult their papers, as well as the references cited in them.

1-D CONDUCTORS WITH DISORDER

There is an important class of organic conductors whose electrical properties are dominated by disorder. The disorder may be intrinsic, as in the case of many TCNQ salts with asymmetric donors [21], induced chemically by alloying one chain with similar symmetric molecules [22], or induced by irradiation [22, 23]. On a microscopic scale the effect of the disorder can be thought of as creating interruptions in 1-d metallic chains, with a subsequent temperature dependent localization of the conduction electrons. There are many consequences for the electrical and magnetic properties of such materials, including: (a) a T dependence for σ similar to that of a semiconductor [21], (b) a frequency dependence to σ [5], (c) a giant dielectric constant which has a strong dependence on T and ω [5], (d) nonlinear electrical conduction [5], (e) an unusual specific heat at low T [24], and (f) a quasi-universal low T χ which varies as [7,23]

$$\chi(T) = AT^{-\alpha} \qquad (7)$$

where the amplitude factor A and exponent $\alpha \simeq 0.8$ are constants which depend somewhat on the material and its method of preparation. In addition, (g) the disorder inhibits or eliminates [7,25] the Peierls transition which commonly occurs in 1-d conductors with little or no disorder [1].

Here, we discuss how ESR measurements of $\chi(T)$ have played a substantial role in establishing the last two points. Most of the materials investigated have a narrow enough ESR absorption that low field (H = 10 Oe, $\omega_e/2\pi$ = 36 MHz) measurements can be made. By comparing the area under the absorption with that of the proton absorption in the same sample and using the same spectrometer settings (H = 8.46 kOe), the absolute value of χ is obtained [26]. This ESR method has permitted measurements of χ over an extremely broad range of T, from 0.4 mK to 300 K in Qn(TCNQ)$_2$ [6,25,27], and from 30 mK to 300 K in several others [28]. It is typically observed that the smoothly divergent χ of Eq. (7) is obeyed in materials with intrinsic disorder between about 20 K and 5 mK. The relation to disorder comes from the fact that the physical origin of Eq. (7) is a 1-d Heisenberg antiferromagnet with random exchange [7]. The 1-d exchange is due to the strong intrachain electronic interactions, and its randomness comes from the stochastic variation in the spacing of the electrons localized by disorder.

The measurements of $\chi(T)$ also demonstrate the absence of a phase transition, even down to very low T. This is in sharp contrast to 1-d metals, such as TTF-TCNQ, which show a large drop in χ when the Peierls gap opens at low T. In fact, by introducing disorder into TTF-TCNQ via neutron irradiation, the emergence of a term in χ following Eq. (7) and a suppression of the Peierls transition are observed.

A recent ESR investigation [23] of the g-value, χ, and linewidth ($\Delta H_{1/2}$) in Qn(TCNQ)$_2$ as a function of T and defect concentration induced by neutron irradiation has shown several interesting points regarding this disordered 1-d conductor. At low T, A [Eq. (5)] increases linearly with, and α is nearly independent of, the induced defect concentration, as expected on the basis of recent renormalization calculations [29]. This work indicates an intrinsic defect concentration for as-grown material that is about the same as that inferred from the effect of neutron irradiation on the dielectric constant of Qn(TCNQ)$_2$ by using an interrupted metallic strand model. It also shows no change in the g-value of the ESR with irradiation, which indicates that all of the unpaired spin density observed via ESR occurs on the conducting TCNQ chain.

We also point out here that the very narrow ESR line ($\Delta H_{1/2}$ as low as 70 mG) seen in Qn(TCNQ)$_2$ is consistent with what is expected of a 1-d organic conductor. If the unpaired electron spin were localized on a single TCNQ molecule with no intrachain exchange, the hyperfine and dipolar interactions would generate a $\Delta H_{1/2}$ much larger than what is observed [23,27]. But the strong intrachain and/or delocalization lead to a line with a Lorentzian shape which is very narrow.

In addition to the ESR measurements of χ just described, a substantial number of pulsed NMR and ESR experiments have been used to study 1-d conductors with disorder. Some of them carried out at high T on Qn(TCNQ)$_2$ have already been discussed here in connection with establishing that the electronic motion is 1-d [10,15]. Pulsed ESR experiments at low T have been interpreted successfully in terms of the same random exchange model used to establish Eq. (7) [6, 27]. Low T proton T_1 measurements, on the other hand, show an unusual behavior in Qn(TCNQ)$_2$ which has not yet been explained [14].

PHASE TRANSITIONS

Magnetic resonance methods have been used successfully to study certain microscopic aspects of the CDW transition exhibited by many 1-d conductors. A beautiful example of this is the recent work of Suits et al. [30] in 2H-TaSe$_2$.

Although this material is not a 1-d metal in the strict sense, the Fermi surface properties responsible for the CDW transition at 120 K are those of a 1-d metal, i.e., large parts of it form nearly parallel sheets.

Below 90 K, the CDW distortion in 2H-TaSe$_2$ was known to be commensurate with a wave vector $\vec{k}$ exactly 1/3 that of the lattice. Between 90 K and 120 K $\vec{k}$ is incommensurate by about 1%. The work of Suits et al. was to test the hypothesis of McMillan [31] that in the incommensurate regime the CDW was commensurate over short distances, with short phase slip breaks ("discommensurations") between these regions to generate the overall incommensurability seen in x-ray measurements [32]. The probe used by Suits et al. was the Knight shift (K) of ^{77}Se in a single crystal sample. At high T a single NMR line is seen as all of the ^{77}Se are in equivalent positions. Below 90 K, in the commensurate CDW phase, there are three equivalent positions, with Knight shifts which give a three line spectrum whose splitting is proportional to the amplitude of the CDW distortion. In the incommensurate phase discommensurations retain the three line Knight shift spectrum, whereas a purely incommensurate $\vec{k}$ would give a single broad line. The experimental observation was that the three line spectrum persisted with a spread which collapsed continuously on going from 90 K to 120 K. From this, they were able to confirm the discommensuration model and measure the T dependence of the CDW amplitude.

Similar work on 2H-NbSe$_2$ has been done by Berthier et al. [33]. Also, ESR studies of the CDW transition in TTF-TCNQ have been discussed earlier in this paper [17].

EXOTIC CHARGE TRANSPORT IN 1-D CONDUCTORS

One of the facts that makes 1-d conductors so interesting is that they support unusual mechanisms of charge transport, such as moving CDW's in NbSe$_3$, etc. In this section we describe magnetic resonance work on the conducting polymer trans-(CH)$_x$ which has led to the idea that in the lightly doped regime its electrical conduction is via charged solitons [3,4]. This is one of the most exotic electronic conduction processes known. It is also a subject of intense controversy [34] and activity; therefore, the picture presented here may have to be modified in the future.

First, we briefly outline what is meant by a soliton in trans-(CH)$_x$. This material is formed from the cis-isomer at room temperature by what is probably a nucleation process followed by propagation along the polymer chains [35]. The structure of trans-(CH)$_x$ is shown in fig. 1. Because of the high symmetry, segments with the double bond in the positions indicated by parts A and B are degenerate. Where such pieces come together there is a discontinuity as indicated by the arrow in fig. 1a. This discontinuity is called a neutral soliton, and carries one unpaired spin. Calculations [3] and the experiments described below indicate that it is actually spread out, as shown in fig. 1b, and that it is mobile. If the isomerization is a random nucleation process from the cis-isomer, it is expected that solitons will always be present in pure trans-(CH)$_x$. Electrical conduction in trans-(CH)$_x$ is obtained by doping it with AsF$_5$ [36]. It has been proposed that in the lightly doped material this dopant accepts an electron from the soliton to form a soliton which is now charged and mobile, but with no unpaired spins [3,4].

Now we describe the variety of magnetic resonance experiments which have helped to establish the above picture. They cover a wide range, including ESR, proton relaxation, and the Overhauser and solid state effect. Several conclusions have been drawn with the help of standard ESR absorption experiments. The amplitude of the ESR in undoped material shows an intrinsic defect concentration of about 1 in 3,000, which is consistent with the idea that A and B type segments are generated at random in the isomerization process [4]. Upon light doping with AsF$_5$, the conductivity increases rapidly, but the ESR signal

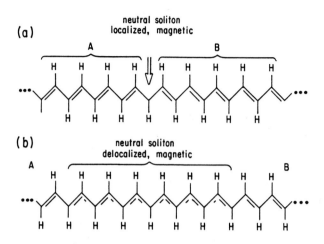

Fig. 1. Schematic representation of the neutral soliton in trans-(CH)$_x$.
(a) The defect separating equivalent chain semgnets A and B is indicated by the arrow. The defect is neutral and has one unpaired electron spin. (b) The energetically favorable [3] spreading of the defect
over several polymer units is indicated.

intensity decreases [4]. This is interpreted as conversion of neutral magnetic
solitons to charged, non-magnetic ones. The linewidth of the ESR is much less
than that expected of the hyperfine interaction of a point soliton (fig. 1a)
with the nearest proton, thereby indicating that the soliton is spread out, as
in fig. 1b, and/or that the soliton is mobile [4]. Similar measurements on
deuterated samples show that the hyperfine interaction is an important source
of the linewidth. The fact that the line has a Lorentzian shape is taken to
show that there is rapid motion of the soliton, as a purely static, spread out
soliton would exhibit a Gaussian shape [4].
    Additional tests of soliton motion have been carried out using proton
relaxation and double resonance methods [37]. Measurement of the proton T_1
in trans-(CH)$_x$ shows $1/T_1 \propto \omega^{-1/2}$ over an extremely wide range of ω. As indicated earlier in this paper, such a result is evidence of 1-d diffusive motion
by the magnetic entities which relax the protons. In the case of trans-(CH)$_x$,
this 1-d character is perhaps more pronounced than in any mother material so
far studied. Finally, a study of the enhancement of the proton NMR signal in
mixed cis/trans-(CH)$_x$ samples with saturation of the ESR signal shows a crossover from the solid state effect in cis-(CH)$_x$ to the Overhauser effect in
trans-(CH)$_x$ [37]. This is interpreted that the solitons become more mobile
as conversion to the transisomer becomes more complete.

CONCLUSION

    In this paper we have mentioned many of the magnetic resonance methods
which have been brought to bear on the myriad properties exhibited by 1-d conductors. These examples are summarized in Table 1, where the numerical entries

indicate the appropriate reference. This field is at present very active, and new applications will undoubtedly appear in the future. That part of the work reported here from the group of W. G. Clark at UCLA is based on research supported by the National Science Foundation on Grant DMR 77-23577.

TABLE I Some 1-d conduction problems and materials studied by magnetic resonance.

TYPE OF PHYSICAL PROBLEM STUDIED	MAGNETIC RESONANCE METHOD															
	NMR							ESR						DOUBLE		
	Knight shift	intensity, χ	linewidth	simple T_1	3 vs T_1	T vs T_1	vs deuteration	intensity χ	linewidth	line shape	g-shift	3 vs T_1	T vs T_1	Day shift	Overhauser	Solid state
Is cond. 1-d? Qn(TCNQ)$_2$, (CH)$_x$ TTF-TCNQ					9,10 13 15	13 15	10 13									
Carrier location Qn(TCNQ)$_2$ TTF-TCNQ							10				23 6 17			6 20	6 20	
Cond. el. χ Qn(TCNQ)$_2$ TTF-TCNQ	19	27 6						17 6 27			17					
Motion, deloc. of solitons (CH)$_x$		4			37			4	4	4					37	37
CDW trans. TTF-TCNQ	19			19				17			17					
el. localization by disorder Qn(TCNQ)$_2$		23 6						23 6	23			23 6	23 6			
Intrinsic defect conc. Qn(TCNQ)$_2$		23						23	23							
Discommensuration in CDW TaSe$_2$	30															
el.-nucl. coupling const's TTF-TCNQ + $\cdots$	12		12													

REFERENCES

1. T. D. Schultz in: The Physics and Chemistry of Low Dimensional Solids, L.
    Alcácer, ed. (D. Reidel, Dordrecht, 1980) p. 1.

2. G. Grüner, L. C. Tippie, W. G. Clark, and N. P. Ong, Phys. Rev. Lett. 45,
    935 (1980); R. M. Fleming and C. C. Grimes, Phys. Rev. Lett. 42, 1423
    (1979); N. P. Ong and P. Monceau, Phys. Rev. B 16, 3443 (1977). M. Weger,
    G. Grüner, and W. G. Clark, Solid State Commun. 35, 243 (1980).

3. W. P. Su, J. R. Schrieffer, and A. J. Heeger, Phys. Rev. Lett. 42, 1698
    (1979); M. J. Rice and E. J. Mele, Solid State Commun. 35, 487 (1980).

4. B. R. Weinberger, E. Ehrenfreund, A. Pron, A. J. Heeger, and A. G.
    MacDiarmid, J. Chem. Phys. 72, 4749 (1980).

5. G. Grüner, Chemica Scripta, to be published (Proceedings of the Internation-
    al Conference on Low Dimensional Synthetic Metals, Helsingor, Denmark,
    1980).

6. W. G. Clark, J. Hammann, J. Sanny, and L. C. Tippie in: Lecture Notes in
    Physics 96, Quasi One-Dimensional Conductors II, Proceedings of the Inter-
    national Conference, Dubrovnik, 1978, S. Barisić et al., eds. (Springer,
    New York, 1979) p. 255.

7. W. G. Clark in: Proceedings of the International Conference on Physics in
    One-Dimension, Fribourg, 1980 (Springer, Berlin), to be published.

8. In this paper, only electronic conduction is considered. Note, however,
    that ionic conduction can also be highly 1-d [see J. Bernasconi, H. V.
    Beyeler, S. Strässler, and S. Alexander, Phys. Rev. Lett. 42, 819 (1979)].

9. M. A. Butler, L. R. Walker, and Z. G. Soos, J. Chem. Phys. 64, 3592 (1976).

10. F. Devreux and M. Nechtschein in: Lecture Notes in Physics 95, Quasi One-
    Dimensional Conductors I, Proceedings of the International Conference,
    Dubrovnik, 1978, S. Barisić et al., eds. (Springer, New York, 1979) p. 145.

11. L. N. Bulaevskii, A. I. Buzdin, and D. I. Khomskii, ibid., p. 135.

12. F. Devreux, Cl. Jeandey, M. Nechtschein, J.M. Fabre, and L. Giral, J. Phys.
    (Paris) 40, 671 (1979).

13. G. Soda, D. Jerome, M. Weger, J. Alizon, J. Gallice, H. Robert, J.M. Fabre,
    and L. Giral, J. Phys. (Paris) 38, 931 (1977).

14. L. J. Azevedo, W. G. Clark, E. O. McLean, and P. F. Seligmann, Solid State
    Commun. 16, 1267 (1975).

15. F. Devreux, M. Nechtschein, and G. Grüner, Phys. Rev. Lett. 45, 53 (1980).

16. A. J. Epstein and J. S. Miller in: Lecture Notes in Physics 96, Quasi One-
    Dimensional Conductors II, Proceedings of the International Conference,
    Dubrovnik, 1978, S. Barisić et al., eds. (Springer, New York, 1979) p. 286.

17. Y. Tomkiewicz, A. R. Taranko, and J. B. Torrance, Phys. Rev. B 15, 1017
    (1977).

18. Y. Tomkiewicz, J. R. Andersen, and A. R. Taranko, Phys. Rev. B 17, 1579
    (1978).

19. E. F. Rybaczewski, L. S. Smith, A. F. Garito, A. J. Heeger, and B. G.
    Silbernagel, Phys. Rev. B 14, 2746 (1976).

20. J. Sanny, Ph.D. Thesis, University of California at Los Angeles, 1980.

21. K. Holczer, G. Mihály, G. Grüner, and A. Jánossy, J. Phys. C 11, 4707
    (1980).

22. M. Miljak, B. Korin, J. R. Cooper, K. Holczer, G. Grüner, and A. Jánossy, J. Magn. and Magn. Mater. 15-18, 219 (1980); M. Miljak, B. Korin, J. R. Cooper, K. Holczer, and A. Jánossy, J. Phys. (Paris) 41, 639 (1980).

23. J. Sanny, G. Grüner, and W. G. Clark, Solid State Commun. 35, 657 (1980).

24. L. J. Azevedo and W. G. Clark, Phys. Rev. B 16, 3252 (1977).

25. H. M. Bozler, C. M. Gould, and W. G. Clark, Phys. Rev. Lett. 45, 1303 (1980).

26. R. T. Schumacher and C. P. Slichter, Phys. Rev. 101, 58 (1956).

27. L. C. Tippie and W. G. Clark, Phys. Rev. B, to be published.

28. J. Hammann, W. G. Clark, A. J. Epstein, and J. S. Miller in: Lecture Notes in Physics 96, Quasi One-Dimensional Conductors II, Proceedings of the International Conference, Dubrovnik, 1978, S. Baristić et al., eds. (Springer, New York, 1979) p. 310; J. Hammann, L. C. Tippie, and W. G. Clark, ibid., p. 309.

29. C. Dasgupta and S.-k. Ma, Phys. Rev. B 22, 1305 (1980); S. R. Bondeson and Z. G. Soos, Phys. Rev. B 22, 1793 (1980); J. E. Hirsch and J. V. José, J. Phys. C. Letts. 13, L53 (1980) and Phys. Rev. B, to be published.

30. B. H. Suits, S. Coutrié, and C. P. Slichter, Phys. Rev. Lett. 45, 194 (1980).

31. W. L. McMillan, Phys. Rev. B 12, 1187 (1975), and 14, 1496 (1976).

32. D. E. Moncton, J. D. Axe, and F. J. DiSalvo, Phys. Rev. B 16, 801 (1977).

33. C. Berthier, D. Jerome, and P. Molinie, J. Phys. C 11, 797 (1978).

34. Y. Tomkiewicz, T. D. Schultz, H. B. Broom, T. C. Clarke, and G. B. Street, Phys. Rev. Lett. 43, 1532 (1979); A. J. Epstein, H. W. Gibson, P. M. Chaikin, W. G. Clark, and G. Grüner, Phys. Rev. Lett. 45, 1730 (1980).

35. J. C. W. Chien, F. E. Karasz, G. Wnek, A. G. MacDiarmid, and A. J. Heeger, J. Polym. Sci. Polym. Lett. Ed. 18, 45 (1980).

36. C. K. Chiang, C. R. Fincher Jr., Y. W. Park, A. J. Heeger, H. Shirakawa, E. J. Louis, S. C. Gau, and A. G. MacDiarmid, Phys. Rev. Lett. 39, 1098 (1977); C. K. Chiang, Y. W. Park, A. J. Heeger, H. Shirakawa, E. J. Louis, and A. G. MacDiarmid, J. Chem. Phys. 69, 5098 (1978).

37. M. Nechtschein, F. Devreux, R. L. Greene, T. C. Clarke, and G. B. Street, Phys. Rev. Lett. 44, 356 (1980).

NMR TECHNIQUES FOR STUDYING IONIC DIFFUSION IN SOLIDS

DAVID C. AILION
Department of Physics, University of Utah, Salt Lake City, UT 84112

ABSTRACT

A survey of NMR relaxation time techniques for studying
ionic diffusion in solids is presented. Particular
emphasis is placed on discussing the kinds of information
obtainable from T_1, $T_{1\rho}$, T_{1D}, and $T_{1D'}$ measurements.
Applications to the study of local and nonlocal diffusion,
diffusion in weakly magnetic systems, and motions between
unequal potential wells are described. Relaxation due to
fluctuating dipolar, quadrupolar, and chemical shift
anisotropy interactions is discussed.

INTRODUCTION

Nuclear magnetic resonance (NMR) has long been used in the study of atomic
diffusion and molecular reorientations in solids. These investigations have
been applied to ionic diffusion in insulators [1], metals [2], superionic
materials [3], and amorphous materials [4]. NMR is not restricted to cases
where there is mass transport as in translational diffusion, but has been
extensively applied to the study of reorienting molecules [5] and to localized
(bound) diffusion [6]. In this paper, the principal NMR techniques for studying atomic motions will be discussed. Examples will be given which illustrate
the potential and limitations of these techniques.

The great strength of NMR is that the resonance signal is characteristic
of the particular nucleus being studied. This features arises from the fact
that the nuclei of many elements are characterized by intrinsic magnetic
moments; accordingly, in a magnetic field H_0, the resonance frequency f_0
equals $\gamma H_0/2\pi$ [7], where the gyromagnetic ratio γ is a constant which is
different for different nuclei [8]. For this reason, NMR can normally distinguish the diffusion of a specific nuclear species from that of other
nuclei. In contrast, techniques like dielectric loss and ionic conductivity
measure properties which do not belong exclusively to the nuclei whose
motions are of interest. Furthermore, the immediate surroundings of a nucleus
may affect measurable NMR properties like linewidth, relaxation time, and
even the resonance frequency. Accordingly, NMR techniques can be used to
study the environment of the nuclei and thereby elucidate the microscopic
details governing atomic motions.

There are two problems facing a researcher who wishes to use magnetic
resonance to study ionic diffusion in solids. The first is that NMR is a
bulk technique whose sensitivity is proportional to the number of identical
nuclei being studied and to their γ. Accordingly, its application to the
direct study of nuclei of low concentration (e.g., impurities and surface
atoms) and to nuclei of low γ is more limited. As a result, these nuclei are
usually studied indirectly, by observing their effect on the resonance of

other atoms [9]. A second problem which occurs to some extent in all solids is that the linewidth is usually quite large (~ 1-100 gauss), in contrast to the narrow line characteristic of liquids. This large linewidth arises in solids from magnetic dipole-dipole and electric quadrupole interactions. As a result, NMR signals may be difficult to observe, especially for nuclei having a large quadrupole moment which are in non-cubic or inhomogeneous environments.

MOTIONAL NARROWING AND SPIN-LATTICE RELAXATION

Linewidths and Motional Narrowing. Probably the most straightforward method for studying ionic diffusion in solids is to observe motional narrowing of the linewidth. In order to understand this phenomenon, we must briefly review the origin of line broadening in solids.

Consider a solid consisting of identical atoms, each of which has spin 1/2 and thus no quadrupole moment. If we place this system in a large static magnetic field $H_o\vec{k}$ and in a perpendicular alternating field $H_1 \cos\omega t$, we will observe a resonance absorption line as ω is swept through the resonance frequency $\omega_0 (= \gamma H_0)$. The linewidth arises from the fact that each nucleus sees a magnetic field which is the vector sum of H_0 and the average local dipolar field arising from the neighboring spins. Since the instantaneous value and direction of this dipolar field depends on the actual orientations of the neighbors, there will be a spread in the value of the dipolar fields through the crystal. This spread gives rise to variations in the resonance frequencies of the individual nuclei of the order of 1-10 gauss typically. This linewidth can be observed conveniently in a pulsed experiment in which H_1 is turned on just long enough to tilt the magnetization into the x-y plane (a 90° pulse). After the rf pulse is turned off, the magnetization in the x-y plane will precess about H_0 which is in the z direction. Since the individual nuclei are precessing at slightly different frequencies, their magnetization contributions will get out of phase and the resultant x-y magnetization will decay to zero with a time constant T_2, called the spin-spin relaxation time. This magnetization response to a 90° pulse is called the free induction decay (FID). Since the FID is the Fourier transform of the line-shape function [10], T_2 is proportional to the inverse of the line-width. In a solid, typical T_2 values are in the range 1-100 μsec.

Now, in order to understand motional narrowing, consider what happens if the nuclei are diffusing between lattice sites having different values of the instantaneous dipolar field at a rate (τ_c^{-1}) more rapid than the rigid-lattice spin-spin relaxation rate (T_2^{-1}). For a symmetric resonance line, the resultant field experienced by any one nucleus will be greater than H_0 as often as it will be less than H_0, thereby resulting in a decreased average spread in the dipolar field. Thus, we get a decrease in linewidth or, alternatively, an increase in T_2. This line narrowing occurs only when the mean atomic jumptime τ_c is less than T_2, in which case it can be shown [11] that the linewidth and thus T_2^{-1} are proportional to the jumptime τ_c. Since τ_c equals $\tau_0 \exp(E_A/kT)$, a measurement of the temperature dependence of T_2 gives the activation energy E_A. By determining the temperature for the onset of motional narrowing, we can determine roughly the pre-exponential factor τ_0. We should emphasize that this motional narrowing method is limited to temperatures for which $\tau_c < T_2$ (~ 10 μsec in solids).

In temperature regions having considerable motional narrowing, the line-width and thus T_2 may be limited by inhomogeneities in the applied H_0 field rather than local field effects. In this case, T_2 will become independent of temperature and will no longer provide information about diffusion. It is still possible, fortunately, to obtain diffusion information in this region by the use of the spin-echo technique [7]. The essential idea is that dephasing due to local field effects is irreversible, whereas dephasing due to static inhomogeneities in H_0 is essentially reversible. So, if we apply a 90° pulse and at a later time τ (after the FID is complete) apply a 180° pulse we will observe a signal (the spin-echo) at a time 2τ after the original pulse. The 180° pulse reverses the sense of the dephasing and thus refocuses the magnetization which was lost due to H_0 inhomogeneity. By then measuring the amplitude of the spin-echo signal as a function of τ, we can measure the actual T_2 due to diffusion effects.

It is possible in some cases to use motional narrowing to obtain dynamic information about the actual motional process, specifically where the atom of interest is part of a reorienting molecular group. Since the second moment of the resonance line can be measured and calculated [7] in both the rigid lattice and at temperatures for which a particular reorientational motion results in an averaging of some of the dipolar linewidth, a comparison of experimental and theoretical second moments can determine the actual motional process [12] as well as the structure in the motionally narrowed region [13].

Spin-Lattice Relaxation. A second well-established method for studying diffusion is to look for a minimum in a temperature plot of the spin-lattice relaxation time T_1. Spin-lattice relaxation characterizes the process by which a spin system reaches thermal equilibrium with its surroundings and can be due to any process which can transfer energy between the spins and the lattice. In a solid, T_1 is usually much longer than T_2 and typically may range in value from a few milliseconds in metals like Cu to thousands of seconds in insulators of high purity. Furthermore, it is normally dependent on temperature and on frequency ω_0. It is easily measured since it charac-terizes the rate of buildup of the z-component of the magnetization after the x-y dephasing following the FID. One may think of spin-spin and spin-lattice relaxation in a solid as a two-step process--the first (T_2) describes the process of establishing a spin-temperature (i.e., a Boltzmann distribution among the Zeeman levels) whereas the second (T_1) describes the process by which this temperature changes as it approaches the lattice temperature.

It is possible to understand in various ways the process by which ionic diffusion affects T_1. One way is to use the language of perturbation theory [14]. If the Zeeman Hamiltonian is the unperturbed Hamiltonian $\mathcal{K}_0$ and the dipolar is the perturbing Hamiltonian $\mathcal{K}'$, then ionic diffusion results in fluctuations in the dipolar Hamiltonian since it causes the distances between pairs of spins to change. We may then use the formalism of time-dependent perturbation theory to calculate the transition probability for flipping a spin ($T_1{}^{-1}$) due to the time-dependent dipolar Hamiltonian. Since this probability will be a maximum (and T_1 a minimum) for frequency components in the perturbation which equal the energy level spacing divided by $\hbar$, the condition for the T_1 minimum is that $1/\tau_c$ equals ω_0 or $\omega_0 \tau_c \approx 1$. Since only τ_c depends upon temperature, this condition describes the temperature of the minimum for fixed ω_0. (One should note that a plot of T_1 vs. ω_0 for fixed temperature will likewise show a minimum when $\omega_0 \tau_c \approx 1$.)

58

Bloembergen, Purcell and Pound (BPP) showed, assuming an exponential correlation function [15], that T_1 due to diffusion is given[†] by

$$\frac{1}{T_1} \sim \frac{(\omega_d)^2 \tau_c}{1 + \omega_0^2 \tau_c^2} \quad . \tag{1}$$

We see that the slope of $\ln T_1$ vs. reciprocal temperature will be proportional to $-E_A$ and to E_A at temperatures above and below the minimum, respectively. Furthermore, a fit to this curve (or simply a measurement of the temperature of the minimum, at which $\omega_0 \tau_c = 1$) will determine the pre-exponential factor τ_0. Finally, on the low temperature side of the minimum ($\omega_0 \tau_c \gg 1$), Eq. (1) predicts that $T_1 \propto \omega_0^2$; thus a measurement of the frequency dependence of T_1 can determine the applicability of Eq. (1) or the validity of assuming an exponential correlation function.

The T_1 method will provide information about diffusion provided the probability of a spin flip due to diffusion is greater than the probability of a spin flip due to other mechanisms (i.e., T_1 due to diffusion is less than T_1 due to other mechanisms). Since the minimum occurs when $\tau_c \sim \omega_0^{-1} \sim 10^{-8}$ sec, we see that, typically, a T_1 minimum occurs in the motionally narrowed region and that the range of atomic jump rates which can be seen by this method is comparable to that of the motional narrowing method ($\tau_c \lesssim 10^{-4}$ sec).

Ultraslow Motions. The temperature range over which diffusion can be studied by NMR can be extended to appreciably lower temperatures by the ultraslow motion techniques [16] which allow the observation of less frequent atomic motions than can be observed by T_1 measurements. The basic idea can be understood by considering what happens to a T_1 minimum if ω_0 is greatly reduced. We see from Eq. (1) that, if we reduce ω_0, the minimum will correspond to a longer value of τ_c and thus to lower frequency motions which occur normally at lower temperatures. There are experimental problems associated with attempting to study relaxation in very weak fields where the NMR signal might be very small. Also there are theoretical difficulties, since the BPP theory underlying Eq. (1) is a perturbation theory which treats the fluctuating dipolar Hamiltonian due to diffusion as a perturbation on the Zeeman Hamiltonian; such an assumption may not be valid in very weak applied fields. We shall now discuss solutions to both these problems.

One approach is to measure $T_{1\rho}$, the spin-lattice relaxation time in the rotating frame of the rf field H_1. In a strong rf field the spin system will form a canonical distribution among energy levels separated by γH_{eff} rather than γH_0, where the effective field H_{eff} in the rotating frame is given by

$$\vec{H}_{eff} = H_1 \vec{i} + (H_0 - \omega/\gamma) \vec{k} \quad . \tag{2}$$

[†]Actually, there are additional terms of similar form. For interactions involving identical spins there is an extra term proportional to $1/(1 + 4\omega_0^2 \tau_c^2)$; for heteronuclear systems, there are terms having $1 + (\omega_I - \omega_S)^2 \tau_c^2$ and $1 + (\omega_I + \omega_S)^2 \tau_c^2$ in the denominator. (See Ch. VIII of Ref. 10.)

In this case, the spins are characterized by a "spin temperature in the rotating frame" [17], which means that Curie's law will hold, with $\bar{M}_{eq} = C \ \bar{H}_{eff}/T$. The significance is that, at exact resonance, the equilibrium magnetization $\bar{M}_{eq}$ will be parallel to $\bar{H}_1$ rather than to $\bar{H}_0$. Furthermore, since H_1 is typically of order 1-50 gauss, we can study the relaxation in a much smaller effective field than H_0 and, accordingly, can observe slower motions. The $T_{1\rho}$ minimum will then occur when $\gamma H_1 \tau_c \cong 1$, for H_0 at exact resonance. If H_1 is much larger than the dipolar local field, the dipolar fluctuations can still be treated by perturbation theory (this approach is sometimes called the "weak-collision" theory [18]).

In order to get even more effective relaxation so as to observe ultraslow motions, one can reduce H_1 (e.g., by adiabatic demagnetization in the rotating frame (ADRF)[19]) to a value comparable to the dipolar local field or even to zero. In this case, the weak-collision perturbation theory is not valid and a "strong-collision" theory [20] is required to determine the jump time τ_c from the experimentally measured relaxation time $T_{1\rho}$. Even though Eq. (1) is not valid in this case, we see that the T_1 minimum occurs when $1/\tau_c$ is of order of the energy level splitting (in units of $\hbar$) which, for large field, is ω_0. In zero field the energy level splitting is due to dipolar interactions. Thus, the minimum occurs when $1/\tau_c \sim \omega_{dip}$ or $\tau \sim T_2$, which is at the onset of motional narrowing [21].

The strong collison theory assumes that sufficient time elapses between diffusion jumps to enable the dipolar and rotating frame Zeeman interactions to cross-relax to a common spin temperature between each jump. This temperature is disturbed by the sudden change in dipolar energy resulting from a jump, but reequilibrates to a new value prior to the next jump. This spin-temperature assumption is equivalent to requiring that $\tau_c \gg T_2$, which means that the strong-collision theory will be valid in the "rigid-lattice" below the $T_{1\rho}$ minimum and the temperature for motional narrowing. The ADRF process transfers long-range Zeeman order due to spins aligned preferentially along $\bar{H}_{eff}$ to short-range dipolar order arising from spins aligned along their individual local fields. A single diffusion jump per spin will have a major effect on the relaxation, so that in zero H_1 we would expect $T_{1\rho}$ to be of order τ_c. The actual strong-collision result for $T_{1\rho}$ in a field H_1 is

$$\frac{1}{T_{1\rho}} = \frac{2(1 - p)}{\tau_c} \ \frac{H_D^2}{H_1^2 + H_D^2} \quad , \tag{3}$$

where H_D is the dipolar local field and 1-p is a geometrical factor of order unity which characterizes the fractional change in energy resulting from a diffusion jump. (This factor shows a 10-20% dependence on the orientation of the field H_0 relative to the crystal axes and, furthermore, this anisotropy depends on diffusion mechanism [22].) We note that, when H_1 equals zero, $T_{1\rho}$ is of order τ_c, as we have seen intuitively. Furthermore, for $H_1 = 0$, $T_{1\rho}$ is identical to the dipolar relaxation time T_{1D} [23] characterizing energy exchanges between the dipolar system and the lattice. Diffusion can be studied by this technique provided the relaxation rate due to diffusion is greater than the relaxation rate due to other T_1 processes. The resultant condition for observability of diffusion effects is then that $\tau_c < T_1$. Since T_1 values are typically 10^{-3} to 10^3 sec, this condition is much less stringent than the condition for motional narrowing ($\tau_C < T_2$). Actual pulse sequences used in measuring $T_{1\rho}$ and T_{1D} are described in Ref. 16.

It may appear surprising, at first glance, that the process of ADRF results in increased sensitivity in diffusion experiments, since the reduction in heat capacity which occurs because of the ADRF is reversed in the subsequent remagnetization. Thus, one might naively think that the effect of diffusion on the temperature change of the entire spin system would be the same after the second adiabatic process as that which would have resulted had their been no ADRF at all. That this is incorrect can be seen by recognizing that the energy change (and thus the entropy change) due to a diffusion jump will be greater in the demagnetized state than if it had occurred in the fully magnetized state [20].

The most obvious advantage of the slow motion methods is to extend downward in temperature the range over which atomic diffusion can be studied. As a result, diffusion mechanisms occurring only at lower temperatures can also be observed. Moreover, since the effect of diffusion on the dipolar relaxation time will be much greater than its effect on the spin-lattice relaxation time, going to zero field magnifies the relaxation effects, thereby making possible the observation of diffusion whose relaxation effects might otherwise be too weak to be observable. A measurement of the frequency dependence of the relaxation time can determine the relaxation mechanism and can check the validity of the theory used, such as, for instance, the assumption of an exponential correlation function, underlying Eq. (1). $T_{1\rho}$ and T_{1D} measurements greatly increase the frequency range available, since they are effectively at very low frequencies ($\sim 10^3$Hz) in contrast to the typical T_1 measurements ($\sim 10^7$Hz). Finally, by having a wider temperature range available, it is possible to observe the onset of other diffusion mechanisms having different activation energies.

DIPOLAR RELAXATION IN THE ROTATING FRAME ($T_{1D}{}'$)

A problem arises in the interpretation of T_{1D} measurements on heteronuclear systems (i.e., having more than one nuclear species) for which one of the species is diffusing and the others are stationary. For simplicity, consider a system containing two species of comparable abundance and gyromagnetic ratio, whose spin quantum numbers are labelled I and S. The problem is that, if we irradiate the I spins and measure T_{1D}, we will obtain precisely the same result as we would if we had measured T_{1D} by irradiating the S spins, independent of which spin is actually diffusing [24]. Hence, from a T_{1D} measurement alone, we will be unable to determine which species is diffusing. This feature arises because there is normally strong coupling between parts of the dipolar interaction involving different species (i.e., the I-I and S-S dipolar terms are strongly coupled via the I-S interaction). So, if an S spin undergoes a jump, the rapid cross-relaxation between I-I and S-S terms will result in the local heating being transferred rapidly to the entire spin system. Thus, in the strong-collision limit, the entire dipolar reservoir is describable by a single temperature prior to a diffusion jump, but either spin species alone will not. (We should realize that there is no similar difficulty identifying the diffusing species in a T_1 measurement since the I and S Zeeman reservoirs will normally be uncoupled or weakly coupled because their resonance frequences are probably quite different).

In order to determine the diffusing species in a "slow motion" experiment, it is necessary to identify a parameter which is sensitive to which species is diffusing. If this parameter can be varied by the experimenter, then the diffusing species can be identified and studied. A way to perform such an experiment is to observe the dipolar relaxation in the rotating frame of one of the species, say the I spins. In this frame, only part of the original dipolar Hamiltonian will be secular. If we relate this secular part $\mathcal{H}_D^{(oo)}$ to the various terms of the original dipolar Hamiltonian $\mathcal{H}_d^{(o)}$, we obtain [25]

$$\mathcal{H}_D^{(oo)} = -\frac{1}{2}(1 - 3\cos^2\theta_I)\,\mathcal{H}_{DII}^{(o)} + \cos\theta_I\,\mathcal{H}_{DIS}^{(o)} + \mathcal{H}_{DSS}^{(o)} \quad, \qquad (4)$$

where θ_I is the angle between $\vec{H}_{eff}$ of Eq. (2) and $\vec{H}_o$. We see that varying θ_I (which is easily done by varying the resonance frequency) will vary the relative strengths of the dipolar tems in Eq. (4) involving I-I, I-S, and S-S interactions. We will then find that the relaxation time T_{1D}' of the rotating frame dipolar Hamiltonian $\mathcal{H}_D^{(oo)}$ will depend on θ_I; this dependence can be used to determine which species is diffusing. Suppose the I spins are magnetically stronger than the S spins (i.e., $\gamma_I > \gamma_S$). If the I spins, which are irradiated, are also diffusing, then the dependence of T_{1D}' on θ_I will be quite weak. However, if the S spins are diffusing, we will see striking effects on T_{1D}' vs. θ_I [26].

For instance, at $\cos^2\theta_I = 1/3$, the I-I interaction terms vanish and our measurements will be much more sensitive (i.e., T_{1D}' will be shorter) to the motion of the S-spins. At exact resonance ($\theta_I = 0$), on the other hand, T_{1D}' will be insensitive to the motion of the S-spins (T_{1D}' will then be very long). The verification of this behavior is shown in Fig. 1, taken from Ref. 26, for a sample of KF:0.1%CaF$_2$ in which the I spins are ^{19}F and the S-spins (^{39}K) dominate the diffusion. It should be noted that, at $\cos^2\theta_I = 1/3$, the relaxation due to the motion of the S-spins is greatly enhanced. Thus, a measurement of T_{1D}' at this θ_I can be used to study the motions of atoms whose magnetic properties are too weak for these motions to be observed in a normal T_1 or even T_{1D} experiment. The potential application of this technique to impurity diffusion is discussed in another paper in this Symposium [27].

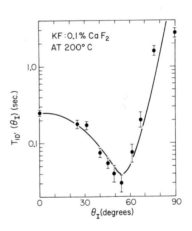

Fig. 1: T_{1D}' vs. θ_I for KF: 0.1% CaF$_2$ at 200°C (taken from Ref. 26).

RELAXATION DUE TO MOTION BETWEEN UNEQUAL POTENTIAL WELLS

Diffusion normally observed in NMR results from an atomic jump from a potential energy well over a barrier to another well of the same depth.

However, if the two wells are of unequal depth, it is possible to see very interesting NMR effects on T_1 and on T_{1D}.

Consider what happens if an atom jumps from a shallow well over an energy barrier of height H to another well whose minimum is Δ below that of the shallow well (Fig. 2). We can see qualitatively what should happen by recognizing that the relative concentration of atoms in the shallow well should be reduced by the Boltzmann factor $\exp(-\Delta/kT)$. Accordingly, the spin-lattice relaxation rate in Eq. (1) should be replaced by

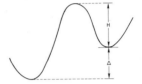

Fig. 2. Two unequal potential wells.

$$\frac{1}{T_1} \sim e^{-\Delta/kT} \frac{(\omega_d)^2 \tau_c}{1 + \omega_o^2 \tau_c^2} \quad . \tag{5}$$

This result has important ramifications in an NMR experiment. The most obvious effect is to reduce the relaxation rate and, correspondingly, raise T_1 by the Boltzmann factor. Thus, this process will give very long T_1's, even at the minimum. A second striking effect is to make the T_1 minimum asymmetric.

Since $\tau_c \propto \exp(-H/kT)$, we see from Eq. (5) that at temperatures above the minimum the slope of $\ln T_1$ vs. $1/T$ is proportional to the energy difference $H - \Delta$, whereas, below the minimum, the slope is proportional to $H + \Delta$.[tt] Additional effects are that the temperature of the minimum is shifted slightly and the frequency dependence of the T_1 minimum is also reduced. These effects have been observed [28] for molecular reorientations in an organic compound [trans, trans-muconodinitrile (TMD)] and are reproduced in Fig. 3. These observations make possible the independent measurement of the parameters H and Δ which characterize the detailed shape of the potential energy profile. However, such observations are not common in NMR because the shallow T_1 minimum due to motion between unequal wells can be masked easily by other relaxation mechanisms. These effects are probably not observable in metals whose T_1's are dominated by the strong conduction electron relaxation. To be seen in insulating materials would require very high purity.

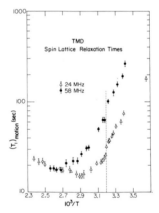

Fig. 3. T_1 vs. $10^3/T$ at two frequencies in TMD (from Ref. 28). Note the unequal slopes above and below the minimum and the large values of T_1.

[tt]For processes which have a temperature-dependent factor involving an energy of formation E_F, H is replaced by the activation energy $E_A = E_F + H$. Also there may be a binding energy factor for some defects. However, the above results will still hold, except that H is replaced by E_A.

LOCALIZED (BOUND) DIFFUSION

One of the great advantages of NMR is that it can observe localized motions of atoms, such as molecular reorientations or the motion of atoms bound to an impurity, as easily as it can observe non-local translational diffusion. This feature contrasts with many non-NMR techniques. Some techniques (e.g., ionic conductivity and redioactive tracers) measure the effects of long-range translational motions whereas others [e.g., anelastic relaxation and ionic thermocurrent (ITC)] observe localized reorientational motions. Very few of these non-NMR techniques have the capability of measuring <u>both</u> local and non-local motions over a wide temperature range using the same technique. Since NMR is not so limited, it has the capability of studying many different diffusion mechanisms which dominate the diffusion in different temperature regions. Thus, one can in principle use NMR to measure the binding energy of a mobile nucleus to a charged impurity by comparing activation energies in temperature regions of local and non-local motions (provided there are no major structural differences between these two regions.)

As an example, consider a substance like CaF_2 or SrF_2 which is doped with a substitutional trivalent impurity. In order to maintain charge neutrality, an excess number of fluorine interstitial ions are formed. (Thus, the concentration of these fluorine interstitials equals the concentration of the impurities.) At low temperatures the fluorines are bound to the impurity atoms and their motion is between neighboring sites. At higher temperatures the fluorines can break the bonds and diffuse over large distances. In an NMR experiment, the transition from the temperature region where the motion of bound (local) fluorines dominates the relaxation to the higher temperature region where the motion of unbound (non-local) fluorine dominates is characterized by a sudden increase in activation energy. This increase is equal to the binding energy plus the difference in barrier heights for the local and non-local potential wells. Figure 4, taken from Ref. 6, shows T_{1D} measurements on SrF_2 doped with various concentration of Y^{3+}. Up to dopant concentrations greater than 0.1% where clustering of impurities may be important, the relaxation rate (T_{1D}^{-1}) is proportional to dopant concentration, thereby supporting the idea that the relaxation is due to fluorine interstitials created to compensate for the charge of the impurity. The sharp increase in slope above 435K (10^3/T = 2.3 K^{-1}) is attributed to the non-local motion of unbound fluorine interstitials (Region I), whereas below this temperature the relaxation is due to the motion of F^- interstitials bound to the Y^{3+}.

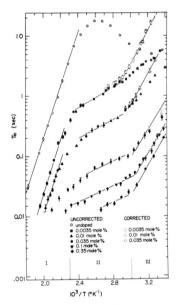

Fig. 4. T_{1D} vs. 10^3/T for SrF_2 doped with various concentrations of Y^{3+}. The low temperature data (Region III) was corrected by subtracting off the background relaxation measured for the undoped sample. This data was taken from Ref. 6.

A somewhat curious feature of these data is the sudden decrease in slope at 333K ($10^3/T = 3.0$ K^{-1}) as we go from Region III to Region II. The activation energy in Region III agrees with other measurements [29] for F$^-$ in a nearest neighbor (nn) site. The transition to Region II may result from motions by next-nearest-neighbor (nnn) fluorines [6] or, possibly, from the formation of clusters. It should be emphasized that the relaxation in Region I is in the strong-collision regime since it involves interstitialcy jumps of all the F spins. In Regions II and III where the diffusion involves the local motions of only a small fraction of the spins, the relaxation of the bulk spins is much weaker (T_{1D} is longer) and occurs by the transfer of dipolar energy from the neighboring atoms to those more distant by mutual spin flips (spin diffusion). (See Ch. V of Ref. 10 for a discussion of spin diffusion.)

DIRECT DETERMINATION OF THE DIFFUSION CONSTANT

An important method for determining the diffusion mechanism is to determine the correlation factor f, which is the ratio of the actual diffusion coefficient to the value it would have if jumps were random and uncorrelated [30]. In order to determine f by using a comparison of NMR with radioactive tracer results, it is necessary to measure D with some precision directly by NMR.

From the temperature dependence of T_1, an activation energy can be determined. With the further aid of an appropriate theory [31], τ_c and even D can be determined, though possibly with insufficient accuracy to determine f reliably. A method, however, does exist for the direct measurement of D by NMR [32]. The basic idea is to apply an external field gradient. Then, if atomic diffusion occurs along this gradient, the atoms will dephase more rapidly and T_2 will be shortened. The decrease in dephasing time is directly dependent on D and can be used to measure D directly. This method has been improved [33] to allow the measurement of smaller values of the diffusion coefficient by applying a pulsed (rather than static) field gradient which is turned off during the rf pulse. Thus, the linewidth will be small at the time of the rf pulse, thereby allowing the use of H_1's which are not particularly large and detection systems which have moderate bandwidth. This technique has been applied to F$^-$ self-diffusion in the superionic conductor PbF$_2$ [34] and very recently to the study of divacancy diffusion in metallic sodium [35].

MOTIONAL STUDIES INVOLVING INTERACTIONS OTHER THEN DIPOLAR

Quadrupolar Relaxation. Many nuclei (of spin greater than 1/2) possess an electric quadrupole moment as well as a magnetic moment. As a result, if these nuclei experience an electric field gradient, their Zeeman energy levels are shifted and are no longer equally spaced. For this reason extra resonance lines (satellites) appear at frequencies slightly different than the Larmor frequencies. The position of these lines depends upon electric field gradients which in turn are very sensitive to the crystal structure in the vicinity of the quadrupolar nuclei [36]. (Such measurements have been widely used in structure determinations and have recently been able to determine ferroelastic domain structure in the elastically ordered phase of NaCN [37].) In a powder sample an effect of these lines is to broaden substantially the apparent NMR linewidth.

Atomic motions can give rise to fluctuating field gradients which, in turn, can contribute to the spin-lattice relaxation time. Thus, the motion of <u>charged</u> atoms which have very small magnetic <u>moments</u> can be studied because of their contribution to the quadrupolar relaxation of nearby nuclei. As an example, Fig. 5 shows T_1 of ^{23}Na in NaCN [38]. This relaxation arises from the fluctuating field gradient at a Na site due to head-to-toe reorientations of neighboring CN⁻ ions. Another interesting example is an observation of the localized motion of off-center Ag⁺ defects in RbCl [39]. This experiment, which was performed by measuring in samples containing only 350 ppm Ag⁺ quadrupolar relaxation of ^{85}Rb, ^{87}Rb, and ^{35}Cl, would have been impossible to perform by direct observation of the very weakly magnetic nucleus Ag⁺.

A problem involved in quadrupole studies is that the effective linewidths may be very large (hundreds of gauss) and, accordingly, the signal-to-noise ratio may be very low. Furthermore, the theory for quadrupolar relaxation and for thermal mixing is not so well-worked out as for dipolar interactions.

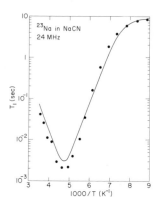

Fig. 5. T_1 vs. $10^3/T$ for ^{23}Na in NaCN. These data will appear in Ref. 38.

<u>Chemical Shift Anisotropy (CSA) Relaxation.</u> Another interaction, which is of considerable importance in chemistry, is the so-called "chemical shift." This interaction is derived from the fine structure interaction between the nuclear magnetic moment and the magnetic field due to the atomic electron's orbital angular momentum. In a molecule or a solid in which electric charges are nearby, the electron orbit precesses so that the average angular momentum is "quenched" [7] and the average interaction is zero. If, however, a magnetic field H_0 is applied, as in an NMR experiment, then the electron's angular momentum is unquenched to some extent and there will be a small additional field at the site of the nucleus. This chemical shift field results in a small shift in the Larmor frequency which depends on the orbital wave function which in turn is strongly dependent on the electron's immediate environment. We thus can obtain for a molecule a spectrum of lines; the study of such spectra in liquids has been widely used to determine the structure of molecules. In solids, one doesn't easily observe these spectra since the chemical shifts (typically 1-100 mgauss) are usually much smaller than the dipolar linewidths. However, by applying a series of rf pulses in different directions, it is possible to simulate motional narrowing and to artificially reduce dipolar linewidths in solids so that very small chemical shifts can be measured [40].

Since the chemical shift typically is anisotropic, it can contribute to T_1 for a molecule reorienting between orientations having different chemical shifts or an atom diffusing between chemically shifted sites. Since the chemical shift is so small, its relaxation contribution is normally not observed in solids and is usually masked by the larger dipolar relaxation. Furthermore, the chemical shift is often invariant for a 180° rotation [38]; hence, such a jump will not then contribute to CSA relaxation even if the chemical shift is larger than the dipolar interaction. For these reasons,

66

observations of CSA relaxation can provide detailed information about the
relative symmetries of the initial and final orientations or sites.

Before proceding further, we should ask what are the characteristics of CSA
relaxation which would identify it as the primary source of relaxation in an
experiment. The answer is that the frequency dependence of T_1 is different
from that of dipolar relaxation shown in Eq. (1). In particular, since the
extra field due to the chemical shift
is proportional to H_0 and to ω_0, the
proper expression for T_1 due to CSA is
given [41] by

$$\frac{1}{T_1} \propto \omega_0^2 \frac{\tau_c}{1 + \omega_0^2 \tau_c^2} . \tag{6}$$

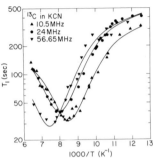

Figure 6 shows measurements of ^{13}C relaxation at three different frequencies in KCN
enriched with 90% ^{13}C. (^{13}C has spin 1/2
and a large magnetic moment, in contrast
to the abundant isotope ^{12}C which has zero
magnetic moment.) Since the neighboring
potassium has a very small magnetic moment,
the dipolar relaxation is relatively weak
and arises primarily from the intramolecular
interaction with the nearest ^{14}N nucleus.
The actual curves in Fig. 6 are due to a

Fig. 6. T_1 vs. $10^3/T$ for
^{13}C in KCN. These data
will appear in Ref. 38.

combination of this dipolar relaxation and CSA relaxation, both resulting
from the reorienting CN⁻ molecule. The low frequency (10 MHz) data is due
primarily to dipolar relaxation, whereas CSA relaxation is favored at the
high frequency (56.65 MHz). Since both the intramolecular dipolar interaction
and CSA interaction are invariant for 180° jumps, these data tells us that the
CN⁻ ion must undergo jumps through angles other than 180°. We should realize
that CSA relaxation is observable here mainly because the neighboring nuclei
have small magnetic moments and correspondingly small dipolar interactions.

ACKNOWLEDGMENTS

The author is grateful to the U.S. National Science Foundation for its
support of much of the research described in this paper.

REFERENCES

1. F. Reif, Phys. Rev. 100, 1597 (1955).

2. D.F. Holcomb and R.E. Norberg, Phys. Rev. 98, 1074 (1955).

3. I. Chung, H.S. Story, and W.L. Roth, J. Chem. Phys. 63, 4903 (1975).

4. G.E. Jellison, Jr., Solid State Commun. 30, 481 (1979).

5. R. Bersohn and H.S. Gutowsky, J. Chem. Phys. 22, 651 (1954).

6. S.H.N. Wei and D.C. Ailion, Phys. Rev. B19, 4470 (1979).

7. An excellent introductory text is Principles of Magnetic Resonance, 2nd ed. (Springer-Verlag, Berlin 1978) by C.P. Slichter.

8. Convenient charts which show basic NMR features for each nucleus (e.g., gyromagnetic ratio, quadrupole moment, and NMR sensitivity) are readily available from instrument companies like Varian Associates and Brucker Magnetics Inc.

9. S.R. Hartmann and E. L. Hahn, Phys. Rev. 128, 2042 (1962).

10. A. Abragam, The Principles of Nuclear Magnetism (Clarendon Press, Oxford 1961) p. 33.

11. See pp. 180-181 of Ref. [7].

12. E.R. Andrew and R.G. Eades, Proc. Roy. Soc. A218, 537 (1953).

13. H.T. Stokes, T.A. Case, D.C. Ailion, and C.H. Wang, J. Chem. Phys. 70 3572 (1979).

14. A. Messiah, Quantum Mechanics, Vol. II (Wiley, New York 1966), Chs. XVI and XVII.

15. N. Bloembergen, E.M. Purcell, and R.V. Pound, Phys. Rev. 73, 679 (1948).

16. D.C. Ailion in Advances in Magnetic Resonance, Vol. 5, J.S. Waugh ed. (Academic Press, New York 1971) pp. 177-227.

17. A.G. Redfield, Phys. Rev. 98, 1787 (1955).

18. D.C. Look and I.J. Lowe, J. Chem. Phys. 44, 2995 (1966).

19. C.P. Slichter and W.C. Holton, Phys. Rev. 122, 1701 (1961).

20. C.P. Slichter and D.C. Ailion, Phys. Rev. 135, A1099 (1964).

21. D.C. Ailion and C.P. Slichter, Phys. Rev. 137, A235 (1965).

22. D.C. Ailion and P. Ho, Phys. Rev. 168, 662 (1968).

23. J. Jeener and P. Broekaert, Phys. Rev. 157, 232 (1967).

24. H.T. Stokes and D.C. Ailion, Phys. Rev. B16, 4746 (1977).

25. M. Goldman, Spin Temperature and Nuclear Magnetic Resonance in Solids (Clarendon Press, Oxford 1970), p. 37.

26. H.T. Stokes and D.C. Ailion, Phys. Rev. B18, 141 (1978).

27. J.R. Beckett, J. Pourquié and D.C. Ailion, in Proc. of MRS Symposium on Nuclear and Electron Resonance Spectroscopies Applied to Materials Science (North-Holland, Boston 1980).

28. M. Polak and D.C. Ailion, J. Chem Phys. 67, 3029 (1977).

29. E.L. Kitts, Jr., M. Ikeya, and J.H. Crawford, Jr., Phys. Rev. B8, 5840 (1973).

30. N. Peterson, in Diffusion in Solids: Recent Developments, A.S. Nowick, J.J. Burton eds. (Academic Press, New York 1975), pp. 115-170.

31. D. Wolf, Spin Temperature and Nuclear Spin Relaxation in Matter (Clarendon Press, Oxford 1979).

32. H.Y. Carr and E.M. Purcell, Phys. Rev. 94, 630 (1954).

33. E.O. Stejskal and J.E. Tanner, J. Chem. Phys. 42, 288 (1965).

34. R.E. Gordon and J.H. Strange, J. Phys. C: Solid St. Phys. 11, 3213 (1978).

35. G. Brünger, O. Kanert, and D. Wolf, Solid State Commun. 33, 569 (1980).

36. T.P. Das and E.L. Hahn, in Solids State Physics: Supplement 1, F. Seitz, D. Turnbull eds. (Academic Press, New York 1958).

37. A. Tzalmona and D.C. Ailion, Phys. Rev. Lett. 44, 460 (1980).

38. H.T. Stokes, T.A. Case, and D.C. Ailion, to be published.

39. O. Kanert, R. Küchler, and M. Mali, J. Phys. (Paris) 41, C6-404 (1980).

40. J.S. Waugh, L.M. Huber, and U. Haeberlen, Phys. Rev. 20, 180 (1968).

41. See p. 316 of Ref. 10.

Electron Resonance

Published 1981 by Elsevier North Holland, Inc.
Kaufmann and Shenoy, editors
Nuclear and Electron Resonance Spectroscopies Applied to Materials Science

ELECTRON PARAMAGNETIC RESONANCE OF MATERIAL PROPERTIES AND PROCESSES*

K. L. BROWER
Sandia National Laboratories[†], Albuquerque, New Mexico 87185, USA

ABSTRACT

The purpose of this paper is to demonstrate, primarily for
the non-specialist and within the context of new and recent
achievements, the diagnostic value of electron paramagnetic
resonance (EPR) in the study of material properties and processes. I have selected three EPR studies which demonstrate
the elegance and uniqueness of EPR in atomic defect studies
and exemplify unusual achievements through the use of new
techniques for material measurement and preparation. A brief
introduction into the origin, interaction, and detection of
unpaired electrons is included.

INTRODUCTION

This series of papers focuses on EPR studies of materials such as semiconductors, glasses, catalysis, amorphous Si:H, ... which are of relevance in
today's technology. Our purpose and approach are intended to demonstrate, primarily for the benefit of the non-specialist and within the context of new and
recent achievements, the diagnostic value of EPR in the study of material properties and processes. Following a brief introduction into some of the basic
aspects of EPR in defect studies, I will highlight the results of EPR studies
on 1) nitrogen in silicon by Brower and Peercy, 2) the boron interstitial in
silicon by Troxell and Watkins--here the ideas of Anderson's negative U energy
and possibly the Bourgoin mechanism are exemplified, and 3) Si-SiO$_2$ interface
states by Poindexter, Caplan, Deal, and Razouk.

EPR IN DEFECT STUDIES

Origin of Unpaired Electrons

It is convenient to characterize the electronic structure of a semiconductor
or insulator in terms of a band structure in which, ideally, the valence band
is filled with electrons and the conduction band, which is separated from the
valence band by an energy gap, is empty. The chemical bonding of these $\sim 10^{23}$
valence electrons/cm^3 is usually such that they are spin paired according to
the Pauli principle. Under these conditions the solid is diamagnetic and yields
a null EPR spectrum.

Imperfections such as vacancies, interstitials, impurities, ... in solids
often have localized states with one or more localized energy levels within the
bandgap. Whether an unpaired electron is trapped in one of the localized states
depends in the case of semiconductors on the position of the Fermi level and in
the case of insulators on the availability of electrons or holes. The effects
of ionizing irradiation, n- or p-type doping, light illumination, temperature,
etc., will sometimes induce paramagnetism in existing defects. Also, defects

*This article was sponsored by the U. S. Department of Energy, Division of
 Basic Energy Sciences, under Contract DE-AC04-76-DP00789.

[†]A U. S. Department of Energy facility.

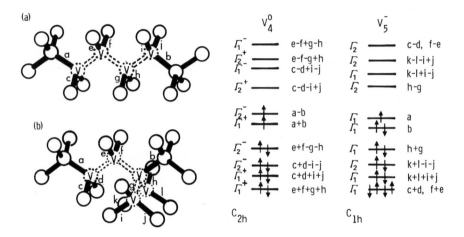

Fig. 1. (a) Model of the four-vacancy for the Si-P3 center (Ref. 3). The distance between end silicon atoms from which broken bonds a and b dangle is ≈ 9.6 Å. (b) Model of the five-vacancy for the Si-P1 center according to Lee and Corbett (Ref. 4). The V's symbolize vacancies and a, b, ..., ℓ represent dangling bonds.

Fig. 2. Electronic structure of the four- and five-vacancy defects (Ref. 3 and 4). The lower four levels, which are filled with electrons, correspond to bonding orbitals (valence band-like), and the upper four levels which are empty correspond to antibonding states (conduction band-like). The middle two levels, which may contain paramagnetic electron(s), are in the bandgap.

can be produced during crystal growth and material preparation, and by displacement or ionizing radiation, ion implantation, (photo-) chemical reactions, thermal or laser annealing, fracturing, catalytic surface reactions, shock waves, etc.

It is often convenient to represent a center in terms of a defect-molecule which is comprised of the defect plus a few neighboring host lattice atoms [1, 2]. Two giant defects which have been identified in neutron-irradiated and ion-implanted silicon are illustrated in Fig. 1. Usually, the defects observed by EPR are somewhat smaller than those in Fig. 1. Macroscopic defects such as interstitial loops, vacancy loops, precipitates, dislocations, bubbles, etc. which have been inferred from other measurements (e.g., electron microscopy) usually arrange themselves so that they are diamagnetic. Even then, in the case of macroscopic imperfections, any dangling bond would still be localized on a very small portion of the total imperfection (e.g., see Sec. V). The link between microscopic and macroscopic imperfections, that is, how they evolve into each other, still remains to a great extent an unexplored frontier.

For simple defect models represented by defect-molecules, it is useful to represent the electronic structure in terms of one-electron molecular orbitals which are consistent with the symmetry of the defect (Fig. 2). Although this simple approach does not give the position of the localized levels in the bandgap or even their relative splittings, it does give the experimentalist insights into many of the electronic properties of the defect as observed by EPR, such as Jahn-Teller effects, energy changes under uniaxial stress, effects of light, etc. First principle calculations of the electronic structure of even the simplest deep-level defects in silicon still present a profound challenge to the theorist [5]. Sometimes the position of the level in the bandgap (see Ref. 5 for a discussion of the concept "level") can be determined by probing with monochromatic light which affects the charge or spin state of the defect. In this connection, the position of electrically active levels in silicon have been measured by deep-level-transient-spectroscopy (DLTS) and correlated with specific defects observed with EPR [6]. Correlations between electrical conductivity and specific defects in silicon have also been made [7]. Thus, EPR is one of the important tools for correlating specific atomic defects with the physical properties of a material.

Interactions of Unpaired Electrons

The relative splittings of the localized defect level containing an unpaired electron can be described by a spin Hamiltonian of the rather general form [8]

$$H_{spin} = \mu_B \, \vec{S} \cdot \overleftrightarrow{g} \cdot \vec{B} \quad + \quad \vec{S} \cdot \overleftrightarrow{D} \cdot \vec{S} \quad + \sum_j \vec{S} \cdot \overleftrightarrow{A}_j \cdot \vec{I}_j$$

$$\text{(electronic Zeeman)} \quad \text{(spin-spin)} \quad \text{(hyperfine)}$$

$$-\sum_j \left(\frac{g_{n,j} \, \mu_N}{I_j} \right) \vec{I}_j \cdot \vec{B} + \sum_j \vec{I}_j \cdot \overleftrightarrow{Q}_j \cdot \vec{I}_j \quad .$$

$$\text{(nuclear Zeeman)} \quad \text{(quadrupole)}$$

It is possible to account for many of the details (e.g., line position, relative intensity, etc.) in an observed spectrum to a very high degree of accuracy with a suitable spin Hamiltonian. The various magnetic interactions are represented by explicit terms in the spin Hamiltonian; whereas, the spatial parts of the wave functions reflecting Coulomb interactions, exchange, symmetry, etc. are buried in the $\overleftrightarrow{g}$, $\overleftrightarrow{D}$, $\overleftrightarrow{A}$, $\overleftrightarrow{Q}$, etc. coupling tensors. Much of the information used in modeling defects comes from interpreting these coupling tensors in quantum mechanical terms while the defect dynamics (migration, reorientation, electrical levels, etc.) can be probed through annealing, stress, light, electric field, ionizing irradiation, etc. Although these are intriguing but difficult tasks, there have been considerable insights gained by these approaches, and they have been extremely fruitful. Clearly, the more interactions a spectrum exhibits, the more information that becomes available. In this respect, it is possible to "load" a defect, for example, by isotopic doping, so it will reveal more.

$\overleftrightarrow{g}$ Tensor. The $\overleftrightarrow{g}$ tensor gives the overall symmetry of a defect. In the case of silicon, general characteristics in the anisotropy of $\overleftrightarrow{g}$ have been systematized by Lee and Corbett in terms of bonding characteristics of the paramagnetic electron [4]. It is yet very difficult to calculate anisotropies in $\overleftrightarrow{g}$ for many defects because this calculation requires information about excited states which is usually unknown; however, in some cases ratios in the anisotropy have been qualitatively predicted [9]. The numerical values for $g_{k\ell}$ are usually unique to each defect—very minor perturbations (charge state, neighboring impurities, etc.) tend to affect it to a measurable extent.

$\overleftrightarrow{A}$ Tensor. The hyperfine (hf) interaction, which is the magnetic dipole-dipole interaction between the unpaired electrons(s) and nuclei with nuclear spin I, is essential for purposes of positive defect identification and modeling. The number of hf lines corresponds to $2I + 1$, and the relative intensity indicates directly the relative abundance of each isotope. This information usually allows a positive identification of the impurities in a defect. In addition, isotopic doping by equilibrium (e.g., thermal-chemical doping [10]) or non-equilibrium (e.g., ion implantation [11]) techniques can also be used to expand one's insights.

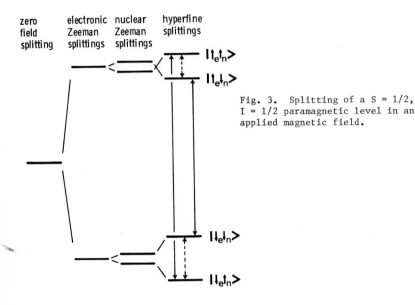

zero   electronic nuclear   hyperfine
field   Zeeman   Zeeman   splittings
splitting   splittings splittings

$|\uparrow_e\uparrow_n\rangle$

$|\uparrow_e\downarrow_n\rangle$

$|\downarrow_e\downarrow_n\rangle$

$|\downarrow_e\uparrow_n\rangle$

Fig. 3. Splitting of a S = 1/2,
I = 1/2 paramagnetic level in an
applied magnetic field.

This interaction also gives a semiquantitative description of the paramagnetic one-electron molecular orbital in terms of s and p character and localization. It is usually possible to fit this together with a more general molecular orbital description of the bonding for a specific defect model based on the
symmetry considerations mentioned in Sec. II A.

The spatial region over which the unpaired electron is mostly localized
varies with the solid and defect. In silicon, the paramagnetic electron usually
spreads out over the defect as seen through the ^{29}Si and impurity hf spectra,
but for insulators, such as SiO_2 which have wider bandgaps and more ionic bonding, the resolved hf spectra barely span the defect so that ones view is more
myopic. A deeper view into the surrounding lattice, which is under the tail
of the wave function, can sometimes be extended by electron-nuclear-double-
resonance (ENDOR). For example, in silicon $\approx$ 30 shells of neighboring silicon
atoms can sometimes be resolved [12]; but in MgO, only the 1st, 3rd, and 5th
shell of Mg atoms around the oxygen vacancy have been observed with ENDOR [13].
The 2nd and 4th shells containing oxygen were not observable.

$\overleftrightarrow{D}$ Tensor. The spin-spin interaction occurs for defects and impurities with
$S \geq 1$. For cases in which this interaction is dominated by the magnetic dipole-
dipole interaction between two spins (S = 1) in a defect, it is possible to deduce the spacing between the unpaired electrons. This information has been very
useful in identifying the length of multiple vacancy (oxygen) centers in silicon
[3] and diamond [14]. There are also defects in which $\overleftrightarrow{D}$ is dominated by other
interactions (e.g., spin-orbit, exchange) which are not easily interpreted in
terms of structural features in the defect.

$\overleftrightarrow{Q}$ Tensor. The quadrupole interaction may only occur for nuclei with I > 1/2
and is due to the interaction between the electric quadrupole moment of a nucleus and the electric field gradient of the crystal at the nucleus. This interaction is usually not perceptable in EPR measurements, but may need to be
considered in ENDOR experiments [15].

Detection of Unpaired Electrons

Typical energy level splittings for a simple S = 1/2, I = 1/2 spin system
are illustrated in Fig. 3. The solid vertical lines indicate the levels between
which magnetic dipole transitions observable by EPR are allowed. Experimentally, one places the sample in a microwave cavity where the $\vec{B}_1$ cosωt of the microwave field is the strongest and perpendicular to the applied quasi-static magnetic field $\vec{B}$. Usually, E_1 cosωt is the weakest in this region. It is desirable to pick sample configurations and locations as well as a cavity (cylindri-
cal vs. rectangular) and mode which minimize ohmic losses since the sensitivity
of a spectrometer is proportional to the cavity Q. Since the microwave frequency is tuned to the cavity, resonances are found by sweeping the applied
magnetic field, $\vec{B}$. Resonance may occur when the microwave quantum, ℏω, equals
the energy difference between levels. Macroscopically, the spins as a whole
exhibit a complex magnetic susceptibility, $\chi = \chi' - i\chi''$, at resonance which
causes a detectable change in the Q of the cavity due to changes in χ'' corresponding to absorption of microwave energy and a related change in resonant
frequency of the cavity due to changes in χ' (dispersion effect). The two main
spectrometer designs used for detecting resonances are the superheterodyne
spectrometer, which is very sensitive at low microwave powers ($\lesssim$ 1 µw), and the
100 kHz homodyne spectrometer, which operates with higher microwave powers
(~ 1 mw). Sensitivities of EPR spectrometers are $\gtrsim 10^{10}$ spins/G [16].

ENDOR involves inducing nuclear magnetic dipole transitions (dotted lines in
Fig. 3, ΔM = 0, Δm = 1) by an applied rf field, $\vec{B}_{rf}$cosω$_{rf}$t, with $\vec{B}_{rf}$ perpendicular to $\vec{B}$. This resonance is detected indirectly with EPR by detecting (transient) changes in a partially saturated EPR resonance due to unbalances in the
spin population of the energy level common to the electronic and nuclear magnetic dipole transitions [17]. The main advantage of this technique is that the
resolution in hf structure is about 10^3 times greater than with EPR.

EPR OF NITROGEN IN SILICON

Although the group V elements P, As, Sb, and Bi are shallow, substitutional
donors in silicon, N, which is also a group V, has been a curious exception.
Crystal growers found it impossible to make silicon n-type by doping with N
using conventional thermal-chemical procedures; instead, silicon nitride precipitates tended to form [18]. Later, experimentalist approached this difficulty using non-equilibrium techniques such as ion implantation and thermal
annealing. Although they did succeed in producing an n-type layer [19], other
studies showed a complexity of electrical and insulating behavior depending on
N fluence and annealing history [20]--properties vastly different from those of
the other group V dopants. In fact, nuclear reaction and channeling experiments
were unable to detect any substitutional N; instead, the N appeared to be trapped on random sites like nitride precipitates or damage clusters [21].

Recently, we discovered a paramagnetic N center in silicon while doing EPR
studies on laser-annealed, ion-implanted silicon [22]. In particular, we were
looking with EPR and Rutherford backscattering (RBS) at the effects of laser
annealing on the recrystallization of amorphous silicon. The silicon had been
made amorphous by implantation with typically 4×10^{15} 160 keV $^{28}Si^+$/cm^2. After
a pulsed ruby laser anneal (λ = .6954 µm) of $\approx$ 2 J/cm^2 in 20 nsec, the amorphous
layer was restored to crystallinity as measured by EPR and RBS, but invariably
a residual paramagnetic spectrum was observed (Fig. 4). We identified this
spectrum as being due to ^{14}N, which has a nuclear spin of 1 and is 99.6% naturally abundant. Our EPR studies indicate that N is introduced during ion implantation as a result of contamination of the $^{28}Si^+$ beam with $\approx$ 1/2% $^{14}N^+_2$
(same mass and charge). In particular, the depth distribution of the N was measured by EPR and matches the calculated depth distribution for $^{14}N^+_2$ implanted
with the same energy as $^{28}Si^+$.

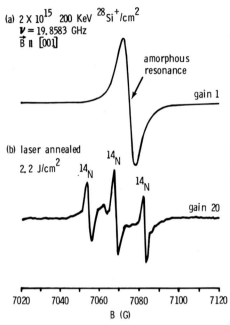

(a) 2×10^{15} 200 KeV $^{28}Si^+/cm^2$
$\nu = 19.8583$ GHz
$\vec{B} \parallel [001]$

amorphous
resonance

gain 1

(b) laser annealed
2.2 J/cm^2

^{14}N ^{14}N ^{14}N ^{14}N

gain 20

7020   7040   7060   7080   7100   7120
B (G)

Fig. 4. (a) EPR spectrum of amorphous silicon, (b) EPR spectrum of the N center in laser-annealed, ion-implanted silicon (Ref. 22).

Our model for this N center [23] is shown in Fig. 5. The N is essentially substituting for a silicon, and the extra electron of the N gives rise to its paramagnetism. Unlike the EPR spectra of the other group V donors, which are isotropic (T_d symmetry) and show no resolved ^{29}Si hf structure, this N spectrum at low temperatures corresponds to a defect having <111> axial (C_{3v}) symmetry. Also, there is a strong ^{29}Si hf interaction [23] which indicates that $\approx$ 73% of the paramagnetic orbital is localized on the $Si_{<111>}$ in Fig. 5 and only ~ 9% is localized on the N atom. These features are characteristic of deep donors. This peculiarity may arise if the donor electron is associated with an orbitally degenerate level rather than a singlet A_1 level as observed for P, As, Sb, and Bi. Consequently, the symmetry of the N center is lowered by virtue of a Jahn-Teller distortion which is indicated in Fig. 5 by the arrows. Nitrogen in diamond exhibits similar effects [24,25]. Paramagnetic N centers have also been observed in the various polytypes of silicon carbide [26].

Under uniaxial stress it is possible to induce a preferential alignment in the N center observed in silicon. Stress-relaxation measurements indicate that the electronic reorientation of this N center follows the Arrhenius relationship $\tau = \tau_0 \exp (E/kT)$ with activation energy E = .084 $\pm$ .005 eV and $\tau_0 \sim 4 \times 10^{-9}$ sec. This implies that the paramagnetic orbital and distortion are thermally activated from one <111> to another with a half-life of $\approx$ 1 sec at 50 K. If the sample is cooled under [110] uniaxial stress from 50 to 7 K and then the stress removed, our EPR spectrum shows a preferential alignment of the paramagnetic orbital perpendicular to the direction of applied stress. This preference in alignment means that the paramagnetic or donor electron is in an antibonding orbital which would tend to displace the N and $Si_{[111]}$ in Fig. 5 away from each other. We observe that $\gtrsim$ 70% of these N centers reorient under stress, which

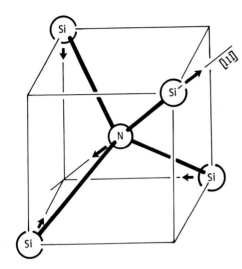

Fig. 5. Model of substitutional N in
silicon with directions of possible
displacements due to a Jahn–Teller
<111> axial distortion indicated by
arrows (Ref. 23).

suggests that they are reasonably isolated in the lattice. The other $\lesssim$ 30%,
depending upon the sample, do not realign under stress, which suggests that
perhaps they are pinned by the strain field of nearby imperfections.

One of the points of practical interest is the location of this donor level
in the bandgap. By exciting the sample with monochromatic light with wavelength
which can vary from 0.4 to 0.9 eV, we observe that stress induced polarizations
are quickly randomized at 7 K for hv $\gtrsim$ .58 eV. This energy is interpreted as an
ionization energy and suggests that the electrical level is near the middle of
the bandgap. The deepness of this donor level is in contrast to that of the
other group V donors which have ionization energies of $\approx$ .050 eV.

Although isolated N atoms are quenched into substitutional sites by pulsed
laser annealing, thermal annealing near 430°C results in the disappearance of
isolated N centers and the brief appearance of another distinct ^{14}N hf spectrum
with a smaller hf interaction. It appears that these isolated N centers are
involved in the formation of other N impurity related complexes for T $\gtrsim$ 430°C.
These centers might be the precursors to those centers or complexes which give
rise to shallow donor levels observed by others in electrical measurements [20].

EPR OF THE INTERSTITIAL BORON IN SILICON

The EPR and DLTS studies of Troxell and Watkins on interstitial boron are of
special interest because they exemplify experimentally for the first time
Anderson's negative U concept for a specific defect [15,27]. Also, the boron
interstitial is of interest because it displays strong ionization enhanced
migration. Watkins found from his EPR studies that substitutional boron is
displaced into interstitial sites by 1.5 MeV electron irradiation at $\leq$ 20 K.
Boron interstitials, like other group III impurities, are believed to be created
by the replacement mechanism.

According to this model [2], Si interstitials, which are produced in the
primary radiation damage event, undergo long-range migration even at 4 K. Upon
encountering a substitutional group III impurity, the Si interstitial replaces
the impurity in the lattice ejecting the impurity into an interstitial site.

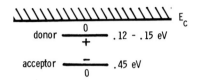

Fig. 6. Proposed electrical level
structure for interstitial boron based
on DLTS measurements. The inverted order
implies a "negative U" system (Ref. 27).

The mechanism by which the self-interstitial migrates at low temperature has
not yet been determined, which is why studies on the enhanced migration of impurity interstitials in silicon under ionizing irradiation and minority carrier
injection are of particular relevance to this intriguing phenomenon [28].

Only the neutral charge state of the B interstitial ($B_i{}^0$) has been observed
with EPR. In order to see this spectrum, it is necessary to illuminate the
sample with near-bandgap light. This result indicated that the neutral, paramagnetic state of the B interstitial is metastable. Furthermore, the failure
to observe the $B_i{}^0$ spectrum (S = 1/2) in the absence of light in either n- or
p-type silicon (in thermal equilibrium) suggested that the B interstitial ordinarily is in the diamagnetic negative or positive charge state. In fact, the
loss of the $B_i{}^0$ spectrum upon warming to 50 K in high resistivity, p-type silicon was attributed to the thermal excitation of the trapped electron from the
donor state at ~ E_c - .15 eV to the conduction band.

More recently, DLTS measurements have indicated a level at E_c - .45 eV which
is identified as the B interstitial acceptor level [27]. Also, the donor level
for the B interstitial has now been detected at ≈ E_c - 0.12 eV by DLTS measurements [29]. In order to account for the metastable state of $B_i{}^0$ and the existence of the stable, diamagnetic $B_i{}^+$ and $B_i{}^-$ for all positions of the Fermi
level, the electrical levels for the B interstitial were arranged as shown in
Fig. 6. Such an arrangement of electrical levels is possible if the configuration of the defect changes significantly for different charge states.

In other words, a different defect exists for each charge state, but each
defect configuration involves the same B atom. Consequently, it is energetically
more favorable for two-electron transitions to occur such as $B_i{}^- \leftrightarrow B_i{}^+ + 2e^-$
than for one-electron transitions such as $B_i{}^+ + e^- \rightarrow B_i{}^0$ or $B_i{}^- \rightarrow B_i{}^0 + e^-$.
Such two electron transitions are the essence of Anderson's "effective negative
U" concept [30] which says, in effect, that an "energy-lowering structural distortion may be sufficiently enhanced by the presence of a second electron so
that the energy gain more than compensates for the electron-electron repulsive
energy" [5]. This concept was originally proposed by Anderson to explain the
properties of defects in chalcoginide glasses. In the case of the B interstitial, further evidence that two electrons are involved in a charge transition
has been inferred directly from DLTS measurements [27].

One model which Troxell and Watkins suggested that could account for the
migrational behavior of the B interstitial is illustrated in Fig. 7. They have
also indicated an alternative model which is also consistent with the known
experimental data (Fig. 8). In particular, their EPR results indicate that the
$B_i{}^0$ is puckered off from a <111> into one of six equivalent distortions, and the
$B_i{}^+$ has <111> axial symmetry. EPR stress-relaxation measurements indicate that
the activation energy for jumps from one <111> axis to another <111> axis is
0.6 eV; this also corresponds to the activation energy for long range migration
of the B interstitial as measured by EPR and DLTS. Troxell and Watkins propose

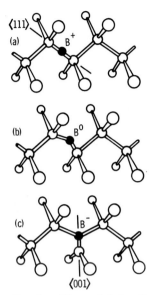

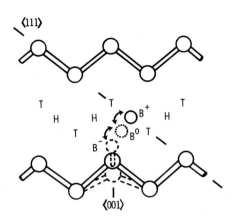

Fig. 8. Alternate model for Bourgoin migration with B_i^+ in a hexagonal interstitial site and B_i^- in a B-Si <001> split interstitialcy (dashed lines). The metastable B_i^0 is in an intermediate position (dotted circle). "T" and "H" represent tetrahedral and hexagonal interstitial sites in a {110} plane (Ref. 27).

Fig. 7. Possible model for Bourgoin migration of interstitial boron in silicon: a) B_i^+ is bond-centered interstitialcy, b) B_i^0 is metastable, distorted bond-centered interstitialcy, and c) is a B-Si <001> split interstitialcy (Ref. 27).

that the B_i^- might correspond to a <001> B-Si interstitialcy as illustrated in Figs. 7 and 8. (The B_i^- configuration has not yet been verified experimentally.) The interesting feature of these models is that migrational jumps may occur as B_i^0 decays to either B_i^- or B_i^+. In fact, DLTS measurements indicate that the annealing rate of the boron interstitial is considerably enhanced under minority carrier injection in either n- or p-type silicon [27]. Furthermore, the quadratic dependence on injected current suggests a two step process: the capture of two electrons (holes) in p-type (n-type) material. Consequently, the enhanced migration of the B interstitial involves the alternation between charge states which differ by two electronic charges [27].

Prior to these findings, Bourgoin and Corbett [31] had proposed that the equilibrium position in the lattice for some interstitials might depend upon their charge state and that successive changes in charge state might lead to motion of the interstitial through the lattice. The ionization enhanced migration of the B interstitial has been considered in terms of the Bourgoin mechanism [27].

EPR OF Si-SiO$_2$ INTERFACE STATES

Si-SiO$_2$ structures, which are an essential element in today's MOS technology, are produced by the thermal oxidation of silicon in either water vapor or oxygen at 800 to 1200°C. Imperfections in stoichiometry and chemical bonding at the Si-SiO$_2$ interface give rise to localized states [32]. The positions of these localized levels with respect to the band structure, in terms of bonding configurations, bond-angle distortions, and defects at the interface, have been the subject of various theoretical studies. Unfortunately, some of these defects can trap charge and affect the electrical performance of an MOS device. Therefore, the identification of specific defect structures and their relationship to particular processing techniques and electrical properties is of interest, but it is an exceedingly difficult task because the density of specific interface states is very small--typically $\lesssim 10^{12}/cm^2$ - eV.

Recently, Poindexter, Caplan, Deal, and Razouk (PCDR) [33,34] have applied EPR to the study of interface states in Si-SiO$_2$, and it appears that they have identified one of the dominant, electrically-active interface states--namely ·Si $\equiv$ Si$_3$. In order to help clarify the relationship between electrical properties and defects, both EPR and capacitance-voltage (CV) measurements were made on (001) and (111) silicon wafers oxidized and annealed by a variety of device-relevant procedures.

Although PCDR observed several different spectra, the spectrum of principal interest and study has been the P$_b$ spectrum, which was first reported by Nishi [35]. In the case of (111) Si-SiO$_2$ interfaces, this spectrum is characterized by a $\overleftrightarrow{g}$ tensor having axial symmetry about the <111>. Ordinarily, if this center were embedded in bulk silicon, it would be randomly oriented among the four equivalent <111> directions, and a unique resonance would be associated with each defect orientation for various orientations of the applied magnetic field. The EPR measurements of PCDR were particularly interesting because the P$_b$ spectrum they observed from a (111) Si-SiO$_2$ interface consisted of only one of the four resonances. This resonance corresponded to P$_b$ centers with only one orientation--namely with their axis of axial symmetry perpendicular to the Si-SiO$_2$ interface. These measurements indicated that the P$_b$ center was on the silicon side of the Si-SiO$_2$ interface and reflected crystalline order. It may be surprising to some that this interface can be as planar on an atomic scale as these measurements indicate; however, this quality is also consistent with the results of high resolution electron microscopy [36].

In the case of (001) Si-SiO$_2$ interfaces, PCDR observed a slightly different P$_b$ spectrum which they labeled P$_{b0}$. In this case, all four resonances were observed. In addition they also observed another distinct spectrum which they labeled P$_{b1}$. The $\overleftrightarrow{g}$ tensors for these spectra are specified in Table 1.

Because of extensive and detailed EPR studies on a variety of defects in silicon as well as SiO$_2$, it is possible to infer overall features in the structure of the P$_b$ center. For the P$_b$ center, PCDR [33] indicate that the anisotropy in the $\overleftrightarrow{g}$ tensor is consistent with the model of a paramagnetic electron localized on a silicon atom which is bonded to only three other silicon atoms (·Si $\equiv$ Si$_3$). In the case of the (111) Si-SiO$_2$, the features in the P$_b$ spectrum tell us that this ·Si $\equiv$ Si$_3$ defect is on the silicon side of the Si-SiO$_2$ interface with the paramagnetic electron perpendicular to the (111) interfacial plane [33]. This model is illustrated in Fig. 9a.

TABLE I. Summary of paramagnetic defects and respective $\overleftrightarrow{g}$ tensors; θ is the angle between g_1 and the [001] direction, and g_3 is parallel to [$\bar{1}$10].

Center	Model	Interface	g_1	g_2	g_3	θ
P$_b$	·Si $\equiv$ Si$_3$	(111) Si-SiO$_2$	2.0013	2.0086	2.0086	54.7°
P$_{b0}$	·Si $\equiv$ Si$_3$	(001) Si-SiO$_2$	2.0014	2.0080	2.0087	62°
P$_{b1}$	·Si $\equiv$ Si$_2$O	(001) Si-SiO$_2$	2.0012	2.0076	2.0052	32°

O oxygen

● silicon

(a) {111} Si-SiO$_2$ interface

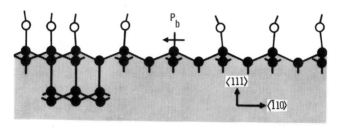

(b) {001} Si-SiO$_2$ interface

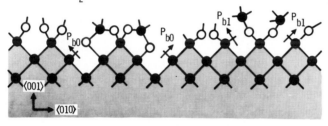

Fig. 9. a) Schematic of (111) Si-SiO$_2$ interface with P_b centers (Ref. 33).
b) Schematic of (001) Si-SiO$_2$ interface with P_{b0} and P_{b1} defects (Ref. 34).

Although a ^{29}Si hf interaction has not been detected, it should be part of
the P_b spectrum; however, it may not be easily detected due to excessive broadening. This broadening is expected to arise because of the sensitivity of the
contact part of the hf interaction to lattice disorder [37], which in this case
exists at the interface.

In the case of the (001) Si-SiO$_2$ interface, the P_{b0} corresponds to essentially
the same model--namely a ·Si ≡ Si$_3$, but now the paramagnetic dangling bond is
canted with respect to the (001) interfacial plane. This breaks the <111> axial
symmetry of the defect, and this may account for the slight perturbations in the
g tensor of the P_{b0} center (see Table I). Furthermore, all four <111> broken
bonds at the (001) interface are equivalent, which accounts for the observation
of four resonances in the P_{b0} spectrum. In the case of the (001) Si-SiO$_2$ interface, PCDR also observe the P_{b1} spectrum which they speculate corresponds to a
·Si ≡ Si$_2$O defect [34]. This defect appears to be a natural consequence of the
structure of the advancing (001) Si-SiO$_2$ interface; however, it does not appear
to be needed for the growth of the (111) Si-SiO$_2$ interface where it is not observed. These models are illustrated in Fig. 9.

Calculations of the local density of states versus energy indicate that
broken bond defects, such as these, lead to distinct interface states within
the silicon bandgap [38]. PCDR have, in fact, measured the density of P_b centers with EPR and the density of electrically active interface states with CV

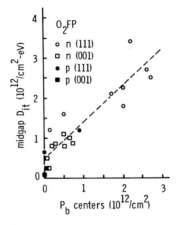

Fig. 10. Midgap interface-states density, D_{it}, versus number of spins in (111) and (001) Si-SiO$_2$ interface with n- or p-type silicon substrate after oxide growth and quench in O$_2$ (O$_2$FP) (Ref. 34).

techniques for a variety of processing and cooling procedures. For the case of (001) and (111) Si-SiO$_2$ interfaces oxidized in dry O$_2$ at various temperatures and quickly cooled in O$_2$ (O$_2$FP) on either n- or p-type silicon, the density of P$_b$ centers is approximately equal to the midgap interface density, D_{it}, as shown in Fig. 10 [34]. Additional EPR and CV measurements have indicated the sensitivity in the density of D_{it} and P$_b$ centers to annealing and cooling in various gases (H$_2$, N$_2$, Ar), impurities, and oxide thicknesses. Also the effects of electric fields in "pulling" charge on and off the P$_b$ center are under investigation by this group.

CONCLUSIONS

The studies reviewed in this paper demonstrate the utility of EPR in probing complex problems in solid state physics. The nitrogen work is novel and illustrates how materials prepared by new and innovative methods can be investigated with EPR. In the case of the boron studies, the complex properties of the boron interstitial are now better understood in terms of Anderson's negative U concept through the combined use of EPR and DLTS. EPR is perhaps presently the only tool which can look at specific atomic defects at the Si-SiO$_2$ interface. The potential for other discoveries through the combined use of EPR and other experimental techniques on interfacial structures and materials appears promising.

ACKNOWLEDGEMENTS

Appreciation is expressed to Professor G. D. Watkins and Dr. E. H. Poindexter and their colleagues for making available preprints of their latest results prior to publication and granting permission to use their figures or adaptations of their figures for this paper.

REFERENCES

1. C. A. Coulson and M. J. Kearsley, Proc. Roy. Soc. A241, 433 (1957).

2. G. D. Watkins, in Radiation Damage in Semiconductors, (Dunod, Paris, 1965) p. 97.

3. K. L. Brower, Radiat. Eff. 8, 213 (1971).

4. Y. H. Lee and J. W. Corbett, Phys. Rev. B 8, 2810 (1973).

5. G. A. Baraff, E. O. Kane, and M. Schlüter, Phys. Rev. B 21, 5662 (1980).

6. L. C. Kimerling, in Radiation Effects in Semiconductors, 1976, edited by N. B. Urli and J. W. Corbett (Institute of Physics, London, 1977), p. 221.

7. H. J. Stein, Radiat. Eff. 9, 195 (1971).

8. B. Bleaney and K. W. H Stevens, Rept. Progr. Phys. 18, 108 (1953).

9. G. D. Watkins and J. W. Corbett, Phys. Rev. 121, 1001 (1961).

10. K. L. Brower, Phys. Rev. B 9, 2607 (1974); Phys. Rev. B 17, 4130 (1978).

11. K. L. Brower, Phys. Rev. B 5, 4274 (1972).

12. E. G. Sieverts, S. H. Muller, and C. A. J. Ammerlaan, Phys. Rev. B 18, 6834 (1978).

13. L. E. Halliburton, D. L. Cowan, and L. V. Holroyd, Phys. Rev. B 12, 3408 (1975).

14. J. N. Lomer and A. M. A. Wild, Radiat. Eff. 17, 37 (1973).

15. G. D. Watkins, Phys. Rev. B 12, 5824 (1975).

16. C. P. Poole, Jr., Electron Spin Resonance (Interscience Publishers, New York, 1967).

17. L. Kevan and L. D. Kispert, Electron Spin Double Resonance Spectroscopy (John Wiley, New York, 1976).

18. W. Kaiser and C. D. Thurmond, J. Appl. Phys. 30, 427 (1959).

19. P. V. Pavlov, E. I. Zorin, D. I. Tetel'baum, and Yu S. Popov, Dokl. Akad. Nauk. SSSR 163, 1128 (1965) [Sov. Phys. Dokl. 10, 786 (1966)].

20. J. B. Mitchell, J. Shewchun, D. A. Thompson, and J. A. Davies, J. Appl. Phys. 46, 335 (1975).

21. J. B. Mitchell, P. P. Pronko, J. Shewchun, D. A. Thompson, and J. A. Davies, J. Appl. Phys. 46, 332 (1975).

22. K. L. Brower and P. S. Peercy, in Laser and Electron Beam Processing of Materials, edited by C. W. White and P. S. Peercy (Academic, New York, 1980), p. 441.

23. K. L. Brower, Phys. Rev. Lett. 44, 1627 (1980).

24. W. V. Smith, P. P. Sorokin, I. L. Gelles, and G. J. Lasher, Phys. Rev. 115, 1546 (1959).

25. R. P. Messmer and G. D. Watkins, Phys. Rev. Lett. 25, 656 (1970), and Phys. Rev. B 7, 2568 (1973).

26. M. F. Deigen, I. M. Zaritskii, and L. A. Shul'man, Fiz Tverd. Tela 12, 2902 (1970); [Sov. Phys. - Solid State 12, 2343 (1971)].

27. G. D. Watkins and J. R. Troxell, Phys. Rev. Lett. 44, 593 (1980); Phys. Rev. B 22, 921 (1980).

28. J. R. Troxell, A. P. Chatterjee, G. D. Watkins, L. C. Kimerling, Phys. Rev. B 12, 5336 (1979).

29. G. D. Watkins, A. P. Chatterjee, and R. D. Harris, 11th International Conference on Defects and Radiation Effects in Semiconductors (1980), Oiso, Japan.

30. P. W. Anderson, Phys. Rev. Lett. 34, 953 (1975).

31. J. C. Bourgoin and J. W. Corbett, Phys. Lett. 38A, 135 (1972).

32. B. E. Deal, J. Electrochem. Soc. 121, 198C (1974).

33. P. J. Caplan, E. H. Poindexter, B. E. Deal, R. R. Razouk, J. Appl. Phys. 50, 5847 (1979).

34. E. H. Poindexter, P. J. Caplan, B. E. Deal, and R. R. Razouk, J. Appl. Phys., to be published.

35. Y. Nishi, Jap. J. Appl. Phys. 10, 52 (1971).

36. O. L. Krivanek, D. C. Tsui, T. T. Sheng, and A. Kamgar, in The Physics of SiO$_2$ and Its Interfaces, edited by S. T. Pantelides (Pergamon, New York, 1978), p. 356.

37. J. G. de Wit and C. A. J. Ammerlaan, in Ion Implantation in Semiconductors, edited by I. Ruge and J. Graul (Springer, Berlin, 1971), p. 39.

38. R. B. Laughlin, J. D. Joannopoulos, and D. J. Chadi, Phys. Rev. B 21, 5733 (1980).

ELECTRON SPIN RESONANCE STUDIES OF AMORPHOUS SILICON

DAVID K. BIEGELSEN
Xerox Palo Alto Research Centers, Palo Alto, CA   94304

ABSTRACT

Electron spin resonance and related spin dependent measurements have been used to make key contributions to the understanding of amorphous silicon, specifically as probes of the dominant states in the gap, recombination processes and doping. In this paper we give a cursory description of the techniques as they apply to this problem. We then review what has been learned in a-Si:H usually from coupled results of spin resonance and other complementary experimental techniques. The results lead us to a surprisingly simple picture of the equilibrium and non-equilibrium behavior of defects and carriers in this prototypical amorphous semiconductor.

INTRODUCTION

Amorphous semiconductors have been an area of technological and fundamental interest for many years. It is only recently, however, that theoretical and experimental results have been achieved which make the field tractable for study. Theoretical notions of charge localization, negative effective correlation energies, chemical short-range order, and characteristic defects in bonding have provided a basis for a microscopic description. In this paper we will consider only the tetrahedrally-bonded material system known as amorphous silicon, a-Si. In early work researchers had tried to grow thin films of pure a-Si by evaporation or sputtering. A large electron spin density [1] of order 10^{20} cm^{-3} and high conductivity showed that this material was far from an ideal "continuous random network" semiconductor. In 1975 Spear and Le Comber [2] demonstrated that thin films of a-Si could be grown (by glow discharge deposition from a silane plasma) with a relatively low density of states in the gap. This then allowed them to dope the material and form p-n junctions. The technological promise of radiation-hard thin film transistors, large-area, inexpensive solar photovoltaic cells, photoreceptors, etc. was thus indicated.

The states in the gap of a semiconductor are dominant in determining the electronic properties, e.g. trapping and recombination. In a-Si the intrinsic gap states are a result of disorder of either the weak bonds in the band tails, or bonding defects. In a-Si it has been found that the density of states can vary by over four orders of magnitude, depending on the deposition conditions. To ascertain the microscopic nature of a-Si (or more properly, a-Si:H, because hydrogen on the order of 10% is necessary to reduce the gap state density) and to supply feedback in the optimization of deposition conditions, a probe sensitive to the localized gap states is necessary. In this article we try to show how resonance techniques have been and will be useful in the study of this important material system.

Electron spin resonance (ESR) of paramagnetic centers (impurities, defects, conduction electrons) in crystalline materials has been a rich and, in many cases, unique source of information about the localization and site symmetry of the electronic wavefunction. Information from ESR in amorphous semiconductors is severely limited due to the random distribution of site angles relative to the applied magnetic fields. The first detrimental effect is the loss of information obtainable in crystals from the explicit angular dependence of the resonance parameters. Secondly, the presence of all possible orientations severely broadens the ESR lines leading to a quadratically reduced sensitivity as well as increased difficulty in resolving nearby signals. Variations in the spin Hamiltonian parameters from site to site further obliterate detailed structure in signals. (Of course, the intrinsic distribution in these parameters is of basic interest and some information can be inferred from the experimental results.) Because the explicit information is lost, ESR results must be interpreted in conjunction with other complementary experiments. In this way much information has in fact been obtained.

TECHNIQUE

The phenomenon of electron spin resonance has been detailed by Brower in a preceding article [3]. We, therefore, present here a brief description of ESR including only those interactions which apply to the problem of amorphous silicon.

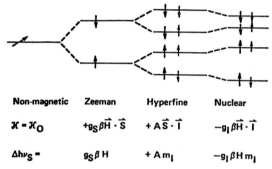

Non-magnetic	Zeeman	Hyperfine	Nuclear
$\mathcal{H} = \mathcal{H}_0$	$+ g_S \beta \vec{H} \cdot \vec{S}$	$+ A \vec{S} \cdot \vec{I}$	$- g_I \beta \vec{H} \cdot \vec{I}$
$\Delta h \nu_S =$	$g_S \beta H$	$+ A m_I$	$- g_I \beta H m_I$

Figure 1   Simplified Hamiltonian and energy levels for an electron and nucleus with spin = 1/2.

The Hamiltonian is given by a series of terms decreasing in strength as shown in Fig. 1. Consider a single unpaired electron. In the high magnetic fields of usual resonance experiments, the dominant perturbation on the non-magnetic electron energy is due to the Zeeman interaction of the electronic dipole in the externally applied field, H_0. The difference in the quantized energies for the electron spin parallel and anti-parallel to the field is given by $g\beta H_0$, where β is the Bohr magneton and g is a dimensionless number (actually a tensor component) which gives the dipole moment. For a free electron g_e = 2.0023. When the electron couples to states of non-zero angular momenta, g is shifted. In silicon, for which the spin-orbit coupling is quite small, the g-shift is also relatively small.

The next term to be considered is the nuclear hyperfine interaction. In essence, an electron overlapping a nucleus with non-zero spin senses an extra local field. In Fig. 1 we show a spin 1/2 nucleus (e.g., Si^{29}, which is ~5% abundant in nature, or H^1). It is readily seen that an electron centered on a Si^{29}, say, will have its resonance (i.e., energy difference between spin up and spin down) split into a pair of lines at $g\beta H_0 \pm A/2$. A is predominantly determined by the nuclear species and the magnitude of the electron wave function at the nucleus. One thus, in principle, can use this information to determine the location and localization of the paramagnetic electron. Finally, the very small nuclear Zeeman term is included for reference in understanding electron-nuclear double resonance (ENDOR) experiments.

The dominant interactions discussed above determine the absorption line position.

The line shape in general also depends on many mechanisms [4], but we will limit our discussion to the few aspects relevant to a-Si:H with "reasonably good" semiconducting properties (spin densities $< 10^{18}$/cc). The homogeneous coupling mechanisms of dipolar broadening, exchange narrowing and motional effects are important only for $n \geq 10^{19}$/cc. They will not be discussed here more than to note that we can conclude that spin clustering does not seem to occur [5]. Spin echo measurements [6] and passage effects on saturation [7] show that the equilibrium ESR signature in undoped a-Si is inhomogeneously broadened (i.e., the line is an envelope of "spin packets" which are nearly uncoupled) with the homogeneous spin packet width $< 1\%$ of the total width. The inhomogeneous mechanisms are related to the line position determining interactions discussed above. The dominant mechanism is g-value anisotropy. Consider as an example a dangling bond on a three-fold coordinated silicon atom. If H_o is parallel to the sp^3 dangling bond, the g-value ($g_{\parallel}$) can be shown by symmetry arguments to be unshifted (to first order) from g_e. If H_o is perpendicular to the dangling bond, the g-value is shifted to $g_{\perp}$. In an amorphous system, these defects will have random orientations with respect to H_o. The resultant line (a "powder pattern" - cf. [13]) spans the range from $g_{\parallel}$ to $g_{\perp}$ with highest density near $g_{\perp}$. A related mechanism arises from, say, the variations in the g-tensor components associated with environmental variations due to bond angle and length fluctuations in amorphous materials. Finally, the hyperfine interaction can produce unresolved line broadening; (a) from the slight resonance shift if the electron wave function ψ_e overlaps a neighboring nuclear spin, and/or (b) from a much-reduced splitting if ψ_e is significantly delocalized (i.e., less electron amplitude at the nucleus implies smaller splitting). The dominance of g-value anisotropy has been confirmed from measurements at various values of H_o [8]. For a factor of four in H_o, the line width (which for a pure powder pattern should be proportional to H_o) varied by approximately three. The origin of the residual inhomogeneous broadening is still experimentally uncertain.

ESR IN UNDOPED AMORPHOUS SILICON

The ESR line shape in undoped a-Si is (surprisingly) very nearly independent of deposition conditions. Again we have chosen not to include the very interesting correlations of linewidth with hopping conduction at high spin densities [9]. Initially seen in evaporated Si [1] and guessed to be due to dangling bonds, the featureless line at g = 2.0055 has also been observed in sputtered Si (with or without hydrogen), glow discharge and chemically vapor-deposited Si, crystalline Si self-implanted to amorphicity, etc. Even in materials known to grow with columnar microstructure [10], the signal is isotropic. This striking blindness to the large variations in local defect structures seems to imply a population of isolated, strongly localized paramagnetic centers. Recent ESR experiments on the crystalline Si/amorphous SiO_2 interface [11] demonstrate similar features for what are undoubtedly Si dangling bonds. (This interface is one of the only systems in which isolated dangling bonds can occur, while still retaining the symmetry ordering demanded by the crystal.) The ESR line is cylindrically symmetric with $g_{\parallel} \sim 2.001$ and $g_{\perp} \sim 2.008$. In Fig. 2 we show a powder pattern formed from such centers, assuming reasonable broadening-functions. W is the half-width at half-height of an assumed Gaussian distribution of g-tensor components. A value of $W = 0.5 \times (g_{\perp} - g_{\parallel})$ gives a peak (i.e., "g-value") at ~ 2.006 - very similar to the g = 2.0055 of a-Si. Of course, detailed agreement is not expected because the relaxation of the silicon back bonds is different in the two cases and a uniform broadening of $g_{\parallel}$ and $g_{\perp}$ is unlikely. Nevertheless, the similarity makes still more plausible the identification of the defect center as a dangling bond. In a-Si, resolved hyperfine splitting has not been observed. In a-SiO_2 on the other

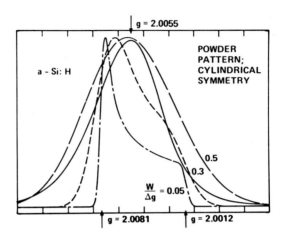

Figure 2   Powder pattern for dangling bond of reference [12] using Gaussian broadening
with half-width at half-height, W. The solid curve is the ESR absorption in a-
Si:H.

hand, hyperfine for the Si^{29} dangling bond (E' center) is seen with a 420G splitting and
~35 G broadening [12]. It is well known that Si bond angle variations in a-Si are ~10
times as great as in $a-SiO_2$. This greater parametric variation may effectively smear out the
resolved peaks in a-Si. Experiments on Si^{29} enriched samples would likely clear up this
question. Another point to be noted here is that the a-Si line width is independent of
hydrogen content (except indirectly, when $n > 10^{19}/cc$). This is further evidence for strong
localization of ψ_e on the paramagnetic site.

Now, whereas the line shape does not depend on deposition conditions, the dangling
bond spin density can vary over many orders of magnitude, from what seems to be a
maximum of ~3 x $10^{19}/cc$ - limited probably by the competing process of bond
reconstruction - to an undetermined low density. Our lowest measurement is ~3 x
$10^{15}/cc$. Problems with defect creation in collecting the large samples can place an
experimental limit on measurements of lower densities. In Fig. 3 we show typical results
for the variation of spin density with two deposition parameters [13]. ESR can thus be
used to characterize the "quality" of the sample and thus feedback to understand and
improve the deposition process. For example, here we see that the lowest defect material
is grown with the lowest power in pure silane. This also corresponds to the lowest growth
rates.   More information comes from comparing ESR measurements with
photoluminescence, i.e., the radiative recombination of optically excited carriers. Figure 4
is such a plot of samples prepared under a wide range of conditions [13]. Within the
scatter a simplifying relation is revealed. The implication is that luminescence of electron-
hole pairs can be quenched when the density of spins is greater than ~$10^{18}/cc$. It is then
more probable for the electron to tunnel to the defect than to return to its geminate hole.
Stuke et al. [14] have found evidence in He^+ bombarded samples of spinless non-
radiative recombination centers, but in as-prepared samples it seems that almost all
electrically active centers are dangling bonds. Thus ESR in a-Si:H is an especially
important tool for understanding the states in the gap.

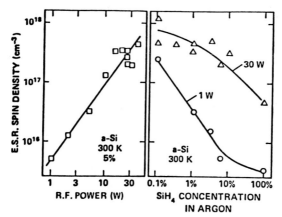

Figure 3   Spin density versus deposition parameters.

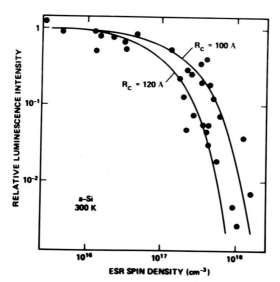

Figure 4   Photoluminescence efficiency versus spin density for samples prepared under a wide range of deposition conditions.

ESR IN DOPED AMORPHOUS SILICON

Up to this point we have been concerned with intrinsic defect centers in *undoped* a-Si. When the density of gap states is low enough, doping can move the Fermi level, E_F. The doping process is still not understood, but substitutional doping by ~10% of the dopant atoms is consistent with all results to date. We note in passing that no signal has been observed from un-ionized dopants. This is no doubt due to the capture of the donor (acceptor) electron (hole) by deeper lying gap states even at T → 0.

The initial effect of doping is to compensate the dangling bond states [15]. Thus the neutral dangling bond energy level is somewhat below mid gap. Therefore, in lightly doped samples (either p-type or n-type) when the number of active dopants is greater than the number of dangling bonds, there is almost no equilibrium ESR signal. At high doping levels, as E_F enters the band tails (~0.1% dopant/Si, depending on deposition conditions), a new ESR line emerges [16]. The line (different for n- and p-type doping) is assigned (see below) to singly occupied states in the band tails (charged weak bonds). The temperature dependent linewidths and short lifetimes at temperatures above ~100°K are indicative of the expected relaxation of more mobile carriers.

There is now emerging strong evidence for the fact that doping also induces defect formation. This may arise indirectly from a change in the deposition process when the doping gas is present; or, more directly, a mechanism such as autocompensation may exist which makes it energetically favorable to induce a dopant-defect pair in lieu of simple doping. This would limit the range that E_F can be varied and would be detrimental to transport properties by introducing concomitant recombination centers. This is an important area of research in which ESR will be a useful tool.

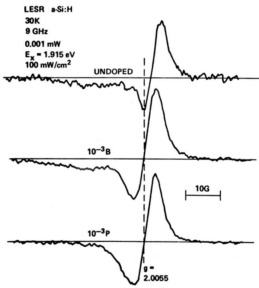

Figure 5   LESR spectra at 30°K in doped and undoped samples.

LIGHT INDUCED ESR

So far we have considered only the equilibrium ESR signals. Irradiation with absorbed light creates non-equilibrium carriers which can give rise to changed ESR spectra [17]. In its simplest (and most frequently used) form, the experiment consists of taking the difference between spectra with and without light on (with due care taken that, (a) little heating occurs, and (b) any dark ESR is not in saturation). Time resolved measurements [18] show an unmeasurably short rise time and a wide distribution of decay times which at low temperatures can have considerable strength out to 1 sec. In Fig. 5 we show LESR lines measured at 30°K for samples of various doping [19]. The n-type sample shows a line like the standard equilibrium signal in undoped material. In the p-type sample, an added broad line at higher g-value is observed. In undoped samples a much weaker signal is observed. The broad line seems to be the same as in doped samples, but the narrow line is in fact narrower and closer to g_e than the standard line.

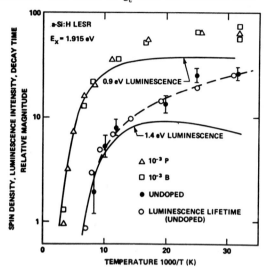

Figure 6   Temperature dependence of LESR, PL lifetime and steady state PL in doped and undoped a-Si.

An understanding of the LESR again comes from comparison with luminescence. Photoluminescence (PL) is a similar non-equilibrium process in which two lines are observed. In materials with low defect concentrations a line is observed at 1.4 eV and is assigned to recombination between a band tail electron and a self-trapped hole with ~100% efficiency. In defective (spin density ~10^{18}/cc) undoped or doped materials, the 1.4 eV line is quenched and a weakly-radiative, defect-related (i.e., independent of n- or p-type doping) band at 0.9 eV is revealed [20]. This line arises from recombination of the electron trapped at the defect with a self-trapped hole. If we assume that the LESR and the PL recombination channels are the same, then the PL lifetime should vary with temperature in the same way as the LESR intensity (since $n_{LESR} \alpha$ $G\tau$, where G is a generation rate and τ a mean lifetime). From Fig. 6 we can see that this is in fact the case. In short, the broad LESR line is associated with the self-trapped hole, the narrow (~6G peak-peak) line is the band tail electron, and the standard signal is due to singly occupied defects. We can further conclude that the defects have a positive effective correlation energy and, from the evidence described above, are most likely dangling bonds. Figure 7 is a schematic drawing of an energy diagram for these states. On the left is shown the background density of states of a completely coordinated a-Si:H alloy (cf. reference [2]). To it is grafted the one and two electron levels of the dangling bond (i.e., a single kind of defect center). The density of these states depends on deposition, hydrogenation, etc. With E_F in the middle of the gap, all dangling bonds are neutral (singly-occupied) and, depending on the defect density, a standard ESR signal is observed. Band gap light induces electron-hole pairs which, after thermalization and localization, either recombine

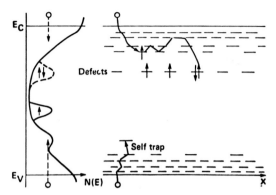

Figure 7   Density of states and spatial variation of states versus energy in a-Si.

radiatively, or the electron tunnels to the neutral defect and recombines non-radiatively or radiatively from the doubly-occupied level to the self-trapped hole level. Obviously, the greater the dark ESR, the more likely is: (a) the quenching of the band tail line, and (b) the increase of the defect luminescence. The LESR signals arise from the geminate band tail electron and hole. No defect LESR is seen here because recombination from the doubly-occupied defect state is too fast ($G\tau_{defect} \to 0$).

We will consider only one example of LESR in doped samples. Consider n-type material in which E_F is just above the doubly-occupied defect band. Clearly the dark dangling bond ESR is zero. Here electrons are the majority carrier so photo-induced holes immediately recombine with the excess electrons. Therefore we expect no hole line in LESR, in agreement with experiment. When the hole capture is by the doubly-occupied dangling bond, a singly-occupied, paramagnetic defect is produced which survives until an electron tunnels to it. In materials of low defect density and high doping, we can pull E_F into the band tails (or self-trapped band for holes) and the origin of the characteristic dark ESR signals in heavily doped samples is explained.

SPIN DEPENDENT RECOMBINATION

Because of the fortuitous occurrence that in a-Si electrically active centers are predominantly paramagnetic and because recombination lifetimes are quite long due to localization and low mobilities, spin dependent recombination is anomalously large and can be used in principle to study the recombination processes. We will not go into detail here but give only an example of an explanation for the observations in one regime. If one observes the cw luminescence [21] or photoconductivity [22] of a sample in a microwave cavity as one scans H_o, a resonant decrease ($\Delta L/L$ and $\Delta \sigma/\sigma < 0$) occurs which is similar to the LESR line. For temperatures above ~40°K these are the only signals observed and, for example, $\Delta L/L$ is approximately independent of temperature. An explanation of this resonant quenching invokes the resonant *enhancement* of the *non-radiative* channel [23]. A photo-excited electron tunneling to a singly-occupied defect finds itself in a triplet pair state three times as often as in a singlet state. (See Fig. 7, right side.) The recombination cross section for the triplet state is much lower than for the singlet state because of Pauli exclusion. Thus, an excess of pairs in the triplet state immediately builds up. Resonating either of the carriers flips a spin and increases the recombination rate. Thus, the band edge PL and the photoconductivity are reduced. We, therefore, obtain insight into the recombination mechanism for free carriers.

ENDOR

Defects in amorphous semiconductors are generally quite strongly localized. One might expect, therefore, that one could resolve the hyperfine signal associated with the nucleus of the dangling bond center. In ENDOR one drives an ESR line nearly into saturation. If the spin-lattice relaxation rate is increased, for example, by turning on a new relaxation channel, then the signal will come out of saturation and the spin signal will increase. By wrapping a radio frequency coil around the sample (in the ESR cavity) and scanning the rf frequency through ω_L for nuclear spins at H_o (e.g., Si^{29} or H^1) a nuclear spin system can be saturated. It turns out that at certain temperatures, this saturation can affect a ~1% increase in the ESR signal. We have attempted such an experiment by modifying our ESR apparatus [24]. No resolved lines could be observed, in agreement with our surmise above. Lines arising from dipolar coupling between the dangling bond and Si^{29} and H^1 were observed. These signals are in agreement with a uniform distribution of nuclear spins, but, so far, no direct information about the extent of the defect wavefunction has been obtained.

FUTURE USES

ESR and LESR are material-intensive measurements. Off the shelf spectrometer systems have spin sensitivities of ~10^{12} spins for the a-Si dangling bond line if no signal averaging is carried out. For "good" thin film samples with thickness ~$1\mu m$ and spin densities $\leq 10^{16} cm^{-3}$, the usefulness of these techniques becomes limited. For evaluation of materials in new growth processes or regimes (e.g., magnetron sputtering, Si:H:O, Si:H:F, etc.) these techniques will be particularly useful. Important fundamental ESR studies which remain are: (1) relating defects to transport (trapping and recombination); (2) understanding the microscopic nature of doping (and any associated defect induction); and (3) studying the nature of tail states.

Spin-dependent mechanisms, SDX, are proportional to $\Delta X/X$, where X is photoluminescence or photoconductivity. Since these are essentially volume independent, they are effectively considerably more sensitive than ESR. For example, Solomon [25] has observed SDPC in the depletion layer of p-i-n diodes when no ESR signal could be observed. Although the mechanism is not yet fully understood, this could be a technique for studying material in device configurations.

SUMMARY

We have presented examples of the usefulness of ESR and related resonance techniques to the study of amorphous silicon. We have shown how, from the results, a picture for electrically active defects in terms of dangling bonds can be inferred. We have concluded with some areas of future application of spin resonance.

ACKNOWLEDGEMENTS

Much of the work reported here has been done in collaboration with Drs. J. C. Knights and R. A. Street. This work has been supported in part by DOE Contract 03-79-ET-23033.

REFERENCES

1. M. H. Brodsky and R. S. Title, Phys. Rev. Lett. 23, 581 (1969).

2. W. Spear and P. G. LeComber, Solid State Comm. 17, 9 (1975).

3. K. Brower (this conference proceedings).

4. B. Movaghar, L. Schweitzer and H. Overhof, Phil. Mag. B, 37 683 (1978).

5. P. A. Thomas and D. Kaplan, in G. Lucovsky and F. L. Galeener (eds.) *Structure and Excitations of Amorphous Solids*, AIP, New York, 1976, p. 85; R. Bachus, B. Movaghar, L. Schweitzer and U. Voget-Grote, Phil. Mag. B. 39 27 (1979).

6. W. B. Mims and D. K. Biegelsen (unpublished).

7. S. Hasegawa and S. Yazaki, Thin Solid Films, 55 15 (1978); and D. K. Biegelsen (unpublished).

8. U. Voget-Grote, Thesis, Marburg, 1977 (unpublished).

9. B. Movaghar and L. Schweitzer, Phys. Stat. Sol. (b), 80 491 (1977); G. A. N. Connell and J. R. Pawlik, Phys. Rev. B, 13 787 (1976).

10. J. C. Knights and R. A. Lujan, Appl. Phys. Lett., 35 244 (1979).

11. P. J. Caplan, E. H. Poindexter, B. E. Deal and R. R. Razouk, J. Appl. Phys., 50 5847 (1979).

12. D. L. Griscom, Phys. Rev. B, 20 1823 (1979); and D. L. Griscom (in this conference proceedings).

13. R. A. Street, J. C. Knights and D. K. Biegelsen, Phys. Rev. B, 18 1880 (1978).

14. R. A. Street, D. K. Biegelsen and J. Stuke, Phil. Mag. B, 40 451 (1979).

15. S. Hasegawa, T. Kasajima, and T. Shimizu, Phil. Mag. 8 (in press).

16. J. Stuke, in W. Spear (ed.), *Proc. 7th Intl. Conf. on Amorphous and Liquid Semiconductors*, U. of Edinburgh, Edinburgh, 1977, p. 406; S. Hasegawa, T. Kasajima and T. Shimizu, Sol. State Comm., 29 13 (1979).

17. J. C. Knights, D. K. Biegelsen and I. Solomon, Sol. State Comm., 22 133 (1977); J. R. Pawlik and W. Paul, in W. E. Spear (ed.), *Proc. 7th Intl. Conf. on Amorphous and Liquid Semiconductors*, U. of Edinburgh, Edinburgh, 1977, p. 437; A. Friederich, D. Kaplan and N. Sol, J. Elec. Materials, 1979 (to be published).

18. D. K. Biegelsen and J. C. Knights, in W. E. Spear (ed.), *Proc. 7th Intl. Conf. on Amorphous and Liquid Semiconductors*, U. of Edinburgh, Edinburgh, 1977, p. 429.

19. R. A. Street and D. K. Biegelsen, Solid State Comm. 33, 1159 (1980).

20. R. A. Street, Phys. Rev. B, 21, 5775 (1980); and references cited therein.

21. D. K. Biegelsen, J. C. Knights, R. A. Street, C. Tsang and R. M. White, Phil. Mag. B, 37 477 (1978); K. Morigaki, D. J. Dunstan, B. C. Cavenett, P. Dawson and J. E. Nicholls, Sol. State Comm., 26 981 (1978).

22. I. Solomon, D. K. Biegelsen and J. C. Knights, Sol. State Comm., 22 505 (1977).

23. D. Kaplan, I. Solomon, and N. F. Mott, J. de Physique-Lettres 39, L51 (1978).

24. R. Kerns and D. K. Biegelsen (unpublished).

25. I. Solomon, Sol. State Comm., 20 215 (1976).

Published 1981 by Elsevier North Holland, Inc.
Kaufmann and Shenoy, editors
Nuclear and Electron Resonance Spectroscopies Applied to Materials Science

DEFECTS IN III-V SEMICONDUCTORS STUDIED THROUGH EPR

T. A. KENNEDY
Naval Research Laboratory, Washington, DC 20375

ABSTRACT

EPR studies of Cr impurities in GaAs and intrinsic defects
in GaP are reviewed to illustrate the contributions that
EPR has made to the materials science and technology of
III-V semiconductors. EPR has shown how the Cr impurity
acts to compensate residual shallow donors or acceptors to
produce semi-insulating GaAs which is used for device
substrates. EPR work has identified two intrinsic
defects, the P_{Ga} antisite and the V_{Ga} vacancy, and
clarified their role in electrical and stoichiometric
properties of GaP.

INTRODUCTION

There is a strong, world-wide effort to improve the III-V semiconductor
compounds for better electronic devices. For example, high-resistivity
or "semi-insulating" GaAs is needed for substrates for microwave
field-effect-transistors (FET's) and high-speed digital integrated
circuits. Often Cr-doping is employed in order to achieve the desired
semi-insulating property. As a second example, III-V materials are
required for light-emitting-diodes (LED's) and semiconductor lasers. In
many cases, such as the GaP:N green-emitting LED, non-radiative
recombination by poorly-understood defects limits the performance of the
device. In both cases, a better understanding of the III-V material, in
particular of the defects in the material, would have an impact on the
device technology.

Electron paramagnetic resonance (EPR) is particularly suited to the
study of localized defects in materials. Indeed, the essential
requirement for EPR is an unpaired electron spin, such as is often found
at a defect in a crystal. The perfect crystal with all chemical bonding
complete provides no EPR and thus forms a clean background to the defect
signals. Defects may be divided into two groups. Impurities may have
an unpaired spin either because of a valence difference from the atom
replaced or because of an incomplete inner shell--as in the case of a 3d
transition element. See Fig. 1 - upper portion. Intrinsic or native
defects may be of three kinds in a compound semiconductor: missing
atoms or vacancies (See Fig. 1 - lower portion), extra atoms or
interstitials, and cation-anion replacements or antisites. Unpaired
spins again occur for intrinsic defects because of the mismatch to the
perfect lattice. In many cases, the defects exist in more than one
charge state - thus nearly guaranteeing paramagnetism and the
possibility for EPR. It is through this ability to trap an electron or
a hole that the defect affects the electrical or luminescent qualities
of the material.

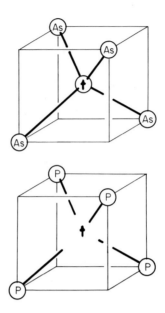

Fig. 1   Point Defects in III-V Lattices. In the upper portion of
          the figure, an impurity substitutes for Ga in GaAs. EPR is
          possible when there is an unpaired spin, such as in an
          incomplete d-shell. In the lower portion of the figure, a
          Ga vacancy is shown in GaP. Again EPR is possible when
          there is an unpaired spin in the broken chemical bonds.

If EPR studies would be so useful to the III-V compounds, why were they
not performed years ago? The EPR linewidths in the III-V's are much
larger than those of Si or the II-VI compounds and make the spectroscopy
more difficult.   Thus sensitivity is limited to concentrations of
$10^{14}cm^{-3}$ rather   than $10^{12}cm^{-3}$ in   Si.   The   larger   linewidths
arise from hyperfine interactions - all the III-V nuclei have magnetic
moments and the covalency of the materials ensures that a substantial
amount of the electron wave function will be at first and second nearest
neighbors in the crystal.

Recent EPR studies in III-V compounds [1] have identified important
defects and clarified their role in affecting technologically important
properties of the materials. Two examples are given in this paper.
Work on Cr in GaAs has demonstrated fully its action to produce
semi-insulating material. Work on GaP has identified two intrinsic
defects and determined their concentrations in bulk material. These
results give a better understanding of intrinsic defects and
self-compensation in GaP.

CHROMIUM IN GALLIUM ARSENIDE

Gallium arsenide grown in quartz crucibles contains a high concentration
of Si, which acts as a donor to produce n-type conductivity. It has
been known that in order to obtain semi-insulating material for device
substrates, Cr can be added to compensate the Si donors [2]. Questions
remained as to the exact nature of the Cr defect. Recent EPR work has
determined that it resides on the Ga site in the lattice without any
nearby defect, exists in three charge states and acts both as an
electron trap and a hole trap.

The EPR spectra observed in semi-insulating, Cr-doped GaAs are shown in
Fig. 2 [3]. The multiplicity of lines indicates that the defects
responsible have spin S > 1/2 and occupy sites of symmetry less than
cubic. Complete analysis of the spectra led to the identification of
Cr^{3+} ($3d^3$) and Cr^{2+} ($3d^4$) ions in the crystal [3,4]. The 3+
charge state corresponds to the Cr substituting for Ga in the lattice
with no excess or deficiency of charge. The EPR arises from the
incomplete d-shell. In the 2+ charge state the Cr has trapped an extra
electron into the d shell, thus compensating a shallow donor in the
material. Since Cr^{3+} and Cr^{2+} are the only equilibrium charge
states observed in semi-insulating material, it is precisely the
trapping of an electron by Cr^{3+} to become Cr^{2+} which produces the
semi-insulating quality.

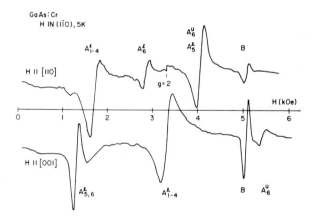

Fig. 2   EPR spectra in GaAs:Cr. The lines labelled A arise from
            Cr^{3+} and those labelled B from Cr^{2+}. The figure is
            taken from Krebs and Stauss [3].

The experiments also show that the local symmetry around each Cr^{2+} or Cr^{3+} is lowered from the tetrahedral symmetry of the Ga site in the perfect crystal. This was attributed to Jahn-Teller distortions [3,4]. That the lowered symmetry did not arise from a subsidiary defect near the Cr was proved through uniaxial stress experiments at low temperature [5,6]. In these experiments relative populations of defects oriented in different directions could be altered rapidly by the uniaxial stress.

EPR from a third charge state of Cr has also been observed. Careful co-doping studies with shallow acceptors have shown that the charge state is Cr^{4+} $(3d^2)$ [7,8], indicating that the Cr acts as a hole trap as well as an electron trap. This charge state is stable only in p-type material.

The concentration of Cr in each charge state has been measured [7]. Thus a complete, quantitative picture of the compensating action of Cr in GaAs has been achieved using EPR. See Table 1. In Sn-doped, n-type material, the Cr is predominately in the Cr^{2+} state, acting as an electron trap. In semi-insulating material, both Cr^{2+} and Cr^{3+} are present as some electron trapping occurs. Finally in Zn-doped, p-type material, Cr^{4+} appears as the Cr acts as a hole trap.

Table 1. Concentrations of the Cr
charge states in different GaAs samples
in units of $10^{16} cm^{-3}$.
The table is taken from G. H. Stauss et al. [7].

Sample	Nominal Doping			EPR Calibration 4.5 K - dark			ρ at 295K (Ω cm)
	[Cr]	[Zn]	[Sn]	[2+]	[3+]	[4+]	
1	8	44	-	0	0	7.9	0.074-p
2A	> 8	> 5	-	∿ 0	16.7	4.3	224-p
2B	8	5	-	∿ 0.1	7.2	≲0.3	1950-p
3	10	-	7	7.8	5±3	0	5×10^8-SI
4	6	-	11	obs.	∿0	∿0	0.090-n

INTRINSIC DEFECTS IN GALLIUM PHOSPHIDE

A considerable amount of effort has gone into the search for the non-radiative defect or defects which limit the performance of GaP LED's. The red photoluminescent efficiency of a series of GaP:Zn, O

samples grown at different temperatures has been studied [9]. The data
can be interpreted to show a decreasing minority carrier lifetime with
increasing Ga-vacancy (V_{Ga}) concentration. Large V_{Ga} concentrations
are deduced from this analysis, from 10^{16} to $10^{19} cm^{-3}$. However,
other intrinsic defects are possible in a III-V semiconductor, such as
the antisite defects, a P-atom on a Ga site (P_{Ga}) or a Ga atom on a P
site (Ga_P), and P and Ga interstitials. Associates of two or more
point defects may be the dominant defects occurring during bulk crystal
growth. Large concentrations of isolated Ga vacancies in crystals
cooled from growth temperature were shown to be unlikely on the basis of
thermodynamic arguments [10]. However, no experimental data existed to
verify the existence of vacancies or antisite defects and to determine
their concentrations.

Recently two intrinsic defects in GaP have been identified using EPR.
The P_{Ga} antisite was first observed in semi-insulating GaP and
identified from central and ligand hyperfine interactions [11]. The Ga
vacancy was observed in electron-irradiated GaP and also identified
through hyperfine splittings [12]. See Fig. 3. The intensity ratios
for the magnetic field in the [001] direction, 1:4:6:4:1, are
characteristic of a interaction with four equivalent nuclear spin
one-half atoms, in this case the nearest neighbor phosphorus (See Fig.
1). These identifications demonstrate the power of EPR, as there is
really no other way to positively identify intrinsic defects in III-V
semiconductors.

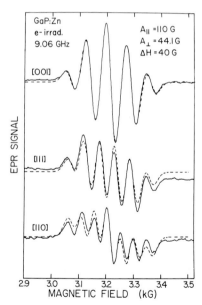

Fig. 3   EPR of V_{Ga} in electron-irradiated GaP:Zn. The
multiplicity arises from hyperfine interactions. The
dotted lines are a data simulation with the parameters
given in the upper-right-hand portion of the figure. The
figure is taken from Kennedy and Wilsey[12].

In order to address the materials problems of GaP LED's, measurements of the concentrations of P_{Ga} and V_{Ga} in a selection of liquid-encapsulation-Czochralski samples were performed [13] with results given in Table 2. Antisites were found in semi-insulating and p-type samples in concentrations sufficient to affect the electrical properties of the material. Since P has two extra electrons relative to Ga, the P_{Ga} antisite acts as a double donor. It thus compensates two of the Zn acceptors in p-type material. The antisite thus may act as a self-compensator. Self-compensation has been observed in a series of n-type GaAs samples over a wide range of donor doping [14]. Isolated vacancies were not observed in as grown crystals. Since the EPR-active charge state may not have been sufficiently populated in the experiments, it can only be safely stated that the total isolated vacancy concentration is less than $10^{16}cm^{-3}$ in p-type material.

Neither isolated P-antisites nor isolated Ga-vacancies are present in the very high concentrations indicated by the photoluminescent efficiency/minority-carrier lifetime study [9]. Thus the prediction that intrinsic defect associates are the cause may be correct [10].

Table II. Concentrations of V_{Ga} and
P_{Ga} in units of $10^{16}cm^{-3}$
The table is taken from Kaufmann and Kennedy [13].

Sample	RT carrier conc.	V_{Ga} $^{2-}$ conc.	P_{Ga} $^{4+}$ conc.
1. GaP:S	30	< 0.1	< 0.1
2. GaP:Te	15	< 0.1	< 0.1
3. GaP:S	$\sim$ 10	< 0.1	< 0.2
4. GaP	undoped	< 0.02	< 0.1
5. GaP:Cr	semiins.	< 0.04	< 0.2
6. GaP:Cr	semiins.	< 0.02	0.6
7. GaP:Zn as grown	13	< 0.02	3.5
8. GaP:Zn Vac. ann.	$\sim$ 13	< 0.04	0.8
9. GaP:Zn e$^-$ irr.	compens.	0.9	4.8

Further EPR work is now being done on associates formed of an antisite plus a second point defect. Such associates have been observed in electron-irradiated GaP [15] and as grown GaP [16].

CONCLUSION

For both Cr in GaAs and the intrinsic defects in GaP, two properties of the electron paramagnetic resonance techniques made possible the contributions to materials science. First, EPR has the ability to positively identify point defects and their charge states. Second, quantitative measurement is possible so that defect concentrations of the particular charge states can be made.

Current EPR work continues to address III-V materials problems of technological importance. Iron in InP [17,18] and the As_{Ga} antisite in GaAs [19] are being studied. Thus although the knowledge of III-V materials still lags behind our knowledge of Si and II-VI compounds, EPR has contributed and will contribute to their understanding and improvement.

ACKNOWLEDGMENTS

I am grateful to U. Kaufmann, J. J. Krebs, J. Schneider, G. H. Stauss and N. D. Wilsey for helpful discussions.

REFERENCES

1. For a general review, see U. Kaufmann and J. Schneider in Festkörperprobleme (Advances in Solid State Physics), Vol XX, p. 87, Vieweg, Braunschweig (1980).

2. G. R. Cronin and R. W. Haisty, J. Electrochem. Soc. 111, 874 (1964).

3. J. J. Krebs and G. H. Stauss, Phys. Rev. B 15, 17 (1977).

4. J. J. Krebs and G. H. Stauss, Phys. Rev. B 16, 971 (1977).

5. J. J. Krebs and G. H. Stauss, Phys. Rev. B 20, 795 (1979).

6. G. H. Stauss and J. J. Krebs, Phys. Rev. B 22, 2050 (1980).

7. G. H. Stauss, J. J. Krebs, S. H. Lee and E. M. Swiggard, Phys. Rev. B 22, 3141 (1980).

8. U. Kaufmann and J. Schneider, Appl. Phys. Lett. 36, 747 (1980).

9. A. S. Jordan, A. R. Von Neida, R. Caruso and M. DiDomenico, Jr., Appl. Phys. Lett. 19, 394 (1971).

10. J. A. Van Vechten, J. Electrochem. Soc. 122, 423 (1975).

11. U. Kaufmann, J. Schneider and A. Räuber, Appl. Phys. Lett. 29, 312 (1976).

12. T. A. Kennedy and N. D. Wilsey, Phys. Rev. Lett. 41, 977 (1978).

13. U. Kaufmann and T. A. Kennedy, J. Electron. Mater., to be published.

14. C. M. Wolfe and G. E. Stillman, Appl. Phys. Lett. 27, 564 (1975).

15. T. A. Kennedy and N. D. Wilsey, Inst. Phys. Conf. Ser. No. 46, 375 (1979).

16. U. Kaufmann, private communication.

17. G. H. Stauss, J. J. Krebs and R. L. Henry, Phys. Rev. B $\underline{16}$, 974-977 (1977).

18. W. H. Koschel, U. Kaufmann and S. G. Bishop, Solid State Commun. $\underline{21}$, 1069 (1977).

19. R. J. Wagner, J. J. Krebs, G. H. Stauss and A. M. White, Solid State Commun. $\underline{36}$, 15 (1980).

Published 1981 by Elsevier North Holland, Inc.
Kaufmann and Shenoy, editors
Nuclear and Electron Resonance Spectroscopies Applied to Materials Science

ESR STUDIES IN AMORPHOUS INSULATORS

DAVID L. GRISCOM
Naval Research Laboratory
Washington, DC 20375

ABSTRACT

Some applications of electron spin resonance to the study
of amorphous insulators are briefly described. Specific
topics addressed include radiation-induced defect centers,
redox equilibria, phase separation, photochromics, and
ferromagnetic precipitates.

INTRODUCTION

Among the insulating materials of current technological interest are a great
many which are employed widely, and sometimes exclusively, in their amorphous
forms. Some examples include glasses for windows, lamp jackets, laser media,
and fiber optics, and thin amorphous films which form the insulating and
passivating layers on metal-insulator-semiconductor (MIS) devices. In many
of these cases, materials properties affecting device performance are found
to be strongly related to details of the atomic scale structure, including
defect structure, and to the nature and disposition of impurities and dopants.
Electron spin resonance (ESR) is perhaps the most powerful single technique
for identifying and characterizing point defects in insulating solids. In
addition, ESR provides an effective means for determining the valence states
of many transition-group impurities. The chief limitation of the method is
the, requirement that there be unpaired electrons at the defect or impurity
site. However, the necessary "unpaired spins" are routinely present at the
sites of transition-group ions with partially filled d or f shells, and they
are readily produced by the metastable trapping of radiation generated electrons or holes at defect sites. In either case, if the spin system is dilute
and relatively non-interacting, the ESR experiment is commonly termed electron
paramagnetic resonance (EPR). The terminology "ESR" is adopted here so as to
include the phenomenon of ferromagnetic resonance (FMR), which is observed when
the spins are clustered and strongly interacting.

The standard ESR spectrometer is an apparatus which records the resonant
absorption of fixed-frequency microwaves by the sample as a function of the
magnitude of an applied magnetic field. If the sample should be a single
crystal, the resonance spectrum would be sensitive as well to the orientation
of the laboratory field with respect to the crystallographic axes, but in
powders and amorphous materials only an angularly averaged "powder pattern" is
recorded. The ESR technique in general is developed in many textbooks (e.g.,
[1-5]) and the means of dealing with powder patterns have been covered in a
number of review articles (e.g., [6-12]). While the focus of the present
paper is restricted to practical applications of the method to amorphous
insulators, the brief format precludes a comprehensive coverage of even
this subset of the literature. Building upon a minimal review of the needed
theory in Sec. 2, the six case studies of Sec. 3 hopefully provide a reasonable cross section of the types of problems which can be solved, together with
some insights into the methods of analysis. The interested reader with no
prior background in magnetic resonances is strongly encouraged to consult

additional references, which might include Refs. $\begin{bmatrix} 9 \end{bmatrix}$ and $\begin{bmatrix} 12 \end{bmatrix}$ plus the article in the present volume by Brower $\begin{bmatrix} 13 \end{bmatrix}$.

ESR THEORY

ESR spectra are customarily analyzed in terms of a so-called "spin Hamiltonian", which contains the energies of interaction of the unpaired electron spins with all of the significant microscopic and macroscopic electric and magnetic fields in their environments. For present purposes, a spin Hamiltonian comprising only two terms will suffice:

$$\mathcal{H} = \beta \; \vec{S} \cdot \overset{\leftrightarrow}{g} \cdot \vec{H} + \vec{S} \cdot \overset{\leftrightarrow}{A} \cdot \vec{I}. \tag{1}$$

In eq. (1), the first term is the Zeeman interaction of electronic spin $\vec{S}$ with the applied field $\vec{H}$, where $\overset{\leftrightarrow}{g}$ is a matrix which generally can be represented in diagonal form and β is the Bohr magneton. The second term describes the hyperfine interaction of the electron magnetic moment with the magnetic moment of a nucleus with spin $\vec{I}$; the matrix $\overset{\leftrightarrow}{A}$ is the so-called "hyperfine coupling tensor", which can also be represented in diagonal form in most practicle situations.

Considering the Zeeman interaction alone leads to the simple resonance condition

$$h\nu = g \; \beta \; H_{res}, \tag{2}$$

where h is the Planck's constant, ν is the microwave frequency and H_{res} is the magnitude of the applied field $\vec{H}$ at which resonance occurs. In eq. (2), g is a dimensionless number which is a function of the principal-axis (diagonal) components of $\overset{\leftrightarrow}{g}$ and the angles describing the orientation of $\vec{H}$ with respect to the principal-axis directions. In the case of axial symmetry

$$g^2 = g_\parallel^2 \; \cos^2\theta + g_\perp^2 \; \sin^2\theta, \tag{3}$$

where θ is the angle between $\vec{H}$ and the symmetry axis of $\overset{\leftrightarrow}{g}$. For single crystal samples, the angular dependence of H_{res} can be understood in terms of eqs. (2) and (3). When the specimen is in powder form, the spectrum becomes smeared out between extremal field values given in the present example by setting $\theta = 0$ ($g = g_\parallel$) and $\theta = \pi/2$ ($g = g_\perp$). The envelope of this powder pattern has a well defined shape which can be calculated numerically, and in the simple axial case it can also be derived as a closed-form analytical expression. Figure 1(a) gives the ESR absorption envelope calculated from eqs. (2) and (3), assuming zero single-crystal linewidth. Figure 1(b) exhibits the first derivative of this envelope together with an experimental first-derivative spectrum shown in superposition. The experimental spectrum is broadened by a finite single crystal linewidth (see Sec. 3.1).

The hyperfine interaction (second term of eq. (1)) causes the general powder spectrum to split up into multiplets comprising 2I + 1 individual (though possibly overlapping) component powder patterns, where I is the scalar magnitude of the nuclear spin. The splittings between the singular features of these component patterns give the diagonal elements of $\overset{\leftrightarrow}{A}$, the "hyperfine coupling constants" whose values specify the degree to which the unpaired electron spin is localized on the nucleus in question (see, e.g., [9]). Further discussion of hyperfine interactions in the ESR spectra of amorphous insulators is provided in Sec. 3.2.

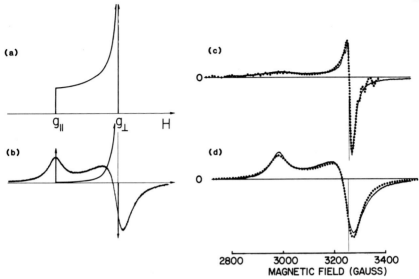

Fig. 1. ESR powder pattern (a) for an axial g tensor and its first derivative (b) are compared to experimental derivative spectra of O_2^- ions (b, c, and d). Experimental curves of (b) and (d) are identical; dotted curves are computer simulations. (After Refs. [8, 9].)

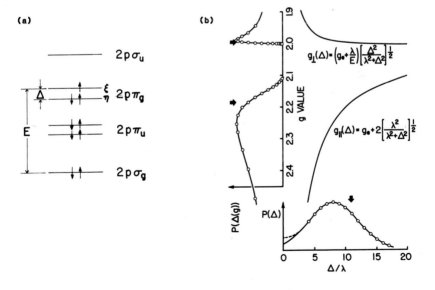

Fig. 2. Energy level diagram (a) and relationship between g values and energy level splittings (b) for the O_2^- molecular ion. (After Refs. [8, 9].)

SOME CASE STUDIES

The O_2^- Ion in Powdered and Amorphous Media

The superoxide anion O_2^- provides an excellent prototype for discussing ESR line shapes in powdered and amorphous samples. The experimental spectrum in Fig. 1(b) is in fact that due to O_2^- inadvertently present in a commercial sodium peroxide powder (Na_2O_2) [9, 14] . The values of $g_\parallel$ and $g_\perp$ measured for this molecular ion can be related to its valence shell electronic structure as schematically illustrated in Fig. 2. Specifically, Känzig and Cohen [15] have shown that these g values depend most sensitively on the splitting Δ of the Π-antibonding levels resulting from the interaction of the ion with its surroundings. The expressions for $g_\parallel$ and $g_\perp$ in Fig. 2(b) are approximations [14] to Känzig and Cohen's formulae; they are plotted here as functions of Δ/λ, where λ is the spin-orbit coupling constant for oxygen. It was found that the spectrum of Fig. 1(b) could be computer simulated on the basis of the powder pattern of Fig. 1(a) by choosing the values of $g_\parallel$ and $g_\perp$ corresponding to $\Delta/\lambda \approx 11$ in Fig. 2(b)(bold arrows) and convoluting by a Lorentzian broadening function. The results are shown as the dotted curve in Fig. 1(d).

An interesting class of amorphous materials can be obtained by careful heating of certain crystalline peroxyborate compounds (e.g., $NaBO_3 \cdot H_2O$) in the temperature range 100-130°C. Measurements of static susceptibility and ESR intensity have indicated the presence in these materials of large numbers of unpaired spins, all of which are reasonably attributed to various combined forms of oxygen [14, 16] . The continuous curve in Fig. 1(c) is the ESR spectrum of one of the paramagnetic species encountered in these amorphous peroxyborate preparations. The dotted curve in superposition is a computer simulation based on O_2^- theory under the assumption that the line shape is entirely determined by a broad Gaussian distribution in the splitting Δ (illustrated by the open circles in Fig. 2 b). Note in this figure that the predicted distributions in $g_\parallel$ and $g_\perp$ have shapes which appear quite different, yet are rigorously interrelated via the formulae of Känzig and Cohen. The ability to achieve a successful line shape simulation under such rigorous theoretical constraints by employing only two adjustable parameters (the peak position and width of the Gaussian distribution function) provides a stringent test of the postulate that the experimental spectrum arises from the O_2^- species. This example also illustrates a canonical influence of the amorphous state of ESR spectra: random site-to-site variations in crystal fields give rise to Gaussian distributions in energy levels generally leading to skewed distributions in spin Hamiltonian parameters.

Defect Centers in Amorphous Silica

Non-crystalline forms of silicon dioxide find important applications as the light transmitting medium in fiber optic waveguides and as the insulating layers in metal-oxide semiconductor device structures. Since both technologies are seriously degraded by exposure to ionizing radiation, considerable interest has centered on the nature of the radiation-induced defect centers in amorphous SiO_2.

The best known defect in vitreous silica, the E' center, has also been studied extensively in the crystalline polymorph α-quartz (for a review, see [17]). Thus, it has been possible to compare the orientations of the principal axes of the spin Hamiltonian with known aspects of the crystal structure. A significant finding was that the g and hyperfine tensors each possessed approximate axial symmetry, with the unique axes of each being coparallel and lying along the direction of a normal Si-O bond [18] . The magnitudes of the ^{29}Si hyperfine coupling constants indicated the unpaired spin to be strongly localized in a tetrahedrally hybridized orbital of a single silicon. These data formed much of the basis for the currently accepted model of the E' center as a hole trapped at an asymmetrically relaxed oxygen vacancy [19,20] as illustrated in Fig. 3(a). The existence of E' centers in irradiated glassy silica was first inferred on the basis of the observation of an ESR powder pattern similar to that of Fig. 1 but corresponding to the same principal-axis g values as determined for α-quartz [21]. Hyperfine structure due to the magnetic isotope ^{29}Si (I = 1/2, 4.7% abundant) proved difficult to observe in the glass, however, since the lines were weak not only due to the low natural abundance but also due to their greater experimental linewidths. Furthermore, as seen in Fig. 4(a), not one but two pairs of hyperfine lines were recorded in some cases. By studying samples enriched to 95% in ^{29}Si (Fig. 4b) it was possible to show that the doublet having a splitting of ∿ 420 G was the true hyperfine structure of the E' center [22,23], while the 74-G doublet turned out to be due to a defect involving impurity protons (see, e.g., Ref. [12]). By computer line shape simulation it was shown that the E' center in silica glass is characterized by a distribution in coupling constants which is nearly Gaussian in shape (Fig. 4c) [22]. In turn, the latter distribution has been related by elementary theory to an inferred distribution of bond angles at the defect sites, as also illustrated in Fig. 4 (c). This study provides one more example of the influence of vitreous disorder on ESR spectroscopic properties. More remarkable is the narrowness of the inferred bond angle distribution: just 0.7 deg in halfwidth.

Two other intrinsic defect centers have also been indentified in amorphous SiO_2 by means of ESR. These are the peroxy radical [24] and the nonbridging oxygen hole center [25], whose inferred modes of formation are illustrated in Fig.3(b) and (c), respectively. The ESR analyses which led to the models for these "oxygen-associated hole centers" (OHCs) are described in Refs. [9],[24], and [25]. In both cases, important use was made of samples enriched in the magnetic isotope of oxygen, ^{17}O. As illustrated in Fig. 3(b), the bridging peroxy

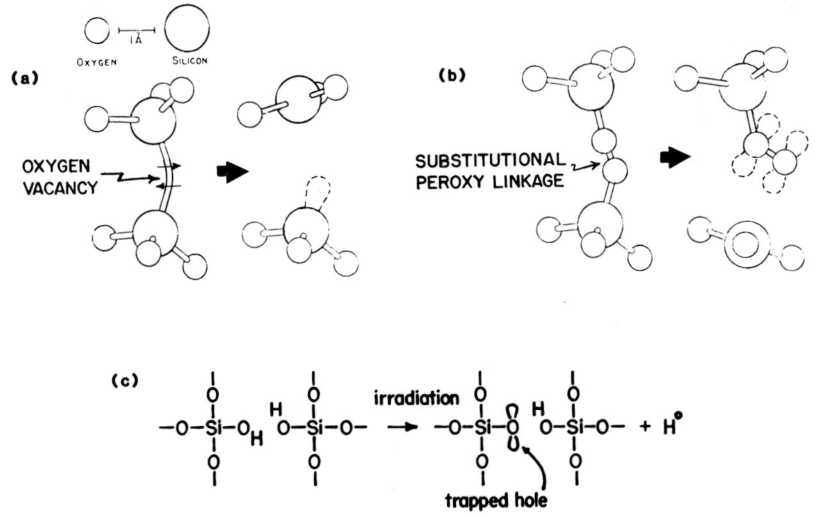

Fig. 3. Models for defect formation in fused silica: (a) the E' center, (b) the peroxy radical, and (c) the nonbridging oxygen hole center. (After Refs. [12] and [25].)

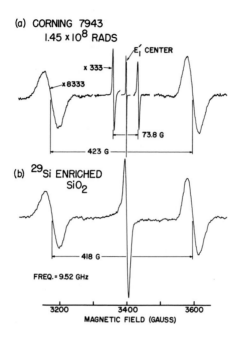

(a) CORNING 7943
1.45 x 10^8 RADS

E_1' CENTER

x 333

x8333

73.8 G

423 G

(b) ^{29}Si ENRICHED SiO$_2$

418 G

FREQ. = 9.52 GHz

3200   3400   3600

MAGNETIC FIELD (GAUSS)

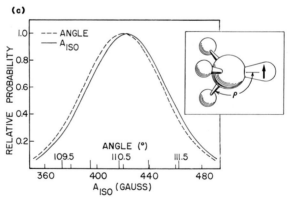

(c)

--- ANGLE
— A$_{ISO}$

RELATIVE PROBABILITY

1.0

0.8

0.6

0.4

0.2

ANGLE (°)

109.5   110.5   111.5

360   400   440   480

A$_{ISO}$ (GAUSS)

Fig. 4. ESR spectra of γ-irradiated fused silica samples (a and b) and distribution in bond angles at E' site (c) derived from an analysis of the 420-G ^{29}Si hyperfine doublet. (After Ref. [22].)

linkage is now suspected to exist as a defect in silica glass prior to irradiation. The peroxy linkage and the oxygen vacancy may be viewed as complementary members of a Frenkel defect pair. By contrast, the nonbridging oxygen hole center is best ascribed to hole trapping at the sites of hydroxol impurities.

Figure 3(c) shows the hydrogen atom as a product of radiolysis of the hydroxol group. Although H^o is a highly reactive species at room temperature, it can be stabilized below $\sim$100K, where its ESR signature is a hyperfine doublet with a splitting of $\sim$ 500 G. Making use of this identification together with the spectral signatures of the other species described above, it becomes possible to study the kinetics of radiolysis by irradiating at cryogenic temperatures and observing the various defect populations as functions of time and temperature. Figure 5 presents some ESR intensity data as a function of 5-min isochronal anneals for a high purity fused silica sample containing $\sim$1200 ppm OH which had been x-irradiated at 77K [26]. Since the inital concentration of atomic hydrogen is seen here to exceed the OHC population by a factor of $\sim$ 3, one must consider the possibility that some of the H^o results from radiodissociation of dissolved molecular hydrogen. The "133-G doublet" of Fig. 5, like the 74-G doublet of Fig. 4(a), has been shown by deuterium substitution to be due to a defect involving a proton [26]. A tentative interpretation of Fig. 5 is that the decay of H^o "feeds" the abrupt increases in E' center and 133-G-doublet populations between 100 and 125K. In glasses containing $\lesssim$ 5 ppm OH, neither H^o nor the 133-G doublet is observed, and the E' and OHC populations decay "hand in hand". Thus, the ESR technique is bringing to light the critical role of hydrogen in the radiation damage kinetics of high-OH vitreous SiO_2.

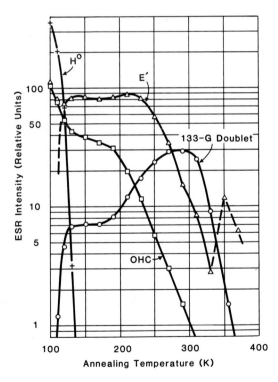

Fig. 5.   Decay of ESR centers in a high-purity fused silica (Suprasil 1) following x-irradiation at 77K and 5-min. isochronal anneals. (After Ref. [26].)

Redox Equilibria in Oxide Glasses

Knowledge of the oxidation-reduction equilibria in oxide melts and glasses is of particular importance in the fields of glass technology, geochemistry, and metallurgy (for specific references, see, e.g. Ref. [27]). Since glasses quenched rapidly from the melt are unable to reachieve chemical equilibrium, the valence states of multivalent elements in these glasses are generally expected to reflect the melt-temperature equilibria. While wet chemical techniques have been developed which work well even with small samples and low doping levels, the use of ESR is an attractive alternative [27, 28] which may in fact become the preferred method when several multivalent transition-group elements are present in the same melt. The reason for this preference is that the ESR spectra of many transition-group ions in glass are distinctive and easily resolved from one another (for details of the spectra and their interpretations, see, e.g., Ref [12]), whereas titrations become unreliable when the number of multivalent elements is increased beyond one.

Given a single multivalent element M with two valence states m + and (m + 1)+, the redox equilibrium can be represented

$$4M_{(glass)}^{(m + 1)+} + 2O_{(glass)}^{2-} \rightleftarrows 4M_{(glass)}^{m+} + O_{2(gas)}. \qquad (4)$$

A simplified theory for this situation yields

$$-\log fO_2 = 4 \log R_M - E_M^*, \qquad (5)$$

where fO_2 is the oxygen fugacity, R_M is the concentration ratio $[M^{m+}]/[M^{(m+1)+}]$, and E_M^* can be viewed as a relative reduction potential for M in a particular melt at a particular temperature [27]. Thus, for any element M with two valence states differing by one unit of charge, a plot of $-\log fO_2$ versus R_M should

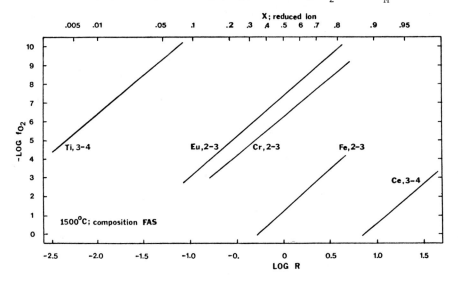

Fig. 6. Oxidation-reduction equilibria of multivalent ions in a Ca-Mg-Aℓ-silicate melt as determined by ESR and wet chemical methods. (After Ref. [29].)

yield a straight line with a slope of 4. Figure 6 presents a family of experimental curves obtained by ESR and redox titrations for a variety of one-electron
redox couples individually doped into a common silicate host and equilibrated
at a common melt temperature [29]. It can be seen that a scale of relative
redox potentials is established, such that Ce(IV)→Ce(III)>Fe(III)→Fe(II)>>
Cr(III)→Cr(II)>Eu(III)→Eu(II)>>>Ti(IV)→Ti(III) [27, 29].

ESR studies of glass samples simultaneously containing two multivalent species
M and N have demonstrated the existence of mutual interactions between two
different one-electron couples, even when both species are relatively dilute
($\lesssim$ 1 wt %) [27]. Thus, a plot of log R_M versus log R_N yields a straight line
of slope 1, as expected from eq. (5), but with a y intercept which is <u>displaced</u>
from the value predicted on the basis of data gathered separately for glasses
containing only N and only M.

Phase Separation in Glass
Since the advent of electron microscopy, many glasses formerly thought to be
homogeneous have been found to possess submicrostructure due to the phenomenon
of liquid-liquid immiscibility (e.g., Ref. [30]). It is now recognized that
nearly all glass properties are affected by such "phase separation", with the
influence being particularly strong in the case of transport properties.
Accordingly, considerable interest has centered on the development of sensitive
experimental methods, such as small angle x-ray scattering, capable of determining the existence and extent of "miscibility gaps" in the phase diagrams of
various multicomponent glass systems. A recent paper [31], summarized below,
has demonstrated the usefulness of ESR in this application.

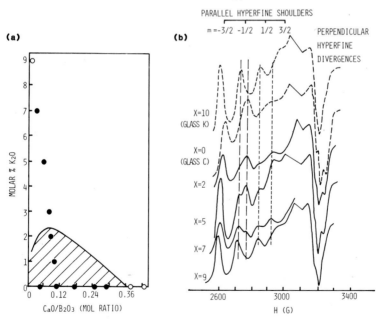

Fig. 7. Glass forming region (a) and ESR spectra of Cu^{2+} ions (b) in the
system $K_2O-CaO-B_2O_3$. (After Ref. [31].)

It has long been known that the CaO-B$_2$O$_3$ system exhibits a broad immiscibility dome for mole ratios [CaO]/[B$_2$O$_3$] extending from $\sim$ 0.01 to $\sim$ 0.36. Indeed, quenched glasses in this composition range are visibly opalescent. Based on such visual inspection techniques it was noted (e.g. [32]) that substitution of K$_2$O for a small fraction of the CaO tended to diminish the region of immiscibility, confining opalescence to the shaded composition region in Fig. 7(a). Kawazoe et al [31] selected Cu^{2+} as an ESR probe ion because of the sensitivity of its spin Hamiltonian parameters to its ligand environment and its ability to be incorporated into at least two of the separated phases. Potassium-free CaO-B$_2$O$_3$ glasses of various mole ratios were studied, together with a series of glasses specified by xK$_2$O-(10-x)CaO-90 B$_2$O$_3$, $0 \leq x \leq 10$. When potassium was left out, a single Cu^{2+} spectrum was observed irrespective of composition, consistent with the conclusion that copper tends to be rejected from the high-B$_2$O$_3$ liquid and resides mostly in the homogeneous Ca-rich phase. However, with the substitution of K$_2$O (x < 10), a second spectrum was observed in superposition with the first, eventually supplanting it entirely for x = 10 (Fig. 7b). When both "end-member" spectra (those characteristic of x = 0 and x = 10) were observed in superposition the glass could be inferred to be phase separated [31]. As seen in Fig. 7, the actual composition region subject to phase separation as determined by ESR (filled circles in (a)) is more extensive than could be inferred from visual inspection.

Photochromic Glasses

The most commonly used photochromic glasses comprise a suspension of silver-halide particles $\sim$ 100 Å precipitated in an oxide glass matrix (see, e.g. [33]). Such glasses darken upon exposure to ultraviolet light and thermally bleach at room temperature to their original state of transparency when the uv illumination is removed. In analogy to the photographic process, photochromism is due to photolysis of the silver halide phase resulting in silver colloid formation. Thus, the simplest conceivable description of the photochromic mechanism is given by

$$Ag^+ + C\ell^- \xrightarrow{h\nu} Ag^o + C\ell^o, \tag{6}$$

where reversibility would be attributed to the glass matrix preventing the escape of $C\ell_2$. However, it has long been known that copper doping (at sufficiently low levels) sensitizes the photochromic process, presumably by providing Cu$^+$ ions which act as deep hole traps:

$$Ag^+ + Cu^+ \xrightarrow{h\nu} Ag^o + Cu^{2+} \tag{7}$$

Marquardt [34] gave the first report of an ESR spectrum due to photolytic Cu^{2+} in a photochromic glass. This relatively weak resonance (Fig. 8c) was disentangled from the spectrum of Fig. 8(b) by computer subtraction of a strong background signal due to non-photolytic Cu^{2+} in the glass matrix (Fig. 8a). Spin Hamiltonian parameters derived by computer simulation of the spectrum of Fig. 8(c), when compared to the parameters for Cu^{2+} in other media, indicated the photolytic Cu^{2+} to be located in highly distorted sites in the silver halide precipitates [34]. As seen in Fig. 8(d), the photo-induced darkening has been found to correlate linearly with the photolytic Cu^{2+} concentration determined by ESR for a series of glasses with wide ranges of copper and silver contents [35]. Although the room-temperature equilibrium darkening plotted in Fig. 8(d) contains contributions from both colloidal silver and photolytic Cu^{2+} [34,35], the zero intercept of the least-squares fit to the data is definitive evidence that hole trapping occurs only at cuprous ion sites [35]. [A positive y intercept would be observed, e.g., if some holes were to be trapped on chlorines according to eq. (6).] Thus, ESR experimentation has demonstrated that the photochromic process is properly described by eq. (7) alone.

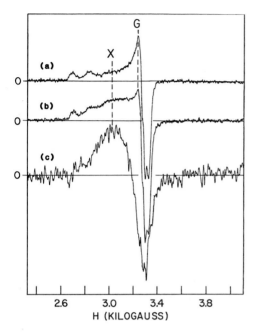

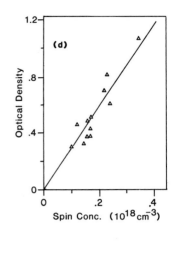

Fig. 8. ESR of Cu^{2+} ions in a photochromic glass: (a) "background" due to Cu^{2+} in matrix glass, (b) background plus contribution of photolytic Cu^{2+} in AgCℓ, (c) difference [(b) minus (a)] x 4. (d) Plot of optical darkening versus concentration of photolytic Cu^{2+}. (After Ref. [35].)

Ferromagnetic Precipitates in Glasses

It is a well known phenomenon that glasses containing transition group ions can be made to precipitate crystalline phases wherein the unpaired spins are strongly coupled (e.g., [36-38]). Such phases are characterized as ferro, ferri-, or antiferromagnetic depending on whether the saturation magnetic moment which is developed is the maximum possible, intermediate in magnitude, or small. Precipitated glasses containing such phases are of high intrinsic interest because the individual particles tend to be spherical, well dispersed, and uniformly sized and because their compositions and properties can be engineered by controlling melt composition, redox conditions, and thermal history. Moreover, precipitated magnetic phases are evidently ubiquitous in natural glasses (e.g., [39]) and can occur inadvertently as undesirable scattering centers in optical glasses containing transition group impurities (e.g., [40,41]).

ESR is a particularly useful technique for characterizing magnetic precipitates in glass. Because of the large magnetic moments developed by ferro- or ferri-magnetic particles ≳ 50 Å, the FMR intensity at room temperature can be ~ 10^3 times larger than the EPR intensity of the same total number of spins behaving paramagnetically [42,43]. Thus, detection of magnetic precipitates at the parts-per-million level is routinely accomplished. In the usual case of spherical particles having a single magnetic domain, the FMR line shape and width are determined by the magnetocrystalline anisotropy field $H_a \equiv 2K_1/M_s$, where K_1 is the first order anisotropy constant and M_s is the saturation magnetization. The dashed curve of Fig. 9(b) shows the theoretical powder pattern for a ferrite with negative H_a (the exact shape depends slightly on the ratio H_a/H_o, where $H_o \equiv h\nu/g\beta$ [41]). The heavy unbroken curve of Fig. 9(b) is the result of convoluting the dashed pattern by a Gaussian broadening function; Fig. 9(a) exhibits the first derivative of this broadened pattern. In Fig. 9(c) experimental data are presented for the temperature dependence of certain field intervals (proportional to H_a) for an unknown ferrite phase encountered in a BeF_2 glass containing only a few ppm transition group impurities [41]. The temperature dependence of the area under this absorption curve (the integrated intensity) is shown by the data points in Fig. 9(d). These behaviors are qualitatively typical of many ferro- or ferri-magnetic precipitates in glass: the integrated intensity varies weakly with temperature (approximately as M_s) below the Curie point while H_a decreases with increasing temperature (approximately as M_s^n where $n \sim 10$). By contrast, the intensity of typical paramagnetic resonance signals varies as $1/T$ over the temperature range examined in Fig. 9.

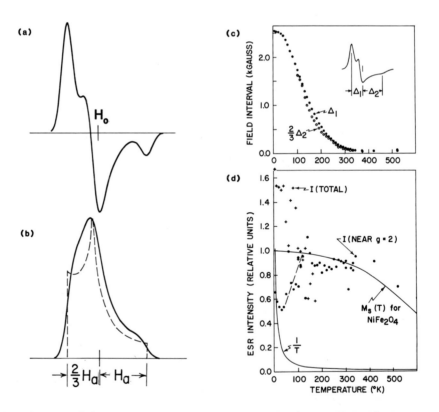

Fig. 9.   FMR of ferrimagnetic precipitates in a BeF_2 glass. Computed spectrum (a) is based on theoretical powder pattern (b). Field intervals (c) and intensities (d) are experimental data. (After Ref. [41].)

Another example of the application of FMR to precipitated phases in glass involves the case of metallic iron precipitates in fused silica [40]. Body-centered-cubic iron and its dilute alloys with Si or Ni are characterized by positive anisotropy constants K_1, giving rise to powder pattern shapes similar to that obtained by reflecting the dashed curve of Fig. 9(b) about a vertical line passing through H_o. Identical iron spectra are exhibited by silica glasses produced in the laboratory by melting under reducing conditions [40] or in nature by lightning strikes in the desert [44]. Further aspects of the FMR phenomenon in glasses are reviewed in Ref. [43].

REFERENCES

1. G. E. Pake and T. L. Estle, The Physical Principles of Electron Paramagnetic Resonance, 2nd ed. (Benjamin, Reading, Mass, 1973).
2. J. E. Wertz and J. R. Bolton, Electron Spin Resonance Elementary Theory and Practical Applications (McGraw-Hill, New York, 1972).
3. P. B. Ayscough, Electron Spin Resonance in Chemistry (Methuen, London, 1967).
4. P. W. Atkins and M.C.R. Symons, The Structure of Inorganic Radicals, (Elsevier, Amsterdam, 1967).
5. A. Abragam and B. Bleaney, Electron Paramagnetic Resonance of Transition Ions (Oxford, London, 1970).
6. P. C. Taylor and P. J. Bray, J. Magn. Res. 2, 305 (1970).
7. D. L. Griscom, J. Non-Cryst. Solids 13, 251 (1973/74).
8. D. L. Griscom, in Defects and Their Structure in Nonmetallic Solids, eds. B. Henderson and A. E. Hughes (Plenum, New York, 1976) p. 323.
9. D. L. Griscom, J. Non-Cryst. Solids 31, 241 (1978).
10. J. Wong and C. A. Angell, Glass Structure by Spectroscopy (Marcel Dekker, New York, 1976).
11. P. C. Taylor, J. F. Baugher, and H. M. Kriz, Chem. Rev. 75, 205 (1975).
12. D. L. Griscom, J. Non-Cryst. Solids 40, 211 (1980).
13. K. L. Brower, this volume.
14. D. L. Griscom, Ph.D. Thesis, Brown University, Providence, RI (1966).
15. W. Känzig and M. H. Cohen, Phys. Rev. Lett. 3, 509 (1959).
16. J. O. Edwards, D. L. Griscom, R. B. Jones, K. L. Watters, and R. A. Weeks, J. Am. Chem. Soc. 91, 1095 (1969).
17. D. L. Griscom, in the Physics of SiO_2 and Its Interfaces, ed. S. T. Pantelides (Pergamon Press, New York, 1978) p. 232.
18. R. H. Silsbee, J. Appl. Phys. 32, 1459 (1961).
19. F. J. Feigl, W. B. Fowler, and K. L. Yip, Solid State Commun. 14, 225 (1974).
20. K. L. Yip and W. B. Fowler, Phys. Rev. B11, 2327 (1975).
21. R. A. Weeks and C. M. Nelson, J. Appl. Phys. 31, 1555 (1960).
22. D. L. Griscom, E. J. Friebele, and G. H. Sigel, Jr., Solid State Commun. 15, 479 (1974).
23. D. L. Griscom, Phys. Rev. B20, 1823 (1979).
24. E. J. Friebele, D. L. Griscom, M. Stapelbroek, and R. A. Weeks, Phys. Rev. Lett. 42, 1346 (1979).
25. M. Stapelbroek, D. L. Griscom, E. J. Friebele, and G. H. Sigel, Jr., J. Non-Cryst. Solids 32, 313 (1979).
26. D. L. Griscom, M. Stapelbroek, and E. J. Friebele, Am. Ceram. Soc. Bull. 59, 386 (1980); to be published.
27. H. D. Schreiber, T. Thanyasiri, J. J. Lach, and R. A. Legere, Phys. Chem. Glasses 19, 126 (1978).
28. R. V. Morris and L. A. Haskin, Geochim. Cosmochim. Acta 38, 1435 (1974).
29. H. D. Schreiber, M. S. Sobota, and H. V. Lauer, Jr., Am. Ceram. Soc. Bull. 58, 877 (1979); to be published.

30. D. R. Uhlmann and A. G. Kolbeck, Phys. Chem. Glasses 17, 146 (1976).
31. H. Kawazoe, H. Hosono, and T. Kanazawa, J. Non-Cryst. Solids 29, 249 (1978).
32. L. Shartsis, H. F. Shermer and A. G. Bestul, J. Am. Ceram. Soc. 41, 507 (1958).
33. W. H. Armistead and S. D. Stookey, Science 144, 15 (1964).
34. C. L. Marquardt, Appl. Phys. Lett. 24, 209 (1976).
35. C. L. Marquardt, J. F. Giuliani, and G. Gliemeroth, J. Appl. Phys. 48, 3669 (1977).
36. R. R. Shaw and J. H. Heasley, J. Am. Ceram. Soc. 50, 297 (1967).
37. D. W. Collins and L. N. Mulay, J. Am. Ceram. Soc. 53, 74 (1970).
38. M. P. O'Horo and J. F. O'Neill, in Amorphous Magnetism II, eds. R. A. Levy and R. Hasegawa (Plenum Press, New York, 1977) p. 651.
39. D. L. Griscom, E. J. Friebele, and C. L. Marquardt, Proc. Lunar Sci. Conf., 4th, Geochim. Cosmochim. Acta, Suppl. 4, Vol. 3, p. 2709.
40. D. L. Griscom, E. J. Friebele, and D. B. Shinn, J. Appl. Phys. 50, 2402 (1979).
41. D. L. Griscom, M. Stapelbroek, and M. J. Weber, J. Non-Cryst. Solids, in press.
42. F.-D. Tsay, S. I. Chan, and S. L. Manatt, Geochim. Cosmochim. Acta 35, 865 (1971).
43. D. L. Griscom, Proc. "Frontiers in Glass Science" Conf., J. Non-Cryst. Solids, in press; also, to be published.
44. R. A. Weeks, M. Nasrallah, S. Arafa, and A. Bishay, J. Non-Cryst. Solids 38 and 39, 129 (1980).

CATALYSTS EXAMINED BY ELECTRON SPIN RESONANCE-EXAMPLES FROM
HYDRODESULFURIZATION CATALYSIS

BERNARD G. SILBERNAGEL
Corporate Research-Science Laboratories, Exxon Research and Engineering Co.,
Linden, New Jersey 07036

ABSTRACT

The utility of electron spin resonance (ESR) for catalyst
characterization is illustrated for the desulfurization
catalysts used to remove sulfur, nitrogen, and organically
bound metals from petroleum. These catalysts consist of
active metals (Mo, W, Co, Ni) in mixed oxide and sulfide
phases on high surface area alumina supports. Coordinated
studies of unsupported sulfide and oxide model systems
determine the chemical form and number of defect sites on
the actual catalysts. Parallel catalysis studies correlate
ESR defects and catalytic activity for many reactions.

INTRODUCTION

Heterogeneous catalysis is widely used throughout the chemical and petroleum
industries. Table I illustrates the broad range of current applications,
from producing important chemical intermediates like styrene and ammonia, to
cleaning and upgrading petroleum-based feedstocks, by desulfurization,
cracking, and reforming processes, to producing synthetic fuels by such
techniques as Fischer-Tropsch chemistry. The catalyst forms employed in
these processes represent almost every area of materials science: elemental
metals and alloys, transition metal oxides, high surface area ceramics, and
novel solid forms like zeolites and metal sulfides [1]. The future need
for non-petroleum based energy sources will require significant improvements
in our ability to do such catalytic chemistry.

Increasingly sophisticated spectroscopic and surface characterization techniques are being employed in catalysis research. When used in conjunction
with an active effort in solid state chemistry and in direct contact with
experiments in catalytic chemistry, they can be extremely powerful probes
of the chemistry of the catalyst and the nature of its interaction with the
chemical species it transforms. A systematic application of these three
tools: solid state chemistry, catalytic chemistry, and characterization
techniques, provides a new dimension in the design of advanced catalyst
systems.

Electron spin resonance (ESR) is a particularly appropriate illustration of
this interaction because it has been applied in one form or another to
catalytic problems for the last two decades. The rationale for using ESR
is apparent. The technique is sensitive to several of the most important
constituents in catalytic systems: paramagnetic transition metal species
with partially filled electron shells, and electronic defects in both inorganic and organic components of the catalyst systems. Such partially occupied
electron states are good candidates for many of the electron transfer pro-

TABLE I   HETEROGENEOUS CATALYSTS OF INDUSTRIAL IMPORTANCE[a]

Process	Example	Catalyst
Dehydrogenation	Ethylbenzene → styrene	$Fe_2O_3 + Cr_2O_3$
Hydrogenation	$N_2 + 3H_2 \rightarrow NH_3$	$Fe + (K_2O, CaO, MgO)/Al_2O_3$
Partial Oxidation	Naphthalene + Air → Phthalic Anhydride	Supported V_2O_5
Acid-Catalyzed Reactions	Catalytic Cracking	Zeolites in $SiO_2-Al_2O_3$
Catalytic Reforming		Pt, Pt-Ir, Pt-Re/Al_2O_3
Heteroatom Removal	Hydrodesulfurization	Co-Mo/Al_2O_3 or Ni-Mo/Al_2O_3
Synthesis Gas Reactions	$CO + H_2 \rightarrow$ Paraffins (Fischer-Tropsch)	Fe (promoted)

(a) per Ref. 1.

cesses expected during catalysis. The form of the ESR absorption signal: its <u>position</u> (or g-value), its <u>width</u> and <u>shape</u>, and <u>hyperfine interactions</u> with atomic nuclei can provide specific information about the <u>valence</u> of the paramagnetic species, the types of its <u>atomic neighbors,</u> and its <u>site symmetry.</u>

Like any characterization technique, ESR has certain limitations. Not all catalytically active species will be paramagnetic. Many transition metal ions, such as Ni^{2+} and Co^{3+} can occur in "low-spin" configurations in which all d electrons are paired and the ion is diamagnetic. Finally, the relationship of a particular transition metal or defect site to a given chemical reaction is not obvious <u>a priori.</u> A clear picture of the chemistry and behavior of a given catalyst requires the systematic application of several complementary characterization techniques, as illustrated below.

HYDROTREATING CATALYSIS: SPECIFIC EXAMPLE

Since the literature on ESR and catalysis is so extensive, the present discussion will focus on one specific case, the use of supported catalysts for the hydrotreating of heavy crude oils. As Table II indicates, such crude oils contain high levels of heteroatoms (especially sulfur and nitrogen) as well as organically complexed metal species (especially vanadium and nickel) [2]. These contaminants must be removed prior to more elaborate catalytic treatments (such as cracking or reforming) to avoid deactivation of the catalysts used in those processes. Even for direct end uses, like the combustion of heavy fuel oils, environmental considerations require that the level of these contaminants be made as low as possible. Finally, as progressively heavier feeds are being refined, hydrotreating processes provide an initial opportunity for molecular weight reduction.

TABLE II   TYPICAL RESIDUA FOR HYDROTREATING CATALYSIS[a]

(Crude oil components with boiling points $>350^{\circ}C$)

	Kuwait	Iranian Heavy	Arabian Heavy
Sulfur (wt.%)	4.21	2.73	4.31
Vanadium (wppm)	53	145	88
Nickel (wppm)	12.7	33.6	26.9
C_7 Asphaltenes (wt.%)	2.7	3.6	6.5
Ramsbottom Carbon (wt.%)	9.8	10.0	13.1

[a] per Ref. 2.

Desulfurization of such feeds places severe constraints on the types of catalysts which can be used. For a typical desulfurization unit processing 25,000 barrels/day of crude oil with 5 wt.% sulfur, approximately 200 tons/day of sulfur would be removed from the oil. It is essential that a sulfur-tolerant catalyst be used. In fact, catalysts composed of transition metal sulfides, typically supported on a high surface area alumina, are effective and have long lifetimes in such applications [3]. A typical starting catalyst consists of a mixture of molybdenum and cobalt oxides impregnated on alumina. The metal oxides are converted to sulfides during the course of the desulfurization process. Mixtures of nickel and tungsten oxides and nickel and molybdenum oxides are also used for these purposes.

While the precise form of molybdenum and transition metal on the catalyst is not known, a priori, a great deal can be learned because the unsupported oxides and sulfides have been studied extensively by solid state physicists and chemists [4]. A major research theme during the last decade has been to exploit the connection of catalysts to these well-characterized model systems. The microscopic information obtained from such relatively simple models is useful because the number of variables in the actual catalyst system is very large: (1) the chemical form (or forms) of the molybdenum (or tungsten) species, (2) the chemistry of the promoting transition metal (usually Ni or Co) and its spatial and chemical relationship to the Mo and W forms, (3) the role which the alumina support plays, either in anchoring the microcrystallites of the active metal compounds or including metal ions within its crystal structure or in defects on its surface, (4) the way in which carbon and metals from the crude oil deposited on the catalyst during the desulfurization process ultimately cause it to deactivate.

The best strategy for dealing with such a problem is to begin with systems which contain only a subset of these variables. For example, the effect of the sulfides can be observed by studying unsupported tungsten and molybdenum sulfides first. The effects of promotion can be examined by adding the promoting metal to the unsupported sulfides. Support effects can be proved by examining simple systems like MoO_3/Al_2O_3 and CoO/Al_2O_3. Deactivation can also be surveyed by recourse to model systems. The present discussion will begin with these simple models. The references given here are not exhaustive. They also are not intended to indicate priority for the discovery of a given concept. They have been chosen because they are illustrative of the combined materials-ESR-catalysis approach outlined above.

SULFIDE CHEMISTRY AND HYDROTREATING CATALYSIS

Transition metal sulfides with moderately high surface areas can be prepared by precipitation from aqueous solution or thermal decomposition of an appropriate precursor [5]. For example, $(NH_4)_2 WS_4$ decomposes thermally to yield WS_3 as the product at temperature around 200°C. WS_2 is produced at higher temperatures (~330°C). These changes can be traced in situ by high temperature x-ray (Guinier) and differential thermal analysis (DTA) techniques. The resulting products of this decomposition have a high surface area (~50-70 m^2/g).

Paramagnetic defects which occur on these sulfides are easily observed by ESR since the starting materials and end products are diamagnetic. WS_4^{2-} contains hexavalent tungsten, with no unpaired electrons. In WS_3 and WS_2 tungsten occurs in the tetravalent state. The crystal field properties of the tungsten site favor a "low spin" configuration for these

species, again resulting in diamagnetism. By contrast W^{5+} and W^{3+} have
an odd number of spins and are observable, at least in principle. The
$5d^1$ configuration of the W^{5+} species typically yields an orbital singlet
which is readily observable by ESR and possesses a g value near the free
electron value (g $\sim$ 2). The $5d^3$ configuration of the W^{3+} species is
significantly more complicated. The unquenched orbital angular-momentum
of these states can lead to highly anisotropic ESR absorption spectra
and short spin lattice relaxation times. Since changes in crystalline
field strongly effect these spins, it is much more difficult to anticipate
the W^{3+} ESR properties a priori.

The initial phase of the ESR analysis of such systems is an exercise in
pattern recognition. The ESR absorption curve has a certain shape which
is the "fingerprint" of the type of defect site which occurs. This can be
illustrated by comparing three tungsten samples prepared in different ways
[5] , as shown in Fig. 1. In this figure, the derivative of the ESR
absorption with respect to magnetic field is plotted, a particularly
advantageous representation since the derivative is sensitive to subtle
changes in the shape of the absorption spectrum. Samples prepared by

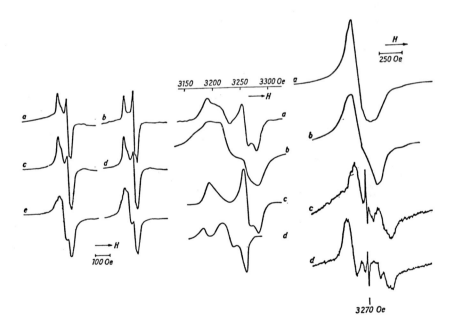

Fig. 1. The Tungsten ESR signal is similar for sulfides produced by precipitation (a) or low temperature ($\sim$200°C) thermal decomposition (b). High
temperature ($\sim$320°C) decomposition yields a different product (after
Ref. 5).

precipitation and low temperature (200°C) thermal decomposition yield similar signals: a highly articulated absorption of approximately 85 gauss in width. The g factor is in the vicinity of the free electron value (g ∿ 2.0023). Slight variations in the preparation conditions can cause minor changes in the derivative spectra, suggesting that there may be more than one paramagnetic species in this sample. Condensed oxygen and sulfur radicals will produce ESR signals in this region and may account for the observed variations.

The position for resonance serves as a quantitative means of identifying the paramagnetic species. In general, the precession frequency of a para-magnetic spin in a magnetic field can be expressed in terms of the following tensor equation

$$\hbar\omega = \beta_e \ (\vec{H} \cdot \overset{\leftrightarrow}{g} \cdot \vec{S}), \qquad (1)$$

where ω is the resonance frequency, $\vec{S}$ is the electron spin, $\vec{H}$ the applied magnetic field, $\overset{\leftrightarrow}{g}$ the g tensor, and $\hbar$, β_e red Planck's constant and the Bohn magneton respectively. In the isotropic case, this expression reduces to the usual Larmor expression for the electron precession frequency: $\hbar\omega = g \ \beta_e (\vec{S} \cdot \vec{H})$. In most cases, a suitable transformation of axes can reduce the tensorial expression to a scalar form.

$$\hbar\omega = \beta_e (g_x S_x H_x + g_y S_y H_y + g_z S_z H_z) \qquad (2)$$

If the system has axial symmetry, $g_x = g_y = g_\perp$; $g_z = g_{11}$. The resonance lineshape of Fig. 1a, suggests non-axial symmetry and values of g_x, g_y and g_z determined from the maxima and zero crossings (the derivative curve are shown in Table III. These g values are very different from the ones determined for W^{5+} in optically irradiated WO_3, for which g ∿ 1.5-1.6 [6]. Conversely, similar g values are found for organometallic species in which tungsten is coordinated to sulfur ligands in a prismatic symmetry similar to the sulfides [7,8]. The ligands of the W^{5+} species influence the resonance position, i.e. cause a g shift by means of spin-orbit coupling effects.

TABLE III   G-VALUES FOR W^{5+} SPECIES

Sample	g-values			Ref.
WS_3	$g_x = 1.995$	$g_y = 2.005$	$g_z = 2.050$	5
WS_2	$g_x = 2.010$	$g_y = 2.030$	$g_z = 2.060$	5
WO_3	$g_x = 1.505$	$g_y = 1.661$	$g_z = 1.532$	6
$W[S_2C_2(CF_3)_3]_3^-$		$g_{11} = 1.987$	$g_\perp = 1.993$	7
$W[S_2C_2(Ph)_3]_3^-$		$g_{11} = 1.987$	$g_\perp = 1.996$	8

The shape of the ESR absorption changes when the samples are heated above
320°C, as shown in Fig. 1b. While the average g value does not change
drastically, the absorption broadens from ∿85G for the low temperature
signal to ∿250G for the high temperature one. This signal has been attributed to a W^{3+} species on the basis of indirect chemical arguments [5].
An analogous signal is claimed for NaW_2S_4, and is seen in sulfur deficient
products like $WS_{1.95}$. The low temperature W^{5+} signal has nearly completely
vanished before this new signal appears, which is again associated with
the sulfur deficiency. Such an indirect line of reasoning is typical of
many of the inferential arguments which appear in the catalysis literature.

These defects are surface species, as verified by the loss of ESR intensity
upon the absorption of molecules like cyclohexene on the sulfides [9].
However, the real connection to catalysis is furnished by the correlation
between defect site density, determined by integration of the ESR absorption
curve and the activity for certain model catalytic reactions. In Fig. 2,
the rate constant for benzene hydrogenation over a $Ni_{0.5}WS_2$ catalyst varies
linearly with the intensity of the "W^{3+}" signal [9]. These correlations
are the justification of the ESR observation strategy. However, each
type of chemical reaction is different and may involve different sites on
this catalyst. For example, cyclohexene hydrogenation does not correlate
with ESR intensity in this $Ni_{0.5}WS_2$ system.

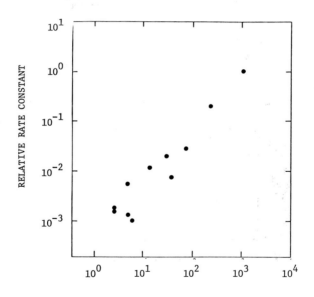

RELATIVE ESR INTENSITY

(Per gram of Catalyst)

Fig. 2. Catalytic activity for benzene hydrogenation correlates
linearly with Tungsten "W^{3+}" ESR signal intensity for
$Ni_{0.5}TiS_2$ (after Ref. 9).

124

PROMOTION EFFECTS

The catalytic activity of molybdenum or tungsten sulfides is greatly enhanced
by the addition of relatively small amounts of transition metal atoms like
cobalt or nickel. The precise mechanism of promotion for particular
chemical reactions is still open to some question. The metal atoms would
be intimately mixed, forming alloys or intermetallic compounds. The change
in catalytic activity would then result from a modification of the bulk
electronic properties of the sulfide crystallites. Alternatively, the
transition metal species might "decorate" the surface of the sulfide
crystallites, forming electronically modified surface states. A third
possibility is that the materials form separate chemical phases and that
catalysis occurs by transfer of species from one crystallite to the next.
ESR studies provide data which often bear upon such questions.

For example, mixed cobalt-molybdenum sulfides have been formed from co-
precipitation of the oxides by $(NH_4)_2S$ in an aqueous solution [10]. The
ratio of cobalt to molybdenum species in the resulting products was
determined by the ratios of CoO and MoO_3 in the starting solution. The
precipitates were subsequently sulfided in H_2S at 500°C.

For low cobalt concentrations, a Mo^{5+} signal with g $\sim$ 2.007 dominates
the room temperature ESR spectrum. As in the tungsten sulfide case, this
g value is appreciably higher than for the oxide. Single crystal
measurements on reduced MoO_3 indicate g_z = 1.878, g_y = 1.942, g_x = 1.953:
$<g>$ = 1.924 [11]. Enhanced spin orbit coupling from sulfur ligands is
assumed to account for this positive g shift. Integrated intensity measurements of the ESR spectrum indicate that roughly 10^{-2}% of all Mo atoms
occur in this Mo^{5+}-sulfide form.

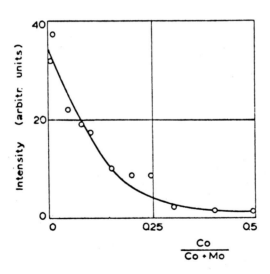

Fig. 3. Mo^{5+} ESR intensity drops rapidly after cobalt addition
to unsupported Co-Mo sulfides, (after Ref. 10).

The intensity of the Mo^{5+} spectrum drops sharply with the addition of Co as shown in Fig. 3. By the time the Co/Mo atomic ratio reaches 0.33, the Mo^{5+} intensity has fallen to $\sim$1/10 of its value for zero Co concentration [10]. A new ESR signal with g $\sim$ 2.017 appears near 140K and reaches a constant intensity below 100K. The temperature dependence of this new signal is not consistent with that of known oxides or sulfides. A low temperature (93K) survey of samples with varying composition shows a very pronounced maximum in the intensity of this new signal for Co/Mo $\sim$ 0.2. Catalytic activity for some reactions correlates with the intensity of this g = 2.017 signal [10]. A series of model compound reactions were examined, and are shown, together with the ESR intensities in Fig. 4. While all three reactions exhibit an activity maximum for Co/Mo $\sim$0.2, only the cyclohexane isomerization closely tracks the ESR signal. For the other reactions additional factors must be contributing to the observed conversion chemistry.

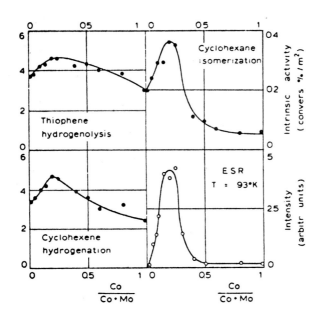

Fig. 4. Cyclohexane isomerization closely tracks the new g = 2.017 signal observed in unsupported Co-Mo sulfides. The correlation is less pronounced for other processes (after Ref. 10).

SUPPORT EFFECTS

While studies on unsupported sulfides have the virtue of simplicity, it is also clear that they must be viewed with some caution in attempting to describe catalytic processes for metals deposited on ceramic supports. Simple H_2/H_2S

reduction experiments of unsupported and supported MoO_3 illustrate strong support effects [12]. For example, bulk MoO_3 sulfided between 300 °C and 500 °C yields a mixed product of MoO_2 and MoS_2. MoO_3 supported on a SiO_2 catalyst behaves in much the same way. By contrast, MoO_3 on Al_2O_3 exhibits a continuous change in weight upon sulfiding without evidence for a MoO_2 intermediate [12]. The sulfidation process is easily traced with ESR, since the calcined MoO_3/Al_2O_3 starting catalyst has no ESR signal. After the sulfidation, an intense Mo^{5+} signal occurs with a g value $\sim$1.929, corresponding to an oxide-coordinated Mo^{5+} form [11]. Similar defects can be generated in MoO_3/Al_2O_3 samples by heating in vacuum, in the presence of hydrogen, or hydrocarbons like propene [13]. Intensity measurements of the ESR absorption spectra indicate that $\sim$2% of the total Mo atoms present occur in the Mo^{5+} form.

This is approximately two orders of magnitude greater than for the unsupported sulfides mentioned above. While no evidence for Mo^{3+} species is observed, radicals, corresponding to sulfur and oxygen species are often observed [12, 14] and are usually identified by their highly articulated spectra and distinctive g values. For MoO_3/Al_2O_3 catalysts, the Mo^{5+} ESR intensities after moderate treatment (9% H_2S in H_2, 260 °C, 2 hr) vary linearly with initial MoO_3 loading up to values $\sim$10 wt% Mo. Higher loadings and more stringent treatments reduce the Mo^{5+} level.

The effect of transition metal promoter atoms on Al_2O_3 supports is still a subject of conjecture. There is ESR evidence that Mo/Al_2O_3 catalysts to which Co or Ni have been added are much more readily affected by sulfidation treatments [15]. Attempts to probe the chemical state of the deposited Co have led to inconclusive results [16]. Different Co ESR signals are seen, depending upon the preparation scheme employed. There is no evidence for the existence of bulk cobalt oxides, such as CoO, Co_2O_3 and Co_3O_4, which exhibit strong exchange narrowed ESR signals [16]. There is similarly no evidence for known compounds such as $CoAl_2O_4$. Some of the Co^{2+} ESR signals observed correspond in g values to Co^{2+} impurities in Al_2O_3 [17].

CATALYST DEACTIVATION

During the active life of a catalyst, components from the feed being treated react with the catalyst, changing its chemical behavior and ultimately causing its deactivation. Molecules are absorbed, reducing the ESR intensity of certain defects [9]. Molecular species are transformed resulting in a carbonaceous deposit on the catalyst which is customarily called coke. Heteroatoms, particularly sulfur, will react with the deposited metal species, changing the chemical form of these metals. If the feedstock contains metals, as most heavy fractions of petroleum crudes do, these metals will be deposited on the catalyst as well.

It is obvious that deactivation studies are very complex, since active metals, support, carbon, and deposited metals are encountered simultaneously. However, present studies are beginning to clarify some of the features of these systems. An ESR derivative spectrum of one such deactivated catalyst is shown in Fig. 5. In this broad (2KG) scan, one observes an extremely sharp peak arising from the carbon radicals. At somewhat higher fields

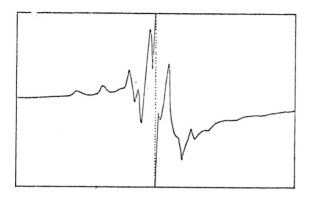

Fig. 5. A discharged Co-Mo/γ-Al₂O₃ hydrodesulfurization catalyst has
ESR signals from oxygen coordinated Mo^{5+} species, deposited
coke and VO^{2+} species.

one observes a signal from the Mo^{5+} oxygen-coordinated species. In addition,
a highly articulated, 16 component spectrum arising from deposited VO^{2+}
species are also observed [18]. Analysis of the g-values and hyperfine
coupling constants derived from these spectra indicates that the chemistry
of the VO^{2+} species has been changed from that of the VO^{2+} porphyrin
encountered in the feed. Comparison of oxidized VO^{2+} forms with those seen
after the same catalyst has been sulfided suggests that sulfur has been
incorporated into the ligand sites of the VO^{2+}.

Intensity measurements suggest that the capacity of Al₂O₃ catalyst for
such VO^{2+} forms is ~ 1.0wt%), vanadium sulfides are formed which are
observable by NMR. These crystalline sulfides are believed to deactivate
the catalyst by plugging its pores [19].

CONCLUSION

The present article is intended to illustrate the context in which ESR
is applied to catalytic problems. The ESR literature in this area is
substantial and in some instances conflicting, but the basic picture which
emerges is as follows. The presence of Mo and W ESR defects on supported
catalysts has been well documented. Their chemical forms and local
environment have been established by comparison with unsupported model
systems. For certain reactions, direct correlations between defect
density and catalytic activity are observed. The effects of the catalyst
support are major: the process of sulfidation proceeds differently in
the presence of supports. Hundredfold higher defect density levels are
observed in the supported samples even though the surface area is only $\sim$5
times greater.

Promotion by the addition of transition metal impurities is still not
fully understood. A new ESR absorption occurred upon the addition of Co to
unsupported molybdenum sulfides, and its intensity correlated with certain
catalytic processes. The issue is still less clear for promoted, supported

catalysts, although several ESR absorptions attributed to the Co promoter have been observed. Kinetics studies suggest that promoters may influence the rate of reduction or sulfidation of these catalysts.

During actual use, these catalysts accumulate deposited molecules, carbon aceous coke, and often metals from the feed. ESR has traced the loss of "$W3^+$" defect intensity associated with molecular absorption. Carbon free radical signals are prominent, and paramagnetic species like VO^{2+} are observed to change chemical form upon their deposition. Correlations with actual catalytic activity are currently in progress.

In conclusion, ESR has some clear limitations. It is insensitive to dia-magnetic chemical forms which may be very active chemically. Strong spin lattice relaxation processes or exchange coupling may render the paramagnetic forms unobservable. Thus the technique is most fruitfully applied in the context of a range of characterization tools. Second, for these observations to be significant, they must be performed in the context of a unified pro-gram. Appropriate model systems must be synthesized. The catalyst studies must systematically probe the "variables" of the system: support chemistry, active metals composition, catalyst preparation, and sulfidation (or reduction) techniques. Furthermore, correlations with significant catalytic activity experiments are essential. Such a comprehensive approach is the key to modern catalysis research.

REFERENCES

1. C. N. Satterfield, Heterogeneous Catalysis in Practice (McGraw-Hill, 1980). Ch. 1.

2. S. Sie, in Catalyst Deactivation, Vol. 6, B. Delmon and G. Froment, eds. (Elsevier, 1980) p. 556.

3. O. Weisser and S. Landa, Sulfide Catalysts, Their Properties and Applications (Pergamon, New York 1973).

4. J. Wilson and A. B. Yoffee, Adv. Phys. 18, 193 (1969).

5. R. J. H. Voorhoeve and H. B. M. Wolters, Z. anorg. u. allgem. Chemie 376, 165 (1970).

6. R. Gazzinelli and O. F. Schirmer, J. Phys. C. 10, L145 (1977).

7. A. Davison, N. Edelstein, R. H. Holm and A. K. Maki, J. Amer. Chem. Soc. 86, 2799 (1964).

8. E. I. Stieffel, R. Eisenberg, R. S. Rosenberg and H. B. Gray, J. Amer. Chem. Soc. 88, 2956 (1966).

9. R. J. H. Voorhoeve, J. Catalysis 23, 236 (1971).

10. G. Hagenbach, P. Menguy and B. Delmon, Bull. Soc. Chim. Belg. 83, 1 (1974).

11. R. Juryska and H. Bill, Nuovo Cimento 38B, 369 (1977).

12. K. S. Seghadri, F. G. Massoth, L. Petrakis, J. Catalysis 19, 95 (1970).

13. G. Martini, J. Magnetic Resonance 15, 262 (1974).

14. M. Lojacona, J. L. Verbeek, and G. C. A.Schuit, Proc. 5th. Int. Cong. Catalysis, (Elsevier, 1973), v. 2 p. 409 ft.

15. R. Galiasso and P. Menguy, Bull. Soc. Chim. France 4, 1331, (1972).

16. H. Ueda and N. Todo, J. Catalysis 27, 281 (1972).

17. G. M. Zverev and A. M. Prokhorov, Zh. Eksp. Teoret. Fig. 36, 647 (1959).

18. B. G. Silbernagel, J. Catalysis 56, 315 (1979).

19. B. G. Silbernagel and K. L. Riley, in Catalyst Deactivation Vol. 6, B. Delmon and G. Froment, eds. (Elsevier, 1980) p. 313.

Mössbauer Effect

Published 1981 by Elsevier North Holland, Inc.
Kaufmann and Shenoy, editors
Nuclear and Electron Resonance Spectroscopies Applied to Materials Science

THE MÖSSBAUER EFFECT AND SOME APPLICATIONS IN MATERIALS RESEARCH*

G. K. SHENOY
Solid State Science Division
Argonne National Laboratory
Argonne, IL 60439

ABSTRACT

A brief introduction to the Mössbauer effect is presented.
The hyperfine interactions associated with the electric monopole, magnetic dipole and electric quadrupole moments of the
nuclear states involved in the Mössbauer transition are described. Their use in materials research is illustrated
through examples dealing with phase analysis, binary solubility, defect interaction and surface properties.

INTRODUCTION

Since its discovery, over the past 20 years the Mössbauer effect has become
an important technique as applied to physics, chemistry and materials science.
There are inumerable instances where the Mössbauer effect has played a significant role in microscopic probing of materials. We list here some examples:
phase analysis, study of solid solubilities, defect interaction, damage
studies, catalysis, corrosion, amorphous solids, steel technology, atomic
diffusion, coal, storage hydrides, battery materials, implantation studies,
etc. There are many contributions related to some of these problems in this
volume.

The present article is intended to be a broad introduction to familiarize a
novice to the technique of Mössbauer spectroscopy, while his primary interest may be in understanding materials. In no way will we deal with all the
above applications of this technique; but a few selected cases will be presented to give a deeper appreciation of the tool.

In the next section the basic ideas behind the Mössbauer effect are briefly
presented along with the interactions that one measures with this nuclear
gamma ray resonance technique. It is hoped that this non-mathematical description will prepare the reader to understand the usual jargon of the method, and
allow him to appreciate some of the applications discussed in the remaining
sections of this article and in many other contributions to this volume.

There are many general [1] and specialized books on the subject of Mössbauer
spectroscopy. In particular, the books edited by Gonser [2], Cohen [3] and
Stevens and Shenoy [4] deal with many material science applications.

BASIC IDEAS

Principles
The phenomenon of the emission of nuclear gamma rays and their resonance absorption by identical nuclei without recoil energy loss is known as the Mössbauer effect. The recoil energy of an individual nucleus is of the order of
10^{-3}-10^{-2} eV in a typical low energy gamma ray transition, sufficient to

destroy the resonance. The Mössbauer effect is realized when the entire lattice shares this recoil and in effect produces no loss in gamma ray emission
and absorption energy. In a microscopic description, a certain fraction of
the gamma rays are emitted and absorbed by the nuclei without exciting any
phonons (zero-phonon processes). This fraction is usually termed the resonance fraction, f. It is a measure of the phonon spectrum of the solid (usually characterized by the Debye temperature of the solid).

These gamma rays with exactly the resonance energy will have an energy spread
determined by the lifetime τ of the excited nuclear state. The energy dependence of the emitted resonant intensity $I(E)$ has a Lorentzian shape with full-
width at half intensity being $\Gamma_0 = h/\tau$. The resonant absorption cross-section
also has a similar line shape. The measurement of the transmitted intensity
of the resonant radiation through a resonant absorber constitutes the usual
Mössbauer experiment. In order to trace the resonance line shape in a Mössbauer experiment, the transmitted intensity is measured as a function of the
Doppler shifted gamma ray energy. This is accomplished by imparting to the
radioactive source relative to the absorber. The line shape is then

$$T(\nu) = \frac{f_s f_a \; n \; \sigma_0 \; d \; (\Gamma/2)}{(\nu-\nu_0)^2 + (\Gamma/2)^2} \qquad (1)$$

where $\Gamma = 2\Gamma_0$ is the measured resonance line width, ν_0 is the position of the
resonance, f_s and f_a are resonance fractions for the source and the absorber,
n is the number of resonant nuclei per cm^{-3}, d is the geometrical absorber
thickness, and σ_0 is the resonance absorption cross-section.

What makes the Mössbauer resonance most useful is its ability to measure the
narrow line width 2Γ ($\sim 10^{-8}$ eV) of the gamma ray energy (~ 10 keV). This resolution of 1 part in 10^{12} makes it possible to measure accurately various interactions of the nucleus moments with the extranuclear electrons. These interactions, called the hyperfine (hf) interactions, are usually larger than 2Γ
thus permitting their precise measurement, and are discussed in the following
section.

Interactions

Isomer shift. The Coulomb interaction between the electronic charge present at the nucleus and the nuclear monopole charge distribution in the ground
and the excited state of the nucleus produces a small energy shift in the resonant gamma ray energy, E_0. This is measureable only when the electronic density at the nucleus, ρ, between the source and the absorber are not identical.
In velocity units, the isomer shift (IS) is given by

$$IS = \frac{2}{3} \frac{\pi c}{E_0} Ze^2 \; \Delta\langle r^2 \rangle \; \Delta\rho(o) \qquad (2)$$

where $\Delta\langle r^2 \rangle$ represents the change in the second moment of the nuclear charge
distribution between the nuclear excited and ground states, $\Delta\rho(o) = \rho(absorber$
$- \rho(source)$. The primary contribution to the IS is from the change in the
total s-electron density, and reflects the change in the s contribution of the
electronic structure of the material. In addition, differing non-s orbital
distributions between the two matrices can also significantly contribute to
$\Delta\rho(o)$ through the screening of s-electrons. Experimentally, the shift in the
centroid of the resonance pattern constitutes the IS. There is however an
additional (and usually small) shift due to the second-order Doppler (SOD)

effect of the resonant radiation from lattice vibrations. The separation between the IS and SOD shift is sometimes difficult, and a detailed treatment is essential [5].

   Zeeman interaction. The interaction of the magnetic dipole moment of the nucleus with the magnetic field at the nucleus splits the nuclear excited and ground states through the Zeeman interaction. The resonance pattern is then made up of hf transitions between various excited and ground state sublevels. The intensities of different transitions are determined by the multipolarity of the gamma ray and the difference in the magnetic quantum numbers of the hyperfine levels involved in the transition [1].

One normally observes a hyperfine magnetic field at the nucleus of a transition atom in a lattice which shows either short-range or long-range magnetic order. The contributions to the hf magnetic field arise from core-polarization, orbital currents, spin dipolar interactions, and conduction electron polarization (in metallic systems) [6]. The hf fields can also be observed in the spectrum of a paramagnetic material if the electronic spin relaxation frequency is small compared to the hf frequencies [7].

The hf magnetic field at the nucleus deduced from the Zeeman spectrum allows us to discuss the magnetic behavior of the material. In the case of rare-earth atoms (except Eu^{2+} and Gd^{3+}), the measured hf field is proportional to the local magnetic moment on the atom. In many applications, mere finger printing of the material using the Mössbauer magnetic hf pattern aids in understanding materials and processes.

   Quadrupole interaction. Nuclei with spins greater than 1/2 have quadrupole moments which can interact with the electric field gradient (EFG) at the nucleus. This is a tensor interaction and details have been discussed elsewhere [8]. The quadrupole interaction can partially or fully lift the degeneracy of the nuclear states involved in a Mössbauer transition. The spectrum then consists of hf transitions, the separation between them being governed by the quadrupole moment and the EFG.

In materials applications, the knowledge of the EFG is important and primarily reflects the lattice symmetry of the atom in the lattice. There are many interesting aspects of the EFG measurement and they can be found in other contributions to this volume.

   Mixed interactions. Usually all the above three hf interactions are present simultaneously and they should be separated through a careful analysis of the Mössbauer spectrum. This spectroscopy is now a fairly mature field, and use of computer programs in the lineshape analysis is essential in obtaining information useful to materials scientists.

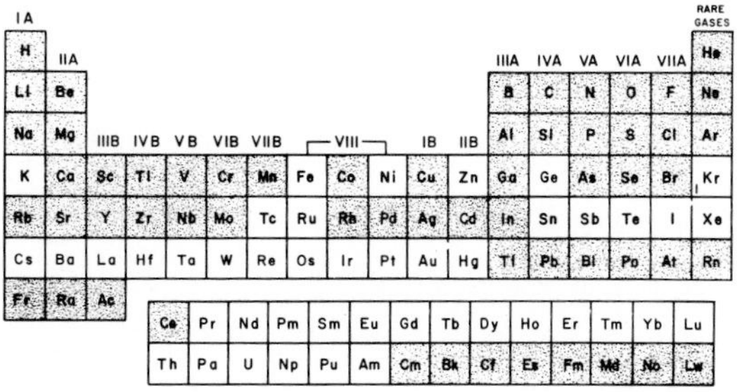

Fig. 1. Mössbauer periodic table. One or more isotopes of the unshaded elements have Mössbauer transitions.

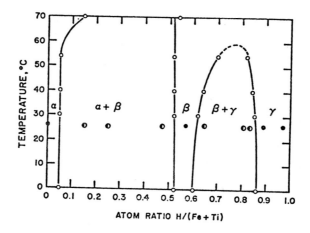 136

Before we leave this section it should be pointed out that there are a large number of Mössbauer resonances spanning the entire periodic table (see Fig. 1), many of which are suitable for materials research. The choice of a particular resonance for a specific problem should be done judiciously, since the resolution factor $(2\,\Gamma_0)$ and the nuclear parameters determining the hf interactions ($\Delta\langle r^2\rangle$, magnetic and quadrupole moments) are not the same for all the Mössbauer transitions.

We shall now proceed to present a few simple examples to illustrate the use of hf interactions deduced from the Mössbauer measurements in microscopically probing the materials.

PHASE ANALYSIS

To illustrate the use of the Mössbauer effect in phase analysis we consider the phase diagram of the system FeTi-H_x. The phase diagram of this system, shown in Fig. 2, has been deduced using conventional methods [9]. There are three phases, viz., α-phase describing the solid solution of hydrogen in FeTi, and the two ternary hydride phases β and γ. When the ternary hydride phases are formed the material breaks down into particles of. 10-50 μ size. It is not unusual that when hydrogen has reacted with FeTi the resulting hydride is in-homogeneous-the surface of the particle forming hydrides with higher hydrogen concentration than deep inside the particle.

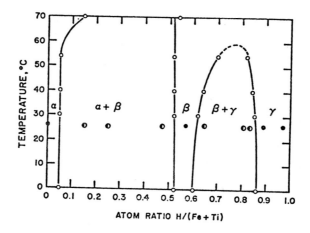

Fig. 2. The phase diagram of FeTi-H_x [9].

If electron and X-ray diffraction techniques are used in a routine fashion to investigate such inhomogeneous particles, we might only identify the hydride phases in the outer regions of the particle. On the other hand, the 14.4 keV resonance gamma ray of ^{57}Fe will penetrate more deeply and will give information on much of the particle. In Fig. 3 we present the spectrum of an inhomogeneous sample of FeTiH$_x$ due to Shäfer et al. [10] The spectrum is resolved into subspectra characterizing the three hydride phases. The γ-phase presents a single absorption line. For the β and γ phases one observes two-line quadrupole pattern. The centroid of each of these three sets of patterns are distinct, due to the different isomer shifts for the three phases.

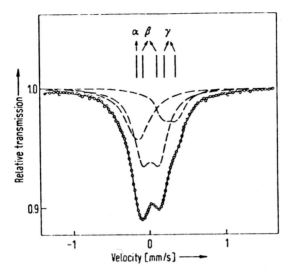

Fig. 3. Mössbauer spectrum of an inhomogeneous sample of FeTiH$_x$ measured with 14.4 keV resonance in ^{57}Fe. The subspectra represent various hydride phases [10].

With increasing hydrogen absorption, the electron density at the ^{57}Fe nucleus decreases [11] and hence the isomer shift consequently increases (since $\Delta\langle r^2\rangle$ for the 14.4 keV transition in ^{57}Fe is negative). In addition, the quadrupole interactions are different in various phases. The areas under the subspectra provide an approximate composition of the three phases in this inhomogeneous sample.

RULES OF ALLOY FORMATION AND STABILITY

One of the long standing problems in metallurgy concerns the degree of solubility of one element in another and the energetics of alloy formation. Microscopic techniques such as the Mössbauer effect would be ideal in probing the environment of an impurity atom residing in another host. Numerous metastable phases have been produced by techniques such as ion implantation and have been investigated with the Mössbauer effect. The hf parameters such as the IS and the quadrupole interaction can be profitably utilized to evaluate the local structure surrounding an atom. That is to say, one can distinguish from the hf parameters those atoms entering the lattice substitutionally, associated with a vacancy, residing interstitially, residing in a grain boundary, producing a cluster, etc. A detailed paper by Nasu et al. [12] deals with the stability of Fe in Al.

Similarly, the configuration of Fe atoms implanted in various elements has been investigated using the Mössbauer effect [13]. In some of the hosts, Fe enters the lattice substitutionally (Fe, Co, Ni, Ru, Rh, Pd, Os, Ir, Pt). It forms random solid solution consisting of dimers and monomers in V, Cr, Nb, Mo and Ta, agglomerates into clusters in Al, Cu, Zn, Ag, Cd and Au, and enters interstitially in Sc, Lu, Hf, Y and Zr. These tendencies of stability can broadly be predicted on the basis of a Darken-Gurry plot (a plot of electro-negativity vs metallic radius) [14]. Miedema and his co-workers [15] have developed the stability rules of binary systems in terms of the heats of for-mation. In this scheme one plots the chemical potential ρ^* as a function of electron density, at the Wigner-Seitz boundary $n_{ws}^{1/3}$. The heat of formation is then given by

$$\Delta H = -P(\Delta\rho^*)^2 + Q(\Delta n_{ws}^{1/3})^2 \qquad (3)$$

where P and Q are positive constants. Such a plot for the solubility of Fe in various hosts is presented in Fig. 4 [13]. It is interesting to note that the four groups of hosts given above lie in Fig. 4 in a remarkably distinct way, and are completely described by Eq. (3).

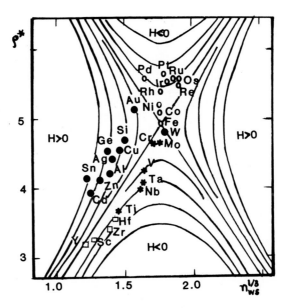

Fig. 4. The plot of two Miedema parameters described by Eq. (3). The data points indicate iron implants in various hosts and different symbols represent the groups discussed in the text [13].

DEFECT STUDIES

The behavior of defects in metals has long been studied from bulk measurements like the resistivity. More recently the Mössbauer effect has been extensively employed in microscopic probing of defects [16]. Here we present an example where Eu atoms which are divalent in a strain-free Mg lattice transform to a trivalent state when associated with defects produced by cold-work [17].

In Fig. 5, we present the Mössbauer spectra of 6300 ppm Eu impurities dispersed in Mg. The resonance at -11.8 mm/s is characteristic of divalent Eu.[17]. The intensity of trivalent Eu peak at -0.1 mm/s increase at the cost of the divalent peak with an increase in the percentage of cold-work. The cold-work primarily produces dislocations in the Mg lattice and some of them pin on the Eu impurity. In the dislocation strain field the Eu atoms transform their valence, the energy needed in such a transformation being typically a fraction of an electron volt. These ideas are further supported by detailed isochronal annealing studies. The large difference in the IS between the divalent (4f^7) and trivalent (4f^6) Eu, produced through the difference in the shielding by 4f electrons, permits one to monitor the impurity atoms in these Mössbauer studies of defect interaction.

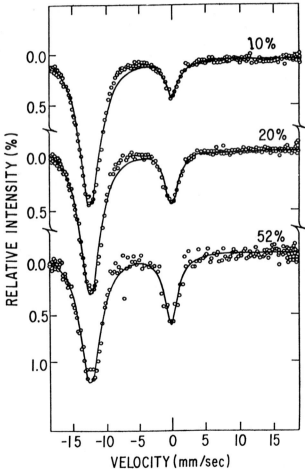

Fig. 5. Mössbauer spectra of Eu:Mg system measured with 21 keV resonance in ^{151}Eu as a function of cold-work. Increasing cold-work transforms the divalent Eu atoms to trivalent atoms [17].

SURFACE STUDIES

The detection of the Mössbauer resonance is usually done by counting transmitted gamma rays through the resonant absorber. The detection can be done in
other ways, such as by scattering of the gamma rays. This is advantageous when
the material containing the resonant nuclei under study cannot be made thin
enough for transmission measurements and for some in situ investigations [18].
When the resonant nuclei de-excites in the absorber, one usually has in addition to the gamma ray, the production of conversion electrons as well as
X-rays. These can also be used to obtain the signature of resonant spectra.
In the case of the [57]Fe resonance, there are 3 K-X-rays of 6.3 keV energy
and 9 K-conversion electrons of 7.3 keV energy produced for every re-emitted
14.4 keV gamma ray.. While the gamma rays transmit through the bulk of the
material, the X-rays provide information through a thickness less than about
10^5Å. On the other hand the electrons have a maximum depth penetration of
only 4000 Å. Thus by proper selection of re-emitted radiation, one can make
some depth selectivity in the studies. In particular, conversion electron
Mössbauer spectroscopy (CEMS) has been employed for various surface studies
[19].

To illustrate the power of the tool, we select an example of the surface catalytic behavior of FeTi in hydrogen absorption [20]. The absorption of hydrogen molecules in FeTi takes place in two stages. Firstly, the hydrogen
molecules from the gas phase break into atoms and secondly, these atoms enter
the lattice. The energy involved in breaking the hydrogen molecule into atoms
is about 4 eV and is expected to occur at the surface through its catalytic
activity. It is suggested that a disproportionation of the surface of FeTi

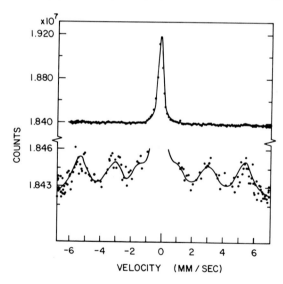

Fig. 6. The conversion electron Mössbauer spectrum of FeTi representing
its surface. The six-line Zeeman pattern measured with [57]Fe resonance shows
the presence of magnetic clusters of Fe atoms on Pauli paramagnetic FeTi surface [20].

takes place at early stages of exposure to hydrogen. This produces catalytically active small metal clusters of Fe covered by a thin layer of disproportionated Ti. To test these thoughts a Mössbauer experiment was carried out using the 14.4 keV resonance gamma rays and the conversion electrons from ^{57}Fe. The gamma rays examined the bulk behavior of the Pauli paramagnetic FeTi, yielding a single resonance absorption with a characteristic IS. The conversion electron spectrum (Fig. 6) consisted of a single line representing FeTi and a six-line hf Zeeman spectrum characteristic of ferromagnetic (or super paramagnetic) Fe clusters at the surface. These clusters residing in a layer of about 200Å thickness carry a magnetic moment of 2.2 μ_B per atom, as deduced from the Mössbauer hf magnetic fields. These studies thus directly offer support to the interpretation that the Fe-rich layer acts as a catalytically active surface responsible for dissociating the hydrogen molecule in the absorption process.

CONCLUSIONS

In this article we have attempted to present a simple picture of the Mössbauer effect to attract the reader's attention to its simplicity and usefulness in materials research. We have presented a few examples which amply illustrate the hf parameters that one measures, and their use in providing microscopic information on materials.

REFERENCES

*Work supported by the U. S. Department of Energy.

1. For example, G. K. Wertheim, Mössbauer Effect: Principles and Applications (Academic Press, New York, 1964).

2. U. Gonser, Mössbauer Spectroscopy (Springer-Verlag, Berlin and New York, 1976).

3. R. L. Cohen, Applications of Mössbauer Spectroscopy, Vol. 1 and 2, (Academic Press, New York, 1976 and 1981).

4. J. G. Stevens and G. K. Shenoy, Chemical Applications of Mössbauer Spectroscopy, Advances in Chemistry Series (American Chemical Society, Washington, D.C., 1981).

5. See for example, G. K. Shenoy and F. E. Wagner, Mössbauer Isomer Shifts (North-Holland Publ., Amsterdam, 1978).

6. A. J. Freeman and R. B. Frankel, Hyperfine Interactions (Academic Press, New York, 1967).

7. H. H. Wickman in: Mössbauer Effect Methodology, Vol. 2, I. J. Gruverman ed. (Plenum Press, New York, 1966), p. 39.

8. M. H. Cohen and F. Reif, Solid State Phys. 5, 321 (1957).

9. G. D. Sandrock (private communication, 1978).

10. W. Shäfer E. Lebsanft and A. Bläsius, Z. Phys. Chem. N.F. 115, 201 (1979).

142

11. L. J. Swartzendrauber, L. H. Bennett and R. E. Watson, J. Phys. F. Metal Physics $\underline{6}$, L331 (1976).

12. S. Nasu, V. Gonser, and R. S. Preston, J. Physique $\underline{41}$, C1-385 (1980).

13. B. D. Sawicka, J. Physique $\underline{41}$, C1-429 (1980).

14. L. S. Darken and R. W. Gurry, Physical Chemistry of Metals (McGraw-Hill, New York, 1973), p. 87.

15. A. R. Miedema, J. Less-Common Met. $\underline{32}$, 117 (1973).
    A. R. Miedema, R. Boom and R. R. de Baer, J. Less-Common Met. $\underline{41}$, 283, (1975), and $\underline{46}$, 67 (1976).

16. G. Vogl, Hyperfine Int. $\underline{2}$, 151 (1976), and references cited therein.

17. Ron G. Pirich, G. R. Burr, G. K. Shenoy, B. D. Dunlap, B. Suits, and J. D. Phillips, Phys. Rev. Letters $\underline{38}$, 1142 (1977).

18. P. A. Flinn and T. O'Connell, U.S. Atomic Energy Commission Rept. WASH-1220 (1973).

19. M. J. Tricker, in Ref. 4, and G. Longworth in this volume.

20. G. K. Shenoy, D. Niarchos, P. J. Viccaro, B. D. Dunlap, A. T. Aldred and G. D. Sandrock, J. Less-Common Metals $\underline{73}$, 171 (1980); A. Blässius and U. Gonser, Appl. Phys. $\underline{22}$, 331 (1980).

Published 1981 by Elsevier North Holland, Inc.
Kaufmann and Shenoy, editors
Nuclear and Electron Resonance Spectroscopies Applied to Materials Science

MÖSSBAUER STUDIES OF ION IMPLANTED ALLOYS

G. LONGWORTH

Nuclear Physics Division, Atomic Energy Research Establishment,
Harwell, Oxfordshire, U.K.

ABSTRACT

A review is presented of the use of Mössbauer spectroscopy
to study the mechanisms responsible for the improvements in
wear resistance in ferrous alloys implanted with either
light or heavy ions. The phases present in the near surface
layers of iron and steel samples implanted with either
nitrogen, carbon or tin ions are identified using Conversion
Electron Mössbauer Scattering (CEMS) spectra. Samples of
wear debris from tin-implanted iron wear discs are analysed
in Mössbauer absorption measurements.

INTRODUCTION

It is now well established that the surface durability of a variety of
metals may be improved by ion implantation [1,2] . In this present article
several examples are given of the reduction in the wear rate of ferrous alloys
after implantation with either light or heavy ions. In order to study the
wear processes involved we need to characterise the implanted layer in terms
of both elemental and chemical composition. The use of nuclear reaction
techniques to measure the total retained dose, and X-ray photoelectron
spectroscopy (X.P.S.) to determine the approximate depth profile of the
various elements present, are well known [3] . Here we discuss the use of [4]
Mössbauer spectroscopy to identify the phases present in the implanted alloy
and to determine how these are modified as wear proceeds.

ION IMPLANTATION

Ion implantation allows the introduction of a controlled amount of one
atomic species into the surface layers of a substrate. The atoms are ionised
and then accelerated to energies in the region 40-200 keV in a reasonably good
vacuum (base pressure $\sim 10^{-7}$ Torr). When the ions penetrate the target material
they lose energy, mainly in elastic collisions, before coming to rest. The
statistical nature of the collision process leads to a distribution of ion
ranges within the target, which is roughly Gaussian in shape for polycrystalline materials [5,6,7,8] . As an example 50 keV N^+ ions are expected to have
a mean projected range (normal to substrate surface) of about 0.055 μm and a
standard deviation of about 0.025 μm.

When an incident ion makes an elastic collision with a target atom the
latter will be displaced from its lattice site if an energy greater than about
20 eV is transferred. When this occurs a vacancy is left at the lattice site
together with the formation of an interstitial atom. Frequently the energy of
the atom is sufficient to induce further displacements and in this way a
collision cascade is built up. Each ion toward the end of its path will
create a large number of vacancies in a small region. Here the probability is
high that the ion will occupy a substitutional site. The production of large
numbers of point defects ensures that the surface regions are heavily

144

disturbed. For a dose of 1 x 10^{17} N$^+$ ions cm^{-2} at 50 keV, at the depth for
which the greatest damage occurs, each target atom will have been displaced on
average about 60 times. This quantity is usually referred to as the number of
displacements per atom (d.p.a.). Some of the defects will aggregate to form
clusters while others will migrate to sinks, such as dislocations and grain
boundaries. These processes may promote diffusion of the implanted atoms-
radiation enhanced diffusion, at temperatures for which thermal diffusion is
minimal.

Ion implantation is a non-equilibrium process and the normal solubility
restrictions may be exceeded. However, an upper limit is set by sputtering,
whereby an incident ion ejects several target atoms from the surface. Usually
the number of atoms removed per incident ion, the sputtering ratio (S), is
greater than one, so that the surface is continually being eroded during
implantation. The effect of increasing dose is to flatten the implantation
profile and to move the depth at which the peak concentration occurs nearer
the surface (fig.1) [9]. Eventually the peak becomes situated at the surface
and a maximum in the retained dose is reached. The maximum atomic concentration of solute is given approximately by the reciprocal of the sputtering
ratio. The solid solutions formed in this way by implantation are metastable
and will generally decompose when the sample is annealed, either by solute
precipitation or by formation of intermetallic compounds.

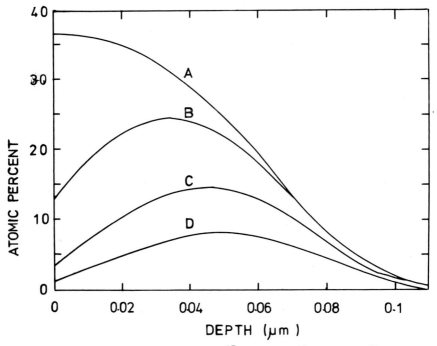

Fig. 1 Implantation profiles for 6 x 10^{17}(A), 2 x 10^{17}(B), 1 x 10^{17}(C) and
5 x 10^{16}(D) N ions cm^{-2} on iron at 50 keV, S = 1.7.

The nature of the wear process at the surface of an implanted alloy may be studied using CEMS spectra measured before and after given periods of wear. In addition, if a few milligrams of the wear debris is collected, this can be analysed in the more conventional Mössbauer absorption measurement.

WEAR ON IMPLANTED METALS

Measurements of the wear resistance of ion implanted metal surfaces have been performed at Harwell using a modified Avery-Denison T62 wear tester of the pin-on-disc type [13]. In such measurements a standard pin of constant cross-section is allowed to wear against a rotating disc which forms the test surface. The volumetric wear rate (K_v) is determined from the displacement of a loaded arm carrying the pin in a given time under known conditions of load and speed. Here K_v is a dimensionless parameter given by V_p/xA, where V_p is the volume of material removed from the pin, x is the sliding distance and A the apparent area of contact. The effectiveness of the implantation in improving the wear rate is assessed from measurements in turn on implanted and unimplanted regions of the same disc. The disc is lubricated with a kerosene flow to remove wear debris and to provide cooling. These measurements are expected to fall within the regime of adhesive wear in which the dominant process is due to the local temperature rise at asperities leading to welding, microadhesion and rupture of the surface [2].

It was observed that the wear rate could be reduced by 1 to 2 orders of magnitude for surfaces of mild steel implanted with either nitrogen or boron ions, 440 C steel implanted with molybdenum ions[13] and nitriding steel implanted with nitrogen ions[14]. In this latter case the greatest improvement was noted for a dose of about 10^{17} N ions cm^{-2}. Similar reductions were measured for aluminium pins implanted with unspecified ions rubbing against steel cylinders [15]. A feature of many of these measurements was the long lasting nature of the wear improvement after implantation [1,15,16]. There was no apparent breakdown of the wear resistance when the wear track reached a depth of 12 µms, which is two orders of magnitude greater than the range of the implanted ions [1].

At this stage, in 1977, it was decided to use Mössbauer spectroscopy to investigate the final locations of the implanted ions in the lattice. In order to simplify the analysis of the Mössbauer spectra, samples of pure iron implanted with either nitrogen or carbon ions were used. These measurements were later extended to ones on commercial steels. Similar studies have also been undertaken on samples of iron implanted with tin ions.

MÖSSBAUER MEASUREMENTS

1) Iron implanted with nitrogen or carbon ions.
In fig.2 are shown the ^{57}Fe CEMS spectra for iron foils (125 µms thick) implanted with 100 keV N_2^+ molecular ions, to doses of 1, 2, 4, and 6 x 10^{17} ions cm^{-2} over a circle $\simeq$ 1.5 cm in diameter [17]. The spectrum for the sample with a dose of 1 x 10^{17} N ions cm^{-2} is similar to that of pure iron, while for doses of 2 and 4 x 10^{17} cm^{-2} iron is also present in an additional magnetic phase (6 line pattern), and in a non-magnetic phase (doublet)for doses of 4 and 6 x 10^{17} N ions cm^{-2}. The Mössbauer parameters of these magnetic and non-magnetic phases are similar to those for respectively γ'-Fe_4N and either ε-$Fe_{2+x}N$ with x $\simeq$ o or ζ-Fe_2N.

In the absence of sputtering the expected range for 100 keV N_2^+ ions in iron is 0.055 µms. For a sputtering ratio S $\sim$ 1.7 [3] the implantation profile approaches the surface as the dose is increased (fig.1). There is a maximum retained dose of 2.9 x 10^{17} ions cm^{-2}. In table 1 are shown the retained dose for a given incident dose, the expected peak position and peak concentration of nitrogen as well as the observed relative areas of the nitride components.

MÖSSBAUER CONVERSION ELECTRON SPECTROSCOPY

The surface alloys produced by ion implantation typically extend to a depth of the order of about 0.1 μm and may be studied conveniently by Mössbauer conversion electron scattering (CEMS). In this technique the appropriate gamma rays from the Mössbauer source are incident on the sample and scattered conversion electrons resulting from the decay of the first excited state in the nucleus are detected. For ^{57}Fe about nine out of ten decays result in the emission of a conversion electron. In the work described here the Mössbauer source is either ^{57}Fe or ^{119}Sn with gamma rays at 14.4 keV and 23.8 keV respectively. For iron the detected electrons originate mainly from the K shell with an energy ≃ 7 keV, while for tin they come from the L shell with an energy of ≃ 20 keV. In each case the sample depth analysed in the Mössbauer spectrum is determined by the electron escape depth, which is 0.1 - 0.2 μm.

There are two types of implanted alloy that may be studied in this way, according to whether the Mössbauer atoms are the implanted atoms or are contained in the substrate. Measurements in the first type of alloy are more sensitive since the number of Mössbauer atoms can be large if an isotopically separated beam is used, and the spectrum relates directly to the location of the implanted atoms. When the Mössbauer atom is contained in the substrate, for example in iron or in steels, the presence of implanted ions such as nitrogen is studied indirectly by their effect on the Mössbauer resonance. The natural abundance of the Mössbauer isotopes may be low (≃ 2% for ^{57}Fe) and the spectrum will contain contributions from iron atoms both inside and outside the implanted region. The latter contribution may obscure the former one, leading to a reduction in sensitivity of the technique.

The detector used is typically a proportional counter, through which a helium/methane mixture is allowed to flow [10]. The sample, an area of about 2 cm in diameter, usually in the form of a foil or flat disc, is sealed temporarily on to the back plate of the counter, which is illuminated by the source (50 -100 mCi ^{57}Fe or 10 -20 mCi ^{119}Sn) placed at the opposite side of the counter. The counter gas is largely insensitive to the incident gamma rays but detects virtually all electrons backscattered into a 2π solid angle. Apart from the detector the remaining equipment used is identical to that used for Mössbauer absorption measurements.

The detector has poor energy resolution for electrons and it is advantageous to count electrons having energies down from the conversion energy to essentially zero. Thus in ^{57}Fe measurements the L Auger electrons are also included since they too give rise to a Mössbauer spectrum. A limitation of the procedure is that very little depth information is available. This is in fact a basic weakness of the CEMS technique since depth information is not directly available even if electrons of a given energy are counted, because all electrons detected at one energy do not originate at the same depth in the scatterer. Although this difficulty may be partially overcome by measuring spectra for electrons detected at several energies and deriving from them depth selective spectra [11], the poor counting geometry of the detector used e.g. a magnetic spectrometer, has ensured that measurements have been restricted mainly to those of highly enriched ^{57}Fe scatterers. However, even using detectors with poor energy resolution and natural iron scatterers it is possible to get some depth information. A curve of probability that electrons emitted at a given depth will escape and be detected can be deduced by measuring the CEMS spectra for an iron foil after successive evaporation of layers of ^{56}Fe of known thickness [12]. This shows that about 50% of the spectral area comes from iron atoms within the first 0.1 μm below the surface. This knowledge is useful, for example in the analysis of the spectra for iron implanted with nitrogen ions, in order to distinguish contributions to the spectrum from iron atoms inside and outside the implantation profile.

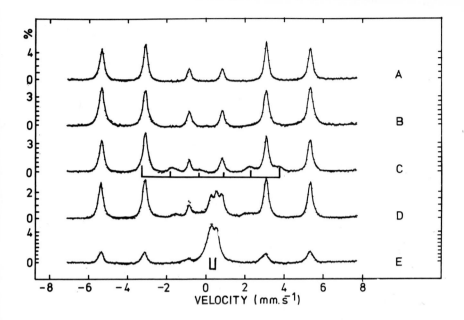

Fig. 2 CEMS spectra for iron (A), and iron implanted with 100 keV N_2^+ ions at doses 1 x 10^{17}(B), 2 x 10^{17}(C), 4 x 10^{17}(D) and 6 x 10^{17} ion cm^{-2}(E). The patterns due to Fe_4N (sextet) and Fe_2N (doublet) are indicated.

Table 1: Implantation parameters for nitrogen implanted iron.

Incident dose (ion cm^{-2})	Retained Dose (ion cm^{-2})	Peak Position (μm)	Peak concentration (at .%)	Nitride area (%)
1 x 10^{17}	0.96 x 10^{17}	0.045	15	0
2 x 10^{17}	1.82 x 10^{17}	0.036	25	19
4 x 10^{17}	2.76 x 10^{17}	0.015	34	33
6 x 10^{17}	2.90 x 10^{17}	0	37	56

The expected peak concentrations are of the right order for nitride formation but any quantitative depth analysis is difficult. Although at first sight the shape of the implantation profile is incompatible with the existence of nitrides of well defined composition, it is conceivable that the relative proportions of iron and iron nitride vary as a function of depth. It is significant that the dose (1 x 10^{17} N ions cm^{-2}) above which nitrides become apparent in the Mössbauer spectra, is that for which the greatest reduction in wear rate was observed for nitriding steel implanted with nitrogen ions. [14]

148

Carbon ions have also been shown to be effective in reducing the wear rate of nitriding steel [2]. ^{57}Fe CEMS spectra of iron foils implanted with 50 keV carbon ions at doses between 5 x 10^{16} and 8 x 10^{17} ions cm^{-2} show components due to iron carbides at all doses(fig.3).These are Fe$_3$C for doses of 5 x 10^{16} and 1 x 10^{17} ions cm^{-2}, and Fe$_5$C$_2$ for higher doses. Again the effect of sputtering (S $\simeq$ 1.0) is to impose a maximum retained dose (4.4 x 10^{17} ions cm^{-2}).

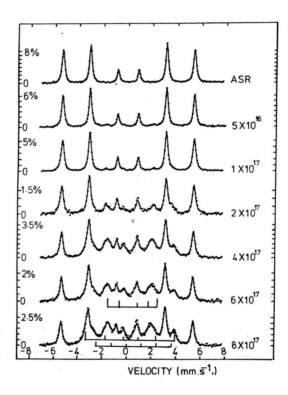

Fig. 3 CEMS spectra for iron implanted with 40kev C ions at stated doses in ions cm^{-2}. The three patterns due to Fe$_5$C$_2$ are indicated.

When the nitrogen implanted iron samples were annealed for one hour at successively higher temperatures the CEMS spectra indicate that the nitrogen atoms were able to diffuse at temperatures above about 275°C and they had largely migrated out of the detection depth at about 500°C. For example, for the sample with a dose of 4 x 10^{17} ions cm^{-2} the non-magnetic phase Fe$_2$N was converted to a magnetic phase ε-Fe$_{2+x}$N, richer in iron, above 275°C and to nitrogen martensite with about 5 at .% nitrogen at 500°C (fig 4). Clearly the phases present are less stable than iron nitrides prepared by conventional techniques, for example γ'-Fe$_4$N is expected to be stable up to 680°C [18]. Similar measurements on a carbon-implanted iron foil (2 x 10^{17} ions cm^{-2}) indicated little migration below 450°C but migration out of the detection depth at about 600°C [12].

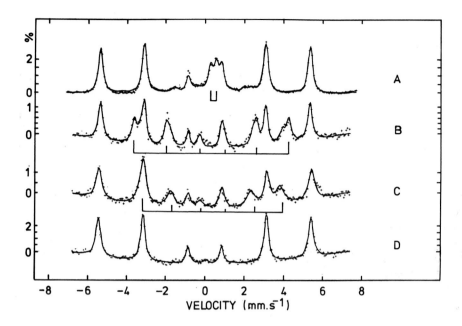

Fig. 4 CEMS spectra for iron implanted with 100 keV N_2^+ ions to 4×10^{17} cm^{-2}
(A), after annealing for 1 hour at $275^{\circ}C$ (B), $400^{\circ}C$ (C) and $500^{\circ}C$ (D).
The patterns are indicated for $\zeta-Fe_2N$, $\varepsilon-Fe_3N$ and $\gamma'-Fe_4N$ (reading
downwards).

There is now a good deal of evidence to suggest that the observed
improvement in wear rates is not simply due to the formation of nitrides near
the surface, since it is maintained when the wear track is many times deeper
than the original implantation depth [1,15,16]. The Mössbauer results indicate
that diffusion can occur at relatively low temperatures, and this has been
used to put forward a possible explanation [1]. It is suggested that the
movement of dislocations is impeded by the trapping of nitrogen atoms thus
hardening the surface layers. As wear proceeds these nitrogen ions are carried
along with the dislocations as they are driven in, thus continually recreating
a hard surface. This model has been extended [2,19,20,21] to one in which the
implanted nitrogen ions produce iron dislocation loops which appear locally as
iron nitride. During the initial wear process the temperature rise at the
local asperities causes the dislocation loops to shrink, releasing the nitrogen
atoms into solution. These nitrogen atoms may then move to further dislocations created in the wear process itself as suggested in the earlier model.

2) Steels implanted with nitrogen ions.
    So far we have been comparing the results for wear measurements on steels
with Mössbauer measurements on pure iron. In order to make a closer comparison
^{57}Fe CEMS spectra were measured for samples of iron, a mild steel with wt.% (0.2C
0.6 Mn), carbon steel (EN8: 0.4 C, 0.8 Mn) and a tool steel (NSOH: 1.0 C,
1.2 Mn, 0.5 Cr) implanted with 40 keV N$^+$ ions to a dose of 2 x 10^{17} ions cm^{-2},
at both a high dose rate (30 μa cm^{-2}) and a low dose rate (3 μa cm^{-2}), the
spectra for the high dose rate are shown in fig. 5. Analysis of the spectra
indicates that the amount of iron nitrides is considerably greater for the
higher dose rates (table 2). The nitride present is γ' –Fe$_4$N in the implanted
iron, mild steel and carbon steel samples and 'Fe$_2$N' for tool steel. Although
the fraction due to nitride decreases from iron to mild steel to carbon steel,
the trend is not continued in tool steel with the highest carbon content.

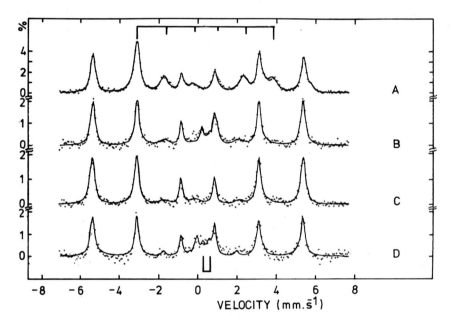

Fig. 5 ^{57}Fe CEMS spectra for iron (A), mild steel (B), carbon steel (C) and
        tool steel (D), implanted with 2 x 10^{17}N ions cm^{-2}, at 30 μa cm^{-2}. The
        patterns are indicated for Fe$_4$N (upper) and Fe$_2$N (lower).

    The fact that no γ' –Fe$_4$N was observed in the spectrum for implanted
tool steel (NSOH) is in agreement with similar measurements on another tool
steel (U.N.I. C80: 0.8 C, 0.4 Si, 0.85 Mn) [22]. Here a quadrupole doublet
was observed which was ascribed to a ε -like carbonitride. The presence of
γ'-Fe$_4$N is unlikely since carbon atoms do not enter this structure.

Table 2: Measured spectral areas due to nitride.

Dose rate $(\mu a\ cm^{-2})$	Iron	Mild steel	Carbon steel	Tool steel
3	21%	5%	0	11%
30	38%	26%	14%	16%

It is difficult to account adequately for the dependence of the nitride area on dose rate. The change in dose rate should give rise to changes in the diffusion rate of the implanted ions due both to radiation enhanced diffusion and to thermal diffusion. During implantation large numbers of vacancies are produced which may cause transport of material out of the implanted region. The temperature rise at the substrate during implantation is not negligible, particularly for the higher dose rate where the power density $\simeq 1$ watt cm^{-2}. The measured rises in temperature were in fact about $50^{\circ}C$ (3 $\mu a\ cm^{-2}$) and $240^{\circ}C$ (30 $\mu a\ cm^{-2}$). The CEMS spectra do indicate a differing amount of nitride within the first $0.1 - 0.2\ \mu m$. However, the overall nitrogen profiles, as determined in XPS measurements, are very similar for high and low dose rates for the mild steel samples. They both show a peak at about $0.2\ \mu m$ with a tail extending to about 1 μm [3]. Thus although some migration has occured it is essentially independent of dose rate. It appears that although the nitrogen ions are mobile, they become trapped to form compounds with the host atoms and at defects, before they can diffuse appreciably. A quite different behaviour is observed in carbon-implanted iron foils, where the effect of increasing the dose rate from 3 $\mu a\ cm^{-2}$ to 30 $\mu a\ cm^{-2}$ is to reduce the amount of iron carbide (Fe_5C_2) present from 69% to 3% of the spectral area. Such behaviour does imply appreciable diffusion. However, it is not possible to compare these results directly with those for nitrogen-implanted iron because of possibly different implantation conditions. For example, the temperature rise during implantation was not recorded for the carbon-implanted samples.

The fact that the amount of nitride in implanted steels, as deduced from the CEMS spectra, appears to decrease with increasing carbon content of the steel may indicate that the presence of carbon interstitials tends to promote diffusion of the nitrogen atoms, but this does not apply to tool steels, with the highest carbon content, particularly for the measurements on U.N.I. C80 steel [22] where the area due to ε - carbonitride ($\simeq 60\%$) was appreciably greater than that of Fe_2N in pure iron at the same dose (4 x 10^{17} ions cm^{-2}) (table 1).

A major source of difficulty in the comparison of results obtained using different techniques on different samples is that there are several parameters that determine the implantation profile and frequently not all these parameters have been measured. Apart from the dose rate and total dose another parameter is the pressure in the implantation chamber, since any residual oxidising gases present will contribute to the build up of an oxide film on the target during implantation [3]. A series of measurements on implanted tool steel and stainless steel has been performed [23] in which several of these parameters have been varied in turn and the samples studied in wear measurements, and in XPS and nuclear reaction studies. The parameters varied were dose rate, sample temperature, target chamber base pressure and the nominal dose of implanted nitrogen ions. The main results are that some reduction in retained dose is found for high dose rates (25 μa cm^{-2}), that the implantation temperature for tool steel should be below about 200°C in order to reduce the effects of out-diffusion of nitrogen and that the base pressure in the implantation chamber, within a range 10^{-7} to 3 x 10^{-6} Torr, has little influence on the wear behaviour. This type of study is being extended to include CEMS measurements to determine the chemical nature of the nitrogen present both before and after periods of wear.

TIN-IMPLANTED IRON

Implantation with heavy metallic ions such as tin is known to reduce the frictional force between steel surfaces [24] and to improve the wear resistance of other metals [25]. The effects of tin implantation on the tribological and oxidation characteristics of pure iron has been investigated using ^{57}Fe and ^{119}Sn CEMS spectra to identify the phases present both before and after wear, while Mössbauer absorption spectra are used to identify the wear debris produced. In this way the type of wear process may be characterised [26,27].

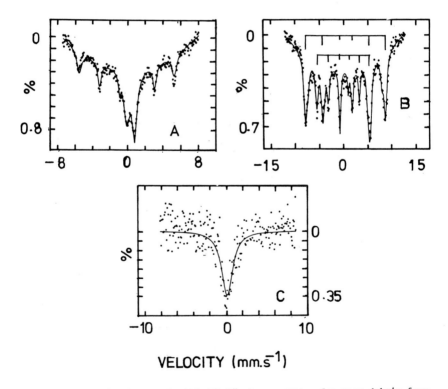

VELOCITY (mm.s^{-1})

Fig. 6 Iron-57 (A,B) and tin-119 (C) Mössbauer spectra for wear debris from tin-implanted iron, at 300K (A), 4.2K (B) and 77K (C). The patterns shown for B are due to Fe_2O_3/Fe_3O_4 (upper) and iron metal (lower).

Samples of pure iron implanted with 200 keV Sn ions to a dose of 5×10^{16} ions cm^{-2} were shown, using the reaction $^{16}O (d,p_1)^{17}O$, to have an oxygen take up at $500^{\circ}C$ about five times lower than that for unimplanted iron [26]. Wear measurements on similarly implanted iron discs made on a pin-on-disc machine with a load of 25N, gave a wear rate, $K_v = 2.1 \times 10^{-8}$, similar to that measured in pure iron, while at a sliding distance of about 1.5×10^5 cm the wear rate decreased and at a distance of about 2×10^5 cm reached a constant value of 2.7×10^{-10}. A similar improvement in wear rate was observed for the lower dose of 1×10^{16} Sn ions cm^{-2}, but not for a dose of 5×10^{15} Sn ions cm^{-2}.

Examination with the electron microscope of the pin, after a prolonged period of wear, showed evidence of transfer of particles from the disc. These were found by energy dispersive X-ray analysis (EDAX), to contain a large proportion of tin as well as iron. ^{57}Fe and ^{119}Sn Mössbauer absorption spectra for the debris revealed the presence of tin oxide (SnO_2) and also iron metal, ferric oxide (α-Fe_2O_3) and a form of magnetite (Fe_3O_4) (fig. 6.). The iron

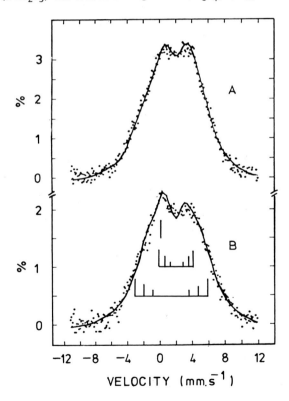

Fig. 7 ^{119}Sn CEMS spectra for iron implanted with 5×10^{16} tin-119 ions cm^{-2} as implanted (A) and after wear for 10^6cm (B). The patterns for SnO_2, $FeSn_2$ and $\underline{FeSn}$ are indicated (reading downwards).

oxides were shown to be finely divided (diameter 100 - 200 Å) by the fact that
the magnetic splitting expected at 300K was suppressed, due to the rapid
relaxation of the iron spins. On cooling the sample to 4.2K the magnetic
splitting was restored and the area could be divided up into 58% ferric oxide,
18% magnetite and 24% iron metal. ^{119}Sn CEMS spectra measured for the surface
of an iron wear disc implanted with 5 x 10^{16}, ^{119}Sn ions cm^{-2} show that the
tin before wear is present as $FeSn_2$ (66% area), FeSn ≃ 4 at .% (34% area)
while after wear for 10^6 cm it is present as $FeSn_2$ (58% area), FeSn ≃ 4 at .%
(39% area) and SnO_2 (3% area) (fig. 7).

As a result of these measurements, using a variety of techniques, the type
of wear may be characterised as mild adhesive wear [28] since the wear rate is
in the range 10^{-8} to 10^{-10} and the wear debris is shown to consist mainly of
finely divided oxides. This wear process occurs when metals of good surface
finish are slid against each other and localised welding and surface forces
cause the transfer of metallic fragments or surface failure. These fragments
may become oxidised before being removed to form the wear debris. The wear
rate is thought to be determined by the rate of oxidation of transferred
material. Thus one reason for the observed reduction in wear rate after
implantation is the reduced oxidation rate of the iron.

A similar transition to a regime of lower wear rate was also observed in
a study of the mild wear of carbon steel [29]. There it was attributed to the
combined effects of phase hardening and to the formation of a protective oxide
layer. In the present experiment the presence of $FeSn_2$ and the solid solution
of tin in iron would be expected to produce phase hardening, while the
observed stannic oxide component may form a protective oxide layer, which
lubricates the surface.

A rough idea of the temperature reached at asperities during wear is
estimated from the appearance of stannic oxide and also from the ^{119}Sn CEMS
spectra for an implanted sample after isochronous anneals at several temperatures. Such measurements suggest that FeSn is formed after annealing at
650K. Since this component is not seen in the spectrum of the worn disc, it
is likely that the temperature during wear does not reach 600K, but must be
greater than about 500K in order to produce stannic oxide.

CONCLUSIONS

The measurements described demonstrate the use of Mössbauer spectroscopy
as a diagnostic technique in the study of wear processes in implanted alloys,
in conjunction with the use of other complementary techniques. Using CEMS
spectra the phases present in the near surface layers of an alloy may be
identified. Although improvements in wear resistance brought about by
implantation may be maintained after the surface has been worn to many times
the implantation depth, it is likely that the mobility of the implanted
species during the initial stages of wear will depend upon their chemical
state. In the past, one of the difficulties has been that different techniques
have not been applied to the same or to well characterised samples. There are
several parameters, such as, total dose, dose rate, sample temperature which
can affect the implanted region, and it necessary to establish what is the
effect of varying each parameter in turn.

ACKNOWLEDGMENTS

The author gratefully acknowledges the collaboration of I. J. R. Baumvol,
L. W. Becker, G. Dearnaley, P. D. Goode, N. E. W. Hartley and R. E. J. Watkins
in these measurements. He would also like to thank R. Atkinson and
T. E. Cranshaw for several fruitful discussions.

REFERENCES

1. G. Dearnaley and N. E. W. Hartley, Thin Solid Films $\underline{54}$, 215 (1978)

2. N. E. W. Hartley, Thin Solid Films $\underline{64}$, 177 (1979).

3. G. Dearnaley, P. D. Goode, N. E. W. Hartley, G. W. Proctor, J. F. Turner and R. E. Watkins, 1979 In Proceedings of Conference on Ion Plating and Allied Techniques: (IPAT, London, July 1979) pp. 243-254.

4. G. K. Wertheim, Mössbauer Effect, Principles and Applications (Academic Press, New York 1964).

5. J. Lindhard, M. Scharff and H. E. Schiott, K. Dan. Vidensk. Selsk., Mat.-Fys. Medd., $\underline{33}$, (1963) No. 14.

6. I. Manning and G. P. Mueller, Computer Physics Comm. $\underline{7}$, 85 (1974).

7. M. D. Matthews, UKAEA Report, AERE-R 7805 (1973).

8. M. D. Matthews, UKAEA Report, AERE-R 9166 (1978).

9. R. Bett and J. P. Charlesworth, UKAEA Report, AERE-R 7052 (1973).

10. G. Longworth, Instrumentation for Mössbauer Spectroscopy, chapter in Advances in Mössbauer Spectroscopy, B. V. Thosar and P. K. Iyengar eds. (Elsevier 1981).

11. B. Bodlund-Ringström, U. Bäverstam and C. Bohm, J. Vac. Sci. Tech. $\underline{16}$, 1013 (1979), and T. Shigematsu, H.-D. Pfannes and W. Keune, Phys. Rev. Lett. $\underline{45}$, 1206 (1980).

12. G. Longworth and R. Atkinson, in Proc. Symp. on Recent Chem. Appl. Mössbauer Spectroscopy, Houston 1980. (Adv. in Chemistry, 1981).

13. N. E. W. Hartley, G. Dearnaley, J. F. Turner and J. Saunders, in: Applications of Ion Beams to Metals, S. T. Picraux, E. P. Eernisse and F. L. Vook eds. (Plenum, New York 1974) pp. 123-138.

14. N. E. W. Hartley, Wear $\underline{34}$, 427 (1975).

15. A. V. Pavlov, P. V. Pavlov, E. I. Zorin and D. I. Tetelbaum, All Union Conf. on the Interaction of Atomic Particles with Solids, Kiev 2, 114 (1974).

16. J. K. Hirvonen, J. Vac. Sci. Technol. $\underline{15}$, 1662 (1978).

17. G. Longworth and N. E. W. Hartley, Thin Solid Films $\underline{48}$, 95 (1978).

18. M. Hansen, Constitution of Binary Alloys (McGraw-Hill, New York 1958).

19. M. Wuttig, J. T. Stanley and H. K. Birnbaum, Phys. Stat. Sol. $\underline{27}$, 701 (1968).

20. P. Haasen, Physical Metallurgy (Cambridge 1978).

21. W.-W. Hu, C. R. Clayton, H. Herman and J. K. Hirvonen, Scr. Metall. 12, 697 (1978).

22. G. Principi, P. Matteazzi, E. Ramous and G. Longworth, J. Mater. Sci. 15, (1980).

23. P. D. Goode and I. J. R. Baumvol, Third International Conference on Ion Implantation: Equipment and Techniques, Kingston, July 1980, and Nucl. Instr. Meth. to be published.

24. N. E. W. Hartley, W. E. Swindlehurst, G. Dearnaley and J. F. Turner, J. Mater. Sci. 8, 900 (1973).

25. R. E. J. Watkins, private communications 1979.

26. I. J. R. Baumvol, R. E. J. Watkins, G. Longworth and G. Dearnaley, Proc. LEIB-II, Inst. Phys. Conf. Ser. (London 1980).

27. I. J. R. Baumvol, G. Longworth, L. W. Becker and R. E. J. Watkins, Proc. V. Int. Conf. Hyperfine Interactions (Berlin 1980).

28. E. F. Finkin, Mats. in Eng. Appl. 1, 154 (1979).

29. N. C. Welsh, Phil. Trans. Roy. Soc 257A, 31 (1964).

MOSSBAUER SPECTROSCOPY STUDIES OF AMORPHOUS METALLIC SOLIDS*

C. L. CHIEN
The Johns Hopkins University, Baltimore, Maryland   21218

ABSTRACT

Mössbauer spectroscopy as applied to the studies of amorphous
metallic, and often magnetic solids will be discussed. The
advantages of using this microscopic technique will be compared
with others in terms of obtaining information on electronic,
magnetic, hyperfine, structural and other properties. Examples
of magnetic ordering structures, concentration dependence of
ordering temperature, hyperfine field distribution, spin-wave
excitations, crystallization behavior, quadrupole interaction,
isomer shift, etc., as obtained by Mössbauer spectroscopy will
be presented.

INTRODUCTION

In the 1960's, Duwez and his co-workers demonstrated that amorphous metals
can be made by extremely rapid cooling from the melt (cooling rates in excess
of 10^5 K/sec) [1]. Since then several other methods have successfully been
used to fabricate amorphous metallic solids. In recent years the field of
amorphous metallic solids or metallic glasses has been one of the most active
areas of research in condensed matter physics.

The term amorphous metallic solid and metallic glass are here taken to be
synonymous although some authors reserve the term glass to mean materials
formed exclusively from the melt. Metallic is used in the general sense to
indicate materials having high concentrations of metallic elements, even
though for example, few have electrical conductivities that approach that of
a good metal. Given sufficient activation energy (e.g. heat), the metallic
glasses invariably and irreversibly transform into crystalline phases, since
the amorphous state is metastable.

Research in the field of amorphous metals has revealed many unusual mechanical,
electrical and magnetic properties. Both because of the ease with which transition metals and rare earths can be incorporated into the metallic glasses and
because of possible applications, amorphous magnetism has perhaps received the
greatest attention. Magnetic bubble memory, magnetic shielding and transformer
core are just few examples of such applications [2,3]. Many other practical uses
will undoubtedly be found for utilizing the superior properties and inexpensive
production of the metallic glasses.

Since the first systemetic Mössbauer studies of metallic glasses by the Caltech
group twelve years ago [4], this technique has been widely used in investigations of various amorphous systems. In this paper, we will briefly describe
the applications of Mössbauer spectroscopy and summarize some of the results in
the literature. Since metallic glasses will be featured, the results on
insulating glasses will not be included [5,6]. In keeping with the emphasis of
this Symposium, general features rather than fine details will be presented.

FABRICATION AND CHARACTERIZATION OF AMORPHOUS METALLIC SOLIDS

There are generally three ways of making amorphous metallic solids:
liquid-quench (piston and anvil, melt spinning) [1,7], vapor deposition (evaporation, sputtering, e-gun, etc) [8], and less commonly electrodeposition [9].
The liquid-quench methods preserve the random atomic arrangement in the liquid
state through the rapid quenching process. In deposition, the random
arrival of atoms is preserved in the solid state by appropriately maintaining
the deposition rate, bias, substrate conditions, etc. Liquid-quench methods
can mass produce materials, but are restricted in composition range. For
example, all transition metal (TM)-metalloid (M) systems made by the liquid-
quench techniques have compositions near that of $TM_{80}M_{20}$, the eutetic composition. Using other techniques, compositions other than $TM_{80}M_{20}$ can be readily
made. Deposition methods have fewer compositional restrictions but suffer
the drawback of yielding only a small quantity of material. The thickness of a
typical liquid-quench sample is about 30 μm. A single layer of an as-quench
sample makes an adequate absorber for ^{57}Fe Mössbauer spectroscopy if the Fe
content is about 10 at.% or higher. The thickness of the vapor-deposited
samples ranges from 1000 A to a few μm, limited usually by evaporation time
and adhesion. For these thin films, multilayers or enriched samples are
desirable.

A word of caution perhaps should be made on the composition of the samples,
especially the commercial ones. The actual composition may be different from
what is advertised or expected. The deviation in composition may not change
appreciably the gross properties such as coercivity, 4πM at room temperature,
or other mechanical properties. But this deviation could substantially alter
the magnetic ordering temperature, crystallization behavior and other properties.
For example, commercial Allied Chemical Metglas 2826 ($Fe_{40}Ni_{40}P_{14}B_6$) has been
shown to vary in composition by as much as 10 at.% [10]. Carbon may end up in
the sample due to the melting process. A few at.% C when present in 2605
($Fe_{80}B_{20}$) can greatly stabilize the unstable Fe_3B phase upon crystallization.
A slight change (a few at.%) of Mo in 2605A ($Fe_{78}Mo_2B_{20}$) can drastically alter
the Curie temperature [11]. Thus deviation in composition often accounts for
different results when samples of supposedly the same composition are studied.

To date, many amorphous metallic systems have been studied by Mössbauer
spectroscopy. Most of them belong to the transition metal-metalloid (TM-M)
systems (e.g. $Fe_{40}Ni_{40}P_{14}B_6$) and the rare earth-transition metal (RE-TM)
systems (e.g. $DyFe_2$, YFe). Many other new systems will undoubtedly be studied
in the future using perhaps isotopes other than ^{57}Fe, ^{161}Dy, ^{119}Sn, ^{151}Eu
and ^{197}Au, which are the ones used so far.

Under x-ray diffraction (or other diffractions), an amorphous solid exhibits
a diffuse scattering pattern which is distinctively different from that of
a crystalline solid [12]. The Fourier transform of the scattered intensity,
or the radial distribution function, describes correlations among the atomic
positions. The model which best describes the experimental results is the
model of dense random packing of hard spheres (or .its relaxed versions).
The average coordination number for most of the metallic glasses is in the
range of 10-13. X-ray diffraction experiments have a resolution of about 20 A
and therefore can only verify a lack of crystalline ordering beyond this range.
Additional information on structure may be inferred from microscopic measurements such as NMR and Mössbauer, which are sensitive to the short range
atomic order surrounding the probe atom.

One of the difficulties in amorphous metallic solids is the problem of experimentally and definitively establishing a solid as amorphous, since all the analytical tools have limited resolution. The minimum criterion one often uses is that the amorphous solid must not reveal any crystallinity under x-ray or other measurements. Samples that show signs of partial crystallinity (such as the ones that are made under poor experimental conditions, those partially crystallized by annealing at high temperatures, radiation damaged crystalline solids) are not called amorphous.

All crystalline solids have definite structure and crystal directions. There are a few inequivalent sites which have a unique number and species of neighbors. The site symmetry, electrical field gradient, crystal field, magnetic exchange interaction and hyperfine interactions are all well defined. For amorphous metallic solids, the atoms are packed in a disordered manner so that there are a large number of inequivalent sites (or a "crystal" of a single unit cell with 10^{23} atoms). All directions are now equivalent. Most quantities which are well defined in crystalline solids are now replaced by distributions. Many unusual properties, some of which have no crystalline counterparts, are a direct consequence of these fundamental differences. In amorphous solids, crystal momentum is no longer a good quantum number. This fact affects most vividly x-ray and neutron diffraction measurements as well as excitations in the solids.

ADVANTAGES OF MOSSBAUER SPECTROSCOPY

The details of Mössbauer spectroscopy have been reviewed by G. Shenoy in this volume and references therein [13]. Here we will only describe briefly the advantages and shortcomings of Mössbauer spectroscopy when compared with other techniques in studying amorphous metallic solids.

A. Microscopic Measurement

Similar to other techniques featured in this volume, Mössbauer spectro-scopy measures the interaction of an isotope embedded in a solid. The inter-action energy is sensitively dependent on the charge distribution and chemical bondings of the probe atom with its surrounding atoms as well as the state of the solid. Since the absorber itself is always macroscopic (cross section of the order of 1 cm^2), all inequivalent sites of the probe atoms contribute to the spectrum. The sensitivity thus is somewhat compromised when the solid is an amorphous one. Fortunately the resolution is never completely wiped out so that useful information can still be obtained.
A useful feature of this technique is the ability to separate out the signals from various inequivalent sites in the solid. They could be inequi-valent sites of a solid (e.g. magnetic inequivalent sites in a ferrimagnet) or sites belonging to several different phases in the sample (e.g. crystalline phases when an amorphous solid is crystallized).

B. Sample Forms

Omitting attenuation of the γ-ray by all atoms in the sample, the important quantity is the number of resonant nuclei per unit area in the γ-ray beam. The physical shape of the absorber can be foil, ribbon, thin film, single crystal, powder, etc. Or course, certain directional information of the sample is lost if an as-prepared foil or single crystal is ground into powder.
Since the resonance cross sections of the favorable Mössbauer isotopes (e.g. ^{57}Fe) are large, only a small quantity of specimen is needed. For a non-enriched Fe containing absorber, typical thickness is of the order of 10 μm. For an

enriched [57]Fe sample, a few hundred Angstrom would suffice. The amount of materials needed is therefore orders of magnitude smaller than those for neutron diffraction, specific heat or vibrating sample magnetometer. For samples where large quantities are impractical (e.g. vapor-deposited samples), Mössbauer spectroscopy is particularly advantageous.

C. Measurements of Magnetic Solids

In conventional magnetization measurements, an external magnetic field is usually needed. Then it is necessary to ascertain effects due to the external field. Ideal geometric shapes of the specimen are often preferred to account for the demagnetizing effect. The strength of the external field as compared with the anisotropy field is crucial in aligning the moments. The measured magnetization includes contribution from all magnetic species in the sample. In Mössbauer spectrosocpy, the measurements can be performed without an external field, and can be applied to ferromagnets, antiferromagnets and spin glasses. Or course, an external field can sometimes provide additional information.

In conventional bulk magnetization measurements, both the change in direction and magnitude of the magnetization are measured. Saturation magnetization is obtained only when all the moments are aligned. In Mössbauer spectroscopy, the directional and the magnitude parts are separated. The former is specified by the line intensities whereas the latter is specified by the magnetic hyperfine splitting.

MOSSBAUER STUDIES OF AMORPHOUS METALLIC SOLIDS

Most of the results reported to date are on amorphous TM-M systems made by the liquid-quench techniques. These systems comprise the bulk of the results presented in the following.

A. Structure

In Figure 1 we show spectra of amorphous and crystallized $Fe_{75}P_{15}C_{10}$. Amorphous Fe-P-C system ($Fe_{80}P_{12.5}C_{7.5}$) was the first amorphous alloy examined by Mössbauer effect [4]. It is clear from Fig.1 that amorphous and crystalline alloys show very different spectra. In the crystallized sample, the sharp lines originated from α-Fe, Fe_3P and Fe_3C [14]. In Fig.2, spectra of two as-prepared samples of $Fe_{84}B_6C_{10}$ are shown. The sample (top spectrum) was first thought to be amorphous since it survived the initial x-ray test. A simple Mössbauer measurement quickly determined that this sample is not amorphous, since its spectrum is very different from that of amorphous $Fe_{84}B_6C_{10}$ (bottom spectrum). Indeed, subsequent lengthier x-ray measurements confirmed that this sample contained crystalline phases. In this regard, Mössbauer effect is reasonably sensitive. It may not certify the amorphicity of a sample, but it can conclude that the sample is not amorphous.

Spectra of amorphous $Fe_{75}P_{15}C_{10}$ and $Fe_{84}B_6C_{10}$ in Fig.1 and Fig.2 resemble the spectra of many amorphous metallic solids. They show broadened but well-defined six-line patterns. Aside from these six peaks, there is no hint of any other peaks. Computer simulations show that if the spectrum were made of as many as 25 sub-spectra, the resultant spectrum would produce 'gliches' that are not observed in actual spectrum with good and obtainable statistics[15].

One competing model of amorphous metallic solids is the micro-crystallite model in which the amorphous solid is made of small crystallites, tens of Angstroms in size which are not revealed by x-ray [12]. However, Mössbauer

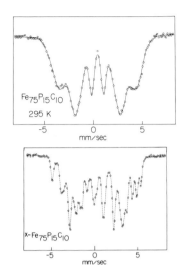

Fig.1 : Mössbauer spectra of amorphous and crystalline $Fe_{75}P_{15}C_{10}$.(Ref.14)

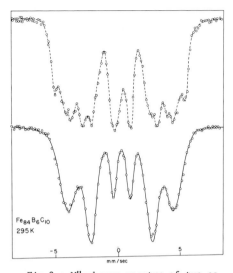

Fig.2 : Mössbauer spectra of two as-prepared $Fe_{84}B_6C_{10}$ alloys. The top spectrum is from a specimen which is partially crystalline. (Chien, unpublished)

measurements on ultra-fine particles show that for particles 50 - 100 A in size, the spectra at low temperatures (to suppress superparamagnetism) show almost all the features of the bulk materials [16]. One then argues that if amorphous Fe-B solids, for example, were made of micro-crystallites, their spectra would show sharp lines corresponding to these crystalline phases. There are only four crystalline compounds of Fe and B, with saturation fields of 340 kOe (α-Fe), 131 kOe (FeB), 250 kOe (Fe$_2$B) and 242, 284 and 305 kOe (Fe$_3$B) [17]. Instead, in the spectra of amorphous Fe_xB_{100-x} ($72 \leq x \leq 86$), or the field distributions [P(H)], one observes a smooth single-maximum P(H) encompassing all the field values of Fe$_2$B, Fe$_3$B and α-Fe, but no sharp lines corresponding to these individual values (Fig.3). Thus these amorphous solids are not made of micro-crystallites of these known crystalline compounds. Furthermore, as shown in Fig.4 and Fig.5, as the Fe concentration is varied, the Curie temperature, average hyperfine field and isomer shift vary continuously by substantial amounts. They are very different from those of crystalline Fe-B compounds [17].

B. Quadrupole Interaction

Although amorphous solids are isotropic macroscopically, the atomic site symmetry is in general lower than cubic. One then expects non-zero electric field gradients (EFG) and quadrupole interaction for all amorphous solids. The quadrupole interaction, however, can be unequivocally measured only in the paramagnetic states of the samples. For ^{57}Fe, the pure quadrupole spectrum is a doublet, whose separation is the averaged value of

$$\Delta E_Q = |e^2qQ|/2, \tag{1}$$

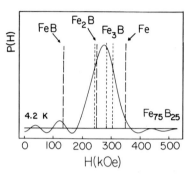

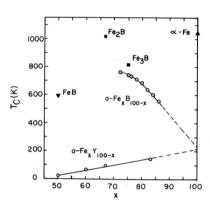

Fig.3 : Hyperfine field distribution of amorphous $Fe_{75}B_{25}$. The vertical dashed lines indicate the field values for crystalline FeB, Fe_2B, Fe_3B and α-Fe.(Ref.17)

Fig.4 : Curie temperatures of crystalline α-Fe, FeB, Fe_2B, Fe_3B and of amorphous Fe-B and Fe-Y as a function of Fe concentration. (Ref.17)

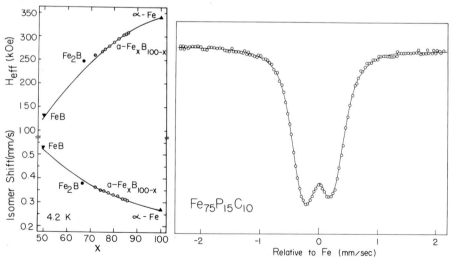

Fig.5 : Effective hyperfine field and isomer shift of crystalline α-Fe, FeB, Fe B and amorphous Fe-B at 4.2 K as a function of Fe content.(Ref.17)

Fig.6 : Mössbauer spectrum of amorphous $Fe_{75}P_{15}C_{10}$ at 640 K.

where eQ is the nuclear quadrupole moment and eq = V_{zz} is the principal EFG tensor. In Eq.(1), we have neglected the factor $\sqrt{1 + \eta^2/3}$ which is specified by the asymmetry parameter η of the EFG tensor [13].

We have measured over 100 different amorphous solids in their paramagnetic states. In _every_ case, we have observed a quadrupole doublet. One such example is shown in Fig. 6. The values of ΔE_Q range from 0.3 mm/sec to 0.9 mm/sec, and they decrease slightly with increasing temperature [18].

The distribution of EFG's is reasonably well defined so that one observes broadened but still resolved doublets. The two peaks of the doublets are slightly asymmetrical (see e.g. Fig. 6). This is likely to be caused by a distribution of isomer shifts. In the magnetically ordered state, since the magnetic hyperfine field direction is in general different from the principal axes of the EFG's, the effective quadrupole interaction is averaged out or nearly so [13]. The line positions of the magnetic hyperfine spectra are thus very symmetric about the centroid of the spectrum, or the isomer shift. Such examples are shown in Fig. 1 and Fig. 2.

An interesting but challenging problem is to relate the measured e^2qQ values to the local environment in an amorphous solid. Ideally then, one can use the quadrupole interaction to probe the amorphous structure on a microscopic level. However, the measured EFG's are "amplified" by the large Sternheimer and antishielding factors, whose values are, among other things, dependent on the atomic state [13]. Thus accurate values of these factors, precise atomic states, as well as reliable calculations of EFG are needed for such an endeavor.

C. Magnetic Ordering Structure

In the magnetically ordered state, the nucleus experiences a magnetic hyperfine field. The direction of the hyperfine field for most isotopes, including ^{57}Fe, is antiparallel to its atomic moment. Thus by determining the direction(s) of the magnetic hyperfine field, one can determine the magnetic ordering structure of the sample.

The six lines of ^{57}Fe magnetic hyperfine spectrum have an intensity ratio of 3 : b : 1, where b is the averaged values of $4\sin^2\theta/(1+\cos^2\theta)$ according to the directions of the magnetic moments in the sample, and θ is the angle between the magnetic moment and the γ-ray direction. The value of b varies from 0 (all moments are parallel to γ-ray) to 4 (all moments are perpendicular to the γ-ray). For a powder sample regardless of the magnetic ordering, or a spin glass sample, a value of b = 2 is observed. For amorphous solids, due to the distribution of hyperfine fields, the linewidths of the six lines are different. A more relevant ratio, therefore, is that of spectral areas.

By physically orienting the samples of as-prepared amorphous ferromagnetic glasses, Chien _et al_ showed that the moments are predominantly in the ribbon plane and along the ribbon long axes [19]. In fact, all the (Fe,Co)-rich TM-M glasses are very soft ferromagnets. This is a consequence of the fact that in amorphous solids all axes are equivalent. The coercivity of these samples is very small so that moments can be aligned by a small field. In the as-prepared ribbons, the shape anisotropy causes the moments to stay along the ribbon axes. Since many of these ferromagnetic glasses are magnetostrictive, the directions of the magnetic moments are susceptible to external stresses.

Heiman et al [20] and Nishihara et al [21] showed that, in sputtered Gd-Fe films, the Fe moments are perpendicular to the film (b = 0). This result, although useful for   magnetic bubble memory application, is rather puzzling and apparently complicated; pair ordering, bias, substrate condition, oxidation, etc. have been suggested as contributing factors.

Coey et al [22] used ^{161}Dy and ^{57}Fe to show that in Dy(De-Co)$_{3.4}$, the TM magnetic network is ferromagnetically ordered but the RE network is randomly oriented as in a spin glass state. This unusual magnetic ordering has no crystalline counterpart.

In other amorphous systems, such as Fe-Pd-Si [23] and Y-Fe [24], the spectra of the as-prepared samples show a value of b = 2 consistent with a spin glass ordering where the moments are locked into random directions. This conclusion is further affirmed from spectra taken under an external field. Indeed, ac susceptibility of these samples exhibits finite cusps, characteristic of spin glasses.

Finally, we discuss the possibility of amorphous antiferromagnets. It is clear that an uniaxial amorphous antiferromagnet cannot exist, for it violates the property that all directions are equivalent (either equally soft as in amorphous ferromagnets, or equally hard as in spin glasses). But an amorphous solid where the exchange interactions are largely antiferromagnetic can be envisioned. The actual ordering though should be the same as that of a spin glass. Mn-rich samples of the (Fe-Mn)-metalloid systems are some of the possibilities [25,26]. The discussions of these and other magnetic structures in amorphous solids have been given elsewhere [27]. Some of these were determined for the first time by Mössbauer spectroscopy.

D. Hyperfine Field Distribution

Mössbauer spectroscopy is among the few techniques capable of obtaining information of the magnetic hyperfine field distribution [P(H)] in an amorphous solid. The P(H) is obtained by deconvoluting the spectrum.

In liquid-quench (Fe-Co-Ni)-metalloid systems, the P(H) shows the following features[28,29,30]:
    a. All P(H)'s are single-maximum and well-defined. An example is shown in Fig. 3.
    b. The shape of P(H) is largely influenced by the metalloid content. For example, the P(H) in (Fe-Co-Ni)$_{80}$M$_{20}$ are rather symmetric when M = B. Asymmetry towards the low field side is observed if M contains P and C.
    c. Within each series (e.g. (Fe-Ni)$_{80}$P$_{14}$B$_6$, (Fe-Co)$_{80}$P$_{17}$Al$_3$), the values of P(H) are mainly determined by the transition-metal content. P(H) shifts to lower H values by about 25% from Fe$_{80}$M$_{20}$ to Ni$_{80}$M$_{20}$, whereas it shifts to higher values by about 5% from Fe$_{80}$M$_{20}$ to Co$_{80}$M$_{20}$. Thus in these Fe-containing systems, Fe nevers loses its moment.

It may be noted that in crystalline Fe-metalloid compounds (e.g. Fe$_3$C, FeB, Fe$_3$P, Fe$_3$C$_{0.9}$B$_{0.1}$, etc), a variety of values have been found for the Fe hyperfine field (H) and its moment (μ_{Fe}). However, the ratios of H/μ_{Fe} are all very close to 135 kOe/μ_B [31]. Thus in these crystalline compounds, the Fe hyperfine field is a good measure of its atomic moment. It is interesting to compare the values for amorphous Fe-metalloid solids. Since bulk measurements give only the average moment per Fe atom ($\bar{\mu}_{Fe}$), we can only obtain the ratio of $\bar{H}/\bar{\mu}_{Fe}$ where $\bar{H} = \int H\, P(H)\, dH$. In a number of systems, a ratio close to 135 kOe/μ_B has also been obtained [17,32]. This suggests that in these Fe-metalloid systems, the P(H) represents approximately a distribution of Fe moments with a width of about 0.5 μ_B.

For TM-M systems containing Mn, Cr, Mo, etc., the P(H) becomes much broader and far from being a well-defined, single-maximum function of H [25]. Complicated P(H) have also been observed in evaporated RE-TM systems (e.g. Y-Fe [24]), Fe-Hf, Fe-Zr, [15] etc. On the RE site of RE-TM systems (e.g. [161]Dy) the field distribution appears narrow: mainly because the hyperfine field itself is very large (a few MOe) [33].

E. Temperature Dependence of Hyperfine Field

One of the useful features of Mössbauer spectroscopy is that in favorable cases, the measured temperature dependence of the hyperfine field reflects that of the spontaneous magnetization (of a ferromagnet) or the subnetwork magnetization (of an antiferromagnet or ferrimagnet). Mössbauer results showed that in a large number of amorphous ferromagnets, the hyperfine field at low temperatures has a temperature dependence of [34,35]

$$H_{eff}(T)/H_{eff}(0) = 1 - B\ T^{3/2} - C\ T^{5/2} \ldots \ldots \qquad (2)$$

$$= 1 - B_{3/2}\ (T/T_c)^{3/2} - C_{5/2}\ (T/T_c)^{5/2} \ldots \qquad (3)$$

In amorphous Fe-metalloid systems where comparisons are available, the results obtained from Mössbauer are in good agreements with those of magnetization, ferromagnetic resonance and neutron diffraction [34,35,36]. Thus in these systems, it has been experimentally established that magnetization and hyperfine field have the same temperature dependence.

The $T^{3/2}$ dependence, indicative of spin-wave excitations, is shown in Fig. 7 for amorphous $Fe_{75}P_{15}C_{10}$. This may seem surprising at first since spin waves are usually discussed in the context of crystalline solids, where spin waves propogate along a regular array of spins. However, the excitable spin waves at low temperatures have long wavelengths, which are many atomic distances across. With these long wavelengths, the spin waves would not be sensitive to whether the spins are regularly arranged. The disordered arrangements of spins and therefore a distribution of exchange interactions in amorphous ferromagnets are reflected in other features; such as the density of spin-wave states which is weighted more on the long wavelength side. The B coefficient, and the reduced value $B_{3/2}$ as defined in Eq.(2) and (3), are much larger in amorphous solids than in the crsytalline solids[34].

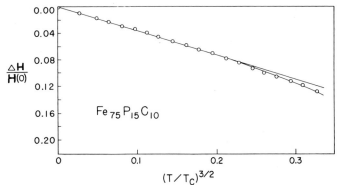

Fig.7 : Temperature dependence of magnetic hyperfine field of amorphous $Fe_{75}P_{15}C_{10}$ as a function of $T^{3/2}$. (Ref.14)

When the reduced magnetization, or reduced hyperfine field is plotted as a function of reduced temperature, as shown in Fig.8, the curve for an amorphous ferromagnet is much "flatter" when compared with those of the crystalline case. This general feature has been observed in many amorphous ferromagnets[17,37]. This is because in amorphous ferromagnets, there are distributions of exchange interactions instead of uniquely defined ones as in the crystalline case. The amount of departure of the magnetization curve gives a measure of the root mean square deviation of the exchange interactions.

F. Magnetic Ordering Temperature

The magnetic ordering temperature of a magnetic solid (ferromagnets, antiferro-magnets or even spin glasses) can be determined by the onset of the magnetic hyper-fine interaction. This is usually accomplished by measuring many spectra across the ordering temperature, or by using the method of constant velocity thermal scan [37].

In TM-M systems, the ordering temperatures have been studied in many pseudo-binary systems, such as $(Fe-Ni)_{80}P_{14}B_6$, $(Fe-Co)_{75}P_{16}B_6Al_3$, etc. In Fig. 9, the concentration dependence of the ordering temperatures of a few representative series are shown. To avoid complications, all of these series have the same metalloid compositions. Not surprisingly, all the Fe-Co samples are ferro-magnetic with high T_c, since both crystalline Fe and Co are strong ferromagnets. The (Fe-Ni) samples are ferromagnetic only on the Fe-rich side. For decreasing Fe content, T_C decreases, goes through a critical concentration. For samples with low Fe concentrations, they are spin glasses. No ordering temperature has been found for Ni-metalloid samples; they are essentially non-magnetic. Very different results are observed in (Fe-Mn)-metalloid alloys, and similarly in Fe-Mo and Fe-Cr systems [11,25]. As Mn is first introduced, T_C decreases drastically and linearly at a rate of about 25 K/at.% of Mn. At about $x = 0.7$, the decrease of T_C abruptly levels off, reminiscent of a percolation limit. The samples with high Mn content show spin glass behavior [26].

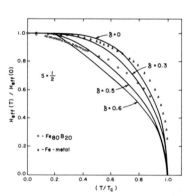

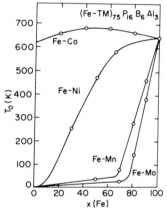

Fig.8 : Reduced hyperfine field versus reduced temperature for crystalline Fe (triangles) and amorphous $Fe_{80}B_{20}$ (circles). The solid curves are calcu-lated results using different distri-butions of exchange interactions. (Ref. 37)

Fig.9 : Concentration dependence of magnetic ordering temperatures of amorphous $(Fe-TM)_{75}P_{16}B_6Al_3$ for Fe-Co, Fe-Ni, Fe-Mn and Fe-Mo sereis as measured by Mössbauer spectroscopy. (Ref. 11, 25, 35)

G. Crystallization and Crystalline Phases

At high enough temperatures, the metastable amorphous state will transform into more stable states, namely the crystalline state. The time for the onset of crystallization is described by the equation

$$\tau = \tau_0 \exp (\Delta E/kT) \tag{4}$$

where ΔE is the activation energy and τ_0 is a characteristic time constant [38]. Thus the crystallization process depends not only on the temperature but also on the duration of heating. For this reason the crystallization temperature is usually measured under specified heating rates. For technological applications, crystallization is of course important since it dictates the usable temperature range and the intrinsic lifetime of the device.

Under usual experimental conditions, amorphous metallic solids, depending on the composition , may crystallize at tens of degree, or remain amorphous at about 1000 K. For most of the TM-M samples near the eutetic composition crystallization usually takes place between 550 K and 750 K. For these glasses ΔE is about a few eV. For these values of ΔE, and because of Eq.(4), a 1 sec lifetime at 700 K implies a lifetime of over 10^{11} years at room temperature !

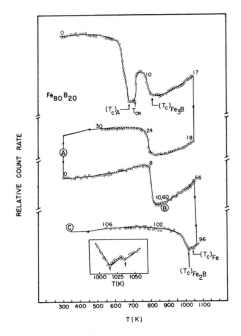

Fig.10 : Count rate measured at the centroid of $Fe_{80}B_{20}$ spectrum as a function of temperatures. Two complete cycles from 300 K to 1050 K are shown. The numbers indicate time in minutes. The arrows indicate the Curie temperatures of $Fe_{80}B_{20}$, Fe_3B, Fe_2B and α-Fe, and the crystallization temperature (T_{CR}). (Ref.37)

Since the data collection time for a conventional Mössbauer spectrum is long
(~hours) compared with those of differential scanning calorimetry, it is not
useful in studying the dynamics of crystallization. It is useful however in
detecting signs of partial crystallization and   crystalline phases. However,
when the thermal scan method is employed, counting time is reduced to seconds.
Crystallization can be followed under high heating rates [37]. One such example
is shown in Fig. 10. T_C of amorphous $Fe_{80}B_{20}(T_C)_A$ crystallization temperature
(T_{CR}) and T_C of the crystalline Fe_3B phase can be clearly located. The instability of Fe_3B is evident since it transforms into more stable Fe_2B and α-Fe
after heating at 1050 K for 30 min. It may be mentioned that if the sample of
$Fe_{80}B_{20}$ contains a small amount of carbon, the stability of the Fe_3B can be
greatly enhanced.

Returning now to Fig.1 where the spectrum of a crystallized $Fe_{75}P_{15}C_{10}$ is
shown. The presence of crystalline Fe_3P, Fe_3C and α-Fe can be easily
ascertained by the thermal scan method as shown in Fig.11, where the ordering
temperatures of these three phases are clearly indicated by the change in the
count rate.

H. Other Applications and Amorphous Systems
Several other applications have been omitted in this short discussion. Isomer
shift, which provides a probe of the atomic state, exhibits a variety of values
for metallic glasses. The systematic changes of isomer shift when the composition is varied suggest the possibilities of charge transfer between metal and
metalloid atoms and hybridization of the valence orbitals [17, 28, 29].
However, isomer shift in crystalline as well as amorphous alloys are complicated by various contributions. Better situations occur in oxide glasses and
rare earth systems. where in many cases, the valence states of the atoms can
be better characterized.

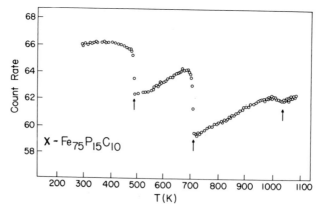

Fig.11 : Count rate versus temperatures or crystallized $Fe_{75}P_{15}C_{10}$. The arrows
indicate the Curie temperatures of Fe_3C (490 K), Fe_3P (715 K) and α-Fe (1040 K).
(Chien, unpublished)

The rf induced sideband effect utilizing Mössbauer spectroscopy has been used
to study magnetostriction of ferromagnetic glasses [37].   Mössbauer spectroscopy
using conversion electrons provides sensitive surface measurements of amorphous
samples as well we samples which are physically very thin [39].   Considerable
studies can be made in RE-containing systems using isotopes other than ^{161}Dy
and ^{151}Eu.   Amorphous Fe-Sn and Co-Sn systems offer the advantage of probing
all species microscopically [40].   Systems made up of fine particles of amorphous
alloys also offer interesting possibilities [41].

REFERENCES

* Work supported by the NSF Grant No. DMR79-10536.
1. P. Duwez, Trans. Amer. Soc. Metals 60, 607 (1967).
2. J. J. Gilman, Physics Today 28(#5), 46 (1975).
3. P. Chaudhari, B. C. Giessen and D. Turnbull, Scientific American 242(#4), 98 (1980).
4. C. C. Tsuei, G. Longworth and S. C. H. Lin, Phys. Rev. 170, 603 (1968).
5. C. R. Kurkjian, J. Non-cryst. Solids 3, 157 (1970).
6. S. P. Taneja, C. W. Kimball and J. C. Shaffer, Mössbauer Effect methodology 8, 41 (1973).
7. H. S. Chen and C. E. Miller, Rev. Sci. Instr. 41, 1237 (1970).
8. e.g. Thin Film Processes (J. Vossen and W. Kern ed.) (Academic, New York, 1980).
9. B. G. Bagley and D. Turnbull, J. Appl. Phys. 39, 5681 (1968).
10. F. E. Luborsky, J. J. Becker and R. O. McCary, IEEE Trans. MAG-11, 1644 (1975).
11. C. L. Chien and R. Hasegawa, J. Appl. Phys. 49, 1721 (1978).
12. G. S. Cargill, Solid State Physics 30, 27 (1974).
13. e.g. Chemical Applications of Mössbauer Spectroscopy, (V. I. Goldanskii and R. H. Herber ed.) (academic, New York, 1968).
14. C. L. Chien and R. Hasegawa, J. de Physique (C-6) 759 (1976).
15. C. L. Chien, unpublished.
16. T. K. McNab, R. A. Fox and A. J. F. Boyle, J. Appl. Phys. 39, 5703 (1968).
17. C. L. Chien, D. Musser, E. M. Gyorgy, R. C. Sherwood, H. S. Chen, F. E. Luborsky and J. L. Walter, Phys. Rev. B20, 283 (1979).
18. C. L. Chien, Hyperfine Int. 4, 869 (1978).
19. C. L. Chien and R. Hasegawa, J. Appl. Phys. 47, 2234 (1976).
20. N. Heiman, K. Lee and R. I. Potter, AIP Conf. Proc. 29, 130 (1975).
21. Y. Nishihara, T. Katayama, Y. Yamaguchi, S. Ogawa and T. Tsushima, Japan J. Appl. Phys. 17, 1083 (1978).
22. J. M. D. Coey, J. Chappert, J. P. Rebouillat and T. S. Wang, Phys. Rev. Lett. 36, 196 (1976).
23. C. L. Chien, Phys. Lett. 68A, 394 (1978).
24. J. Chappert, J. de Physique (C2) 107 (1979).
25. C. L. Chien, J. H. Hsu, J. P. Stokes, A. N. Bloch and H. S. Chen, J. Appl. Phys. 50, 7647 (1979).
26. M. B. Salamon, K. V. Rao and H. S. Chen, Phys. Rev. Lett. 44, 596 (1980).
27. J. M. D. Coey, J. Appl. Phys. 49, 1646 (1978).
28. C. L. Chien, D. Musser, F. E. Luborsky and J. L. Walter, J. Phys. F8, 2407 (19780.
29. C. L. Chien, D. Musser, F. E. Luborsky and J. L. Walter, Solid State Commun. 28, 645 (1978).
30. I. Vincze, Solid State Commun. 25, 689 (1978)
31. H. Bernas, I. A. Campbell and R. Fruchart, J. Phys. Chem. Solids 28, 17 (1967).

32. D. Musser, C. L. Chien, F. E. Luborsky and J. L. Walter, J. Appl. Phys. $\underline{50}$, 1571 (1979).
33. D. W. Forester, R. Abbundi, R. Segnan and D. Sweger, AIP Conf. Proc. $\underline{24}$, 115 (1975).
34. C. L. Chien and R. Hasegawa, Phys. Rev. $\underline{B16}$, 2115 (1977).
35. R. J. Birgeneau, J. A. Tarvin, G. Shirane, E. M. Gyorgy, R. C. Sherwood, H. S. Chen and C. L. Chien, Phys. Rev. $\underline{B18}$, 2192 (1978).
36. S. M. Bhagat, M. L. Spano and K. V. Rao, J. Appl. Phys. $\underline{50}$, 1580 (1979).
37. C. L. Chien, Phys. Rev. $\underline{B18}$, 1003 (1978).
38. F. E. Luborsky, Materials Science and Engineering $\underline{28}$, 139 (1977).
39. O. Massenet and H. Daver, Solid State Commun. $\underline{25}$, 917 (1978).
40. B. Rodmacq, M. Piecuch, G. Marchal, Ph. Mangin and C. Janot, IEEE Trans. $\underline{MAG-14}$, 841 (1978).
41. A. E. Berkowitz, private communication.

COAL AND THE MOESSBAUER EFFECT*

PEDRO A. MONTANO
Department of Physics, West Virginia University, Morgantown,West Virginia 26506

ABSTRACT

The presence of iron bearing minerals in coal makes the
Moessbauer effect extremely useful as an analytical tool.
In this paper we present a general review of the use of
Moessbauer spectroscopy in coal research. We discuss a
simple method to identify iron bearing minerals in coal.
Several researchers have used the Moessbauer effect to
study the transformations of iron minerals during coal
processing. We have studied the stoichiometries of iron
sulfides produced during coal liquefaction. We found
drastic changes in the stoichiometry of the pyrrhotites
as a function of the total sulfur content of the coal,
and the reaction conditions. Some in situ experiments
are reported in this paper. We have observed evidence of
reactions between the coal components and the iron sulfides
at high temperatures (400° C). The possibility of studying
in situ reactions makes the Moessbauer effect a unique
analytical tool.

INTRODUCTION

The diminishing supplies of domestic oil and the high cost of imported fuels
have created great interest in the use of coal. By comparison to other energy
resources coal represents 85% of the total reserves in fossil fuels in the
United States [1]. However, direct coal utilization is not always feasible due
to strong environmental constraints. The use of coal for transportation requires its conversion to a liquid fuel. This conversion method was already
known in pre-war Germany, but the high cost of this process hinders its use in
the present economic structure. Considerable research is necessary to obtain
a better understanding of the coal conversion process, as well as to improve
existing technologies of coal conversion.

In order to completely characterize a coal a careful study of the organic
and inroganic components is necessary. From the materials science point of view
coal is a composite material with a very complex organic matrix. The carbon
structure of coal can be viewed as consisting of hydroaromatic structures with
aromaticity increasing from low-rank to high-rank coals [2]. The organic part
of coal contains also sulfur, oxygen, nitrogen in variable amounts depending
on the coal. There is not a single type of coal but many coals which can
differ considerably in both components, organic and inorganic. In order to
characterize a coal completely a coordinated study with as many techniques as
possible is necessary.

In recent years resonance techniques have become very powerful in the study
of the different organic components appearing in coal [3]. The components
of coal are of central importance from the point of view of Moessbauer
spectroscopy. The first study of coal using Moessbauer spectroscopy was carried out by Lefelhocz et al. [4] in 1967. After a research gap of about 10
years renewed interest started in the use of Moessbauer spectroscopy to coal
research [5,6,7]. This resurgence in the use of Moessbauer spectroscopy is as-

sociated to the great importance of iron as a major constituent of the mineral matter. In all the Moessbauer studies of U.S.A. coals, the iron compounds identified were always inorganic in origin. However, recently Cashion and co-workers [8] have observed a Moessbauer spectrum in Australian coals that can be identified as due to organically bound iron.

The major emphasis in the application of Moessbauer spectroscopy to coal research has been on the characterization of the iron bearing compounds. More important applications lie ahead in the study of mineral transformations during coal utilization. The use of Moessbauer spectroscopy to characterize liquefaction residues, char, coke, ashes, etc. opens new areas of use for this highly sensitive resonance technique.

In this paper we will try to give a review of some of the applications of Moessbauer spectroscopy to coal research. We do not expect this review to be complete, since each of the areas of research covered here is progressing at a very fast rate. However, we expect to give the reader a general picture of the complexity of the study of coal and the areas of research where Moessbauer spectroscopy can be useful.

IRON-BEARING MINERALS

In most of the U.S.A. coals the iron sulfides are the dominant iron-bearing minerals [9]. The iron disulfide, pyrite, is the most abundant. Marcasite, the orthorhombic form of FeS_2, is less abundant. They can be readily identified by XRD when their abundance in coal is 1% or more. The effect of having a mixture of pyrite and marcasite in the Moessbauer spectrum is to give a more negative isomer shift and a slight asymmetry of the pyrite doublet. The iron disulfides appear in various morphological forms: framboids, euhedral crystals, and large pieces of massive pyrite usually of the order of 100 μ in mean diameter [10]. Other iron sulfides are rare in coal. However, we have detected the presence of pyrrhotites (less than 0.1% by total weight of the coal) in several fresh coals. Although this is not a common occurrence, the presence of other iron sulfides is quite important in selecting the coal for utilization. Other sulfides like sphalerite, mackinawite, greigite, smythite, chalcopyrite, troilite, and arsenopyrite were not detectable in our studies. Some of these minerals may be detected by other techniques, like scanning electron microscopy (SEM).

Another group of iron bearing minerals of considerable importance in coal are the clays. The clays are more abundant in coal than the sulfides. The principal clays in coal are illite, kaolinite, chlorite, montmorillonite, mixed clays, etc. [9]. Iron appears as a substitutional impurity in the clays, usually replacing aluminum. The iron ions in the clays can be both divalent and/or trivalent; in many cases more than one crystallographic site is possible [11]. The amount of iron in kaolinite is very small, and consequently, the detection of this mineral using Moessbauer spectroscopy is difficult. There is a great diversity of Moessbauer spectra for the clays, many of them indistinguishable from each other at room temperature [12]. Consequently, in many cases measurements at low temperatures and in the presence of an external magnetic field are necessary to obtain a more clear identification of a clay.

A very important group of iron bearing minerals in coal are the carbonates, siderite ($FeCO_3$) and ankerite [$Ca(Fe,Mg,Mn)(CaCO_3)_2$]. Siderite can appear in coal as nodules and lenticular masses. Siderite and pyrite usually do not occur together in significant amounts in coal. Siderite is easily identified from its Moessbauer spectrum.

Another group of iron bearing minerals appearing in coal are the sulfates. The most important sulfates are szomolnokite ($FeSO_4 \cdot H_2O$), rozenite ($FeSO_4 \cdot 4H_2O$), melanterite ($FeSO_4 \cdot 7H_2O$), coquimbite ($Fe(SO_4)_3 \cdot 9H_2O$), roemerite ($FeSO_4 \cdot Fe(SO_4)_3 12H_2O$), and the jarosites. The presence of sulfates in coal is almost always an indication of weathering. Some of the trivalent sulfates detected by XRD of low temperature ashes (LTA) of coal are generated during the LTA process [13].

Other iron minerals like oxides and hydroxides are rare in deep-mined coal. However, FeOOH is a common constituent of weathered strip-mine coal and its detection by Moessbauer spectroscopy has been suggested as a possible method of detecting excessive coal oxidation by weathering [14].

APPLICATION OF MOESSBAUER SPECTROSCOPY TO COAL CHARACTERIZATION

Moessbauer parameters. The Moessbauer parameters give information about the local environment and chemical bonds of the Moessbauer atom. The isomer shift (IS) gives unique information on the valence states of iron, especially for high spin Fe^{2+} and Fe^{3+}. Besides the IS there exists a shift of the Moessbauer lines due to the second-order Doppler effect. The IS values reported in the literature include the second order Doppler shift. The IS values must always be given with respect to a standard reference material, for example for ^{57}Fe metallic iron is a common standard. Another important parameter is the quadrupole splitting (QS). The QS is produced by the interaction between the nuclear quadrupole moment and the electric field gradient (EFG) at the nucleus. It gives information on the coordination number and crystal field parameters of the Moessbauer atom. The last of the parameters related to the hyperfine interactions is the magnetic hyperfine splitting. The magnetic hyperfine splitting arises from the interaction of the nuclear dipole moment with the internal magnetic field of the atom. For many iron compounds magnetic hyperfine splitting can appear at low temperatures, even in the absence of long range order, this is caused by slow spin relaxation. One can induce a magnetic hyperfine splitting at low temperatures by applying a large external magnetic field. This is a useful technique for the study of the electronic ground state of iron ions in minerals. A more detailed description of the Moessbauer parameters can be found in the paper by G.K. Shenoy appearing in this Proceedings.

The Debye-Waller factor of the Moessbauer atom in a solid is a very important parameter. In order to use the Moessbauer effect as a quantitative analytical tool, one has to know the Debye-Waller factor of the mineral under study. This parameter is not always easy to determine.

Methodology for analyzing minerals in coal. The identification of iron bearing minerals in coal required a very careful analysis of the spectrum. Care must be taken that the samples under study are representative of the coal seam. The coal samples we have studied were collected by strictly following ASTM Procedure D2013-72. Several samples from the same seam should be analyzed in order to check for consistency of the results. In our studies we usually used coal ground to 200 mesh. The samples were either mounted in lucite containers that were hermetically sealed or as pressed pellets. The average surface densities were usually between 150 to 300 mg/cm^2 of coal.

One has to be careful in fitting the experimental points. In general constraints are necessary in order to limit the number of free parameters. To identify the iron minerals one has to compare the parameters obtained from the fit, with those of standard minerals. A serious problem is the overlap of the Moessbauer spectral lines for different compounds. Several clay lines are difficult to distinguish from each other; they also overlap with the ferrous sulfates. The trivalent sulfate lines partially overlap the pyrite spectrum. If any trivalent iron is present in the clays, its identification by Moessbauer spectroscopy is very difficult at room temperature. In many cases we find it necessary to either chemically treat the sample (e.g., with diluted HCl) to eliminate the sulfates or carry out low temperature measurements (in some cases in the presence of an external magnetic field) [15]. The low temperature measurements facilitate the identification of the minerals due to the higher resolution obtained when the magnetic hyperfine splitting is present. In all

174

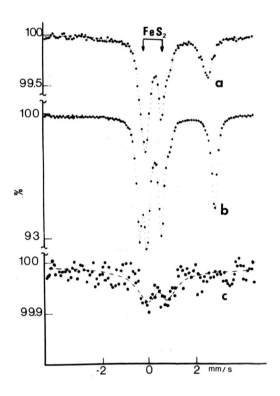

Fig. 1. Moessbauer spectra of (a) Pittsburgh coal (Pa), pyrite, szomolnokite, jarosite and illite identified in the sample; (b) IL 6 coal, pyrite, rozenite and jarosite present; (c) North Dakota lignite, kaolinite present, containing a very small amount of iron.

cases comparison of the spectra with those of standard minerals is necessary. It is a good practice to use other techniques like x-ray diffraction (XRD) and scanning electron microscopy (SEM) to complement the Moessbauer measurements. In Figure 1 the Moessbauer spectra are shown of three coals from different regions in the U.S.A. The mineral content of the three coals differs considerably. The North Dakota lignite has the smallest amount of iron. The spectrum given in Figure 1c was identified as that of kaolinite. These three spectra should be taken only as examples, since for a careful study of the minerals in a coal seam the mine of origin should be specified.

Quantitative analysis. The Moessbauer effect was used by several groups to analyze the amount of pyritic sulfur [5,6,7]. The use of this technique for quantitative analysis presents several problems. First, the Debye-Waller factor of pyrite has to be known accurately. Second, granularity, texture and crystallinity of the pyrite present in the coal are all factors that can create difficulties in analyzing the Moessbauer spectral area. The non-resonant radiation background should be known with good precision. We have observed variations between 10 and 30% in the background counting rate for samples of coals with the same weight per unit area. If all the above factors are taken into consideration one can obtain reliable information about the amount of pyritic sulfur using the Moessbauer effect. We observed that good agreement is obtained between our method of analysis and the standard chemical procedures (ASTM D2492) [16].

Other techniques. The purpose of any study of the mineral matter in coal is to identify all the inorganic components. Consequently, the Moessbauer effect is limited to the investigation of the iron-bearing minerals in coal. Because of this limitation the use of other spectroscopic techniques is necessary in order to identify all the minerals. XRD is the most commonly used analytical tool for mineral characterization. The only disadvantage of XRD is the need to use it on LTA and not on fresh coal. We observed that the LTA process alters the composition of the sulfates and creates new species [5,13]. For very small amounts of iron minerals, the Moessbauer effect is more sensitive than XRD. SEM is a very powerful tool when combined with energy dispersive analysis [17]. It allows the identification of many trace minerals in coal. Infra-red spectroscopy is widely used in studying minerals in coal, in particular clays [18].We find that all the above techniques are complementary.

TRANSFORMATION OF IRON BEARING MINERALS IN COAL

Perhaps the most important application of Moessbauer spectroscopy to coal research is in the study of the transformations in the iron minerals during coal utilization. In this area of research the Moessbauer effect presents great advantages over other techniques, due to the possibility of carrying out in situ measurements. In the following paragraphs some examples are given of the use of Moessbauer spectroscopy to study mineral transformations in coal.

Acid mine drainage. One of the major environmental problems associated with coal mining is the acid mine drainage. It is accepted that FeS_2 is the main precursor for acid formation [19]. When coal and the associated strata are exposed to the atmosphere, the FeS_2 is oxidized to a series of hydrous iron sulfates which subsequently dissolve in water and produce the acid mine drainage. However, there is a lack of correlation between the amount of FeS_2 and the amount of acid subsequently produced. The Moessbauer effect has been applied successfully to study the acid producing capabilities of a coal seam. The Moessbauer effect becomes a very powerful tool for the detection of small amounts of sulfates that escape other analytical techniques (like XRD). In Figure 2 the Moessbauer spectra are shown from a pyrite lens in a Waynesburg coal seam, in a Dipple and Dipple surface mine in West Virginia [20]. The top spectrum (a) is representative of the acid producing part of the lens. One observes the presence of szomolnokite ($FeSO_4 \cdot H_2O$). In the bottom spectrum (b) the unreactive part of the lens is shown; one notices the absence of szomolnokite. It is to be noted that the total amount of pyritic sulfur was larger for the unreactive part (6.21%) than for the reactive part (2.68%).

Combustion. The most common usage of coal is in direct combustion to generate electricity. One of the major problems facing the direct utilization of coal to generate electricity is associated with the pollution of the environment. The Moessbauer effect becomes very useful to study the different cleaning processes since pyrite is usually the most detrimental inorganic component of coal. The Moessbauer effect is also very useful in studying the residual ashes of coal. In the standard method (ASTM D3174), coal is slowly heated in air to 750° C for 24 hours. Huffman and Huggins have carried out extensive studies of the coal ashes [21]. They observed the transformation of

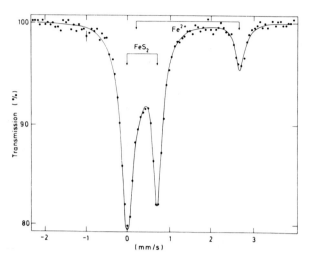

Fig. 2a. Moessbauer spectrum at room temperature of the reactive portion [20].

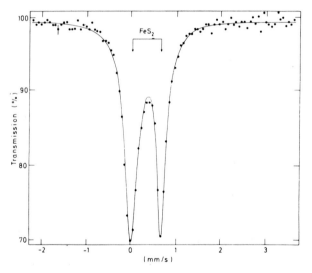

Fig. 2b. Moessbauer spectrum at room temperature of the unreactive portion [20].

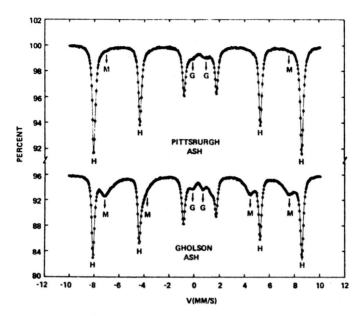

Fig. 3. Moessbauer spectra of ashes prepared from coals from Pittsburgh seam, Pa., and from Gholson seam, Ala. by heating in air to 750° C, for up to 24 hours. Arrows labeled H, M and G indicate Moessbauer peaks due to α- Fe_2O_3 (hematite), $Mg_xFe_{3-x}O_4$ (magnesioferrite) and paramagnetic ferric phases including glass, oxides and silicate phases [21].

$FeS_2 \rightarrow \alpha$- Fe_2O_3, of $FeCO_3 \rightarrow \alpha$- Fe_2O_3, of ankerite $\rightarrow Mg_xFe_{3-x}O_4$ and of Fe^{2+} in clays to Fe^{3+} (in glass). The sulfates are always oxidized to hematite. Two Moessbauer spectra from different coals ashes [21] are shown in Figure 3. Under more realistic conditions the combustion temperatures are usually higher than 750° C. The composition of the ash depends very much on the original minerals and the cooling conditions of the slags. In Figure 4 the Moessbauer spectra for two rapidly quenched slags are shown [21].

Coke. Coke is produced by exposing a properly chosen and prepared coal or blend of coals to sufficiently high temperatures for a long period of time in the absence of air. There is not much knowledge about the process taking place during coking. This lack of knowledge has now become a critical problem. The presence of iron minerals makes the Moessbauer effect a unique tool for the study of coke. Huffman and Huggins have studied extensively the transformation of iron minerals during coking [21,22]. They detected the presence of iron sulfides, FeS and $Fe_{1-x}S$, hematite, magnetite, α- iron, and the characteristic ferrous glass spectrum due to the transformations of the clays. It should be emphasized that much remains to be done in this area of research and that considerable use of Moessbauer spectroscopy should be expected.

178

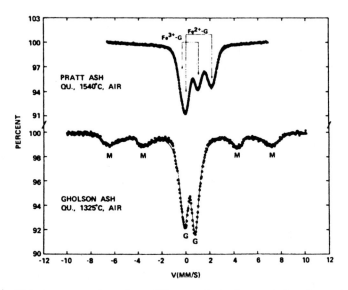

Fig. 4. Moessbauer spectra of rapidly quenched slags arising from mineral matter in coals from the Gholson and Pratt seams, Ala. Spectra show that all the
iron is in glass for the slag from the Pratt coal,whereas substituted and
superparamagnetic magnetite is formed on quenching the slag from the Gholson
coal [21].

Coal gasification.   There are two stages in coal gasification. The first
is the direct coal gasification; in general up to 50% of the original mass can
be gasified. The second stage is the char gasification. The mineral matter
and trace elements can have strong catalytic effects on the gasification process. There is only a small amount of work reported in the literature where
the Moessbauer effect was applied to study coal chars. No attempt to study
systematically the transformations in the iron minerals during gasification
has been carried out. This area of research remains open to the application of
Moessbauer spectroscopy. The possibility of using the Moessbauer effect for
in situ measurements during coal gasification should be pursued.
    Coal liquefaction.   There are considerable reserves of fossil fuels, but
there is a shortage of desirable fuels. Consequently, the production of liquid
fuels from coal becomes a necessity. Liquefaction of coal has been a known process for a number of years: to convert coal to a liquid fuel, one has to
hydrogenate the coal, remove the mineral matter and the hetero atoms (sulfur,
nitrogen, oxygen), and reduce the molecular weight of the products.
    There are two types of liquefaction processes: direct and indirect. In the
direct liquefaction of coal, process derived solvent and hydrogen are mixed
and heated to temperatures in the range of 400 to 450° C for 15 to 60 minutes
in a pressure range of 1500-3000 psi H_2. Currently, several direct liquefaction
processes are under study: the SRC-II process (Solvent Refined Coal) where
the vacuum bottoms are recycled [23]; the EXXON donor solvent, where the sol-

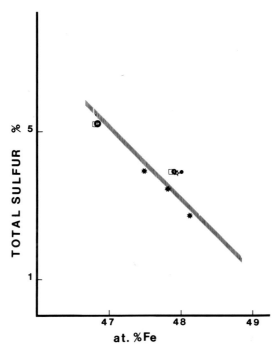

Fig. 5. Total sulfur content of the coal versus atomic % Fe in $Fe_{1-x}S$. The stars, squares and dots represent different liquefaction runs.

vent is catalytically hydrogenated prior to being mixed with the coal [24]; and the H-Coal (Hydrocarbon Research, Inc.) process, where a Co-Mo catalyst is used to improve liquid properties. In all the above processes one has to consider the possible catalytic role of the mineral matter.

In indirect liquefaction the coal is first gasified and syngas is produced (mixtures of H_2 and CO). The syngas is cleaned and catalytically shifted via the water-gas shift reaction to desired H_2 to CO ratio. This mixture is later upgraded to synthetic fuels using the Fischer-Tropsch synthesis. Moessbauer spectroscopy has been applied recently to characterize catalysts used in the synthesis reaction [25].

The Moessbauer effect has been used by several groups to study liquefaction residues [26,27,28]. Under liquefaction conditions FeS_2 is transformed to $Fe_{1-x}S$. Moessbauer spectroscopy has been also used to find out the stoichiometry of the pyrrhotites present in the liquefaction residues [28]. A number of different pyrrhotites can be formed, depending on the reaction temperature and H_2S partial pressure. We have observed that the most important factor controlling the stoichiometry of the pyrrhotite formed from the decomposition of pyrite is the total sulfur present in the coal. This is related to an increase in the partial pressure of H_2S in the reactor during the liquefaction reactions. In Figure 5 the total amount of sulfur is plotted vs. the atomic % iron

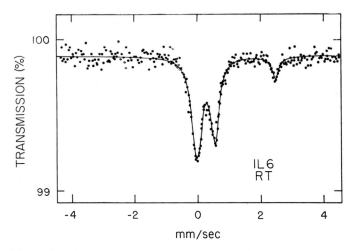

Fig. 6a.IL 6 coal used for the in situ studies (4.05% total sulfur content 1.85 pyritic sulfur). Pyrite and szomolnokite are identified in this coal.

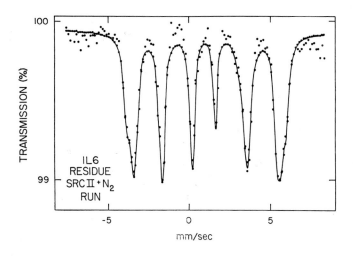

Fig. 6b. Moessbauer spectrum of IL 6 coal after a high temperature run (420° C) in the presence of solvent (SRC-II heavy distillates 850^{+} F).

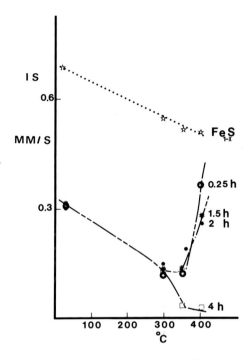

Fig. 7. Isomer shift of the iron species in coal versus temperature (4h means four hours at that temperature). All the IS are given relative to α- iron.

in $Fe_{1-x}S$ for various runs. One observes that the amount of iron in $Fe_{1-x}S$ is reduced when the total amount of S increases. The combination of sulfur and pyrrhotite may be the most important element in any catalytic process associated with the mineral matter present in coal.

The presence of H_2 is not a necessary condition to fully convert pyrite to pyrrhotite. We carried out in situ measurements in a specially designed reactor using an IL 6 coal (Fig. 6) and SRC-II heavy distillates (850+ F) as a solvent under high nitrogen pressure (200 psi). Full transformation of pyrite to pyrrhotite took place under such conditions [29]. The spectrum of the residue of this run is shown in Figure 6; the pyrrhotite formed has 48.0 at. % iron.

We have carried out in situ studies of coal liquefaction. In the liquefaction runs the H_2 pressure was 250 psi and the solvent SRC-II heavy distillates 850+ F (solvent to coal ratio of 2:1). In this study we observed strong evidences of reactions at high temperatures between the coal components and the pyrrhotites formed [29]. In Figure 7 the effect of reaction time and temperature is shown for four different runs. The IS becomes more negative and the lines much broader as the residence time increases. For comparison the IS for a typical pyrrhotite is shown in Figure 7. One can easily observe the great difference in the IS values for the two cases. It is evident from this study

that the pyrrhotites formed during the liquefaction of coal are actively par-
ticipating in the process. It is noted that by cooling to RT all the above
effects disappear. A study of mixtures of pyrrhotites and model compounds has
been started in order to obtain more information about the reactions taking
place at high temperatures. One can expect that the Moessbauer effect will
play a central role in the study of liquefaction residues and elucidate the
importance of the pyrrhotites in the liquefaction process.

SUMMARY

Some applications of Moessbauer spectroscopy to coal research have been
presented in this paper. A limited number of applications have been selected
as examples and no exhaustive review of the subject was intended. In such a
fast-moving area one is always afraid to be outdated by the time the publication finally appears.

ACKNOWLEDGMENTS

I wish to acknowledge the invaluable help of Doctors P. Vaishnava and
A. Bommannavar as well as the technical assistance of Mr. V. Shah.

* Work supported by U.S. Department of Energy and Energy Research Center of
West Virginia University.

REFERENCES

1. H.R. Linden, "The Robert A. Welch Foundation Conferences on Chemical Research. XXII. Chemistry of Future Energy Resources," Houston, Texas,p. 153
(1978).

2. I. Wender, Catal. Rev. Science Eng. 14, 97 (1976).

3. H.L. Ketcofsky, Appl. Spectros. 31, 116 (1977).

4. J.F. Lefelhocz, R.A. Friedel, and T.P. Kohman, Geochim. Cosmochim. Acta. 31,
2261 (1967).

5. P.A. Montano, FUEL 56, 397 (1977).

6. L.M. Levinson and I.S. Jacobs, FUEL 56, 453 (1977).

7. G.P. Huffman and F.E. Huggins, FUEL 57, 597 (1978).

8. J.D. Cashion, P.E. Clark, P. Cook, F.P. Larkins, M. Marshall, B. Maguire,
L.T. Kiss, Proceedings from "Nuclear and Electron Resonance Spectroscopies
Applied to Materials Science," ed. by E.N. Kaufmann and G.K. Shenoy (1980).

9. H.J. Gluskoter and C.P. Rao, Illinois State Geological Survey, Cir 476
(1973).

10. W.C. Grady,"Petrography of West Virginia Coals," in Carboniferous Coal,
Eastern Section of the American Association of Petroleum Geol. (1979).

11. G. Brown, "X-ray Identification and Crystal Structure of Clay Minerals,"
Mineralogical Society,London, England (1961).

12. J.M.D. Coey, Proc. Int. Conference on Moessbauer Spectroscopy VI, p. 33, Cracow (1975).

13. P.A. Montano, "Characterics of Iron-Bearing Minerals in Coal," ch. 22 in Coal Structure, ed. by M. Gorbaty (1980).

14. F.E. Huggins, G.P. Huffman, D.A. Kosmack and D.E. Lowenhaupt, Int. J. of Coal Geology 1, 75 (1980).

15. P. Russell and P.A. Montano, J. of Applied Physics 49 (3), 1573 (1978).

16. P.A. Montano in "Recent Chemical Applications of Moessbauer Spectroscopy," ed. by J. Stevens and G.K. Shenoy (1980).

17. R.B. Finkelman and R.W. Stanton, FUEL 57, 763 (1978);
F.E. Huggins, D.A. Kosmack, G.P. Huffman and R.J. Lee, Scanning Electron Microscopy (1980) I, SEM, Inc., O'Hare, IL, p. 531.

18. H.L. Retcofsky and R.A. Friedel, FUEL 47, 487 (1968).

19. H.L. Barnes and S.R. Romberger, J. Water Pollut. Contr. Fed. 40, 371 (1968).

20. A.H. Stiller, J.J. Renton, P.A. Montano, and P.E. Russell, FUEL 57, 447 (1978).

21. F.E. Huggins and G.P. Huffman, "Analytical Methods for Coal and Coal Products," 13, Academic Press (1980).

22. G.P. Huffman and F.E. Huggins in Recent Chemical Applications of Moessbauer Spectroscopy, ed. by J. Stevens and G.K. Shenoy (1980).

23. B.K. Schmid and D.M. Jackson, Coal Processing Technology 5, 146 (1979).

24. W.R. Epperly and J.W. Taunton, Coal Processing Technology 5, 28 (1979).

25. E.E. Unmuth, L.H. Schwartz, and J.G. Butt, Journal of Catalysis 61, 242 (1980).

26. B. Keisch, G.A. Gibbon, and S. Akhtar, ACS, FUEL Preprints, 263 (1978).

27. I.S. Jacobs, L.M. Levinson and H.R. Hart, J. of Applied Phys. 49(3), 1775 (1978).

28. P.A. Montano and B. Granoff, FUEL 59, 214 (1980).

29. A. Bommannavar, V. Shah, and P.A. Montano, To be published (1981).

Spin Precession

SOME APPLICATIONS OF SPIN PRECESSION METHODS TO PROBLEMS IN MATERIALS SCIENCE

E. N. KAUFMANN
Bell Laboratories, Murray Hill, New Jersey 07974 USA

ABSTRACT

Nuclear and electron resonance and the Mössbauer effect are techniques which observe the interaction of moments with fields directly in the energy domain. An energy splitting, however, also implies the precession of the moment in the field. When a means exists to determine the orientation of the moment then the precession can be observed in the time domain. The direction of radiation emitted in a nuclear, muonic or atomic decay is correlated to the direction of corresponding moments (spins) and can thus act as a detector of spin precession. In the language of nuclear, muonic and atomic physics, these methods are called perturbed angular correlations (PAC), muonic spin rotation, and quantum beats, respectively. Below, these methods will be illustrated by displaying some examples of the application of perturbed angular correlations to a variety of materials systems.

INTRODUCTION

The basis of all spin precession methods [1] is the detection of nuclear radiation emitted by a decaying nucleus or in the case of μSR a decaying muon whose spin is not randomly oriented with respect to the laboratory frame. A classical analogy can be drawn to the radiation pattern generated by a simple antenna. Antennae of various shapes (dipole, quadrupole, etc.) will radiate in a nonisotropic fashion. The nonisotropic radiation pattern is, naturally, fixed to the orientation of the antenna itself. In a similar fashion the radiation pattern from a decaying nucleus whose spin is not isotropically oriented will be anisotropic and will be oriented in space according to the orientation of the nuclear spin. The radiation direction is, of course, in the quantum case coupled to the nuclear spin because of the conservation of angular momentum. Spin precession methods are therefore based on the production in the first instance of a nonisotropic distribution of directions of nuclear spins followed by the detection of emitted radiation.

An anisotropic spin distribution can be achieved in several ways. For example, a nucleus which is cooled to temperatures of a few milliKelvins in the presence of an external field (either an electric field gradient or magnetic field) will minimize its interaction energy with the field by orienting its moment (and therefore its spin) with respect to the field. This is known as the nuclear orientation method [1]. The anisotropy of subsequent decay radiation indicates the degree to which the nucleus was oriented and therefore the strength of the interaction, i.e. a product of moment and field. This is not a spin precession method but is an energy-based method. It is the energy splitting of the nuclear sublevels by the field which causes the orientation to occur. Thus, although the anisotropic nuclear radiation is used as a detection device this method properly belongs in the category of other energy methods, such as nuclear resonance or the Mössbauer effect.

Nuclei in excited states can be found oriented as the result of a nuclear reaction. Because the incident particle from an accelerator beam has a definite direction in space, the spin of the excited state will be correlated with that direction. Therefore, subsequent decay radiation will be anisotropic with respect to the direction of the particle beam. Measurement of this distribution and its precession in a field is termed a perturbed angular distribution experiment. In this case all nuclei created have an orientation which is on the average nonuniform. In a third approach, an ensemble of nuclei, i.e. a subset of the nuclei in a radioactive sample, can be "oriented" through the observation of radiations in cascade. The observation of the first radiation selects an ensemble of nuclei whose spins are anisotropically

A. NO PERTURBATION:

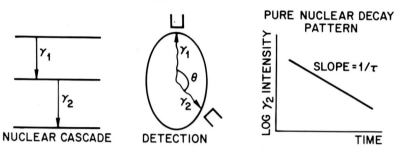

PURE NUCLEAR DECAY PATTERN

B. PRECESSION IN A STATIC FIELD:

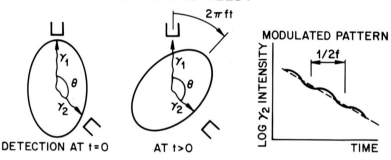

MODULATED PATTERN

Fig. 1. Schematic illustration of the anisotropy in a gamma-gamma cascade perturbed angular correlation and the time dependence of the intensity of the second gamma ray. a) Unperturbed angular correlation. b) An angular correlation perturbed by an external field illustrating the rotation of the correlation pattern and the modulation of the intensity versus time.

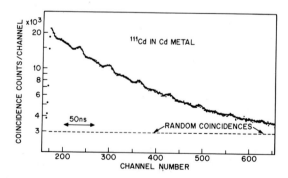

Fig. 2. The perturbed angular correlation pattern for 111-Cd (247keV level) in cadmium metal as observed by gamma-gamma perturbed angular correlations. The parent radioactivity in this case was 111m-Cd with a 48 minute halflife. The quadrupole precession modulations are clearly evident superimposed on the exponential decay curve of the excited nuclear state. (Data from Ref. [3]).

distributed with respect to that emission direction. With suitable electronic coincidence requirements, one can selectively observe only those second decay radiations which originate from the same nuclei as did the first radiation. Thus the second radiation originates from a group of nuclei which are anisotropically oriented. The second radiation will therefore appear anisotropic under these coincidence conditions, notwithstanding the fact that the radiation from the radioactive sample as a whole remains isotropic. No field, i.e., no sublevel splitting, is needed to produce the anisotropy. Observing the precession of the anisotropic distribution in a field is the perturbed angular correlation method [1].

Figure 1 schematically shows the cascade gamma rays which populate and depopulate a nuclear state, the anisotropic radiation pattern and detectors, and a plot of the intensity of the second radiation as a function of time after the first decay. This is effectively a measurement of the lifetime of the intermediate nuclear state. One sees from Figure 1b that in the presence of an extranuclear field which causes a spin precession, the precession of the radiation pattern is evidenced as a modulation in the intensity of gamma radiation versus time. An actual example of such data is shown in Figure 2 where the spin pre-

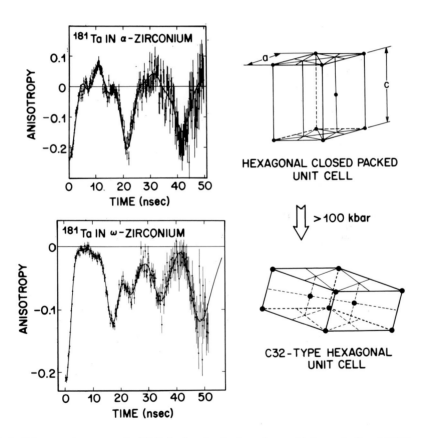

Fig. 3. Spin precession patterns for 181-Ta in zirconium in the α-phase and (after application of 100 kbar hydrostatic pressure) in the ω-phase. The obvious change in the pattern arises from the transition from hexagonal closed-packed to C32-type structure. The parent radioactivity in this case is 45 day halflife 181-Hf and the nuclear excited state is at 482keV in Ta with a halflife of 10ns. (Data from Ref. [4]).

cession of the nucleus 111-Cd embedded in cadmium metal is shown. This represents the interaction of the quadrupole moment of the 247keV state in 111-Cd with the electric field gradient arising from the noncubic nature of the cadmium metal lattice. One sees modulations with a period of approximately 50ns superimposed on the 84ns halflife of this nuclear excited state. It is usual in the presentation of data of this type to remove the lifetime exponential and display only the modulation pattern which will be done in subsequent examples.

For those who are interested in the fundamental quantum mechanical derivation of the general perturbed angular distribution and correlation functions and in the associated angular momentum algebra, several excellent articles are cited in Ref. 1. In addition, a large number of examples of applications of techniques such as this can be found in the proceedings of conferences on hyperfine interactions which are cited in Ref. 2. The review papers contained in this volume by Wichert and Recknagel, deWaard, and Richter cover topics concerned with the detection of defect trapping, the detection of gas trapping, and muon spin precession in materials; thus, we will not dwell on those topics here.

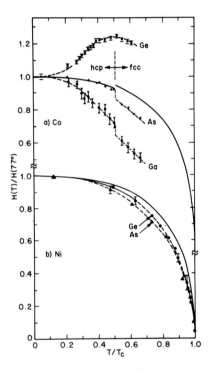

Fig. 4. The magnetic hyperfine field at various probes in a) cobalt and b) nickel host lattices. The hyperfine field is normalized to that observed at liquid nitrogen temperature. These fields, plotted as a function of temperature here, show a trend with bulk magnetization, particularly in nickel, as well as local moment anomalies as evident in cobalt. These data were obtained using the time-differential perturbed angular distribution technique based on the creation of isomeric excited states through nuclear reactions using an accelerator beam. Each magnetic field value was, of course, derived from a single measurement of a spin precession frequency. (Data from Ref. [5]).

PHASE TRANSITIONS

A good example to illustrate how the modulation pattern of gamma radiation intensity versus time is affected by the character of the interaction of the emitting nucleus with its environment is given for the case of the phase transition from α- to ω- zirconium as pressure is applied [4]. The data displayed in Figure 3 show the spin precession pattern obtained using the radioactive parent nucleus 181-Hf which decays to the 482 keV excited state in 181-Ta. Sources were prepared by alloying hafnium with zirconium metal and subsequently irradiating with pile neutrons. In the top of Figure 3, data obtained with the as-irradiated rolled zirconium foil are displayed. One discerns a single set of frequencies in this pattern characteristic of a unique quadrupole interaction. This is consistent with the hafnium impurity residing at regular lattice sites in the hexagonal close-packed unit cell of zirconium. After the application of 100kbars pressure to the sample and its subsequent slow release, the zirconium is found to have transformed to the so-called ω-phase which is retained metastably at atmospheric pressure. The spin precession pattern in this case, which is shown in the bottom of Figure 3, shows two components, one with twice the amplitude of the other. This corresponds to the C32 structure type hexagonal unit cell of the ω-phase structure in which two inequivalent sites are found in a 2:1 occupancy ratio. It is thus clear that the observation of quadrupole precession patterns can be used as an indicator of the occurrence of structural changes in the lattice.

A second illustration of a phase transition and also of the effect of temperature variation is seen in the work of Raghavan et al. [5] on the magnetic fields at impurities in cobalt metal. Figure 4 shows the derived normalized magnetic interaction versus temperature for germanium, arsenic and gallium probes in a cobalt lattice. As the temperature is varied through the hexagonal to face-centered cubic transition and up to the Curie point of the metal, one sees the effect of the phase transition itself. Also seen is the anomalous behavior of the local magnetic fields at these probes, especially when compared to the more normal situation shown in the bottom of Figure 3 for a nickel host. Thus both global structure or magnetization changes as well as local behavior around an isolated impurity atom are accessible through use of radioactive probe atom.

RADIATION DAMAGE

The use of the local probe is also a powerful method to observe the lattice damage created during ion implantation into materials. As an example, Figure 5 shows the results of implanting 111-In into zinc [6]. 111-In is the radioactive parent of 111-Cd, a probe which we encountered in Figure 2. The spin precession pattern of the 247keV gamma ray in Figure 5a shows a strongly damped behavior indicative of a loss of directional anisotropy in short times and therefore implying a randomization of spin directions of the ensemble of oriented nuclei in the same short times. This effect arises from the superposition of nuclear precessions with a wide variety of frequencies characteristic of a damaged lattice or nonunique lattice environment. One sees from the figure (5b) that after annealing of the zinc host matrix, the lattice damage heals and the probe indium atoms find unique substitutional sites where a very well defined electric quadrupole interaction is seen.

We have already shown data in Figure 2 for the gamma-ray cascade available in 111-Cd using a 111m-Cd isomeric parent with a 48min halflife. The electron-capture decay of the 111-In parent with a 2.8 day halflife produced the data of Figure 5. The useful state in 111-Cd may also be populated in the beta decay of 111-Ag and, rather than using the gamma-gamma perturbed angular correlation technique, one can use the beta-gamma technique in which not only the magnitude but the sign of the electric quadrupole interaction becomes available.

An interesting example which demonstrates that the choice of parent radioactivity may be crucial is the ion implantation of 111-Ag and of 111-In into a beryllium metal lattice. Figure 6 shows the decay scheme which populates the 84ns state in 111-Cd [7]. Figure 7 shows the spin precession spectra obtained for the implantation of silver as opposed to implantation of indium [8]. The quadrupole interaction in both cases is with the same 247keV state in 111-Cd. Therefore one must conclude based on the obvious difference in frequencies (factor of approximately 3.4) that the silver parent and the indium parent implant in such a way as to leave the daughter cadmium in different lattice configurations. Ion-beam channeling experiments have demonstrated that whereas silver implants substitutionally into beryllium, indium occupies the tetrahedral interstitial site after implantation. This explains the drastic difference in the quadrupole interaction frequency observed at cadmium when prepared by implantation of the two different radioactive parents.

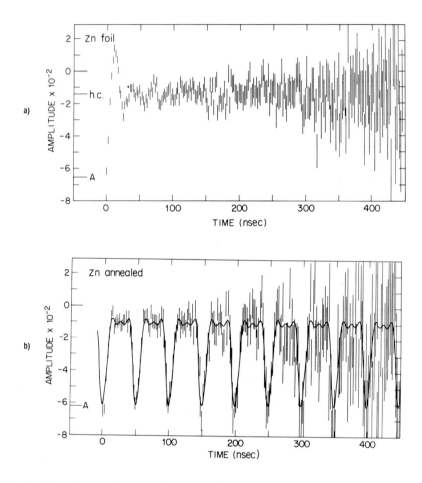

Fig. 5. a) The spin precession pattern for 111-Cd (parent 111-In with 3 day halflife) implanted into zinc metal. The 247keV state of Cd (84ns halflife) is perturbed by a random nonunique quadrupole interaction due to lattice damage arising from the implantation process itself. b) After annealing the zinc at 300°C for 1 hour, a unique quadrupole interaction is recovered corresponding to 111-Cd at a regular lattice site of the zinc. (Data from Ref. [6]).

A subsequent PAC experiment using implantation of 111m-Cd yielded the larger of the two frequencies shown in Figure 7. Thus one deduced that Cd implants tetrahedrally in beryllium without need of a channeling experiment. The uniqueness of the hyperfine interaction is often of value to tag a given configuration in this way.

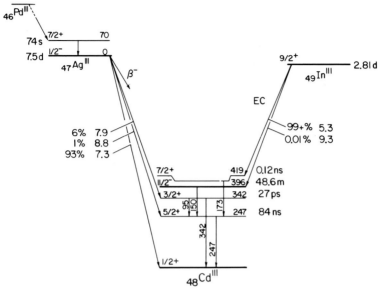

Fig. 6. The decay scheme showing the three parents which populate the 247keV state in 111-Cd. One sees beta decay from 7.5 day silver, the internal 150keV gamma transition from the 48 minute isomer, and the electron capture decay from 2.8 day indium [7].

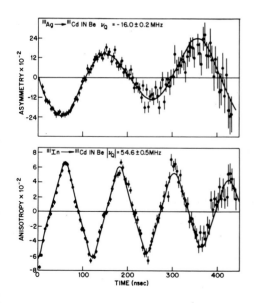

Fig. 7. Upper part) The spin precession pattern using the beta-gamma perturbed angular correlation method for 111-Ag decaying to cadmium in beryllium. (Note that this pattern is a sine function whereas in all other examples given here, cosine functions are typical. This arises because of the polarization component involved in beta decay and makes the sign of the interaction accessible.) Lower part) The spin precession pattern for cadmium as the daughter of indium implanted in beryllium. (Data from Ref. [8]).

194

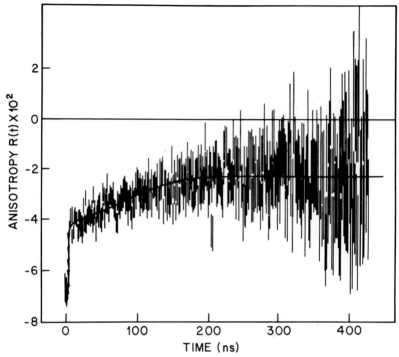

Fig. 8. The spin precession pattern for 111-Cd as the daughter of indium implanted into silicon after annealing for 1/2 hour at 200°C. The various components of rapid damping, slow damping, and constant anisotropy are evident. (Data from Ref. [9]).

AMPLITUDES AND DAMPING FACTORS

In some instances no unique precession frequency is observed for a probe implanted into a host lattice, as was the case for 111-In implanted in zinc shown in Figure 5a. When the loss of anisotropy with time can be resolved into its component parts, these fractions can be used to distinguish groups of impurities in different environments. A good example of this situation is that of 111-In implanted in silicon [9]. Indium is a group III dopant in silicon and displays the same so-called reverse annealing behavior as do boron, aluminum, gallium or thallium implants. That is, as implantation damage is annealed and sheet resistivity decreases, a temperature is reached where the resistivity temporarily increases again before complete recovery. Such behavior has been ascribed to indium which had been electrically active on substitutional sites being displaced from those sites through interactions with defects, in particular with silicon interstitials. This can be investigated from the point of view of the indium probe using perturbed angular correlations.

In Figure 8 spin precession data are shown in which three components can be identified. At the time zero, a rapid loss of anisotropy is observed followed at longer times by a slow decay and followed at still longer times by a constant nonzero anisotropy. These components can be associated with a strongly interacting nonunique environment, a weakly interacting nonunique environment, and an unperturbed environment, respectively. It is also possible that a very strongly interacting environment is present which causes a loss of anisotropy to occur in a time shorter than the instrumental time resolution. This would not be directly observed in the experiment but can be deduced by comparing total observed amplitudes with theory.

With this type of analysis, the fraction of probe atoms in so-called good sites were defined as those being in the unperturbed and weakly perturbed environments. The remaining (bad site) fractions are in strongly perturbed environments where anisotropy decays quickly. When these fractions were plotted against annealing temperature, the results in Figure 9 were obtained. One sees that the good or electrically active fraction is strongly correlated with the sheet resistance of the indium-implanted silicon as a function of annealing temperature including the reversal effect. The reversal can therefore be associated with a change in environment of the indium probe. However, no unique interaction was observed. One cannot then associate the indium which becomes electrically inactive during the reversal with a unique defect configuration or, as might be expected, with a unique interstice. The PAC technique is particularly well suited for this type of experiment where the concentration of impurity probes is necessarily quite small since it is based on the observation of radioactive decays.

The combination of amplitudes, damping constants, and unique frequencies observed in spin precession data clearly provides a powerful tool for the examination of the local probe environment. Frequencies are the unique tag for a given configuration. Damping indicates the degree of uniqueness. Amplitudes yield fractional occupancy of various configurations. For lack of space, we will not touch on the utility of PAC to study time dependent effects such as diffusion or relaxation processes where anisotropy damping of a different character is seen.

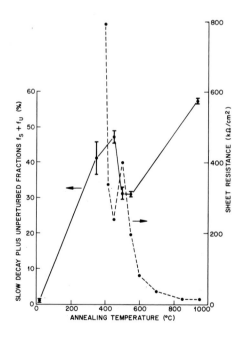

Fig. 9. A plot of the so-called good fraction for indium implanted in silicon, consisting of the probes in unperturbed and mildly perturbed sites, against annealing temperature. Also shown as the dashed lines are resistivity measurements versus annealing temperature. The reverse annealing effect is evident in both curves. (Data from Ref. [9]).

196

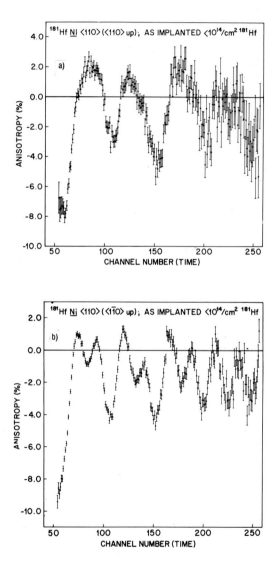

Fig. 10. Spin precession patterns for 181-Ta as the daughter of 181-Hf implanted in a nickel single crystal. a) With the <100> direction up b) With the <110> direction up. In both cases no treatment after implantation was performed. (Data from Ref. [10]).

CRYSTAL ORIENTATION

In the preceeding examples, the sample under study was not assumed to be either polycrystalline or single crystal. In fact, aside from the beta-gamma technique used for the 111-Ag parent previously discussed, only polycrystalline samples are necessary. The spin-precession technique is clearly sensitive, however, to the geometry of the perturbing interaction. This is illustrated as a final example below for the case of 181-Hf implanted in nickel [10]. As in the earlier example of Figure 3, the daughter activity of 181-Hf is 181-Ta which contains the 482keV state useful for PAC. Nickel, of course, is a magnetic host and one expects to see a magnetic Larmor spin precession in this case. With the use of a nickel single crystal, one expects an orientation dependence because in the absence of an external aligning field, the internal magnetization of the sample would be along one set of crystal axes (the easy axis of magnetization). By comparing Figure 10a and 10b which differ only in the orientation of the nickel single crystal sample with respect to the radiation detectors, one sees a clear difference arising from the different orientation of the magnetic hyperfine field in space. This can, in principle, be used to determine the distribution in space of the magnetic interaction, however, in practice without additional information that would be rather difficult. In general, however, the dependence on field orientation can be useful. It has been used, for example, to determine the configuration of trapped defects as is described by Wichert and Recknagel in this volume.

Other features of the data displayed in Figure 10 which arise from the details of the site assumed by hafnium implanted in nickel are discussed in another contribution [10].

SUMMARY

It is clear that for particular materials systems, the spin precession methods are of particular value as a phenomenological indicator of a change of state of the material or to investigate local effects at impurities. Their value is not dependent on a fundamental understanding of the origins of the fields perturbing the nuclei. That is to say, as a diagnostic tool it suffices to observe changes in the hyperfine interaction as a function of extrinsic parameters. Of course, where adequate theoretical understanding of the interactions is available, additional insight into the environment of impurity probes can be obtained. At the present time, however, the fundamental understanding of the electric field gradient produced by non-cubic environments in solids and of the magnetic hyperfine interaction is sufficiently coarse so that they cannot be used in a predictive way for materials research.

REFERENCES

1. H. Frauenfelder and R. M. Steffen in: *Alpha-, Beta-, and Gamma-ray Spectroscopy*, K. Siegbahn ed. (North-Holland Publ. Co., Amsterdam, 1965) p. 997.

    S. R. deGroot, H. A. Tolkoek and W.J. Huiskamp in: ibid, p. 1199.

    R. D. Gill, *Gamma Ray Angular Correlations* (Academic Press, New York, 1975).

    W. D. Hamilton, ed., *The Electromagnetic Interaction in Nuclear Spectroscopy* (North-Holland Publ. Co., Amsterdam, 1975).

2. E. Matthias and D. A. Shirley, eds., *Hyperfine Structure and Nuclear Radiations* (North-Holland Publ. Co., Amsterdam, 1968).

    H. vanKrugten and B. vanNooijen, eds., *Angular Correlations in Nuclear Disintegration* (Rotterdam Univ. Press, Wolters-Noordhoff Publ., Groningen, 1971).

    G. Goldring and R. Kalish, eds., *Hyperfine Interactions in Excited Nuclei* (Gordon and Breach, London, 1971).

    E. Karlsson and R. Wäppling, eds., *Hyperfine Interactions Studied in Nuclear Reactions and Decay* (Almqvist & Wiksell International, Stockholm, 1975).

    R. Coussement, M. Rots and L. Vanneste, eds., Hyperfine Interactions 2 (1976).

    R. S. Raghavan and D. E. Murnick, eds., Hyperfine Interactions 4 (1978).

    F. N. Gygax, W. Kundig and P. F. Meier, eds., Hyperfine Interactions 6 (1979).

    H. Haas and G. Kaindl, eds., Hyperfine Interactions, to be published (conf. in Berlin, July, 1980).

    J. H. Brewer and P. W. Percival, eds., Hyperfine Interactions, to be published (conf. in Vancouver, August 1980).

3. P. Raghavan and R. S. Raghavan, Phys. Rev. Letters *27*, 724 (1971).

4. E. N. Kaufmann and D. B. McWhan, Phys. Rev. B*8*, 1390 (1973).

5. P. Raghavan, M. Senba and R. S. Raghavan, Phys. Rev. Letters *39*, 1547 (1977).

6. E. N. Kaufmann, P. Raghavan, R. S. Raghavan, K. Krien, E. J. Ansaldo and R.A. Naumann, in: *Applications of Ion Beams to Metals*, S. T. Picraux, E. P. EerNisse and F. L. Vook, eds. (Plenum Press, New York, 1974) p. 379.

7. *Table of Isotopes*, C. M. Lederer and V. S. Shirley eds. (John Wiley & Sons, New York, 1978) p. 516 ff.

8. E. N. Kaufmann, P. Raghavan, R. S. Raghavan, E. J. Ansaldo and R. A. Naumann, Phys. Rev. Letters *34*, 1558 (1975).

9. E. N. Kaufmann, R. Kalish, R. A. Naumann and S. Lis, J. Appl. Phys. *48*, 3332 (1977).

10. L. Buene, E. N. Kaufmann, M. L. McDonald, J. Kothaus, R. Vianden, K. Freitag, C. W. Draper, this volume.

Published 1981 by Elsevier North Holland, Inc.
Kaufmann and Shenoy, editors
Nuclear and Electron Resonance Spectroscopies Applied to Materials Science

HYPERFINE INTERACTION INVESTIGATIONS OF HELIUM TRAPPING IN METALS

H. DE WAARD
Laboratorium voor Algemene Natuurkunde, University of Groningen, The Netherlands

ABSTRACT

Helium decorated vacancies may be trapped to a high
degree and with a high activation enthalpy at impurities
in metals. This effect is studied by means of the
hyperfine interaction of the radio-active impurities
^{111}In $\rightarrow$ ^{111}Cd, ^{119}Sb $\rightarrow$ ^{119}Sn and ^{129m}Te $\rightarrow$ ^{129}I, implanted
into a number of f.c.c. metals and into iron and cobalt.
Large changes of the magnetic hyperfine field are
caused at ^{119}Sn impurities in Ni, Co and Fe and high
values of the electric quadrupole interaction strength
are found at ^{111}Cd impurities in Al, Cu, Ag, Ni and Pt
when helium decorated vacancies are trapped at the
impurities. These effects provide a new method for
studying the trapping and desorption mechanism of
helium in metals.

INTRODUCTION

The presence of helium in a metal has a strong influence on its mechanical
properties. Studies of helium embrittlement are important in connection with
the development of suitable materials for fusion reactors. In order to
understand the observed effects in a more than phenomenological manner, we
must try to obtain a microscopic insight in the defect structures produced in
a metal by the penetration of helium and in the trapping and detrapping of
helium by such defects. Also, it is important to study the influence of the
presence of impurities in the metal, because these may sensitively affect the
helium trapping and detrapping stages.
In general, helium may be trapped by isolated vacancies, vacancy clusters,
vacancies associated with impurities or at grain boundaries. If none of these
defects were present, the helium would rapidly evaporate from most metals
through interstitial diffusion already below room temperature.
One useful technique that has been employed to study the behaviour of helium
in metals is by measuring its desorption from the metal as a function of
temperature. As discussed, for instance, by Kornelson and Edwards [1] and by
Reed [2], this desorption often occurs at a number of discrete, sharply defined
temperatures, each corresponding to a particular vacancy-helium cluster.
Systematic studies have allowed the identification of several of such clusters.
We have recently started to explore another method for studying the behaviour
of helium in metals, namely by measuring the hyperfine interaction of impurities
with associated vacancies into which helium may be trapped. To this end we
have implanted radioactive impurities, suitable for time differential perturbed
angular correlation spectroscopy (DPAC) or for Mössbauer spectroscopy (MS) into
various metals and we have looked for changes of their hyperfine interaction
patterns after helium doping and subsequent annealing. Striking effects were
first found in the DPAC spectra of sources of ^{111}In implanted in Ni and Cu which
were post-implanted with helium [3]. When the helium dose is high enough,
helium decorated vacancy clusters are formed with close to 100% efficiency
at the indium impurities, when the sources are annealed. For such clusters,
a very strong quadrupole interaction results at the impurity, corresponding

to a field gradient $V_{zz} \sim 10^{18}$ V/cm^2 at the impurity. These results have stimulated us to extend the investigation to other metals and to some other 5 sp shell impurities, suitable for MS. We are aware that some similar investigations have been started by a group in Konstanz [4] and we will incorporate one of their results in the following review.

2. SURVEY OF SYSTEMS UNDER INVESTIGATION.

In Table 1 we summarize the cases under investigation until now. With the exception of b.c.c. iron, all metals have the f.c.c. structure (for cobalt mixed with h.c.p.)

TABLE I
SURVEY OF HYPERFINE INTERACTION STUDIES OF HELIUM DECORATED VACANCY CLUSTERS; + IS TRAPPING OBSERVED; - IS TRAPPING NOT OBSERVED; (i) INSOLUBLE IMPURITY; (s) SOLUBLE IMPURITY. EXTENSIVE INVESTIGATIONS UNDERLINED.

HOST	IMPURITY		
	^{111}In$\rightarrow$^{111}Cd (DPAC)	^{119}Sb$\rightarrow$^{111}Sn (MS)	^{129}Te$\rightarrow$^{129}I (MS)
Al	+(i)	-(i)	+(i)
Cu	+(s)	+(s)	+(i)
Ag	+(s)	+(s)	
Au[4]	+(s)		
Fe		+(s)	+(i)
Co		+(s)	
Ni	+(s)	+(s)	+(i)
Pd	-(s)		
Pt	+(s)	+(i)	+(i)

Cases where helium decoration effects have been observed are indicated by +, and where not, by -. A distinction must be made between cases where the impurity is soluble in the host (s: solubility $\gtrsim 1^0$/a) and where it is insoluble (i: solubility smaller than 0.05 ^{0}/a). For s-cases we must expect that, after annealing the implanted source at a sufficiently high temperature, the h.f.i. becomes identical to that for a diffused source. For i-cases, a permanent change of the h.f.i. is observed after high temperature annealing because the impurities have clustered, have formed chemical bonds with host atoms and/or acquired a permanent vacancy association. To differentiate between the effects of helium decoration and "normal" annealing, spectra obtained with and without helium post-implantation must be compared. At the moment such a comparison has been made for the cases underlined in Table 1. The quadrupole interaction strengths of ^{111}Cd impurities associated with helium decorated vacancies in various metal hosts are given in Fig. 1. The hyperfine magnetic fields of ^{119}Sn impurities associated with helium decorated vacancies in the 3d metal hosts Fe, Co and Ni are given in Fig. 2 (full lines). For comparison, the field values for substitutional impurities are also given (broken lines).

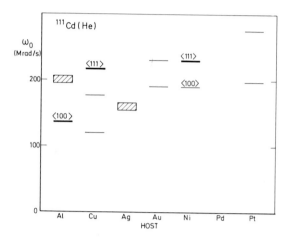

Fig. 1. Quadrupole interaction strengths, given by frequencies
$\omega_0 = (3\pi/10)(3QV_{zz}/h)$, for helium decorated vacancy clusters at ^{111}Cd
impurities in various metals. Field gradient axes, where indicated,
have been obtained from single crystal measurements.

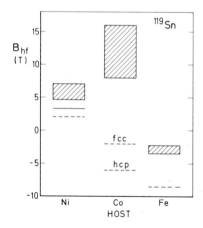

Fig. 2. Magnetic hyperfine field for ^{119}Sn impurities in various metals.
Full lines: in helium decorated vacancy clusters. Broken lines: pure
substitutional

3. TIME DIFFERENTIAL PERTURBED ANGULAR CORRELATION (DPAC) MEASUREMENTS.

All measurements were performed on polycrystalline metal foils and, for Al, Ni and Cu also on single crystals implanted with the 2.8 d ^{111}In activity to a dose of about 5×10^{12} In-ions/cm^2, with an energy of 120 keV, usually at room temperature. In each case, spectra were measured without and with a post-implanted He-dose of 3×10^{16} ions/cm^2 (for Cu $3 \times 10^{13} - 3 \times 10^{16}$ ions/cm^2) The implantation energy was 7 keV (in Cu also 15 keV). At the lower implantation energy, the depth distributions for the In and He overlap as well as possible (mean range for both ~ 200 Å).
DPAC measurements on the well known 171 keV - 245 keV gamma-gamma cascade in ^{111}Cd were usually performed at room temperature after annealing each sample for 15 min at successively higher temperatures. In the case of Ni, measurements were also performed above the Curie temperature, in order to exclude the magnetic hyperfine interaction.
The principles of the DPAC technique have been treated at this symposium by Kaufmann [5]. Our measurements are mainly aimed at establishing quadrupole interaction (q.i.) strengths and directions of electric field gradient axes. For this purpose we only note a few results of the theory: (1) In the q.i. spectrum of each particular site there are 3 Fourier terms with related frequencies. The fundamental frequency is $\omega_0 = (3\pi/10)(eQV_{zz}/h)$. The relative frequencies of the Fourier components may further yield a value for the e.f.g. asymmetry parameter η. A least squares fit of an experimental spectrum to the theory in principle yields values of ω_0, η and the modulation amplitude A_0 for each site. The ω_0 provide us with electric field gradients V_{zz} and the amplitudes with fractional site occupations. (2) For a single crystal the calculated DPAC spectra depend markedly on the angle between the e.f.g. axis and the directions in which the gamma rays are detected. From measurements with the crystal oriented in such a way that different main crystal axes successively point in the detector directions, we can uniquely determine the e.f.g. direction relative to the crystal axes. An example is

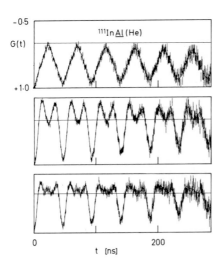

Fig. 3. Time differential perturbed angular correlation spectra for ^{111}InAl(He) source. Top: detector axes along <100> axes; middle: axes along <110> axes; bottom: axes along <111> axes.

shown in Fig. 3, where three spectra are shown for ^{111}InAl(He) (unannealed),
with directions of the 4 detectors pointing along the <1$\overline{0}$0>, <$\overline{1}$00>, <010>
and <0$\overline{1}$0> axes (top), the <110>, <$\overline{1}\overline{1}$0>, <$\overline{1}$10> and <1$\overline{1}$0> axes (middle) and
the <111>, <$\overline{1}\overline{1}\overline{1}$>, <1$\overline{1}$1> and <$\overline{1}1\overline{1}$> axes (bottom). The fits of the measured
spectra to calculated ones uniquely established that the e.f.g. direction
lies along the <100> axis.

3.1. ^{111}InNi.

This is the most thoroughly investigated case. The annealing behaviour without
helium had previously been studied by Hohenemser et al. [6] and by Pleiter [7]
who observed two different vacancy associated sites. One of these (site 2) is
characterized by a practically zero electric quadrupole interaction and a
magnetic interaction reduced to 40% of its normal value, the other (site 3)
by a quadrupole interaction with $\omega_0 \simeq$ 53 Mrad/s and again, a strongly
reduced magnetic interaction. Site (2) exists in an annealing temperature
region from 300-600 K, site (3) in a region from about 300-800 K. Above the
last temperature, all vacancies are detrapped and the In is found in a pure
substitutional state (site 1), with full magnetic h.f. field and zero
quadrupole interaction. The vacancy trapping behaviour of ^{111}InNi with 2 × 10^{16}
(post-implanted) helium atoms/cm^2 is quite different. Here, the vacancy
associated sites 2 and 3 are not clearly identified, but two new types
of lattice defects are found after annealing at 720 K. They are characterized
by very strong quadrupole interactions (ω_0 = 229 Mrad/s and ω_0 = 190 Mrad/s)

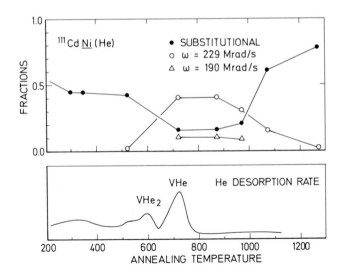

Fig. 4. Top: Annealing behaviour of different sites of ^{111}Cd impurities in Ni
post-implanted with 3 × 10^{16} at/cm^2 of helium. Bottom: desorption of
helium from nickel, implanted with 200 eV energy |1|; temperature scale
converted to our annealing conditions.

204

and together involve up to 70% of the implanted indium atoms. These defects are particularly stable, they resist annealing up to 1270 K. Their annealing behaviour is shown in Fig. 4 (top). In the bottom part of the figure we show the result of a helium desorption experiment carried out by Kornelson and Edwards [1] on a sample of nickel implanted with low energy helium ions, after conversion of their temperature scale to our annealing conditions. It is seen that a strong growth of the high q.i. sites takes place when the helium is released, presumably from vacancies generated by the helium implantation itself. These vacancies are then highly mobile and are apparently trapped by the impurities together with rapidly diffusing helium atoms. The latter immobilize the vacancies up to a high temperature, corresponding to a helium activation enthalpy of about 2.3 eV.

3.2. ^{111}InPd and ^{111}InPt.

We note that Ni, Pd and Pt belong to the same group of the transition elements. Their metal radii increase in the given order but they are all significantly smaller than that of indium. It is, therefore, surprising that helium post-implanted Pd does not show any tendency to capture vacancies: The ^{111}In remains purely substitutional at all annealing temperatures up to 1100 K, as is clear from Fig. 5. Platinum, however, starts capturing helium decorated vacancies at about 700 K and partly retains them up to 1100 K, the highest annealing temperature reached so far.
We interpret this trapping stage as for ^{111}InNi(He), namely by trapping of vacancies resulting from helium desorption that are decorated and thus frozen at the impurities by desorbed helium atoms.

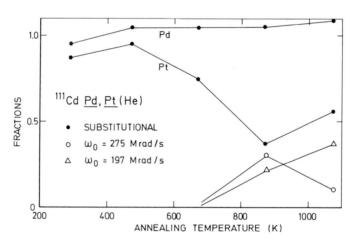

Fig. 5. Annealing behaviour of sites of ^{111}Cd impurities in Pd and Pt, post-implanted with 3×10^{16} at/cm^2 of helium. The errors of the fractions given in this Figure vary between 5 and 10%. This puts the substitutional fraction for Pd very close to 1.

4. MÖSSBAUER SPECTROSCOPY.

The principles of this method have been explained at this symposium by Shenoy [8] and we therefore only briefly review some of the quantities relevant to our measurements.

(1) The magnetic hyperfine field B_{hf} (given in Tesla) derived from the magnetic structure of the spectrum, that results from the Zeeman splitting of the levels between which the recoilless gamma transition takes place and from the relative probabilities of electromagnetic transitions between magnetic sublevels (Clebsch-Gordan coefficients). It turns out that the value of B_{hf} is very sensitive to the presence of vacancies close to the recoilless emitting atom. Our values of B_{hf} include no correction for the Lorentz field.

(2) The isomer shift S (given in mm/s) of the gamma ray energy of some standard absorber relative to the investigated source, which tells us about changes in the outer shell s-electron configuration of the source. It should be remarked that a positive value of S corresponds to a lower gamma ray energy of the source than the absorber.

(3) The electric quadrupole interaction strength $eV_{zz}Q$ (and asymmetry η) derived from the quadrupole splitting of the spectrum. In most of our observations, especially for ^{119}Sn, where the quadrupole moment of the relevant nuclear state (I = 3/2, E = 23.8 keV) is quite small (Q ∿ -0.06 b), this interaction plays a minor part. So far, it has been neglected in our analysis.

(4) An additional feature of Mössbauer spectroscopy, as compared with other h.f.i. techniques, is the dependence of the degree of recoilless gamma emission (and absorption) on temperature. This so-called recoilless fraction f(T) depends strongly on the stiffness of the bonds between the emitting (absorbing) atoms and its surroundings. In particular, the presence of vacancies close to this atom may lead to a substantial reduction of f. Within the accuracy of our measurements we will restrict ourselves to the simple Debye model of the recoilless fraction, which allows us to express the lattice dynamical properties of the impurity with only one parameter: the characteristic temperature θ, which, at T = 0 is related to f by $f = exp(-3E_r/2k\theta)$, with $E_r = E_\gamma^2/2Mc^2$ the free nucleus recoil energy.

We shall again discuss some of the measurements in more detail.

4.1. ^{119}SbNi.

Like ^{111}InNi for DPAC, this is the most thorougly investigated case for Mössbauer spectroscopy. Some typical spectra are shown in Fig. 6. The annealing behaviour of SbNi and InNi, both with and without post-implanted helium, is very similar, but in this case the evidence is mostly obtained from changes in the magnetic hyperfine field of Sn. This field is derived from Mössbauer spectra of the 23.8 keV gamma transition in ^{119}Sn, following the radioactive decay of ^{119}Sb.

In ^{119}SbNi sources without helium, a component with a much larger hyperfine field, $B_{hf}^{(2)}$ = 7.1(3) T, is observed besides the component with normal field, $B_{hf}^{(1)}$ = 2.06(2) T that is due to substitutional Sn-impurities. This second component grows in intensity upon annealing up to about 550 K and then rapidly disappears. It is interpreted in the same way as site 2 in ^{111}CdNi, namely as due to impurities surrounded by a tetrahedron of vacancies (see Fig. 7.A.). Above 600 K, all Sb impurities are substitutional. Again, the annealing behaviour for helium post-implanted ^{119}SbNi sources is quite different. Also here, a site with a strong hyperfine interaction (be it magnetic in this case) grows rapidly after annealing at a temperature where helium desorption from previous traps starts. It reaches an occupation of 95%. The annealing behaviour of this high field site

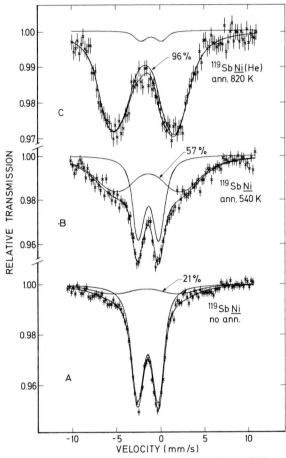

Fig. 6. Mössbauer spectra of the 23.8 keV transition in ^{119}Sn, obtained
for implanted ^{119}SbNi sources and a BaSnO$_3$ absorber both at 4.2 K.
A. No anneal, no helium post-implanted. B. Annealed at 540 K for
30 min, no helium post-implanted. C. 2 × 10^{16} at/cm^2 of helium
post-implanted. Annealed for 30 min at 820 K.

($B_{hf}^{(2')}$ = 6.2 T) is shown in Fig. 8. Between 1000 and 1100 K the high field
collapses to a "medium" large value, $B_{hf}^{(3)}$ = 3.4 T which persists up to
1270 K with 80% site occupation. The normal value of the field for a
substitutional site is only restored after annealing at 1470 K. As for
^{111}CdNi(He), this result proves that the impurity associated vacancies
provide very deep traps for the helium, with activation enthalpies of about
2.6 eV.

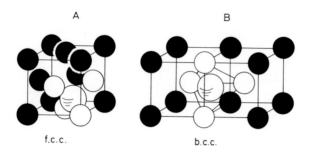

Fig. 7. Symmetric vacancy configurations in f.c.c. and b.c.c. metals.

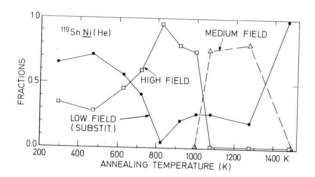

Fig. 8. Annealing behaviour of different sites of ^{119}Sn impurities in nickel post-implanted with 2×10^{16} at/cm^2 of helium.

The isomer shift of the vacancy associated site is 0.5 mm/s more negative than that of the substitutional site, corresponding to an increase of the s-density at the ^{119}Sn atoms, equivalent with the addition of about 0.17 of a 5s electron. At first sight, such a result seems strange, because an increase of s density generally accompanies a compression of the atom whereas vacancy trapping leads to a decompression. The calculations of Williamson et al. [9], however, show that for Sn (and also for Sb and Te) an increase of s-density corresponds to an increase of the atomic volume, contrary to the "normal" behaviour. In fact, on the basis of these cal-culations, we estimate that a 0.5 mm/s more negative isomer shift roughly corresponds to an increase by 30% of the atomic volume.

Measurements were further carried out with ^{119}SbNi and ^{119}SbNi(He) sources at 3 different temperatures (LHe, LN and RT). The recoilless fraction of Sn-atoms in the vacancy associated sites decreases more rapidly with

temperature than that of Sn-atoms in substitutional sites (see Fig. 9).
In terms of characteristic temperatures we have θ(subst. Sn) = 260 K,
θ(vac. ass. Sn) = 200 K. Within the scope of the Debye model, these
temperatures can be related to impurity r.m.s. vibration amplitudes
$a = \langle x^2 \rangle^{\frac{1}{2}} = \hbar \left| (4/3)(m_I m)^{\frac{1}{2}} k\theta \right|^{-\frac{1}{2}}$ at T = 0, with

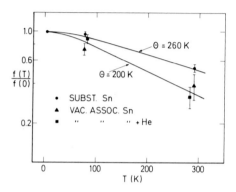

Fig. 9. Relative recoilless fractions vs. temperature for substitutional,
vacancy associated and helium decorated vacancy associated sites
of ^{119}Sn in nickel.

m_I = impurity mass and m = host mass, yielding a(subst.)= 0.030 Å and
a(vac. ass.) = 0.035 Å, which is compatible with an increase in space for
the vacancy associated impurities, both with and without helium decoration.
Apparently, Mössbauer spectroscopy in this case has provided three different
parameters, the hyperfine magnetic field B_{hf}, the isomer shift S and the
recoilless fraction f, from each of which useful information about the impurity
environment can be obtained.

4.2. ^{119}SbCo.

Investigations of this case are somewhat complicated by the co-existence of
f.c.c. and h.c.p. cobalt, with quite different h.f. magnetic fields for
substitutional Sn impurities: $\left| B_{hf}^{(hcp)} \right|$ = 5.7(1) T and $\left| B_{hf}^{(fcc)} \right|$ = 1.96(3) T.
In helium post-implanted sources of ^{119}SbCo, these fields are observed up to
an annealing temperature of 670 K. At 820 K, much larger fields, with an
average value of $\left| B_{hf} \right| \simeq$ 13 T and a large spread, $\Delta B_{hf} \sim$ 14 T (FWHM), appear.
Better statistics of the measurement will be needed to determine individual high
field components. These large fields persist only over a relatively narrow
temperature range: after annealing at 970 K they have vanished, but the
original fields have not yet reappeared. In the spirit of our previous interpretations for ^{111}InNi (He) and ^{119}SbNi (He), we believe that the high field
components are due to impurity atoms that have captured several vacancies
decorated with helium atoms. Again, these remain trapped at the impurities
up to a high temperature, at least up to about 1000 K.

4.3. ^{119}SbFe.

The annealing behaviour of the helium decorated vacancy trapping in this case is very similar to that of ^{119}SbCo, but here the absolute value of the hyperfine field shows a large decrease, from 8.6 T to 2.8 T when decorated vacancies are trapped (at 800 K). Since iron is a b.c.c. metal, the geometry of the vacancy clusters is different from that of the f.c.c metals, but also in this case a (distorted) tetrahedron (see Fig. 7B) is one of the likely nearest neighbour vacancy environments for an impurity. Both fields are found to be negative. Again, the normal substitutional field value is not restored until the sample is annealed at 1300 K.

4.4. ^{129m}TeMe.

Tellurium is insoluble in all metals studied, and in all cases the hyperfine interaction, studied by Mössbauer spectroscopy of the 27.8 keV level in ^{129}I, undergoes permanent changes when the implanted sources are annealed. In the initial annealing stages, up to about 800 K, the behaviour of sources with and without helium is clearly different, but at higher annealing temperatures impurity-host "molecules" are formed with characteristic Mössbauer spectra that are the same for sources with and without helium. It may be that helium is still trapped at the "molecules", but this does not influence the hyperfine interaction significantly. Further analysis is necessary in these cases to reach more detailed conclusions.

CONCLUSIONS.

At the moment, though data taking and analysis is not yet complete for all cases given in Table I, the following general conclusions can be drawn about the annealing behaviour of helium post-implanted sources:
(1) In the f.c.c. metals Ni, Cu and Pt helium decorated vacancies are trapped to a high degree at In and Sb impurities after helium is desorbed from isolated vacancies. Similar strong trapping is observed for In in Ag and Al and for Sb in Co and Fe, but not for In in Pd and for Sb in Al and Ag. These differences in behaviour can not be understood simply in terms of size difference.
(2) Helium decorated vacancies trapped at impurities are often very stable; their dissociation temperatures are several hundred degrees higher than those of undecorated vacancies. The helium activation enthalpy can be as high as 2.6 eV (for SbNi(He)).
(3) The high sensitivity of hyperfine interactions of impurities in metals to the presence of decorated vacancies allows detailed studies of helium trapping and desorption at these impurities.
It is important to combine our h.f.i. measurements with helium desorption experiments in order to understand the mechanism of helium motion in the metal more completely.

REFERENCES

1. E.V. Kornelsen and D.E. Edwards in: Application of Ion Beams to Metals, S.T. Picraux, E.P. EerNisse and F.L. Vook eds. (Plenum New York 1974) p. 521.
2. D.J. Reed, Rad. Effects 31, 129 (1977).
3. F. Pleiter, A.R. Arends and H. de Waard, Phys. Lett. 77A, 81 (1980).
4. M. Deicher, private communication.
5. E.N. Kaufmann, Some Applications of Spin Precession Methods to Problems in Materials Science, these Proceedings, p.

6. C. Hohenemser, A.R. Arends, H. de Waard, H.G. Devare, F. Pleiter and
    S.A. Drentje, Hyp. Int. $\underline{3}$, 297 (1979).
7. F. Pleiter, Hyp. Int. $\underline{7}$, 109 (1979).
8. G. Shenoy, The Mössbauer Effect and Some Applications, these Proceedings,
    p
9. D.L. Williamson, J.H. Dale, W.D. Josephson and L.D. Roberts, Phys. Rev.
    $\underline{B17}$, 1015 (1978).

Published 1981 by Elsevier North Holland, Inc.
Kaufmann and Shenoy, editors
Nuclear and Electron Resonance Spectroscopies Applied to Materials Science

DEFECTS IN METALS - DETECTED BY SPIN PRECESSION METHODS

TH. WICHERT and E. RECKNAGEL
Fakultät für Physik, Universität Konstanz, D-7750 Konstanz,
Fed. Rep. Germany

ABSTRACT

Recent advances in the study of point defects and small
defect clusters by nuclear methods will be reviewed. Suitable
radioactive atoms which fulfill the requirements of spin
precession techniques are introduced into solids, where they
are used as microscopic probes to investigate their immediate
surrounding. Trapping, detrapping or annealing of defects is
reflected by changes of the hyperfine interaction between the
nuclear moments of the probe atoms and the electromagnetic
fields of the defects. The interaction is extracted from a
time differential spin precession pattern or - more obviously - from the Fourier transform of it, which can be regarded
as the resonance spectrum of the involved frequencies.
It allows a distinct recognition of different defects and
comprises detailed information about their structural properties. Experimental data on interstitials, vacancies and
clusters in metals will be quoted as illustrations for the
scope of information, which can be achieved by hyperfine interaction studies.

INTRODUCTION

It is the purpose of this review to demonstrate how nuclear methods can
contribute to the understanding of radiation damage. Being inherently microscopic,
the investigations performed by these hyperfine interaction methods give direct
information about the structure and dynamics of interstitials, vacancies and their
clusters on a scale of a few lattice constants in contrast to the rather indirect
bulk methods where conclusions are drawn from changes in macroscopic physical properties which are influenced by defects. However, both approaches are complementary in many respects. As compared with the resonance spectroscopy the spin
precession methods have not to look for the transition frequencies belonging to
the respective hyperfine interaction, but they directly observe the spin
precession frequencies via the radioactive decay products.

Extensive data, expecially for point defects in metals were already collected
by conventional methods such as residual resistivity and elastic modulus measurements and, in addition, for more extended defects by scattering methods and electron microscopy. Experimental procedures like isochronal annealing were developed, so that at least a more or less consistent picture exists about the formation and recovery of point defects. But also the limitations of definite state-
ments became more and more obvious. Different probe preparations and damage techniques caused controversial results - mainly because the unambiguous recognition
of defects by integral observations was difficult.

At this point the "short-sighted" but highly specific nuclear methods are
able to add conclusive information. They are mainly based on trapping mechanisms
of defects at radioactive impurity probe atoms: Though, in some cases this might

be disadvantageous, it provides on the other hand the possibility to study the role of dilute impurities as trapping centers. The distinct defect is classified via its characteristic electromagnetic field instead of its thermodynamic properties like migration or formation energies. Such a classification is invaluable for the unraveling of the contribution of different defect types to a single recovery stage.

After a short introduction of the relevant hyperfine parameters describing the magnetic dipole and the electric quadrupole interaction, we shall present two experimental techniques to measure the defect induced hyperfine interaction in metals: The time differential observation of the perturbed angular correlation of γ-rays (TDPAC) and the muon spin rotation method (μSR). The feasibility of both the methods will be illustrated in the following two chapters, in the course of which two properties of the applied spin precession methods are recognizable: Its short-sightedness prevents a simultanous sensitivity to all defect types being present in the lattice, so that the measured hyperfine interaction does not represent some mean value due to all defects but characterizes the just thermally activated and trapped defect. Furthermore, the hyperfine interaction is especially specific, when the defect causes a change of the local magnetisation, producing a magnetic dipole interaction or a change of the local symmetry, producing a electric quadupole interaction, i. e. in ferromagnetic and diamagnetic cubic metals.

HYPERFINE INTERACTIONS (HFI)

The study of static and dynamic properties of defects by radioactive probe atoms is performed via hyperfine interactions $|1,2|$. These interactions occur be-

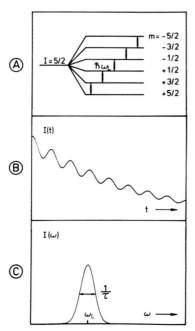

Fig.1: (A) Level splitting of a nuclear state with spin I = 5/2 caused by a magnetic dipole interaction.
(B) The time spectrum measured by a spin precession technique exhibits an intensity modulation governed by the transition frequency ω_L along with its exponential decay due to the finite lifetime τ of the nuclear state.
(C) Fourier transformation of this spectrum results in a single line at the frequency ω_L with a width 1/τ.

tween the nuclear electromagnetic moments of the probe atoms and the electromagnetic fields in their surrounding. If defects appear next to the probe atoms, they will influence the electromagnetic fields thereby causing a change of the HFI.

This already reveals the scope of application of these methods:
- The HFI has to be changed by a measurable amount compared to the unperturbed situation;
- the defect has to be trapped by the probe atom, so that an influence exists during a time interval comparable to the half-life of the radioactive state.

The hyperfine interactions relevant for the present issue are

Magnetic dipole interaction (MI)

The interaction between the nuclear magnetic dipole moment $\vec{\mu}$ and the effective magnetic field at the probe atom leads to the normal Zeeman splitting

$$E_m = - \mu B \frac{m}{I} \qquad (- I \leq m \leq + I) \qquad (1)$$

with the Larmor transition frequency ($\Delta m = 1$)

$$\omega_L = \frac{\mu B}{\hbar I} \qquad (2)$$

In fig.1A this splitting is shown for $I = 5/2$. In case of an unequal population of the magnetic sublevels, i. e. either a polarization or an alignment of the nuclear spins exists, a periodic change of population with respect to a quantization axis caused by the transversal magnetic field is observed in all spin precession methods, the frequency of which is given by the Larmor frequency ω_L while the amplitude of the modulation is determined by the degree of polarization (alignment): In fig.1B a typical time differential spin precession spectrum is shown. After eliminating the experimental part caused by the decay of the nuclear state, the pattern can be fitted by the "perturbation function"

$$G_2(t) = 1 + a \cos \omega_L t \qquad (3)$$

(In case of alignment one observes $\cos 2\omega_L t$). The Fourier transform of eq.(3) as displayed in fig.1C shows a resonance at ω_L of amplitude a and - in absence of relaxations - of width $1/\tau$, with τ being the lifetime of the nuclear state.

Electric quadrupole interaction (QI)

The interaction between the nuclear quadrupole moment and the electric field gradient (efg) leads to the hyperfine splitting of a nuclear energy level as shown in fig.2A $|3|$. Both quantities are tensors, thus the interaction depends on the mutual orientation. Since the quadrupole moment eQ of a nucleus with spin I is usually defined as the maximal component of the quadrupole tensor and since further the number of free components of the traceless efg tensor V_{ij} can be reduced to two by proper transformation to the principle axes of a cartesian coordinate system, the QI energies for the simplest case of an axially symmetric efg ($V_{xx} = V_{yy} = - V_{zz}/2$) are given by

$$E_Q = \frac{3m^2 - I(I+1)}{4I(2I-1)} eQV_{zz} \qquad (4)$$

In the case of a non-axial efg the two tensor components of the principle axes, V_{xx} and V_{yy} are not equal and a second quantity,

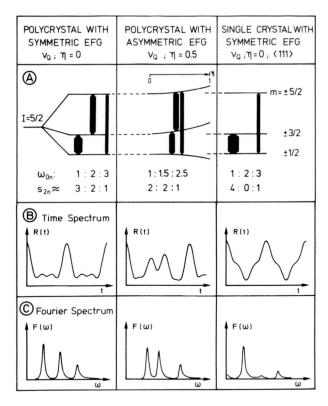

POLYCRYSTAL WITH SYMMETRIC EFG v_Q ; $\eta = 0$	POLYCRYSTAL WITH ASYMMETRIC EFG v_Q ; $\eta = 0.5$	SINGLE CRYSTAL WITH SYMMETRIC EFG v_Q ; $\eta = 0$; $\langle 111 \rangle$
(A) $I = 5/2$		$m = \pm 5/2$ $\pm 3/2$ $\pm 1/2$
ω_{0n}: $\quad 1:2:3$ $s_{2n} \approx \quad 3:2:1$	$1:1.5:2.5$ $2:2:1$	$1:2:3$ $4:0:1$
(B) Time Spectrum $R(t)$	$R(t)$	$R(t)$
(C) Fourier Spectrum $F(\omega)$	$F(\omega)$	$F(\omega)$

Fig.2: (A) Quadrupole splitting of a nuclear state with spin $I = 5/2$ caused by an efg $V_{zz} \sim v_Q$ with ($\eta = 0$) and without ($\eta \neq 0$) axial symmetry . For the single crystal (right diagram) it is assumed that the γ-detectors and the efg are oriented along a $\langle 111 \rangle$ lattice direction. ω_{0n} and s_{2n} are the observable, relative transition frequencies and strengths, respectively. In part (B) the respective time spectra without the exponential decay are shown as measured by the spin precession method TDPAC and in part (C) the Fourier transforms of these spectra are given.

$$\eta = \frac{V_{xx} - V_{yy}}{V_{zz}} \qquad (4')$$

is necessary to describe the interaction (with $|V_{xx}| \leq |V_{yy}| \leq |V_{zz}|$).
The transition frequencies between the non-equidistant hyperfine levels (fig. 2A) are multiples of the fundamental angular precession frequency ω_0:

$$\omega_0 = \frac{6eQV_{zz}}{4I(2I-1)\hbar} \qquad (5)$$

for a half odd integral spin I as I = 5/2 of ^{111}Cd, the only nuclear probe mentioned throughout this review in context with QI.

Since more than one frequency for an unique QI is involved, the spin precession spectra are more complicated than in the magnetic case (fig. 2B). The perturbation function can be expressed as

$$G_2(t) = s_{20} + s_{21} \cos \omega_{o1} t + s_{22} \cos \omega_{o2} + s_{23} \cos \omega_{o3} t \qquad (6)$$

For $\eta = 0$ the frequency ratios are $\omega_{o1} : \omega_{o2} : \omega_{o3} = 1:2:3$ (fig. 2A left side) with $\omega_{o1} = \omega_o$ as given by eq. (5). The coefficients s_{2n} express the transition probabilities between the different hyperfine energy levels. For $\eta \neq 0$ the frequency ratios normally are no longer integral and the ratios and the coefficients s_{2n} depend on η (fig. 2A middle part). The Fourier transforms (fig. 2C) of the spin precession spectra (fig. 2B) directly display the three frequencies and their transition amplitudes s_{2n}.

EXPERIMENTAL PROCEDURE

Spin precession methods
In our case the hyperfine interaction parameters described in the last section are measured by the so-called spin precession method, i.e. the electromagnetic fields interact with the moments of radioactive nuclei that decay after a characteristic lifetime τ by emission of γ quanta or charged particles. In contrast to the so-called resonance methods this property opens a unique way to sense the respective hyperfine splitting: Instead of inquiring the system via a radio-frequency for the frequencies ω_L or ω_o (see eqs. (2) and (5)) these values can be directly read off from the intensity modulation of the emitted radiation, because its directional properties are influenced by the spin orientation of the nuclear state which precesses just with the frequencies ω_L or ω_o about the axis of the MI or QI. Obviously, such a spin precession is only observable for a nonrandom orientation of spins, i.e. the population of the magnetic sublevels m has to be different. When the population changes monotonously with m or with the absolute value $|m|$ we call it a spin polarization or alignment, respectively. With the axis of the HFI perpendicular to the quantization axis of the non randomly populated m-states the spin polarization (alignment) is periodically perturbed as reflected by the modulation of the emitted radiation intensity. Experimental techniques measuring this parameter are the TDPAC or TDPAD (Time Differential Perturbed Angular Correlation or Distribution) and the μSR (Muon Spin Rotation). A second property common to these methods is the small concentration of impurity probes since only a certain absolute number of probes ($10^9 - 10^{12}$) is required which depends on the solid angle of the detectors and the desired statistical accuracy.

The probe ^{111}In/^{111}Cd (TDPAC). The most commonly used spin precession method for studying HFI caused by defects is the TDPAC. In this method metals are doped with radioactive "parent" nuclei, the half-life of which determines the time available to carry out the experiments and to handle the specimen by annealing, irradiation or quenching. The relevant probe discussed in the next sections is the isotop ^{111}In with $T_{1/2} = 2.8$ d. The actual HFI is measured at a short lived excited state of the "daughter" nucleus ^{111}Cd ($5/2^+ - 247$ keV) which decays with a half-life $T_{1/2} = 84$ nsec. The lifetime of the excited state represents a time window, during which the HFI period $1/\omega$ can be observed, or, in case of the Fourier transformed spectra, it determines the line width and hence the resolution at the energy splitting caused by the HFI (see Fig. 1). - In TDPAD experiments the daughter is directly populated by a nuclear reaction performed at an accelerator and there is no time for further handling of the specimen. Therefore, this method seems to be less suited for the detection of defects at the moment.

In γ-γ angular correlation, the observation of the first γ-ray of a transition cascade in a fixed direction selects excited nuclei which have spins in an aligned state, so that the γ-ray of the second transition displays an anisotropic angular correlation with respect to the first one. This angular correlation is approximately given by

$$W(\Theta) = 1 + A_2 P_2 (\cos \Theta) \tag{7}$$

where A_2 is a tabulated anisotropy coefficient, yielding for ^{111}Cd: $A_2 = -0.18$, and Θ is the angle enclosed by the two γ-rays. If an electromagnetic field interacts with the nuclear moment of the excited state during the lifetime the anisotropic γ-distribution precesses with the respective hyperfine interaction frequency ω. In polycrystalline samples internal fields usually occur in all directions so that an average over all precession axes has to be taken into account. The perturbed angular correlation is then given by

$$W(\Theta,t) = 1 + A_2 G_2(t) P_2(\cos\Theta) \tag{8}$$

with the time evolution described by the perturbation factor $G_2(t)$. For the MI and QI the perturbation factor is given by eqs.(3) and (6), respectively. In general, the probe atoms experience different perturbations, so that the experimental data have to be evaluated by a generalized perturbation factor (polycrystalline sample)

for MI:
$$G_2(t) = f_o + \sum_{i=1} f_i \, e^{-\sigma_i t} \cos(\omega_{Li} t), \tag{9}$$

and assuming η = 0

for QI:
$$G_2(t) = \sum_{n=o}^{3} s_{2n} \{ f_o + \sum_{i=1} f_i \, e^{-n\sigma_i t} \cos(n\omega_o^i t) \} \tag{10}$$

where f_o denotes the unperturbed fraction of nuclei, f_i is the fraction associated with unique interaction frequencies ω_o^i or ω_{Li} and σ_i allows for a frequency distribution. The latter parameter corresponds to an increase of line width in the Fourier transforms. In the case of QI, generally, the strength is expressed by the quantity ν_Q, which is independent of the nuclear spin

$$\nu_Q = \frac{eQV_{zz}}{h} = \frac{2I(2I-1)}{6\pi} \, \omega_o \tag{11}$$

for a half odd integral spin I, so that in the following the fundamental frequency ω_o^i is replaced by ν_{Qi}.

A measurement of the QI in a single crystal allows to determine the orientation of the efg tensor. The observation of the γ-rays at various angles with respect to the crystal yields the orientation of the symmetry axis of the efg for η = 0 with respect to the crystallographic axis <hkl> |4|. The coefficients s_{2n} drastically depend on this orientation, whereas the frequencies are not affected (fig. 2 right side).

Thus the relevant information obtained from TDPAC investigations of defects in metals can be summarized as follows:
- ν_{Qi}, ω_{Li} denote the specific defect-probe atom interaction, from which the field quantities V_{zz}, η, B can be derived;
- f_o, f_i denote the fractions of nuclei experiencing the same interaction, and
- <hkl> determine the direction of the efg.

The muon probe (μSR). As a second probe to measure the HFI the muon is
presented, a leptonic particle created in the decay of the π-meson |5|. The
required pions are produced by high energetic protons (E_p > 300 MeV) and decay
after a few nanoseconds into the muon and a neutrino with the consequence that
the muon spin is highly polarized. The muon spin is 1/2 and the lifetime about
2 μsec. Albeit the μSR is an accelerator experiment like the TDPAD a handling
of the specimen during the experiment is possible (see section 3.3).

The orientation of the muon spin is monitored by the intensity distribution of the decay positron , whose maximum energy is 53 MeV. In a magnetic
field perpendicular to the muon polarization the positron rate varies according to eq. (3) (after eliminating the exponential decay function, see fig.1). In
the past the muon has proved itself as a sensitive probe for detecting MI. The
relevant information about defects is drawn from the Larmor frequency ω_{Li}, the
frequency distribution σ_i and the relative amplitude f_i (see eq. (9)).

Doping of materials with radioactive probes
Three processes are commonly used to introduce probe atoms into metals:
Implantation, diffusion and melting |6|.
Implantation has been proved to be the most versatile process, because it offers the following advantages :
- Range and concentration profile can be easily controlled;
- conventional mass separation enables one to implant only the desired radioactive isotope, which usually means low doses;
- the doping can be extremely clean, if a proper ultrahigh vacuum system
is used, with differential pumping of accelerator and target region;
- the influence of the surface region can be mostly avoided;
- the doping can be carried out at any target temperature, which is especially
important for the investigation of the self-produced damage cascades (correlated damage);
- almost all materials can be doped.
There are of course features connected to the implantation process, which under
certain circumstances prove to be disadvantageous:
- Implantation is always accompanied by radiation damage;
- homogeneous doping of thicker probes is not easily accomplished.
These two problems do not hold for the muon, since due to its small mass the
muon is not affected by its own radiation damage (correlated damage) and the
irradiation dose of about 10^{10} particles can be neglected. Secondly,due to its
high kinetic energy of several ten MeV a homogenous doping is easily achieved.
However, in contrast to the TDPAC technique, the muon does not observe the migration of defects but only the result of their mobilisation. The sensitivity to migrating defects is unique to TDPAC or Mössbauer effect and
offers the way to break the symmetrical behaviour of vacancies and interstitials
in annealing experiments as induced by the irradiation process.
Diffusion is in many cases easier than implantation, because it can be performed
with less technical expenditure. It guarantees homogenous doping and causes no
radiation damage, though most of the advantages which distinguish implantation
cannot be reached by diffusion.
Nearly the same holds for melting. There the sample preparation is further restricted by the general limitations of alloying processes.

Creation of lattice defects and isochronal annealing
Handling of the specimens means the controlled irradiation and annealing of
the doped metals. Due to their microscopic character the presented methods are
sensitive to the local energy density deposited by the irradiating particle. Therefore it is very desirable to study the samples under various damage conditions,

ISOCHRONAL ANNEALING

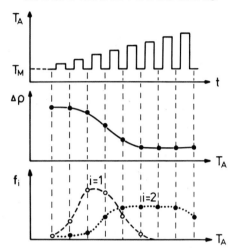

<u>Fig.3</u>: Typical isochronal annealing program to investigate the behaviour of defects: The specimen is heated stepwise to increasing annealing temperatures T_A (upper diagram). The specimen is measured at a temperature T_M lower than T_A. In resistivity measurements (middle diagram), the thermal mobilisation of defects is reflected by a stage-like resistivity decrease (so-called recovery stage). In nuclear methods (bottom), the mobilisation of defects is described by a relative trapping probability f_i with its characteristic HFI. Therefore, separation of different types of defects is possible in the latter case.

i. e. different mass of the projectiles, using electrons, protons or heavy ions (neutrons), or different particle dose. Another possibility to investigate the nature of a distinct defect recovery stage is a comparison of quenched and irradiated samples; whereas the quenching process, i. e. the rapid cooling of the sample from a temperature near the melting point, produces exclusively vacancies, the irradiation process creates vacancies and interstitials with equal concentrations. As in residual resistivity measurements, an isochronal annealing sequence is also applied in defect studies by nuclear methods (fig.3). Usually, the samples are heated stepwise to annealing temperatures T_A for about 10 min., while the measurements are carried out in between at lower temperatures T_M. At certain temperatures T_A a mobilisation of defects occurs, which leads in the case of resistivity measurements to a reduction of $\Delta\rho$, while in nuclear methods the mobile defects may be trapped at the radioactive atoms thus giving rise to a fraction f_i in the HFI. The disappearance of this fraction at higher temperatures T_A may be due to detrapping, annealing by antidefects or changing to other defect configurations. In contrast to $\Delta\rho$ measurements, the HFI methods are able to distinguish, by the defect induced hyperfine interaction, between different defects becoming mobile at the same temperature T_A, which is indicated in the lower part of fig. 3.

DEFECTS IN MAGNETIC METALS

μ^+SR in electron irradiated iron

Defect studies using the muon as probe are known for the diamagnetic metals Al |7,8| and Nb |9|. In these cases, however, the interaction of the muon with the lattice defects was observed via an increase of the line width σ_i (see eq.(9)), whereas a drastic change of the magnetic field at a site of a muon trapped in a vacancy cannot be expected, because there is nearly no change in the local magnetisation.

Recently, defects were investigated in iron, irradiated with 3 MeV-electrons at 10 K |10|. The resulting concentration of Frenkel pairs was about 85 ppm.

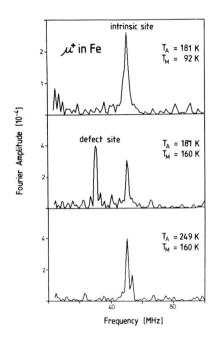

Fig.4: Results of a μSR measurement at electron irradiated iron. Displayed are the Fourier transforms of the experimental time spectra (see Fig.1) which show that the muon spin can precess with two different Larmor frequencies ω_L caused by the magnetisation at an interstitial and a vacancy site.

The preliminary results of μSR measurements performed at the "meson factory" SIN (Switzerland) are shown in fig. 4. According to eq.(9) each field site i, characterized by the Larmor frequency ω_{Li}, appears in the Fourier transform as a single line. Annealing of the Fe sample at $T_A = 181$ K shows one frequency after a measurement at $T_M = 92$ K. This frequency is well known from earlier investigations |11| and is caused by the local magnetisation at an interstitial site, the normal place of the muon due to its size; we call it the "intrinsic" line. Increasing T_M to 160 K, however, results in the appearance of a new line, not known until now. The last spectrum is measured again at 160 K after annealing above the recovery stage III (220 K) at $T_A = 249$ K. The absence of the second line proves that this line is caused by trapping of the muon at a defect, that disappears during the recovery in stage III. It is known that in this stage monovacancies annihilate or are annihilated so that the defect line is caused by trapping of the muon in a monovacancy. It is obvious that at this site the magnetisation should be drastically different from that at a normal interstitial site what is reflected by the experimental result. The fact that below stage III at lower temperature T_M the defect line is not visible (first spectrum in fig.4) is due to the low mobility of the muon at these temperatures, which inhibits an arrival of the muon at the defect within its lifetime.

Nickel implanted with ^{111}In

The first successful investigation of radiation damage using an ion implanted probe atom was carried out by Hohenemser et coworkers |12| in nickel. They observed the defect induced MI at the probe ^{111}In/^{111}Cd in a completely magnetized

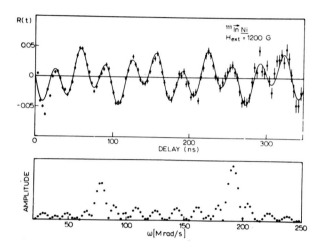

<u>Fig.5</u>: Magnetic interaction at ^{111}In implanted in Ni after annealing at 400 K.
The Fourier transform shows the components with sharp frequencies $2\omega_{L1}$ and
$2\omega_{L2}$. The sample was magnetized perpendicular to the detector plane |12|.

nickel foil applying the TDPAC technique. Usually, in TDPAC experiments, several
angular correlation spectra are measured simultaneously, so that after subtracting the accidental event, the spectra can be combined to give a single TDPAC
spectrum R(t) that is proportional to $A_2G_2(t)$ as described in section 3.1.1. The
time spectrum R(t) in fig. 5 has to be described by two fractions f_1 and f_2
associated with the frequencies $2\omega_{L1}$ and $2\omega_{L2}$. This result is easier recognizable using the Fourier spectrum given below. Similar to the µSR experiment in
iron there is an intrinsic line, now due to a substitutional probe atom,
and a defect line showing a smaller magnetisation than the intrinsic one. In an
isochronal annealing programm the trapping and detrapping of the defect producing the lower frequency were followed up. In contrast to the µSR experiment,
however, the ^{111}In probe directly observes the migration of the defects and not
only the result of the migration. Further the ^{111}In/[111]Cd probe is capable to
measure the QI too. The absence of a QI comparable in strength to the MI points to
a cubic symmetry around the probe atom, so that this In-defect configuration is
interpreted by the authors to be a tetravacancy with the In atom in its center.

DEFECTS IN DIAMAGNETIC METALS

In the preceding section the MI was the relevant HFI, which was used to characterize a defect preferentially in a ferromagnetic material. Now, the QI will be
used, an interaction very sensitive to any change of the local symmetry around
the probe atom. In a cubic metal a probe atom on a substitutional site does not
experience any QI, thus a lattice without defects does not show up any interaction.
The situation is different for diamagnetic hcp-metals: Like in the ferromagnetic
samples an intrinsic lattice signal exists already without defects.

<u>Results in the fcc-metal gold</u>
For a polycrystalline sample with an axial symmetric efg ($\eta = 0$), we expect
the precession frequencies ω_{on} to have the ratio 1 : 2 : 3 and the transition

amplitudes s_{2n} to be about $3 : 2 : 1$, $(n = 1,2,3)$. Thus only the QI frequency ν_{Qi} and an associated fraction f_i can be extracted. A TDPAC example is shown in fig. 6 |13|. ^{111}In doped polycrystalline Au foils are irradiated with electrons at 4.2 K. At an annealing temperature T_A = 77 K, the R(t) spectrum is nearly unperturbed. The slight slope results from a distribution of efg's caused by defects further away from the probe atoms. The second spectrum at T_A = 257 K reveals a periodic structure, the superposition of the frequencies ω_{on} with the respective amplitudes s_{2n}. The result is ν_Q = 91 MHz. At higher annealing temperatures, this structure disappears and a new one turns up at T_A = 320 K with a frequency ν_Q = 40 MHz, evidently due to a different, but also axially symmetric defect configuration. At T_A = 873 K, the probe is annealed out and the full, unperturbed anisotropy is restored. On the right side of fig. 6 the corresponding Fourier transforms are shown. It can be easily seen that in the case of QI and for the spin I = 5/2 each interaction produces three lines (see eq.(6)) and that after annealing at 257 K and 320 K two different QI are measured and hence two different defect configurations are observed.

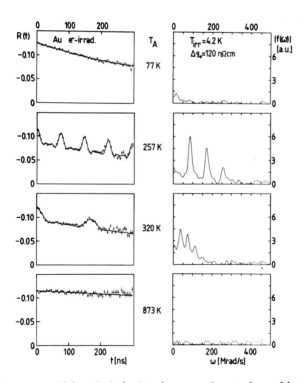

Fig.6: TDPAC spectra R(t) and their Fourier transforms of a gold specimen, doped with ^{111}In, irradiated with electrons, and annealed to temperatures T_A. The solid line is the result of a least-squares-fit of the function $A_2G_2(t)$ with $G_2(t)$ given in eq.(10). The uncorrected experimental value of A_2 is about -0.12 in these measurements.

Identification of defects. In most cases it is not possible to indentify directly a trapped defect by its characteristic parameters determined via the hyperfine interaction at the radioactive probe, since until now the theoretical models are not able to reproduce these values with sufficient accuracy. Therefore we can only use these parameters to uniquely tag the trapped defect and by comparing a number of defect studies one can try to obtain systematics concerning the characteristic parameters of the different defect types.

Taking the monovacancy as an example, this point should be further elucidated. Concerning the absolute value of the efg induced at a probe atom, we do not have any ab initio calculations, but considering the configuration of a trapped monovacancy in a cubic metal one would expect that the perturbation of the cubic symmetry possesses axial symmetry around the line connecting vacancy and probe atom. That means, the induced efg is axially symmetric and points into the direction of a nearest neighbour's site, being a <110> or <111> lattice direction in fcc or bcc-metals, respectively. Measurements of the orientation of the efg produced by defects, which are indentified as a monovacancy by additional experiments or are at least a good candidate, show just this behaviour: In Cu and Au, the efg of the trapped vacancy points into <110> |4| and in Mo into <111> (see section 5.2); in all cases the probe atom is ^{111}In and the efg is axially symmetric. In such experiments one could benefit from knowledge of the efg tensor itself.

As mentioned above the description of the dynamic behaviour of defects is mainly based on the results of the resistivity method. So we know that the defect recovery in metals is governed by several recovery stages, where defects are highly mobile and a mutual annihilation of Frenkel pairs is possible. One interesting question is, which defect type effects the recovery in stage III, which is located for example around 250 K in the case of the noble metals Cu, Ag and Au. The well-known way to attack this question is to compare the experimental results (e.g. the thermodynamical parameters of defects) of an irradiated specimen with those of a quenched one. As after quenching vacancies are preferentially created, the identity of the results proves the vacancy character of the

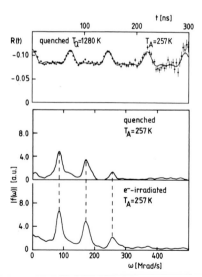

Fig.7: Quadrupole interaction of ^{111}In doped Au observed after quenching and electron irradiation. Comparison of the two Fourier transforms shows that in both cases the identical QI is observed. The vacancy character of the defect v_{Q3} has been proven in this way.

defects and vice versa. Due to the properties of the classical methods, it has not yet been possible to get unambiguous information in this way. Using the defect specific efg as a tag, however, one is able to recognize a particular defect type under several damage conditions. To get an impression of how clear a trapped defect can be recognized in a quenched and an irradiated sample, fig. 7 shows the results of TDPAC experiments in Au after quenching and electron irradiation. From the irradiation experiment it is known that a defect becomes trapped at ^{111}In around 250 K characterized by $\nu_Q = 91$ MHz and $\eta = 0$. In order to create preferentially vacancies, ^{111}In doped Au has been heated to near its melting point in a CO-atmosphere and quenched into a bath cooled to 220 K. At the top of fig. 7, the spectrum along with its Fourier transform is shown after annealing the specimen at 257 K. This Fourier spectrum is compared with the Fourier transform of the R(t) spectrum obtained after irradiation and annealing at 257 K (see fig. 6). Both the Fourier spectra in fig. 7 show three peaks at identical frequencies. In this way it can be unambiguously proved that the trapped defect has to be of vacancy type. It is evident that this method is superior to those using thermodynamic arguments to prove or disprove the vacancy character of a defect.

In several other metals such TDPAC experiments have been performed to identify defects migrating in stage III by trapping at the probe atoms ^{111}In, i.e. Al |14|, Cu |15| and Cd (see section 5.3).

Variation of irradiating particles. It is obvious that the technique to tag defects via their hyperfine parameters, opens the possibility to trace their behaviour under different damaging conditions, yielding changes of the defect structure in the lattice. This can be achieved using several types of irradiating particles like electrons, protons and heavy ions. Such experiments have been performed in Au |13| to investigate the behaviour of stage III around 250 K. In fig. 8 the results are collected which were obtained during an isochronal annealing program: Four distinct ^{111}In-defect configurations have been found (distinguished by the four line types) with

(1): $\nu_Q = 102$ MHz, $\eta = 0.45$,　　　(2): $\nu_Q = 101$ MHz, $\eta = 0$,
(3): $\nu_Q = 91$ MHz, $\eta = 0$,　　　(4): $\nu_Q = 40$ MHz, $\eta = 0$.

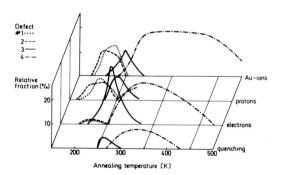

Fig.8: Influence of temperature and mass of irradiating particles on the relative occurence of different types of defects in gold. The different trapped defects are explained in the text.

It is evident, that their formation probabilities strongly depend on the type of the irradiating particle: electron, proton or a heavy particle as the self-ion Au. Complementary results observed in a quenched Au specimen are also displayed. Employing the TDPAC technique again, the efg has been used as the tag in this case. Whereas the defect structure after electron irradiation mainly consists of isolated Frenkel pairs, the structure after damaging with Au ions should be dominated by defect clusters. Consequently, the occurence of the different defect types changes too, indicating their different defect sizes. Configuration No. 3 is preferentially visible after electron irradiation, whereas No. 1 and 2 are nearly absent. Turning to the ion irradiation one sees an increase of configuration No. 1 and 2, and a decrease of No. 3, so that here the three formation probabilities are nearly equal. Therefore one can conclude, that No. 3 should be a trapped point defect and together with the results after quenching it can be identified as the monovacancy. - This result was confirmed by experimental determination of the direction of the induced efg that points into a <110> lattice direction (see above). On the other hand, No. 1 and No. 2 should be produced by the migration of small defect clusters. From the fact that both the configurations appear and disappear nearly at the same temperature and that their relative formation probabilities are nearly constant it is highly probable that the two configurations are produced by only one defect type trapped in two different configurations at the probe atom so that two different efg are produced. Moreover one sees a fourth defect in fig. 8 with the highest formation probability and thermal stability after heavy ion irradiation. This configuration has been indentified as a vacancy loop (see also section 5.1.4) formed at the In probe. From the whole experiment we get the information that in stage III of Au several defect types are mobile with slightly different activation energies, effecting the recovery of this stage and that already at the end of this stage defect clusters are formed. This example shows that hyperfine methods can easily distinguish between several defects with nearly identical migration energies.

Defect-antidefect reactions. As mentioned at the beginning of this report, the investigation of defects by their HFI with probe atoms is based on trapping mechanisms of the defects at the probe atoms. It seems that no such trapping exists for interstitials at ^{111}In in Au. The first Fourier transform in fig. 9 shows the result of a proton irradiation at 9 K. There are no lines visible which would be characteristic for a QI due to a trapped defect. This situation lasts during annealing until 200 K, where the known defects of stage III are observed.

From ME and PAC experiments in Al doped with [57]Co |16| or [111]In |17,18| and Cu doped with [111]In |19| one knows interstitials do produce a small but still measurable efg at the probe's site. However, Au belongs to the metals with an absent recovery stage I, so that the lack of interstitial trapping is of particular interest here: Besides a missing binding energy there is also the possibility that interstitials are not mobile below stage III, a question that is still under discussion. Taking into account the migration of interstitials to be always accompanied by the annihilation of vacancies it should be possible to solve this problem by HFI-techniques.

Extending the term "probe" to an impurity atom with a trapped monovacancy, one can be sure that this probe is sensitive to the migration of interstitials. Preliminary experiments showed that this way is fruitful to detect the migration of defects via defect-antidefect reactions. In the case of proton irradiated Au, the specimen was doped with "In-vacancy" probes by annealing at 255 K (see second spectrum in fig.9) and subsequently was post-irradiated at 9 K. Fig.9 shows the concentration of "probes" to decrease with increasing dose, i.e. monova-

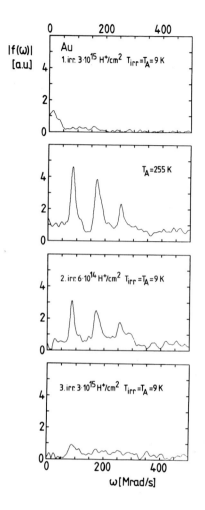

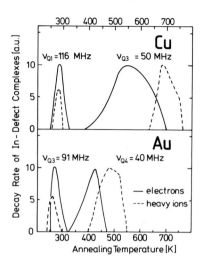

Fig.10: Influence of the damage conditions on the thermal stability of the impurity-cluster configuration in Cu (ν_{Q3}) and Au (ν_{Q4}). In contrast, the stability of the impurity-monovacancy configuration (ν_{Q1} in Cu, ν_{Q3} in Au) is unchanged.

Fig.9: Proton irradiation of Au doped with In-vacancy pairs at 9 K (see text for discussion).

cancies disappear. Since a detrapping of vacancies from the In atoms is impossible at 9 K, one has to assume interstitials to migrate at this temperature in Au |20|.

Thus, the application of defect-antidefect reactions extends the basis for the study of defects, where the binding energy to the particular defect is deficient.

Defect clusters. For the fcc-metals nickel, copper and gold ^{111}In-defect configurations were found, which exhibit a large thermal stability, recognizable by the temperature range where the particular configuration is "visible". These configurations are: ν_{Q2} = 54 MHz in Ni, ν_{Q3} = 52 MHz in Cu, and ν_{Q4} = 40 MHz in Au. They all produce an axially symmetric efg, i.e. η = 0, that points into a

<111> lattice direction. Another common property is their preferential formation after heavy ion irradiation; this point has been explicitly shown for Au in section 5.1.2. An accurate inspection of the experimental data, available for Cu and Au, shows that their thermal stability depends on the type of the irradiating particle, a behaviour, that is not typical for the defects discussed before. After electron irradiation the defect configurations decay at lower annealing temperatures than after heavy ion irradiation |24|.

The normalized decay rate of the fraction f in arbitrary units is displayed in fig. 10, i.e. the high temperature part of $f(T_A)$ has been differentiated with respect to the annealing temperature T_A. The maxima of the decay rate of configuration ν_{Q3} (Cu) and ν_{Q4} (Au) differ by about 100 K between electron and heavy ion irradiation, whereas the configurations ν_{Q1} (Cu) and ν_{Q3} (Au), which are assigned to trapped monovacancies, do not exhibit such a dependence. This property shows that there is an additional parameter that influences the stability of the observed complex, but not the efg at the impurity site. This parameter should obviously be identified with the size of the defect complex. Its influence on the thermal stability is known from transmission electron microscopy; the missing influence on the efg, on the other hand, can be explained by its short-sightedness, provided that the immediate neighbourhood of the probe atom does not change with the cluster size.

Due to their pronounced occurence under correlated damage conditions and as proved by quenching in gold, these clusters should possess vacancy character. Such clusters are known to preferentially collapse into planar loops, so that they form a stacking fault on a {111} plane |21|. The consequence is an alteration of the stacking sequence ABCAB of {111} planes in perfect fcc-metals by one missing plane, yielding a new sequence AB.AB, the stacking sequence of hcp-metals. A probe atom trapped in one of these stacking faults will thus experience a well-defined efg with axial symmetry and an axis pointing into a crystallographic <111> direction. On the basis of their common properties, the configuration ν_{Q2} in nickel should be interpreted in a similar way. A systematic study of its thermal stability would supply more information in this case too.

Results in the bcc-metal molybdenum

In the preceding sections it has been mentioned that in case of QI also structural information about the probe-defect configuration is available besides the strength of the interaction ν_Q being proportional to the tensor component V_{zz} (see eq.(11)): The axial symmetry of the efg tensor should be reflected by the geometry of the probe-defect complex too and the same holds in case of axial asymmetry expressed by the parameter η (see eq.(4')). A further important quantity is the orientation of the efg which can be easily interpreted in case of a trapped point defect producing an axially symmetric efg at the probe's site. In section 5.1 we have shown that in fcc-metals a trapped monovacancy creates an efg pointing towards the next-nearest neighbour, i.e. in a <110> direction. In the fcc-metals copper and gold this identification of the trapped monovacancy based on the orientation of the efg could be confirmed by additional experiments, as e.g. the comparison with the results found after quenching that seems to be an elegant procedure to prove the vacancy character of the trapped defect. Just this technique is strongly hampered in the case of bcc-metals due to their high melting point and their inclination to be contaminated by impurities. Supported by the findings in fcc-metals, however, one should expect for the trapped monovacancy in a bcc-metal like molybdenum an axially symmetric efg in <111> direction.

Monocrystalline molybdenum was implanted with 350 keV ^{111}In and annealed to 470 K, the recovery stage III, in order to mobilize and trap the produced

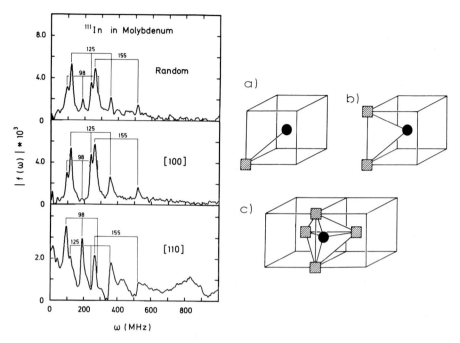

Fig.11: Fourier transform of TDPAC spectra of ^{111}In implanted molybdenum after annealing at 470 K. The spectra measured at a randomized crystal and at a crystal with two different orientations with respect to the γ-detectors show a drastic difference in the intensity of the several defect lines. This behaviour is used to determine the orientation of the respective efg's |22|.

Fig.12: Proposed ^{111}In-defect configurations in irradiated molybdenum. Black circles and cross-hatched squares represent the ^{111}In atoms and vacancies in the bcc-lattice, respectively |22|.

defects |22|. It is known that in this recovery stage different defect configurations are formed at the ^{111}In probe (see fig.11, first Fourier spectrum): v_{Q1} = 125 MHz and η = 0, v_{Q2} = 155 MHz and η = 1, v_{Q3} = 98 MHz and η = 0. Whereas for η = 0 three lines are observed in the Fourier transform (see fig.2), for η = 1 only two lines are observed because ω_{o1} becomes equal to ω_{o2}. Using a single crystal the intensities of the lines vary drastically, as soon as the orientation of the crystal is changed with respect to the detectors. This is shown by the second and third spectrum in fig.11, measured with detectors oriented in [100] and [110] directions, respectively. Comparison of the experimental data with the theoretical predictions yields a <111> and <100> efg orientation for the defects characterized by v_{Q1} = 125 MHz and v_{Q3} = 98 MHz, respectively. Since v_{Q2} possesses no symmetry axis an evaluation of that data is complicated. On the basis of these results the defect No.1 can be identified as a trapped monovacancy (see also fig.12a). Comparison with the predictions of a point charge model suggests a microscopic interpretation of the other configurations too, as shown in fig.12b and c for the defect No.2 and 3, respectively: Configuration b produces an asymmetry parameter η = 1 and configuration c produces an axially symmetric efg in a <100> direction.

Results in the hcp-metal cadmium

In contrast to the cubic metals, a non-cubic metal like cadmium gives an "active" signal, if the probe atom is on a substitutional site (fig.13, first spectrum), a situation similar to that in magnetic metals. On the other hand the experimental spectra are more complicated when trapped defects produce additional efg's, whereby each new efg normally gives three new lines in the Fourier spectra. In fig.13 a sequence of TDPAC measurements at different annealing temperatures is shown |23|. Before annealing the ^{111}In doped cadmium sample was irradiated with 10 MeV protons at 77 K. At 100 K and 290 K mainly the intrinsic signal is observed, whereas at 125 K and 145 K a second signal is visible caused by a trapped defect which is characterized by ν_{QI} = 103 MHz and η = 1. This defect was also observed after quenching proving its vacancy character. Since the de-

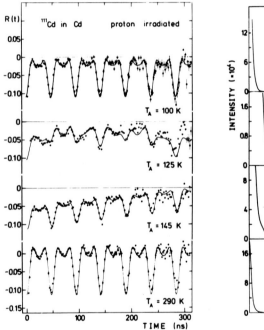
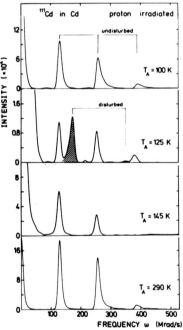

Fig.13: R(t) spectra of ^{111}In doped Cd (left side). Due to the hexagonal structure, a QI is observed in the unperturbed lattice too (first and last spectra). The corresponding Fourier transforms (right side) show the typical three frequencies for an efg with η = 0 in a polycrystalline sample. The rather complicated second spectrum, caused by an additional defect induced QI is quite simple to analyse after Fourier transformation. The defect-lines (dotted area) are clearly visible |23|.

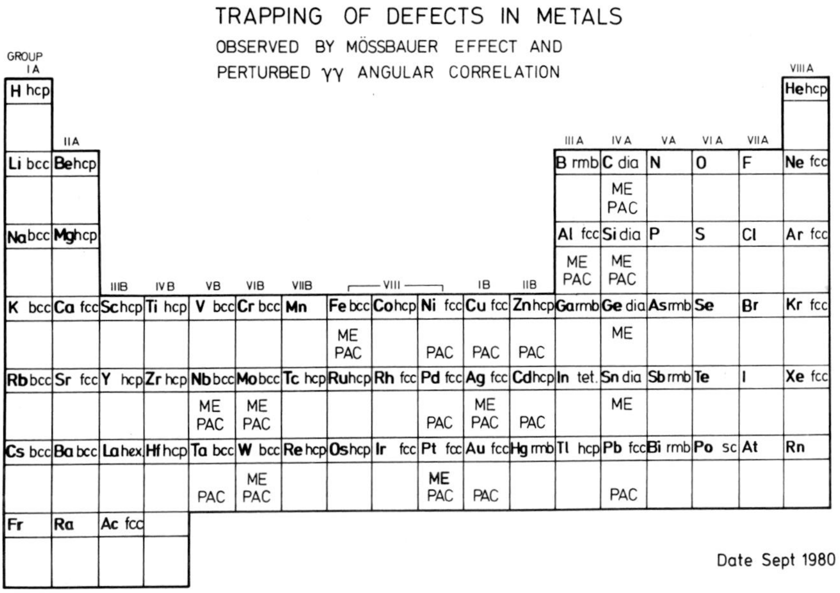

TRAPPING OF DEFECTS IN METALS
OBSERVED BY MÖSSBAUER EFFECT AND
PERTURBED γγ ANGULAR CORRELATION

GROUP IA	IIA	IIIB	IVB	VB	VIB	VIIB	←— VIII —→			IB	IIB	IIIA	IVA	VA	VIA	VIIA	VIIIA
H hcp																	He hcp
Li bcc	Be hcp											B rmb	C dia (ME PAC)	N	O	F	Ne fcc
Na bcc	Mg hcp											Al fcc (ME PAC)	Si dia (ME PAC)	P	S	Cl	Ar fcc
K bcc	Ca fcc	Sc hcp	Ti hcp	V bcc (ME PAC)	Cr bcc	Mn	Fe bcc (ME PAC)	Co hcp (PAC)	Ni fcc (PAC)	Cu fcc (PAC)	Zn hcp	Ga rmb	Ge dia (ME)	As rmb	Se	Br	Kr fcc
Rb bcc	Sr fcc	Y hcp	Zr hcp	Nb bcc (ME PAC)	Mo bcc (ME PAC)	Tc hcp	Ru hcp	Rh fcc (PAC)	Pd fcc (ME PAC)	Ag fcc (PAC)	Cd hcp	In tet.	Sn dia (ME)	Sb rmb	Te	I	Xe fcc
Cs bcc	Ba bcc	La hex	Hf hcp	Ta bcc (PAC)	W bcc (ME PAC)	Re hcp	Os hcp	Ir fcc	Pt fcc (ME PAC)	Au fcc (PAC)	Hg rmb	Tl hcp	Pb fcc (PAC)	Bi rmb	Po sc	At	Rn
Fr	Ra	Ac fcc															

Date Sept 1980

Fig.14: Collection of host-elements (no alloys), where defects are investigated by nuclear methods; in many cases the radioactive atoms were implanted. ME and PAC means investigation by Mössbauer effect and γ–γ perturbed angular correlation technique, respectively.

fect is trapped in the temperature region of stage III, also in this case the contribution of vacancies to the recovery in stage III could be shown. An interesting aspect of this experiment is the detection of another defect that produces vanishing efg in the hcp-lattice. This defect, which removes the characteristic efg of the lattice, displays properties similar to the clusters discussed for the fcc-metals (section 5.1.4), taking into account the hexagonal lattice structure and the lower melting point. Relative to the melting point, the defect possesses a high thermal stability and an alteration of the hcp stacking sequence ABAB effects a local fcc lattice with a vanishing efg; that is confirmed by the experimental value $\nu_{Q3} = 0$ MHz.

CONCLUSION

In the preceding sections the sensitivity of spin precession methods, especially of the μSR and the TDPAC, to detect lattice defects has been illustrated. The application of such methods in this field furnishes information about defects which are complementary to those known from other techniques. Secondly the methods, based on the trapping of defects, promote the understanding of impurity-defect interactions in metals. A collection of host-elements investigated by the Mössbauer effect (ME) and perturbed angular correlation technique (PAC) is given in fig.14. The number of definite results shows that the nuclear methods have overcome the stage of exploring their qualification for the study of lattice defects and are becoming a standard method in this field.

The authors wish to thank M. Deicher, O. Echt, H. Graf, G. Schatz and A. Weidinger for helpful discussions, and are grateful to the Bundesminister für Forschung und Technologie for financial support.

REFERENCES

1. H. Frauenfelder, R.M. Steffen in: Alpha-, Beta- and Gamma-Ray Spectroscopy, edited by K. Siegbahn, Vol. II, p. 997 (North-Holland, Amsterdam 1965).

2. E. Recknagel and Th. Wichert in: Proceedings of the International Conference on Ion Beam Modification of Materials, Albany, USA, 1980, to be published in Nucl. Instr. Meth.

3. E.N. Kaufmann and R.J. Vianden, Rev. Mod. Phys., $\underline{51}$, 161 (1979).

4. M. Deicher, O. Echt, E. Recknagel and Th. Wichert, contribution to this conference, p.

5. A. Seeger in: Hydrogen in Metals I, edited by G. Alefeld, J. Völkl (Springer, Berlin, Heidelberg, New York 1978) p. 349.

6. Site Characterization and Aggregation of Implanted Atoms in Materials, edited by A. Perez and R. Coussement (Plenum Publishing Corporation 1980).

7. K. Dorenburg, M. Gladisch, D. Herlach, W. Mansel, H. Metz, H. Orth, G. zu Putlitz, A. Seeger, W. Wahl and M. Wigand, Z. Physik, $\underline{B31}$, 165 (1978).

8. J.A. Brown, R.H. Heffner, M. Leon, M.E. Schillaci, D.W. Cooke and W.B. Gauster, Phys. Rev. Lett., $\underline{43}$, 1513 (1979).

9. D. Herlach in: Proceedings of the Second International Topical Meeting on Muon Spin Rotation, Vancouver, B.C., Canada, 1980.

10. H. Graf, T. Möslang, E. Recknagel, A. Weidinger, Th. Wichert, R.I. Grynszpan, to be published.

11. H. Graf, G. Balzer, E. Recknagel, A. Weidinger and R.I. Grynszpan, Phys. Rev. Lett., $\underline{44}$, 1333 (1980).

12. C. Hohenemser, A.R. Arends, H. de Waard, H.G. Devare, F. Pleiter and S.A. Drentje, Hyperfine Interactions, $\underline{3}$, 297 (1977).

13. M. Deicher, E. Recknagel and Th. Wichert, Rad. Effects, to be published.

14. H. Rinneberg and H. Haas, Hyperfine Interactions, $\underline{4}$, 678 (1978).

15. Th. Wichert, M. Deicher, O. Echt, E. Recknagel, Phys, Rev. Lett., $\underline{41}$, 1659 (1978).

16. W. Mansel and G. Vogl, J. Phys., $\underline{F7}$, 253 (1977).

17. H. Rinneberg, W. Semmler and G. Antesberger, Phys. Lett., $\underline{66A}$, 57 (1978).

18. W. Semmler, R. Butt, H.G. Müller in: Proceedings of the Fifth International Conference on Hyperfine Interactions, Berlin 1980, edited by H. Haas and G. Kaindl, to be published in Hyperfine Interactions.

19. M. Deicher, R. Minde, E. Recknagel and Th. Wichert in ref. 18

20. Th. Wichert, M. Deicher, E. Recknagel, to be published.

21. M. Wilkens in: Fundamental Aspects of Radiation Damage in Metals, edited by M.T. Robinson and F.W. Young, Jr. (US-ERDA-Conf. 751006, 1976) p. 98.

22. A. Weidinger, R. Wessner, Th. Wichert and E. Recknagel, Phys. Lett. $\underline{72A}$, 369 (1979) and
A. Weidinger, R. Wessner, E. Recknagel and Th. Wichert in ref. 2.

23. W. Witthuhn, A. Weidinger, W. Sandner, H. Metzner, W. Klinger and R. Böhm, Z. Physik, $\underline{B33}$, 155 (1979).

24. M. Deicher, O. Echt, E. Recknagel and Th. Wichert in ref. 18.

Kaufmann and Shenoy, editors
Nuclear and Electron Resonance Spectroscopies Applied to Materials Science

MUONS AS LIGHT HYDROGEN PROBES-DIFFUSION AND TRAPPING

D. RICHTER
Institut für Festkörperforschung, Kernforschungsanlage Jülich, West Germany

ABSTRACT

Contributions of the muon spin rotation technique (μSR)
to metal physics problems are surveyed. The similarity
between the muon and the proton constitutes μSR as a new
tool to investigate the physics of H-like interstitials on
an atomic scale. Experiments on the lattice location and
on local properties like the Knight shift as well as
diffusion measurements are reviewed. In particular the
extreme sensitivity of muons toward impurities and lattice
defects is emphasized. Results on the trapping of muons
by substitutional impurities or vacancies in the ppm range
are displayed.

INTRODUCTION

During the last few years, muon spin rotation (μSR) has emerged as a new
kind of spectroscopy for microscopic investigation of condensed matter. Although μSR as a technique to study solid state phenomence dates back less than
10 years, its application ranges already from fields like kinetics of chemical
reaction to semiconductors and in particular to metal physics which will be the
concern of this paper /1/. Muon sources, which require intense primary medium
energy proton beams, at present are available at the so called meson factories:
SIN in Switzerland, TRIUMF in Canada, LAMPF in USA and KEK in Japan, and at the
synchrocyclotrons at CERN, Dubna and Gatchina the latter both in the USSR. New
facilities at Brookhaven (USA), IKO (Amsterdan) and Rutherford Laboratory (UK)
are under construction.

With a mean lifetime of τ_{μ}=2.2 μs the muon decays into a positron and 2 neu-
trinos. The correlation between the direction of the emitted positron and the
spin direction of the muon allows one to measure the spin precession frequency
and/or the decay of the muon polarization of an ensemble of muons implanted in
a solid. Static as well as dynamic properties can be investigated. In magnetic
materials the precession frequency reveals information on the local field at
the interstitial muon site similarly to Mößbauer and NMR experiments which
yield the local fields on substitutional sites. In nonmagnetic materials Knight
shift measurements cast light on the local electronic structure around the
muon. The investigation of the field dependent decoupling of electronic and
magnetic interactions allows the determination of the electric field gradient
(EFG) created by the muon on its neighboring atoms, as well as the assignment
of the muon lattice location.

The time scale μSR can cover is related to the occuring magnetic field fluctuations $<\Delta B^2>$ and the lifetime of the muon. In nonmagnetic substances,
where $<\Delta B^2>$ is caused by nuclear dipolar fields, the resulting time scale
amounts to $10^{-7} \leq t \leq 10^{-5}$ s, in magnetic materials it can be extended down
to 10^{-12}s. Field fluctuations may occur either due to dynamic processes in the
host material like spin relaxation in spin glasses /2/ or by motion of the muon
itself. Concerning muon diffusion two aspects are of importance: (i) Since the

muon can be considered as a light isotope of the H (m_μ = 1/9m_p) its intrinsic diffusion properties are of great interest. Early experiments on Cu, Nb and Ta /3-6/ yielded simple monotonic temperature dependences of the muon mobility which could be understood in terms of small polaron diffusion /7/. Later experiments on high purity Nb and Al samples, however, revealed that most of the early results were governed by impurity trapping rather than by intrinsic diffusion /8-10/. Furthermore, for a particle at the borderline between small polaron-like localization and band like delocalization the degree of disorder present in the host material can determine the possible diffusion mechanisms /11/. (ii) The extreme sensitivity of the muon toward lattice imperfections makes it a powerful probe for the investigation of small impurity and defect concentrations. Early attempts of defect studies in metals by positive muons, however, have not been encouraging. Heavy neutron irradiation on Cu and Al /12/ showed virtually no effect. Endeavours to observe trapping on thermally induced vacancies close to the melting point in Cu and Al remained unsuccessful /13/. The first evidence for muon trapping on intrinsic defects was found by Kossler et al. /14/ on deformed Al and by Dorenburg et al. /15/ on neutron irradiated Al at low temperatures. Since then, trapping on structural vacancies in nonstoichiometric Al-Cu, Al-Ni compounds /16/, on quenched vacancies and divacancies in Al /17/ and on vacancies created by electron irradiation in Nb /18/ and Al /19/ has been reported. The first clear evidence for impurity trapping has been achieved by Borghini et al. /8/ on Nb doped with varying concentrations of N impurities. They were soon followed by similar studies on V /20/ and oxygen doped Nb /21/. Al can be purified easily and shows high muon mobilities down to very low temperatures. These two properties made it the primary host material for recent investigations on impurities /11, 14, 22, 23/. In particular careful concentration dependent studies on the $AlMn_x$ system demonstrated muon trapping by single impurity atoms. Mn-concentrations as low as 5 ppm were found to trap significant fractions of the implanted muons revealing the extreme sensitivity of muons /24,25/. These experiments also show the necessity for extremely well characterized samples in order to get meaningful results.

THE μ SR METHOD:

Muons (typical properties are displayed in table 1) are created by the weak interaction decay of pions. As a consequence the muons are polarized with respect to their momentum in the rest frame of the pions. In the standard experimental set-up, where pions decay in flight, the extracted muons have about 75% polarization in the laboratroy frame and a linear momentum of 100-300 MeV/c. In order to achieve maximum stopping of muons in the sample they are degraded in energy before the target. High stopping power is then obtained by samples offering a few grams per cm^2 beam cross section. An alternative set up constitutes the surface or Arizona beam, where the pion decays at rest. In this case beam polarization is 100% and the linear momentum amounts to 29 MeV/c.

TABLE I
Properties of the Particles Muon and Proton

particle	mass	spin	magn. moment	lifetime
muon	105.659 MeV/c^2	1/2	3.18334 μ_p	τ_μ = 2.197 $\cdot$ 10^{-6} s
proton	938.28 MeV/c^2	1/2	μ_p = 2.7928 μ_N	stable

At these low energies a sample thickness of typically 0.1 mm is sufficient. The advantage of small samples has to be traded off against difficulties with e.g. windows in the beam and strong aberrations in applied magnetic fields.

In order to determine the time evolution of the muon polarization, knowledge about the time which the muon spends in the solid is required. This can be achieved in two ways: (i) Normally the incoming muon starts a clock and the decay positron stops the clock when it hits a detector telescope at a certain angle with respect to the beam direction. This approach allows only one muon at a time in the sample. (ii) Alternatively in the so called stroboscopic method narrow pulses of muons are implanted in the sample requiring an appropriate time structure of the proton accelerator /26/. During an experiment typically $10^6 - 10^7$ events are recorded and are arranged in a time histogram. The number of positrons counted in a certain time interval measures the direction and magnitude of the muon spin polarization at the time these muons decayed, since the e^+ emission probability is peaked in the μ^+ spin direction.

In the case of transverse field geometry, which is by far the most common one for metal studies, the applied magnetic field is perpendicular to the plane made up by the beam and the detectors. Then, the incoming muon, which loses virtually no polarization during the fast thermalization process in a metal /27/ starts to precess with a Larmor frequency ($\omega_\mu = \gamma_\mu B = 13.55$ KHz/Gauss) and the resulting time histogram is shown schematically in fig. 1. The muon decay function is superimposed by an oscillatory component which has a maximum whenever the μ^+ spin points in the direction of the counter. For the time dependent counting rate we have

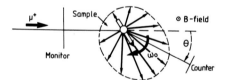

Fig. 1. Schematic sketch of transverse μSR set up, the time dependence of the resulting e^+ counting rate is shown below.

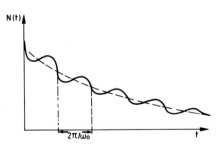

$$N(t) = N_o \exp(-t/\tau_\mu)(1 + a_o P(t) \sin(\omega_\mu t + \phi)) \tag{1}$$

with a_o typically between 0.2 and 0.25. It allows the determination of the precession frequency as well as the damping of the spin polarization $P(t)$. The range of internal fields accessible to the experiment lies between 20 G (2 precessions in a time window of 10 μsec) and 50 KG limited by the time resolution of the electronics. The accuracy can be pushed to a few ppm. The damping of the muon polarization P(t) is caused by random fluctuations of the local field originating e.g. from the nuclear dipolar moments in a nonmagnetic material. For a immobile muon the lineshape of P(t) is Gaussian

$$P(t) = P_o \exp (-\sigma^2 t^2) \qquad (2)$$

where $\sigma^2 = 1/2 \ \gamma^2 \ B^2_{dip}$ is the second moment of the frequency distribution due to internal fields. For a mobile muon the field fluctuations are averaged and motional narrowing occurs. The depolarization than has the form

$$P(t) = \exp (-\sigma^2 \tau_c t) \qquad (3)$$

where τ_c is the average time during which correlations between the frequencies exist. τ_c is proportional to the mean residence time τ of the muons at its interstitial sites. For the whole range of mobilities a formula, introduced originally to evaluate lineshape profiles of diffusing particles in NMR /28/, can be applied

$$P(t) = \exp (-2\sigma^2 \ \tau_c^2 \ \{ \ \exp (-t/\tau_c) - 1 + t/\tau_c \ \} \ ) \qquad (4)$$

For longitudinal relaxation in zero and low field NMR a stochastic theory has been worked out by Kubo and Toyabe /29/ and applied to μSR by Yamazaki /30/. For large enough longitudinal fields, dephasing (T_2) properties caused by nuclear dipoles can be separated from "T_1 relaxation" due to fluctuations of electronic spins. Experimentally the longitudinal relaxation function $P_z(t)$ is obtained in forward backward geometry, where the ratio

$$\frac{N_f - N_b}{N_f + N_b} = AP_z (t) \qquad (5)$$

is measured. For the limiting case of zero external field in a nonmagnetic material the longitudinal relaxation function allows one to distinguish effects of static field broadening from dynamic spin fluctuations: In particular this is important in systems with slowly fluctuating spins such as spin glasses or critical magnetic systems. Another interesting feature of the zero field method has been pointed out by Petzinger /31/: For an immobile muon after an initial decay, $P_z(t)$ increases asymptotically for long times to 1/3 of its initial value, since precession due to z-components of B_{dip} should not change P_z on the average. For a muon undergoing trapping processes this should enable one to distinguish whether a particle diffuses toward a trap or escapes from it. Only in the case when the muon ends up in a trap should the restoration of $P_z(t)$ be observed.

LATTICE SITE AND LOCAL PROPERTIES

A precondition for a quantitative interpretation of μSR results e.g. on Knight shifts, internal fields and diffusion properties is the knowledge of the muon lattice site. In addition, it is expected that the muon and the H atom occupy the same type of site. Thus, muon data might be able to complement the still rather limited data on H-location. The clues to solving the problem of the lattice location are the field and orientation dependence of the static dipolar width. For unlike spins and dipolar coupling in nonmagnetic solids Van Vleck /32/, has evaluated the linebroadening to be

$$\sigma^2 = \hbar^2/6 \, \gamma_\mu^2 \gamma_I^2 \, I(I+1) \sum_j \frac{(1-3\cos^2\theta_j)^2}{r_j^6} \tag{6}$$

where r_j is the vector from the muon to the nuclear spin I_j and θ_j is the angle between $\underline{r}_j$ and the external field $\underline{B}_0$. In a single crystal equ. (6) will depend on the crystal orientation relative to $\underline{B}_0$. For a particular interstitial site characteristic σ-values for the principal directions {100}, {110} and {111} can be calculated, which differ strongly for different sites. Thus, a measurement of these so called Van Vleck values should be enough to determine the interstitial site. First experiments on Cu applying low transverse fields, however, showed very little difference in the σ-values for the three main crystal directions /33/.

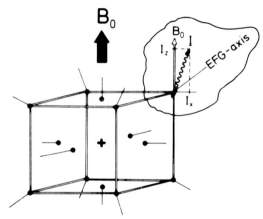

Fig. 2. The EFG produced by a muon (indicated by the +) on its neighboring atoms. B_0: external field.

Subsequently Hartmann pointed out that high transverse fields have to be applied in order to reach the σ-values given by the dipole sums /34/. The muon as a charged impurity creates to first order a radially directed electric field gradient (EFG) at the nearby nuclei (Fig. 2). If these nuclei possess a quadrupole moment, then the direction of the external field is no longer the quantization axis for the nuclear moments and the orientational dependence of the dipolar width is changed drastically. The correct expressions for σ have to be calculated with the nuclear spins I undergoing combined electric and magnetic interactions. The result for the dipolar width depends strongly on the relative interaction strength $b = \omega_B/\omega_E$ where ω_B is the Larmor frequency $\omega_B = \gamma_\mu B_0$ and ω_E is an electric interaction frequency determined by the quadropole moment and the strength of the electric field gradient. For small b the electric interaction dominates and σ will be nearly isotropic. For $b \gg 1$ the Van Vleck values are recovered. The cross over from electric to magnetic behavior occurs around $b=5$. Experiments on the field dependence of σ for the 3 main directions have been undertaken for Cu /35/ and Al doped with 1000 ppm Mn /11/. Figs. 3 and 4 compare the experimental results with the predictions of Hartmann's theory. It is obvious that the muon occupies octahedral sites in Cu and tetrahedral sites in Al $Mn_{0.001}$ around 15 K. In both cases the observed σ-values are lower than the theoretical values, which for example in the {111} -direction in Al is 0.32 μs^{-1} compared to the observed 0.24 μs^{-1}. This effect can be understood in terms of a 5% lattice expansion

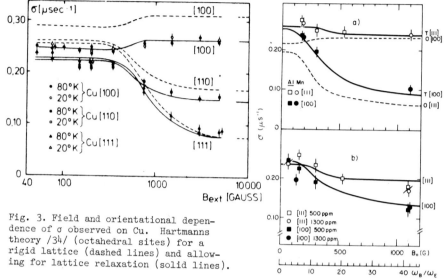

Fig. 3. Field and orientational dependence of σ observed on Cu. Hartmanns theory /34/ (octahedral sites) for a rigid lattice (dashed lines) and allowing for lattice relaxation (solid lines).

Fig. 4. Field dependence of σ for AlMn. a) T = 15 K, solid lines: tetrahedral site, dashed lines: octahedral sites; b) T = 5 K, theoretical curves are a sum of 30% octahedral and 70% tetrahedral occupation /23/.

around the muon site.

From the field dependence of σ, the strength of the electric field gradient created by the muon can be evaluated. For Cu $q = 0.27 \pm 0.15$ Å^{-3} and for $AlMn_{0.001}$ $q = 0.18$ Å-3 have been found at the nearest neighbor atoms. The low accuracies are mainly due to the large uncertainties in the magnitude of the quadrupole moments of $Cu^{61,63}$ and Al^{27}. Still, they are of considerable theoretical interest, since they represent EFG values very close to the charged impurity and allow critical tests of electronic screening calculations. For the case of Al, Jena et al. /36/ using the self consistent formalism of Hohenberg, Kohn and Sham /37/ arrived at 0.17 Å^{-3} for a tetrahedral muon site in Al in good agreement with the experiment. Another experimental observable related to the local electronic structure around the muon is the muonic Knight shift in nonmagnetic materials and the Fermi-contact field in ferromagnets. A series of Knight shift measurements, mainly on alkali and alkaline earth metals, have been performed at SIN /38/. The results, however, seem to show no systematic relation to the electronic properties of these materials and a theoretical explanation is still missing. For a recent review of muon interactions in magnetic materials we refer to reference /39/.

MUON DIFFUSION

The investigation of muonic diffusion is of great interest both for theoretical as well as practical reasons. On the one hand, any quantitative evaluation of μ^+ defect or impurity studies in metals requires knowledge of the μ^+ diffusion. On the other hand, the transport mechanism of light interstitials itself poses challenging theoretical problems. Muon diffusion studies allow one to investigate the isotope effects in H diffusion toward lighter masses, where quantum effects in diffusion are expected to be more pronounced. Certain advantages lie in the fact that μSR works in infinite dilution, and since muons are implanted, temperature ranges and materials are accessible where H diffusion studies are impossible due to low solubilities.

For a light interstitial like the muon or the proton, small polaron tunneling-hopping is regarded to be the basic motional mechanism. In this process the energy difference between adjacent interstitial sites caused by lattice relaxation (polaron effect) are equalized by thermal fluctuations and a sub-barrier tunneling process occurs /7,40/. At low temperatures ($T \ll \theta_D$) 2-phonon processes dominate while at higher T ($T \simeq \theta_D$) multiphonon processes take over. For the temperature depeendence of the rate jump $1/\tau$ one has

$$1/\tau \sim J^2 T^7 \qquad\qquad (T \ll \theta_D)$$
$$1/\tau \sim J^2 T^{-\frac{1}{2}}\; e^{-E_a/kT} \qquad (T \simeq \theta_D)$$

(7)

where J is the tunneling matrix element between adjacent sites. Unlike in classical over-barrier jumps, where the activation energy is related to the barrier height, here E_a is the energy necessary to equalize energetically the initial and final position.

Holstein /40/ and later Kagan and Klinger /41/ theoretically studied the possibility of so called 'coherent transfers' which occur without the assistance of phonons. This band like propagation can be destroyed dynamically by two phonon scattering processes resulting in an inverse T-dependence of the coherent diffusion coefficient

$$D_{coh} \sim J^2_{eff}\; T^{-n} \quad n = 7.....9 \qquad\qquad (8)$$

and by static disorder leading to Anderson localization /42,11/. In a coherent transfer, the muon together with its disortion cloud has to tunnel. Thus the bare overlap J is reduced to $J_{eff} = J \exp (-S)$, where S has a similar structure as a Debye Waller factor /43/. Typical values for exp (-S) at T = 0 lie between 0.1.....0.2. Theoretical estimates for the transition temperature between coherent and phonon assisted tunneling vary widely (e.g. around 30 K for Al, Kehr recently /44/; about 190 K for Cu, McMullen and Bergerson /45/).

Recently the question, whether the muon is always self trapped, has been raised. Stoneham and collaborators have investigated the general conditions for self-trapping /46,47/. At low T it should be determined by the bandwidth $B \sim J$ in the non-trapped state and the energies—thermal and/or elastic-needed to overcome a certain barrier to self-trapping which decreases rapidly with decreasing bandwidth.

As already described in section 2, μ^+ diffusion is conventionally investigated in transverse field geometry. Fig. 5 presents /uSR data on Cu which seemed to be the 'text book' case for a long time /3/. From a low temperature static linewidth, the damping parameter Λ (Λ is the inverse time at which the polarization according to equ. (4) decays to 1/e of its initial value) starts to decrease around 70 K and falls off monotonically until it reaches complete motional narrowing around 300 K. The temperature dependence of the correlation times τ_c obtained from a fit with equ (4) has been analyzed in terms of small polaron hopping theory by Teichler /48/ resulting in J = 17.5 μeV and E_a = 75 meV. The corresponding data on H (E_a = 400 meV, prefactor 10^{14} s^{-1} compared to $10^{7.5}$s^{-1} for the muon) seem to indicate a qualitative change of the diffusion mechanism from subbarrier tunneling in the case of the muon characterized by small activation energy and prefactor to a more classical behaviour for the H with a prefactor of the order of an attempt frequency and a large activation energy characteristic of a high barrier.

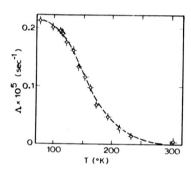

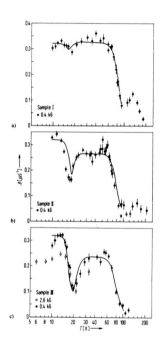

Fig. 5. Muon depolarization data obtained from Cu /3/.

Fig. 6. Temperature variation of the damping parameter Λ for muons in Nb of different purities /8/ (see text).

The history of muon diffusion results on bcc metals, however, advises caution in interpreting line narrowing data as intrinsic diffusion. Taking Nb as an example, the critical influence of impurities on μ^+ diffusion results will be demonstrated. Fig. 6 presents μ^+ diffusion data obtained from Nb containing different amounts of impurities /8/ [sample I with 3700 ppm N; sample II with 60 ppm N+O sample III with 15ppm N+O(in addition, all samples contained about 100 ppm substitutional impurities, mainly Ta)]. While the upper curve (sample I) shows almost the same temperature behavior as the Cu-data, at lower impurity concentrations a more complicated structure appears. The characteristic feature is a dip in the linewidth data around 20 K which deepens with decreasing impurity concentration. The data have been analyzed quantitatively in terms of a two state trapping model /49/ which will be outlined in the next paragraph. The qualitative description is the following. At low T the muons are localized in fixed lattice positions. Above 15 K they start to diffuse and cause the initial drop in linewidth. With increasing temperature the μ^+ mobility increased and the muons are able to reach the traps caused by the interstitial N atoms. An increase of the damping above 20K results. The depth of the dip phenomenon is naturally related to the interstitial impurity concentration, since the trapping probability decreased with trap concentration. In the broad plateau region between 30 K and 60 K, the muons experience the dipolar fields in the traps for most of their lifetime. The sharp drop of damping above 60 K indicates the beginning of escape processes from the traps. The motional narrowing is then caused by repeated capture and release processes. The dissociation energy from a N-trap has been evaluated as E_a^μ = 48 meV and can be compared with the corresponding value for H , E_a^H = 167 meV obtained from a quasielastic neutron scattering experiment /50/. Later experiments on ultra-pure Nb /51/ with an overall impurity concentration below 5 ppm are presented in fig. 7. The surprising result was a high muon mobility over the entire temperature region. In particular the low T plateau, observed in the samples I to III, vanishes. Thus, it cannot be an intrinsic muon property in Nb but has also to be related to impurity effects.

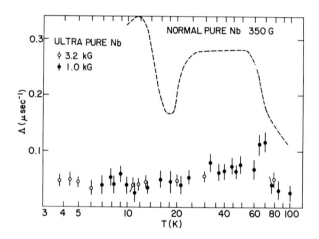

Fig. 7. Temperature dependence of the μ^+ damping parameter Λ in ultra pure Nb. Dashed line: result on sample II (Fig. 6).

For Al which is easily purified to 1 ppm overall impurity concentration, the outcome of μSR experiments was inverse to the situation in Nb. The first measurements have already shown a muon mobile down to the lowest accessible temperatures (presently 25 mK) /23,4/. Adding small amounts of substitutional impurities (e.g., Mn) leads to muon localization up to 20 K /11,23/.

Very recently the muon localization-delocalization phenomena, taking place in Al, were investigated systematically on Al doped with substitutional impurities in the concentration range between 5 and 75 ppm /25/. Mn which causes long range strain fields, and Li which disturbes the lattice only locally, were used as impurities. Fig. 8 presents the results on $AlMn_x$ below 2 K, where the muon depolarization increases with decreasing temperature. A rise of the Mn concentration enlarges the damping at a given temperature. The spectra were fitted with equ (4) and the resulting correlation times are displayed in fig. 9 in the

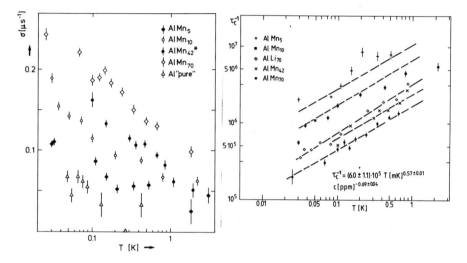

Fig. 8. Damping parameter σ obtained from a fit with eq. (2) for pure Al and doped Al samples below 2 K /25/.

Fig. 9. Inverse correlation time τ_c^{-1} as obtained from a fit of the exp. spectra with eq. (4) /25/.

form of a double logarithmic plot. Three features are noteworthy: (i) τ_c^{-1} follows a simple power law $\tau_c^{-1} \sim T^\nu$ over more than one order of magnitude in temperature; (ii) τ_c^{-1} decreases with increasing Mn concentration; the concentration and temperature dependence of τ_c^{-1} for Mn impurities can be described with good accuracy in a closed form

$$\tau_c^{-1} = (6.\pm1.1)\cdot 10^5 \, (T \, [mK])^{0.57\pm0.1} \, (c \, [ppm])^{-0.69\pm0.04} \, [s^{-1}] \qquad (9)$$

(iii) though causing very different long range distrubances, Li and Mn have essentially the same effect on the muon diffusion behaviour.

These experimental results can hardly be described by an intrinsic diffusion mechanism. In particular the concentration dependence and the lack of sensitivity toward long range strain components are in favour of an interpretation in terms of impurity trapping. In this picture the increasing depolarization with decreasing temperature is ascribed to a trapping rate, growing with falling T. Consequently a larger fraction of muons gets trapped at lower temperatures and depolarizes with the static linewidth. For diffusion limited trapping, the trapping rate $1/\tau_1$ is proportional to cD in qualitative agreement with the experiment. The AlLi$_x$ results are explained naturally since mainly the short range lattice distortion is important for trapping, which is of the same order of magnitude for Mn and Li. For the diffusion coefficient this interpretation involves an inverse temperature dependence

$$D \sim T^{-0.6} \qquad (10)$$

that is, the diffusion accelerates with decreasing T. In principle such a behavior is expected from Kagan's and Klinger's theory on coherent diffusion /41/. However, since two-phonon scattering processes are limiting the coherent transfers, high temperature powers have been predicted $D_{coh} \sim T^{-7...9}$. The much weaker temperature dependence cannot be accounted for and a theoretical understanding of these results is still missing.

Finally, some ideas concerning the importance of Anderson type localization phenomena /42/ for the muon in metals are discussed. In an ideal crystal at zero temperature the muon wave function would spread over the crystal volume, forming a band with a bandwidth of $J \cdot z$, where z is the coordination number of the muon sublattice. According to Anderson's criterion for localization, such a band state is destroyed even at zero temperature if large enough disorder is created in the crystal. For a distribution of random energy shifts ΔE at the muon interstitial sites (width Γ), there exists a critical value $\Gamma_c \geq J$ above which localization occurs. For the muon with a mass much heavier than the electron mass the corresponding bandwidth is much smaller than the electronic one and even very weak distrubances could be sufficient to fulfill the Anderson criterion.

Even in the most purified, isotopically pure, metallic samples like Nb or Al, the most important remaining disburbances originate from the long range elastic interaction between substitutional or interstitial impurities and the muon. It gives rise to small energy shifts ΔE between adjacent sites. The magnitude of ΔE in the intermediate region between impurities can be estimated by elastic theory /52/ and compared with the effective overlapp J_{eff} for the muon. In the following, estimates for ΔE are quoted for two examples: (i) in a very pure Cu sample (overall impurity concentration below 20 ppm) ΔE amounts to ~ 5 μeV. Estimates for J_{eff} from the bare tunneling matrix element obtained from the high temperature diffusion data /43,48/ range in the order of ~ 1 μeV. Thus, at very low temperatures the muon in high purity Cu could be at the brink of delocalization. Experimentally in such a sample a strong decrease of linewidth below 2 K has been observed which could be connected to a delocalization phenomenon /23/. However, due to its isotopic mixture, a Cu matrix may contain additional disorder which exceeds the impurity effect. (ii) for high purity Al (6 N), ΔE is of the order of 0.01 μeV. Doping with Mn at the concentrations of 10, 100 and 1000 ppm leads to disturbances ΔE of 0.2, 5 and 100 μeV, respectively /23/. Taking the μ+ diffusion data for Al which were extracted from vacancy trapping experiments at high temperatures /19/, a bare tunneling matrix element J of the order of a few meV can be evaluated and leads to J_{eff} values of the order of some 100 μeV. Thus, even at impurity concentrations in the range of 100 ppm, low T delocalization seems to be possible in accordance with the low temperature $AlMn_x$ (Li) data discussed before.

The peculiar temperature dependence $D \sim T^{-0.6}$ observed in these experiments might be related to similar observations regarding the electronic conductivity in heavily disordered metals /53/. There, the weak temperature dependence is explained with Anderson type localization theories including the effect of temperature /54/.

MUON TRAPPING

In the previous chapter, the strong interrelation between muon diffusion and trapping became evident. In this chapter, the trapping aspect is emphasized. Commencing with a description of a stochastic model for muon depolarization in the presence of trapping impurities, selected experimental results on impurity and vacancy trapping in Al are discussed.

The starting point of the stochastic trapping model is a random walk description
of the muon diffusion in a metal /49/ which can easily be extended to diffusion
in the presence of randomly distributed trapping centers. This is done in terms
of a so-called two state model which describes repeated capture and release processes and has been considered, e.g., for H-diffusion, in the presence of traps
by Richter and Springer /50/. A muon diffusing in the undisturbed lattice is
caught by a trap after an average time τ_1; a muon in a trap escapes on the average
after a time τ_0. Thus, $1/\tau_1$ is the trapping rate and $1/\tau_0$ the escape rate. The
two states have different polarization decays. The polarization decay in the
free state $P_1(t)$ is taken as the depolarization function of the homogeneous lattice; the depolarization at the trap in the simplest approximation is given by a
Gaussian decay $P_0(t) = \exp(-\sigma^2_t t^2)$ which assumes an immobile muon in the trapped
state. σ^2_t is the second moment of the frequency distribution in the traps.
Finally, $\psi_o(t) = \exp(-t/\tau_o)$ and $\psi_1(t) = \exp(-t/\tau_1)$ are the probabilities for
the muon to remain in the trapped state or in the free state until time t
after its capture or previous release, respectively.

   The basic feature of the random walk description is the separation of the
polarization decay into contributions from a fixed number of state changes

$$Q_1(t) = \sum_{l=0}^{\infty} R_1(t) \qquad Q_o(t) = \sum_{l=1}^{\infty} S_1(t) \qquad (11)$$

where Q_o and Q_1 are the polarization decays from muons disintegrating in the
trapped state or in the free state, respectively. (For simplicity of presentation only muons starting in the mobile state are considered which is a reasonable assumption for low impurity concentrations.) R_1 and S_1 are the
contributions from muons having performed 1 state changes until time t which
have led either to the free state or to a trap. The lowest order terms have
the following structure

$$R_o(t) = P_1(t)\psi_1(t)$$

$$S_1(t) = \int_o^t dt' \psi_o(t-t')P_o(t-t')1/\tau_1 \psi_1(t')P_1(t'). \qquad (12)$$

$S_1(t)$, for example, describes the polarization decay of muons which have stayed
in the free state until time t' (the depolarization there is given by P_1). At
t', a change into the trapped state occurred (its probability is given by
$1/\tau_1 \psi_1(t')$). Thereafter they depolarize in the trap with a depolarization
function $P_o(t-t')$ (the probability that the muons stay in the trap until time t
being $\psi_o(t-t')$). All possible contributions are obtained by integration over
t'. The higher order terms are similar convolutions including larger numbers
of state changes. The whole series of the random walk description (eq. 11)
can be generated from the integral equations,

$$Q_1(t) = P_1(t)\psi_1(t) + 1/\tau_o \int_o^t dt' P_1(t-t')\psi_1(t-t')Q_o(t')$$

$$(13)$$

$$Q_o(t) = 1/\tau_1 \int_o^t dt' P_o(t-t')\psi_o(t-t')Q_1(t').$$

These equations can be solved analytically by Laplace transformation. Finally the muon depolarization function $Q(t) = Q_o(t) + Q_1(t)$ is obtained by numerical inversion of the Laplace transform. As has been shown in Ref. /48/, the integral equations (13) can be approximated by a system of differential equations,

$$\frac{dQ_1}{dt} = 2\sigma^2 \tau_c [\exp(-t/\tau_c)-1] - 1/\tau_1 Q_1 + 1/\tau_o Q_o$$

$$(14)$$

$$\frac{dQ_o}{dt} = 2\sigma^2_t \tau_o [\exp(-t/\tau_o)-1] - 1/\tau_o Q_o + 1/\tau_1 Q_1.$$

The resulting depolarization function is a good approximation for linewidth as well as lineshape to the solution of the integral equation and can be utilized much easier for the purpose of fitting experimental data. It has been applied, for example, to analyze the muon data obtained from Nb with different amounts of impurities, shown in the last paragraph. The solid lines in Fig. 6 represent the result.

Other trapping models which start from the muon correlation function in a Gaussian-Markovian approach /55/ or which do not allow for escape processes /19/ are also in the literature but will not be discussed here.

As examples for the study of muon impurity interaction, recent experiments on Al are chosen. Al, which can be highly purified, offers nearly ideal experimental conditions to investigate trapping properties. (i) The nuclear dipole momemt of Al provides a convenient range of μ^+ relaxation rates and (ii), in pure Al muon diffusion is rapid over the entire accessible temperature range /23/.

Figure 10 presents results for the depolarization rate obtained from Al doped with Mn in the concentration range between 5 and 1300 ppm /24/. The linewidth in all the $AlMn_x$ samples exhibits a pronounced maximum around 17K. The height of the maximum depends on the impurity concentration c_{Mn}. Its position is not affected by c_{Mn}. Measurements on Al samples with other kinds of impurity atoms are shown in Fig. 11. The maxima in linewidth occur at different temperatures for each impurity. Qualitatively, these results can be understood in terms of muon trapping: the fraction of muons caught in a trap depends on the impurity concentration (see $AlMn_x$) and the escape process from the traps takes place at a temperature characteristic for each kind of impurity atom.

Quantitatively the data have been analyzed with the two-state model described above. Thereby fast diffusion in the undisturbed lattice was assumed and the depolarization in the 'free' state was neglected. The solid lines in the Figs. 10

246

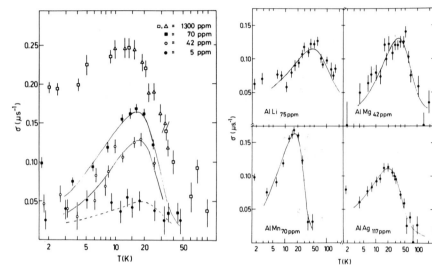

Fig. 10. Gaussian linewidth (σ) for muons in aluminum doped with manganese as a function of concentration and temperature. Solid lines show fit to the two state model. The dashed line for the 5 ppm sample is obtained from an extrapolation of the 42, 57 and 70 ppm Mn alloy results /24/.

Fig. 11. Gaussian linewidth for muons in Al doped with Li, Mn, Mg and Ag. Solid lines: Fits with two state model.

and 11 show the fits to the two-state model. For AlMn$_x$, where experimental data at 57 ppm not shown in Fig. 10 were included, the following results were obtained: (i) the trapping rate $1/\tau_1$ is only weakly temperature dependent, i.e., $1/\tau_1 \sim T^\beta$ with $\beta \lesssim 1$; (ii) the dissociation energy from a Mn trap was found to be 10 meV compared to 48 meV for a N-trap in Nb; (iii) the trapping rate depends linearly on the Mn concentration confirming thereby the consistency of the applied model. In order to illustrate the result with some numbers, the τ_1 and τ_o values at the peak temperature (17K) in AlMn 70 ppm case are

$$\tau_1 = 3.3 \ \mu s, \qquad \tau_o = 40 \ \mu s. \tag{15}$$

This means that (i) only a fraction of the muons finds a trap ($\tau\mu = 2.2 \ \mu s$); (ii) if, however, a muon is trapped, it has a large probability to stay there. Thus the peak height is directly related to the fraction of muons trapped.

It is interesting to note that there is no correlation between trapping capability of a particular impurity atom and the lattice expansion it creates.

Thus, local distortion around the impurity which determines the trapping
behavior cannot be estimated from the long range behavior. This is in agreement with recent diffuse neutron scattering results on AlLi which show large
short-range atomic displacements even with a net volume dilation close to
zero /56/.

Figure 12 presents μ^+ results on electron irradiated Al /19/. The
estimated vacancy concentration was $C_v \simeq 4 \cdot 10^{-6}$. The relaxation rates
increase with temperature toward a saturation value Γ_o^{IV}. Since annealing
of vacancies sets in only well above 100K, the observed increase of damping
is due to diffusion controlled trapping, where the muon mobility increases
with increasing T. The shift of the damping increase observed after annealing reflects the recovery of radiation damage as a function of the annealing
temperature.

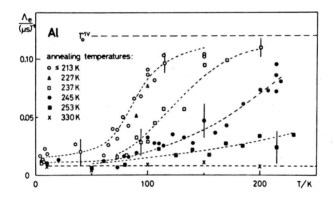

Fig. 12. Muon depolarization Λ in electron irradiated Al
(vacancy concentration before first annealing $c_v \sim 4.10^{-6}$) /19/.

Assuming diffusion controlled trapping, the trapping rate is proportional
to the diffusion coefficient. Thus, applying a trapping model, the temperature
dependence of the diffusion coefficient can be evaluated from the linewidth data.
The temperature dependence can be described by small-polaron hopping theory
(eq. (7)) and the activation energy comes out to 32 meV, considerably lower
than in Cu where 75 meV has been found. For the overlap, a Debye approximation
of the small-polaron theory yields a J of the order of a few meV, two orders of
magnitude larger than in Cu. Finally, the extreme sensitivity of μSR toward
defects should be emphasized by the fact that even vacancy concentrations
below the 1 part/million level can be easily detected with muons. The authors
estimated a vacancy concentration of $\sim 7 \cdot 10^{-7}$ after annealing at 245 K, which
still gives rise to a strong increase in the muon depolarization (see Fig. 12).

SUMMARY

This review on muon diffusion and trapping in nonmagnetic material does not
claim to be complete, but to display the state of the art by way of example.
Two main directions in muon diffusion and trapping studies seem to dominate

(i) There is great interest in the fundamental transport mechanism of light interstials in metals. Here the muon, as a particle at the borderline between band-like propagation and quantum hopping, seems to be the ideal probe to investigate the different aspects of quantum transport theory. (ii) On the other hand, the extreme μ^+ sensitivity toward lattice imperfections makes it a promising tool to investigate impurities and defects in metals. Microscopic information on defect and impurity properties can be obtained at extremely low concentrations. Performing field dependence measurements on single crystals, the structure of the muon traps can be identified and the recovery from radiation damage can be studied in great detail.

Finally, the analogies between the muon and H allow one to extrapolate from μ^+ results on H-properties. Thus, new information on H in metals, such as lattice location, electronic environment and, perhaps, H-diffusion in dense H-storage materials where the muon could be used as a tracer /57/, can be expected.

ACKNOWLEDGEMENTS

Large parts of the work discussed in this review have been carried out within the CERN μSR group. Many discussions with Drs. O. Hartmann, E. Karlsson, K. W. Kehr, T. Niinikoski, L. Norlin, and A. Yaouanc have helped to develop the picture of muon properties in metals presented here.

REFERENCES

1. For recent reviews see
    a) "Muon Spin Rotation" ed. F. N. Gygax, W. Kündig, P. F. Meier, North Holland Publishing Company, Amsterdam (1979).
    b) "Exotic Atoms 79 - Fundamental Interactions and Structure of Matter" ed. K. Crowe, J. Duclos, G. Fiorentini, G. Torelli, Plenum Press, New York (1980).

2. Y. J. Uemura et al., Phys. Rev. Letts. 45, 583 (1980).

3. I. I. Gurevich et al., Phys. Letts. 40A, 143, (1972).

4. O. Hartmann et al., Phys. Letts. 61A, 141, (1977).

5. W. F. Lankford et al., Hyperfine Int. 4, 833, (1978).

6. O. Hartmann et al., Hyperfine Int. 4, 824 (1978).

7. C. P. Flynn, A. M. Stoneham, Phys. Rev. B1, 3966 (1970).

8. M. Borghini et al., Phys. Rev. Letts. 40, 1723 (1978).

9. H. K. Birnbaum et al., Phys. Letts. 65A, 435 (1978).

10. See also Session on "Interstitial Diffusion and Trapping at Impurities" in Ref. 1a.

11. O. Hartmann et al., Phys. Rev. Letts. 41, 1055 (1978).

12. W. B. Gauster et al., J. Nucl. Materials 69+70, 147 (1978).

13. W. B. Gauster et al., Sol. State Comm. 24, 619 (1977).

14. W. J. Kossler et al., Phys. Rev. Letts. 41, 1558 (1978).

15. K. Dorenburg et al., Z. Physik B31, 165 (1978).

16. M. Doyama et al, Hyperfine Int. 6, 341 (1979).

17. J. A. Brown et al., Phys. Rev. Letts. 43, 1513 (1979).

18. H. Bossy et al., SIN Newsletter 12, 81 (1979) (SIN, Villigen, Switzerland, Dok./Int. Dec 79/3500 EDMZ).

19. D. Herlach, Proceedings of EPS 1980 Annual Conf. on Condensed Matter, April 1980, Antwerp, Belgium, to be published.

20. R. H. Heffner et al., Hyperfine Int. 6, 237 (1979).

21. J. A. Brown et al., Hyperfine Int. 6, 233 (1979).

22. W. J. Kossler et al., Hyperfine Int. 6, 295 (1979).

23. O. Hartmann et al., Phys. Rev. Letts. 44, 337 (1980).

24. O. Hartmann et al., Sol. State Comm., in print.

25. K. W. Kehr et al., Proceedings of the 2nd Int. Conf. on μSR, Vancouver, 1980, to be published.

26. M. Camani et al., Phys. Letts. 77B, 326 (1978).

27. D. K. Bryce, Phys. Letts. 66A, 53 (1978).

28. See, e.g., A. Abragam, Nuclear Magnetism (Oxford Univ. Press, Oxford, (1961)).

29. R. Kubo and T. Toyabe in "Magnetic Resonance and Relaxation," ed. by R. Blinc, North Holland Publ. Co., Amsterdam (1967).

30. T. Yamazaki, Hyperfine Int. 6, 115 (1979).

31. K. Petzinger, to be published.

32. J. H. Van Vleck, Phys. Rev. 74, 1168 (1948).

33. O. Hartmann et al., Phys. Letts. 61A, 141 (1977).

34. O. Hartmann, Phys. Rev. Letts. 39, 832 (1977).

35. M. Camani et al., Phys. Rev. Letts. 39, 836 (1977).

36. P. Jena et al., Phys. Rev. Letts. 40, 264 (1978).

37. P. Hohenberg et al., Phys. Rev. 140A, 1133 (1965).

38. M. Camani et al., Phys. Rev. Letts. 42, 679 (1979).

39. P. F. Meier in Ref. 1b, p. 355.

40. T. Holstein, Ann. Phys. (N.Y.) $\underline{8}$, 325, 343 (1959).

41. Yu. Kagan, M. Klinger, J. Phys. $\underline{C7}$, 2791 (1974).

42. P. W. Anderson, Phys. Rev. $\underline{109}$, 1492 (1958).

43. K. W. Kehr in "Hydrogen in Metals," ed. by G. Alefeld and J. Völkl, Springer Verlag, Berlin Heidelberg (1978).

44. K. W. Kehr, Proc. Conf. on Hydrogen in Metals, Japan Inst. of Melts (1979).

45. T. McMullen and B. Bergersen, Sol. State Comm. $\underline{28}$, 31 (1978).

46. A. Browne, A. M. Stoneham, to be published.

47. N. F. Mott and A. M. Stoneham, J. Phys. $\underline{C10}$, 3391 (1977).

48. H. Teichler in Ref. 1b.

49. K. W. Kehr et al., Z. Physik $\underline{B32}$, 49 (1978).

50. D. Richter and T. Springer, Phys. Rev. $\underline{B18}$, 126 (1978).

51. T. O. Niinikoski et al., Hyperfine Int. $\underline{6}$, 229 (1979).

52. See e.g., G. Leibfried, Z. Physik $\underline{135}$, 23 (1953).

53. See e.g., N. Giordano et al., Phys. Rev. Letts. $\underline{43}$, 725 (1979).

54. D. J. Thouless, Phys. Rev. Letts. $\underline{39}$, 1167 (1977).

55. K. Petzinger, Hyperfine Int. $\underline{6}$, 223 (1979).

56. K. Werner, private communication.

57. D. Richter et al., to be published.

Energy Materials

NMR OF SMALL PLATINUM PARTICLES

HAROLD T. STOKES, HOWARD E. RHODES,[*] PO-KANG WANG AND CHARLES P. SLICHTER
Department of Physics and Materials Research Laboratory, University of Illinois,
Urbana, Illinois 61801

J. H. SINFELT
Exxon Research and Engineering Company, Linden, New Jersey 07036

ABSTRACT

     We present ^{195}Pt NMR lineshapes as well as relaxation
data in three different samples of platinum metal particles
(46%, 26%, and 15% dispersion) supported on alumina. We
show that the electronic properties of these particles are
very much different from those of bulk Pt metal. A prominent peak in the lineshape has been identified as a "surface resonance" which arises from Pt nuclei on the surface
of the Pt particles. We find that these surface Pt atoms
are "nonmetallic" when coated with adsorbed molecules.

INTRODUCTION

     Heterogeneous catalysis is a field of great interest because of its impor-
tant technological applications. However, the fundamental understanding of
catalytic phenomena is not well developed [1]. In an effort to learn more
about the microscopic details of heterogeneous catalysis, we have been applying
nuclear magnetic resonance (NMR) to the study of a typical catalyst, platinum
metal. In this paper, we present NMR data [2-4] on ^{195}Pt in small platinum
metal particles supported on alumina. We show that the electronic properties
of these particles are very much different from those of bulk platinum metal.
Furthermore, the Pt atoms on the surface are "nonmetallic" when coated with
adsorbed molecules.

SAMPLES

     We have used three different samples (prepared by one of the authors,
JHS) which we label 1, 2, and 3. These samples consist of small Pt particles
supported on alumina. (All samples are 10% Pt-90% alumina by weight.) The
dispersion (fraction of Pt atoms which are on the surface of the particles)
of samples 1, 2, and 3 was measured by hydrogen chemisorption and was found
to be 46%, 26%, and 15%, respectively. After the samples were prepared, they
were opened to air. The subsequent catalytic reaction of atmospheric oxygen
with adsorbed hydrogen probably caused the Pt particles in these samples to
be coated with water.

KNIGHT SHIFT

     When a material is placed in an applied magnetic field H_0, interactions
between the nuclei and nearby electrons give rise to a displacement of the NMR
frequency ν_0. Polarization of the electron orbital moments gives rise to a
displacement called the chemical shift. In metals, polarization of the conduc-

---

[*] Presently at Eastman Kodak Co., Rochester, New York 14650

tion-electron spins gives rise to an additional displacement called the Knight shift. For a given nucleus, Knight shifts are usually an order of magnitude larger than chemical shifts [5].

In non-metallic Pt compounds, the [195]Pt nuclear resonance is usually observed between $H_0/\nu_0 = 1.093$ and 1.100 kG/MHz, a spread of about 0.6% [6]. This spread of resonances is due to chemical shifts alone, since the Knight shift is zero in an insulator. By convention, the zero chemical shift (and zero Knight shift) of [195]Pt is chosen to be at $H_0/\nu_0 = 1.0996$ kG/MHz which is the position of the [195]Pt resonance in H_2PtI_6, one of the least paramagnetic compounds [6]. In bulk Pt metal, a large Knight shift is observed. The [195]Pt resonance is at $H_0/\nu_0 = 1.1380$ kG/MHz, a Knight shift of -3.37% with respect to H_2PtI_6 [6].

The Knight shift is a local effect, arising from interactions with electrons which are very near the nucleus [5]. Thus the Knight shift is very sensitive to local variations in the state of the conduction electrons. In small metallic particles which contain only a few atoms, the state of the conduction electrons is expected to vary greatly as a function of position in the particle, resulting in a large variation of Knight shifts. This results in a rather broad NMR absorption lineshape.

LINESHAPES

The NMR absorption lineshapes were measured using a pulsed-NMR spin-echo technique, varying H_0 while holding ν_0 constant. The results are shown in Fig. 1. We observe in our samples absorption lineshapes which extend from the full-Knight-shift position of bulk Pt metal to approximately the position of non-metallic Pt compounds, a distance of 4.5 kG at 74 MHz.

In sample 3 (see Fig. 1), a prominent peak in the lineshape can be seen at the position of the [195]Pt resonance in bulk Pt metal ($H_0/\nu_0 = 1.1380$ kG/MHz). This peak arises from [195]Pt nuclei in larger particles where the conduction electrons are in an environment similar to bulk Pt metal. The absence of this peak in samples 1 and 2 indicates that very few of the Pt particles in those samples are large enough to exhibit the electronic properties of bulk Pt metal.

In all three samples (see Fig. 1), another peak in the lineshape can be seen at $H_0/\nu_0 = 1.0897$ kG/MHz. We have identified this peak as a "surface resonance" [2,3] which arises from Pt nuclei on the surface of the particles. From the position of this peak on the line, we see that these nuclei are in a non-metallic environment. The conduction electrons at the surface of the Pt particles are most likely tied up in chemical bonds with various adsorbed molecules. This conclusion is further supported by relaxation measurements discussed below.

From Fig. 1, we see that the intensity of the surface resonance decreases with decreasing dispersion, as expected. Since the area under an NMR absorption line is proportional to the number of nuclei contributing to the absorption, the ratio of the area under the surface resonance peak to the area under the complete lineshape ought to be equal to the dispersion. To calculate the area of the peak, we assume that the left side of the peak is well-resolved from the rest of the line and that the peak is symmetric about its center. Thus we obtain the right side of the peak as shown by the dashed lines in Fig. 1. Measuring the dispersion by integrating the lineshapes, we obtain 46%, 20%, and 12% for samples 1, 2, and 3, respectively, in good agreement with the values measured by hydrogen chemisorption (46%, 26%, and 15%, respectively).

SAMPLE TREATMENTS

We "cleaned" sample 1 by exposing it alternately to hydrogen gas and then gas at 300°C. The final exposure was to hydrogen gas which was then pumped off. The sample was then cooled to room temperature and sealed in a glass ampoule

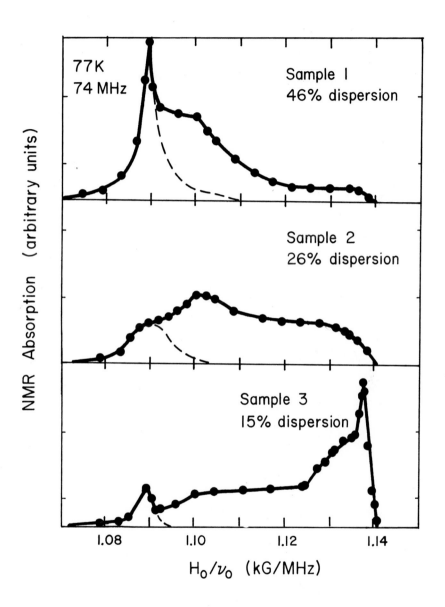

Fig. 1. NMR absorption lineshapes of ^{195}Pt in samples 1, 2, and 3.

256

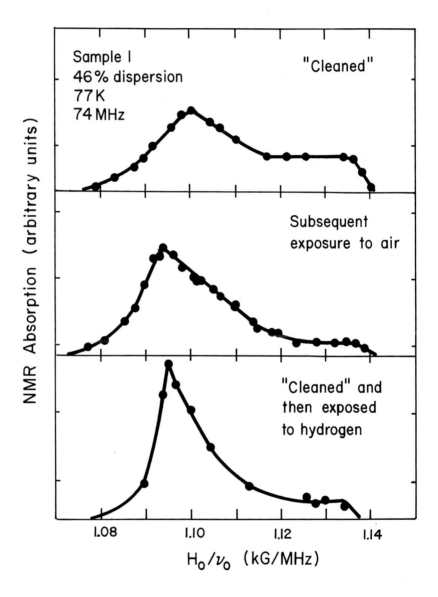

Fig. 2. NMR absorption lineshapes of [195]Pt in sample 1 under various surface treatments

under vacuum. The lineshape for this sample is shown in Fig. 2. We see that the surface resonance peak is gone and that much of the intensity of the line-shape is shifted towards the position of bulk Pt metal. It appears that in our "cleaned" sample, the electrons previously tied up in chemical bonds to adsorbed molecules are now free, and the surface of the Pt particles is more metallic in nature.

We next broke open the ampoule and exposed this sample to air, probably causing oxygen to be adsorbed on the Pt particles. A surface resonance peak reappeared (see Fig. 2), however, somewhat broader and at a different position (H_0/ν_0 = 1.0944 kG/MHz). This displacement of the surface resonance is caused by a chemical shift and thus depends on the nature of the adsorbed molecules.

We repeated the "cleaning" process on sample 1 and the exposed it to hydrogen gas at room temperature. The gas was then pumped off and the sample sealed in a glass ampoule under vacuum. Now, only hydrogen is adsorbed on the surface of the Pt particles and we again see a surface resonance peak (see Fig. 2), this time at H_0/ν_0 = 1.0952 kG/MHz.

RELAXATION

We measured the spin-lattice relaxation time T_1 as a function of position on the absorption line in samples 1, 2, and 3 (see Fig. 3). As can be seen, the data from each of the three samples fall on the same curve. This result suggests that at each position of the resonance line, the electronic environment is the same, without regard to particle size. At H_0/ν_0 = 1.1422 kG/MHz, we find T_1 = 390 µs which is the value of T_1 measured in bulk Pt metal [6]. The relaxation

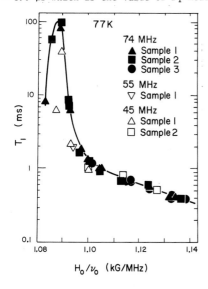

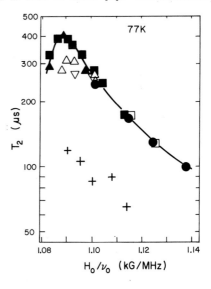

Fig. 3. Spin-lattice relaxation time T_1 as a function of position on the line in samples 1, 2, and 3.

Fig. 4. Spin-spin relaxation time T_2 as a function of position on the line in samples 1, 2, and 3. The symbols have the same meaning as in Fig. 3. The crosses (+) are T_2 data taken in sample 1 after exposure to hydrogen.

here is caused by interaction with the conduction electrons [5]. This is a very effective relaxation mechanism and produces a very short T_1.

Near the position of the surface resonance, T_1 is considerably longer and also field dependent. This is characteristic of relaxation in a non-metal. (Relaxation via conduction electrons is field independent [5]). Thus we conclude that the Knight shift is zero at the surface resonance, and the observed displacement of this resonance from H_2PtI_6 is purely a chemical shift.

We also measured the spin-spin relaxation time T_2 as a function of position on the line (see Fig. 4). In this case, T_2 is mostly limited by T_1 and thus has some of the same general characteristics of the T_1 data in Fig. 3. As with T_1, we see that T_2 is field independent over most of the line except near the surface resonance.

Some T_1 and T_2 data was also taken on the treated samples. Of special interest is the T_2 data on the sample which was exposed to hydrogen (the crosses on Fig. 4). As can be seen, the T_2 in this sample is much shorter than the other samples. Here we see how a particular surface treatment can have a rather dramatic effect on the NMR properties of the ^{195}Pt nuclei. If we place a ^{195}Pt nucleus in the vicinity of other nuclei with strong nuclear magnetic moments, its T_2 can be greatly reduced via the dipolar interaction [5]. Hydrogen nuclei have a very strong nuclear magnetic moment. Sitting on the surface of the Pt particles, they are close enough to the ^{195}Pt nuclei to produce this effect, as we indeed observe.

ACKNOWLEDGMENTS

We gratefully acknowledge Claus Makowka and Serge Rudaz for their helpful discussions and assistance in taking data. This work was supported by U.S. Department of Energy under Contract DE-AC02-76ER01198.

REFERENCES

1. J. H. Sinfelt, Prog. Solid State Chem. <u>10</u>, 55 (1975).

2. H. E. Rhodes, Ph.D. Thesis, University of Illinois, 1980.

3. H. E. Rhodes, C. P. Slichter, and J. H. Sinfelt, Bull. Am. Phys. Soc. <u>25</u>, 272 (1980).

4. H. T. Stokes and C. P. Slichter, Bull. Am. Phys. Soc. <u>25</u>, 273 (1980).

5. C. P. Slichter, <u>Principles of Magnetic Resonance</u> (Springer-Verlag, New York, 1978) pp. 106-121.

6. G. C. Carter, L. H. Bennett, and D. J. Kahan, in <u>Progress in Material Science</u>, B. Chalmers, J. W. Christian, and T. B. Massalski, eds. (Pergamon, New York, 1977) Vol. 20, Part I, pp. 295-302.

SOLID STATE NMR STUDIES OF THE ADSORBED STATES OF FORMIC ACID ON Y ZEOLITES

T. MICHAEL DUNCAN* and ROBERT W. VAUGHAN**
California Institute of Technology
Division of Chemistry and Chemical Engineering
Pasadena, CA   91125

ABSTRACT

Several multiple-pulse double-resonance NMR techniques have
been applied to isolate and characterize the spectra of the
adsorbed states of formic acid on two Y zeolites. The two
surface states, bidentate and unidentate, possess different
motional properties and ^{13}C - ^{1}H cross-polarization
techniques may be used to separate the spectra. The ^{13}C
chemical shift anisotropy is found to correlate with the
symmetry of the formate species. The ^{1}H spectrum of the
carbonyl hydrogen, selectively observed with the dipolar-
difference method, indicates that this hydrogen becomes
more acidic upon adsorption.

INTRODUCTION

An understanding of the reaction mechanisms on heterogeneous catalysts is
essential to optimizing existing processes and designing new catalysts.
This understanding requires a knowledge of the adsorbed state and the
adsorption site of both the reactant and product species. This paper
describes the application of several multiple-pulse NMR techniques to
isolate and characterize the spectra of a catalytic system.

The system studied is the decomposition of formic acid on Y zeolites
(Si/Al = 2.8). Formic acid adsorbs irreversibly on the zeolites through
the loss of the acidic proton. The adsorbed species decomposes almost
exclusively to carbon monoxide and water. The nature of the adsorbed formate
can be described by the type of bonding to the oxide surface. The unidentate
structure is bonded through one oxygen atom with the other oxygen atom double-
bonded to the carbon. The bidentate formate is chelated to the adsorption
site through both oxygen atoms, yielding symmetric carbon-oxygen bonds.

The two zeolites investigated are an ammonium-Y (NH_4-Y) and an ultrastable
hydrogen-Y (u-H-Y) zeolite. A discussion of these molecular sieves can be
found elsewhere [1]. Briefly, the NH_4-Y and u-H-Y (derived from the NH_4-Y
by heating to 775 K) have the same crystal structure, except the u-H-Y has a
fraction of the Al ions that have been displaced from the crystal framework.
Correspondingly, the u-H-Y zeolite has 12-15 new Brönsted sites per unit cell
(unit cell contains 48 Al atoms) and increased catalytic activity with respect
to hydrogenation and cracking.

---

*
Present Address: Bell Telephone Laboratories
                   Murray Hill, NJ   07974
**
  Deceased

The multiple-pulse NMR experiments are used to quantify the dipolar interactions and the chemical shift anisotropy of the ^{13}C nuclei of the adsorbed formate species. The strength of the dipolar couplings is used to determine the extent of motion of the adsorbed molecules and the nature of the adsorption site. The principal components of the chemical shift tensor are analyzed in a correlative fashion to determine the symmetry of the formate. Finally, the ^1H NMR spectrum of the hydrogen bonded to the carbon is observed selectively to complement the analysis.

EXPERIMENTAL METHODS

The NH_4-Y and u-H-Y zeolites have BET surface areas of 460 m^2/g and 515 m^2/g, respectively. Both contain 780 ppm Fe and 12 ppm Mn, by weight. The details of the sample preparation and treatment are discussed elsewhere [1]. After outgassing, each zeolite was dosed with approx. 0.3 monolayers of 90% ^{13}C-enriched formic acid.

The ^{13}C - ^1H double-resonance multiple-pulse NMR spectrometer [2] and probe [3] have been described previously. The single coil probe (5 mm dia.) was tuned to 56.4 MHz(^1H) and 14.2 MHz(^{13}C). Temperatures were regulated with a dewared nitrogen flow system.

The ^{13}C NMR data in this study were obtained by Fourier-transforming the signals observed with three techniques: the 180°-τ-90° [4], ^{13}C-^1H cross-polarization [5], and ^{13}C-^1H dipolar-modulation experiments [6]. The ^1H spectrum of the carbonyl proton of the adsorbed formic acid was isolated via the ^{13}C-^1H dipolar-difference technique [7]. The schematics of these pulse sequences are given in Figure 1. The length and intensity of the pulses are given elsewhere [1].

Description of the techniques may be found in the original references, or in a recent review [8], and monographs [9, 10].

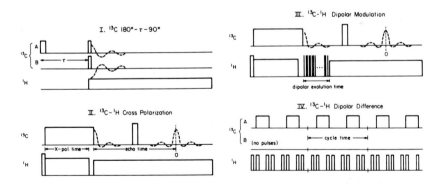

Fig. 1. Schematic representations of the four NMR pulse sequences. The ^{13}C and ^1H pulse sequences are initiated simultaneously. In sequences I and IV, the A and B options of the ^{13}C pulses are applied alternately and the NMR signals are alternately added and subtracted. (From Ref. [1]).

RESULTS AND DISCUSSION

The proton-decoupled ^{13}C NMR spectra of three reference formates; formic acid, ammonium formate, and calcium formate, are shown in Figure 2. The spectrum for formic acid (2a) is not a fully-developed powder pattern, as reflected by the poor agreement with a best-fit theoretical curve (solid line) and a powder pattern with pre-set components (dotted line). The components used in the analysis are taken from the dotted line.

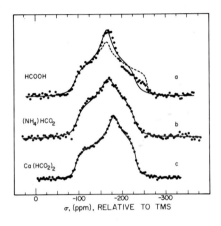

Fig. 2. Proton-decoupled ^{13}C NMR spectra of (a) formic acid at 115 K, (b) ammonium formate at 135 K, and (c) calcium formate at 125 K. (From Ref. [1]).

The chemical shift anisotropies of these compounds and methyl formate [11] are considered with respect to the symmetry of the two C-O bonds in Table I. The average chemical shift and the overall anisotropy ($\sigma_{11} - \sigma_{33}$) do not correlate well with the ratio of the two bond lengths. However, the position of the central component, σ_{22}, relative to the other components, as reflected by the ratio ($\sigma_{11} - \sigma_{22}$)/($\sigma_{11} - \sigma_{33}$) does follow a distinct trend. For unidentate symmetries, σ_{22} is closer to σ_{11}, and the ratio approaches zero. For bidentate symmetry, σ_{22} is closer to σ_{33} and this ratio increases, to as high as 0.65.

The four sequences in Figure 1 were then applied to the zeolite samples. The 180°-τ-90° experiment (Figure 1, sequence T) was used to (1) count the number of ^{13}C nuclei and (2) measure the ^{13}C T_1's by varying τ. The ^{13}C NMR spectra were narrow, relative to static formate tensors, and approximately Lorentzian in shape, indicating motional averaging. The ^{13}C NMR spin counts were low, compared to the amount originally deposited as formic acid on the zeolites, by 11% and 49% for the NH$_4$-Y and u-H-Y samples, respectively. A portion of the unobserved ^{13}C spins may be the result of the formic acid reacting to CO, which desorbed into the dead volume above the sample. Months later, after the formate signal had decreased further, evidence for CO was detected as a sharp peak at -131 ppm.

The ^{13}C spins relaxed inhomogeneously, and in both samples, could be described by the combination of two T_1's. On the NH$_4$-Y sample, 35% relaxed with an average T_1 of 2.6 msec and 65% relaxed with a T_1 of 82 msec. The two groups on the u-H-Y sample were distributed as 25% with a T_1 of 4.4 msec and 75% with a T_1 of 71 msec. The interpretation of the T_1 inhomogeneity could not be unambiguously determined, due to poor signal-to-noise levels in the spectra for long τ.

The ^{13}C - ^1H cross-polarization technique, which utilizes the heteronuclear dipolar coupling, was effective for only a portion of the ^{13}C nuclei of the adsorbed formates observed with the 180°-τ-90° experiment; 52% of the spins on the NH_4-Y zeolite and 83% of the spins on the u-H-Y zeolite. The dipolar coupling in the remaining molecules are too weakened, presumably by motional averaging, and are thus not observed in this experiment. The ^{13}C NMR spectra of the rigidly adsorbed formate species are shown in Figure 3 for two temperatures.

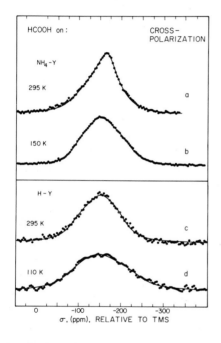

Fig. 3. Proton-decoupled ^{13}C NMR spectra of formic acid adsorbed on: the NH_4-Y zeolite at (a) 295 K and (b) 150 K, and on the u-H-Y zeolite at (c) 295 K and (d) 110 K. (From Ref. [1]).

Though there is substantial residual broadening in the spectra, due to effects such as site inhomogeneity and ^{27}Al dipolar interactions, the lines may be fit with the theoretical chemical shift powder pattern convoluted with a Gaussian broadening function. The chemical shift components are used to compute the values in Table I. At 295 K, the average chemical shift for the adsorbed formic acid on both zeolites, is at least 15 ppm above the normal range of formate compounds. The ratio in the last column of Table I, which indicates the relative position of the central component, is relatively high (0.69 and 0.58) for both samples at 295 K. These values, considered in light of the trends of the reference formates, indicate that the rigid formate species isolated by cross-polarization is a bidentate group. The mobile species which was the narrow portion of the line observed in the 180°-τ-90° experiment, is probably a unidentate compound. The quantitative distributions between the bidentate and unidentate species (52:48 on the NH_4-Y and 83:17 on the u-H-Y) are in agreement with the qualitative results of a recent infrared study [12].

TABLE I
Formate Compounds: C-O Bond Symmetry and Chemical Shift Anisotropy[a]

Formate Compound	$\frac{R(C-O)}{R(C=O)}$	$<\sigma>$	$\sigma_{11}-\sigma_{33}$	$\frac{(\sigma_{11}-\sigma_{22})}{(\sigma_{11}-\sigma_{33})}$
Methyl Formate, $HCOOCH_3$[b]	1.12	-165	146	0.20
Formic Acid, HCOOH	1.05	-168	159	0.44
Ammonium Formate, NH_4HCO_2	1.01	-164	126	0.55
Calcium Formate, $Ca(HCO_2)_2$	1.00	-172	139	0.65
Static Formate Species on:				
NH_4-Y Zeolite (a) at 295 K	-	-147	32	0.69
(b) at 150 K	-	-160	138	0.41
u-H-Y Zeolite (a) at 295 K	-	-150	104	0.58
(b) at 110 K	-	-161	142	0.33

[a] all chemical shift data in ppm, relative to TMS.
[b] from reference [11].

Upon cooling to below 150 K, the average chemical shift decreases to values closer to the normal range of formates and the anisotropy ratio decreases significantly. This large change in both values cannot be totally explained by the quenching of the motion of the other formates (i.e. the unidentates). The large shifts indicate that cooling causes the bidentate groups to become more unidentate-like. However this is a preliminary conclusion and requires further study.

The second moments of the broadening functions which were convoluted into the powder pattern fits are 0.67 G^2 for spectrum 3a and 0.88 G^2 for spectrum 3c. After subtracting contributions due to the estimated chemical inhomogeneity, librational averaging, and lifetime broadenings, the second moments are 0.53 G^2 and 0.79 G^2, respectively. The remaining broadening is attributed to ^{27}Al-^{13}C dipolar interactions. These interactions cannot be calculated exactly for these samples, since the quadrupolar ^{27}Al nucleus may not be in a Zeeman state, which could increase the coupling by a factor of 1.8 [13]. Thus, for a formate group bonded to an Al site, the interaction would be 0.52 to 0.87 G^2 and for a Si site it will be 0.23 to 0.35 G^2. These values indicate that the bidentate formate is bonded to the Al atoms of the zeolites.

Dipolar-modulation experiments (Figure 1, sequence III) were used to quantify the strength of the ^{13}C - ^1H coupling. The rate of decrease of the spectral area as a function of the dipolar evolution time was approximately the same as that of calcium formate, reported previously [6]. Thus, the adsorbed bidentate structure has a ^{13}C - ^1H spin coupling similar to inorganic formates.

The ^1H NMR spectrum of the hydrogen bonded to the adsorbed bidentate formate was isolated from the extraneous signals with the dipolar-difference scheme (Figure 1, sequence IV). This technique yields a dipolar-decoupled (homonuclear and heteronuclear) ^1H spectrum which is approximately 3% of the total spectral intensity. The observed spectrum is 6.0 ppm wide, centered at -12.3 ppm relative to TMS. Neat formic acid is centered at -9.2 ppm. Corresponding protons in calcium formate and ammonium formate have average chemical shifts of -10.9 and -12.8 ppm, respectively. The downfield shift of the adsorbed formic acid suggests that this proton has become more acidic.

hydrogen may be more susceptible to nucleophilic attack or proton-shift reactions.

CONCLUSIONS

Through the application of several multiple-pulse NMR experiments it has been possible to measure the dipolar interactions, spin-lattice relaxation rates, and chemical shift properties of formic acid adsorbed on an NH_4-Y and an u-H-Y zeolite. These quantities have been interpreted to determine the configuration of the adsorbed state and the adsorption site.

In agreement with infrared studies, the formate ion is found to exist in both bidentate and unidentate states on the zeolitic surfaces. The ratio of biden-tate to unidentate groups is 52:48 on the NH_4-Y and 83:17 on the u-H-Y, with 0.3 monolayers of formic acid at 295 K. The bidentate species is adsorbed on the Al atoms of the zeolite, based on conclusions from the ^{27}Al - ^{13}C dipolar interactions. Adsorption into the bidentate state causes the ^{13}C chemical shift to move upfield by almost 15 ppm, whereas the ^{1}H spectrum is shifted down-field, by 3 ppm. In ^{1}H NMR, this shift is commonly associated with a more acidic nature of the hydrogen atom.

The u-H-Y zeolite, which is more active catalytically, has a greater percentage of formates adsorbed in bidentate states. Furthermore, the average ^{27}Al - ^{13}C coupling is stronger, suggesting that the Al atoms may exist in higher electro-static field gradients. Both results are in agreement with the suggestion that the transition from the NH_4-Y to the u-H-Y creates 12 to 15 non-rramework Al ions per unit cell [14].

ACKNOWLEDGEMENTS

This work was supported in part by the Office of Naval Research. We are indebted to J. A. Reimer for the dipolar-modulation and dipolar-difference spectra and to Dr. G. T. Kerr of Mobil Research and Development Corporation for supplying the zeolite samples.

REFERENCES

1. T. M. Duncan and R. W. Vaughan, J. Catalysis, Vol. 67, No. 1, 1981 and references cited therein.
2. R. W. Vaughan, D. D. Elleman, L. M. Stacey, W. K. Rhim and J. W. Lee, Rev. Sci. Instrum. 43, 1356 (1972).
3. M. E. Stoll, A. J. Vega, and R. W. Vaughan, Rev. Sci. Instrum. 48, 800 (1977).
4. H. Y. Carr and E. M. Purcell, Phys. Rev. 49, 630 (1954).
5. A. Pines, M. G. Gibby, and J. S. Waugh, J. Chem. Phys. 59, 569 (1973).
6. M. E. Stoll, A. J. Vega, and R. W. Vaughan, J. Chem. Phys. 65, 4093 (1976).
7. J. A. Reimer and R. W. Vaughan, Chem. Phys. Lett. 63, 163 (1979).
8. R. W. Vaughan, Ann. Rev. Phys. Chem. 29, 397 (1978).
9. M. Mehring, NMR - Basic Principles and Progress, Vol. 11, (P. Diehl, E. Fluck, and R. Kosfeld, Eds.) Springer - Verlag, New York, 1976.
10. U. Haeberlin, Adv. in Magn. Resonance, Supp 1. (J. S. Waugh, ed.) Academic Press, New York, 1976.
11. J. L. Ackerman, J. Tegenfeldt, and J. S. Waugh, J. Amer. Chem. Soc. 96, 6843 (1974).
12. T. M. Duncan and R. W. Vaughan, J. Catalysis, Vol. 67, No. 1, 1981.
13. D. L. Vanderhart, H. S. Gutowsky, and T. C. Farrer, J. Amer. Chem. Soc. 89, 5056 (1967); D. L. Vanderhart, Ph.D. Thesis, U. of Illinois, 1968.
14. G. T. Kerr, J. Catalysis 15, 200 (1969).

THE NATURE OF FLUORINE MODIFIED OXIDE SURFACES: AN NMR STUDY

JOHN R. SCHLUP and ROBERT W. VAUGHAN*
Division of Chemistry and Chemical Engineering, California Institute of
Technology, Pasadena, California 91125

ABSTRACT

Fluorine modified oxide surfaces have received considerable
attention both as research materials and as commercial
catalysts. Pulsed NMR has been used to directly observe
protons and fluorine on fluorine modified silica, alumina,
and aluminosilicates. The center of mass of the ^{19}F spectra
is consistent with values reported for covalent silicon-
fluorine and aluminum-fluorine bonds. The proton and fluorine
concentrations have been investigated as a function of
sample preparation.

INTRODUCTION

The addition of fluorine to oxide catalysts has been shown to drastically
change the properties of oxide catalysts. During the last twenty years,
these properties have been demonstrated with a wide variety of reactions,
catalysts, and reactor conditions.[1] Research on fluorine modified oxide
catalysts has focused on improving product yields and selectivity and on
obtaining a better understanding of the unmodified oxide catalysts as changes
due to the addition of fluorine are observed.
     Fluorine modified oxide catalysts have been studied by measuring the rates
of a variety of reactions and by measuring the surface acidity. Spectroscopic
techniques have included infrared spectroscopy, nuclear magnetic resonance
x-ray photoemission spectroscopy, x-ray diffraction, and inelastic electron
tunneling spectroscopy. In general, the object has been to understand these
catalysts by monitoring changes in their chemical behavior and in their
adsorption of small molecules after the addition of fluorine. It is hoped the
observed changes in behavior may then provide insight into changes at the
catalytically active sites. Although this approach has provided much
information about the catalysts, one would like to have a direct spectroscopic
probe of the local environment of the hydroxyl groups and fluorine atoms.
     In the past infrared spectroscopy has been most widely used as a
spectroscopic probe. However, since the fluorine atom vibrations are buried
among the bands due to the lattice vibrations, one is limited to observing only
the hydroxyl group stretching vibrations. Even then, past research has focused
on the perturbations in these vibrations that result from modifying the surface
or from the adsorption of small molecules. However, nuclear magnetic resonance
provides a very specific probe of the local environments of hydrogen and
fluorine atoms. The chemical shift interactions provides information on
the electronic environment while the dipole-dipole interaction yields
information about the location of neighboring atoms. NMR is also a useful
tool for quantitative analysis.
     Several key questions arise as one investigates fluorine modified oxides.
The type of bonding present must be understood. The existence of fluorine-
hydroxyl group hydrogen bonds has also been proposed.[2,3] The focus of the

---

*
 Deceased.

TABLE 1

Proton and Fluorine Concentrations from NMR Data

	Fluorine Treatment	Pretreatment Temperature (°C)	Protons per 100 Å^2	Fluorine atoms per 100 Å^2	Weight percent Fluorine
Silica	Light	300°	3.2	2.3	1.7
		400°	2.9	2.4	1.7
		500°	2.2	2.4	1.7
		600°	1.2	1.5	1.1
	Heavy	300°	2.3	7.9	5.6
		600°	1.3	1.4	0.9
Alumina	Light	300°	12.8	6.7	4.2
		600°	4.5	4.7	2.5

present discussion is the quantitative analysis of the modified oxide after various pretreatment conditions and the nature of the fluorine and proton bonding on various fluorine modified oxide catalysts.

EXPERIMENTAL

The NMR measurements were performed with a pulse NMR spectrometer described previously[4] operating at 16 kG and at 25 kG. All measurements were made at room temperature. The experiments reported herein were either free induction decays or spin echoes (90-τ-180).

The silica samples were prepared using a commercial dessicant silica gel (Grace Davison Grade 62). The alumina samples were prepared from Alcoa alumina F-20. The aluminosilicates used in this study were xerogel aluminosilicate catalysts prepared by Dr. D. A. Hickson at Chevron Laboratories in Richmond, California. The fluorine modified materials were prepared by soaking the catalysts in aqueous ammonium fluoride solutions (5.4 mM and 17.9 mM NH_4F). The suspensions were gently heated to dryness and then calcined under flowing oxygen in a tube furnace. Nitrogen BET surface area data were taken and then the samples were again calcined under oxygen. The sample tubes were then evacuated and sealed. The surface areas of all of the samples were greater than 125 m^2/g with the fluorine modified silicas having surface areas of 225 m^2/g.

QUANTITATIVE ANALYSIS

Numerous problems exist in obtaining proton and fluorine concentrations of fluorine modified oxides. Precision of 10% is very difficult to obtain for samples with low fluorine concentrations. Many quantitative techniques require preparations which cause irreversible changes in the sample. Vibrational spectroscopic measurements rely on knowledge of the oscillator strengths of the vibrations being considered. NMR is a good quantitative tool for determining proton and fluorine atom concentrations and it is a non-destructive technique. The signal intensity is directly proportional to the number of spins present. Precision of 5% has been demonstrated on samples containing as few as 5×10^{18} spins.

The presence of hydroxyl groups and fluorine atoms as a function of sample preparation is an important question. The relative number of hydroxyl groups and fluorine atoms needs to be known and yet values reported in the literature do not present a clear picture of their relationship. The results of counting the proton and fluorine spins are found in Table I. Protons are present on all the samples regardless of the pretreatment temperature. With silica samples at higher fluorine concentrations no protons were detected. Silica samples with higher fluorine concentrations also decompose to form a volatile silicon-fluorine compound.

The proton concentrations decrease as the sample pretreatment temperature is increased. The fluorine and proton concentrations after pretreatment at 600°C are the same regardless of the fluorine preparation (for this range of fluorine concentrations). The fluorine concentration is equal to the proton concentration for both silica and alumina after pretreatment at 600°C.

SPECTRA OF VARIOUS MODIFIED OXIDES

Various fluorine modified silicas, aluminas, and aluminosilicates have been prepared. In all cases the fluorine spectra has a center of mass which corresponds to fluorine covalently bound to the metal atom of the oxide. Fluorine ions and oxyfluoride species were not detected. The nature of the fluorine-silicon and fluorine-aluminum bond did not change with the

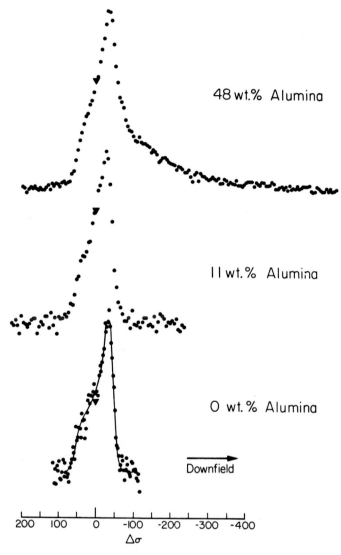

48 wt.% Alumina

11 wt.% Alumina

0 wt.% Alumina

Downfield

200 100 0 -100 -200 -300 -400

Δσ

Fig. I ^{19}F powder patterns of fluorine modified
oxides with varying alumina content. The
signal amplitude is plotted versus chemical
shift (ppm) relative to hexafluorobenzene.

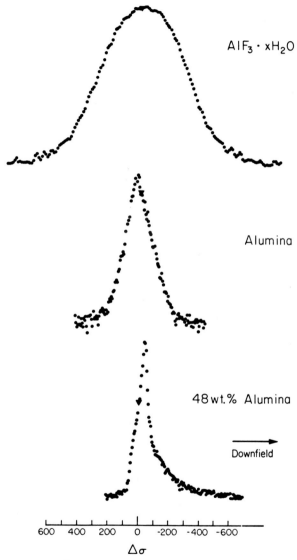

Fig. 2 ^{19}F powder patterns of materials with
aluminum–fluorine interactions. The signal
amplitude is plotted versus chemical shift
(ppm) relative to hexafluorobenzene.

sample pretreatment temperature.

Figure 1 compares the ^{19}F spectra of fluorine modified aluminosilicates with varying alumina contents. The fluorine modified silica exhibits a lineshape resulting from the chemical shift anisotropy and the center of mass is consistent with that of fluorine bound covalently to silicon. The sample containing 11 wt.% alumina exhibits an identical spectrum. This means that the fluorine exists in environments identical to those of a fluorine modified silica. However, the ^{19}F spectrum of an aluminosilicate containing 48 wt.% alumina has two components. The narrow component is identical to that of fluorine bound covalently to silicon. The width of the broad component suggests that there are contributions from aluminum-fluorine dipole-dipole interactions.

Figure 2 contains ^{19}F spectra taken from spin echo experiments with $AlF_3 \cdot xH_2O$, fluorine modified alumina, and a fluorine modified aluminosilicate which was 48 wt.% alumina. The ^{19}F spectrum of aluminum trifluoride is a Gaussian line broadened by dipolar interactions between the neighboring fluorine, aluminum, and hydrogen atoms. The spectra for the fluorine modified alumina is not easy to interpret. It is asymmetric and the asymmetry is not due to chemical shift anisotropy. The spectrum of the fluorine modified aluminosilicate clearly has two components. As was shown in the previous figure, the narrow component results from fluorine bonded to silicon atoms. The center of mass of the broad component is consistent with fluorine bound covalently with aluminum. Therefore, fluorine binds to both silicon and aluminum in aluminosilcates with higher alumina concentrations. The hydroxyl groups of these samples exhibit similar behavior, which is consistent with data reported previously.[5]

CONCLUSIONS

Fourier transform NMR has been shown to be a valuable tool for investigating fluorine modified oxide catalysts. It provides a precise, non-destructive tool for obtaining fluorine atom and proton concentrations. It has been shown that fluorine forms cavalent bonds with the metal atom in silica, alumina, and aluminosilicates. In aluminosilicate materials with low alumina content, the fluorine bonds preferentially to the silicon atoms. Fluorine bonds with aluminum as the alumina content of the aluminosilicate increases.

REFERENCES

1. V. R. Choudhary, Ind. Eng. Chem., Prod. Res. Dev. 16, 12 (1977).

2. A. N. Sidorov and I. E. Neimark, Russ. J. Phys. Chem. 38, 1518 (1964).

3. R. W. Vaughan, D. D. Elleman, L. M. Stacey, W.-K. Rhim, and J. W. Lee, Rev. Sci. Instrum. 43, 1356 (1972).

4. L. B. Schreiber and R. W. Vaughan, J. Catal. 40, 226 (1975).

UTILITY OF THE MÖSSBAUER EFFECT IN THE ASSESSMENT OF CHEMICAL TRANSFORMATIONS
IN UNSUPPORTED CATALYST SYSTEMS AS A FUNCTION OF THE METAL SALT

MARY L. GOOD, M. AKBARNEJAD, M. D. PATIL, AND J. T. DONNER
Division of Engineering Research, Louisiana State University, Baton Rouge,
Louisiana 70803

ABSTRACT

This paper is presented to illustrate the utility of
Fe-57 Mössbauer spectroscopy in the physical and
chemical characterization of metallic catalysts as a
function of preparation and subsequent treatment. The
particular example chosen for investigation is the
Ru-Fe bimetallic salt system which has significant
potential as an active Fischer-Tropsch catalyst. The
interpretation of the resulting Mössbauer parameters
provides information on the chemical state of the metals
as a function of treatment and initial salt properties
and allows certain deductions to be made about particle
size and solid state properties.

INTRODUCTION

Recent reports from our laboratory [1,2] have indicated the unique potential of "duel labeling" Mössbauer studies on the characterization of the
solid state reactions which occur in the preparation and treatment of unsupported, bimetallic catalysts. By combining Ru-99 and Fe-57 Mössbauer
data with supplementary information from ESCA spectroscopy and x-ray diffraction studies it was possible to completely characterize a ruthenium-iron
system from the initial salt mixture through a hydrogen reduction and a subsequent calcination step. Preliminary results for a similar study of
ruthenium-iron on a zeolite support have also been obtained. These investigations have further illustrated the utility of Mössbauer spectroscopy
for the evaluation of heterogenous catalyst systems where one or more of the
catalytic materials involves a metal having a nucleus which is Mössbauer
active [3]. The special capabilities of the technique include the ability
in the transmission mode to penetrate solid substrates which are opaque to
energies in the optical or vibrational spectral ranges and in the back-
scatter mode to provide information about surface species. Chemical
speciation is accomplished through "fingerprint" spectra and structural and
particle size information can be deduced from quadrupole splitting parameters and magnetic spectra. Mössbauer spectroscopy is an especially
powerful tool for evaluating solid state effects when it is combined with
diffraction techniques (both x-ray and electron) and surface measurements
such as ESCA and Auger spectroscopy.

The present study was undertaken to exploit the Mössbauer effect of the
Fe-57 nucleus in the evaluation of the chemical differences of the solid
state reactions which occur in the preparation of Ru-Fe mixed metal catalysts
from a variety of starting materials. Our initial work on the unsupported
Ru-Fe system indicated that a mixture of ruthenium trichloride trihydrate
and hydrated ferrous sulfate produced a hydrogen "reduced" product containing a complex mixture of Ru metal, RuO_2 and $\gamma-Fe_2O_3$. This result was in

contrast to the conclusions drawn from surface analysis techniques on a similar system where the reduced product from an initial mixture of ruthenium trichloride trihydrate and hydrated ferric nitrate was described as "metallic" [4]. A related study using the same initial chloride-nitrate salt mixture impregnated on silica indicated that bimetallic clusters of Ru-Fe were the products of hydrogen reduction [5]. These anion effects on the solid state chemistry of salt mixtures are of vital importance to catalytic chemists who must produce uniform materials of known characteristics. The problem is complex in that one needs to characterize the actual chemical species present and to describe the physical state as well, i.e. particle size, number of phases present, etc. We have begun an extensive study in our laboratory to investigate the mixed iron-ruthenium system as a function of support, salt mixture and reaction conditions. Characterization methods will include x-ray powder diffraction, ESCA/Auger spectroscopy, gas absorption techniques and Mössbauer spectroscopy. In this paper we report the initial results of an Fe-57 Mossbauer survey of the products formed when ruthenium trichloride trihydrate is mixed with a variety of iron salts and the salt slurry is subjected to relatively mild oxidation and reduction conditions. The intent is to show the utility of Mössbauer spectroscopy as a rapid, routine tool for assessing the reaction products of such treatments. The details of the actual catalytic properties of the various product materials will be the subject of a subsequent, much more intensive report.

EXPERIMENTAL

Materials and Sample Preparation

Ruthenium trichloride trihydrate was purchased from Engelhard Industries and used as received. Ferrous sulfate (hydrated) and ferric nitrate (hydrated) were purchased from Fischer Scientific Company and ferric chloride (hydrated) was obtained from Mallinckrodt Chemical Works; all were used as received. 1:2 molar ratio mixtures of $RuCl_3 \cdot 3H_2O$ and the iron salts were slurried with distilled water and a few grains of ascorbic acid and evaporated to dryness on a steam bath. These initial samples were designated as Fe-Ru sulfate, Fe-Ru nitrate and Fe-Ru chloride. Each sample was then heated under vacuum to 400°C for one-half hour and subsequently reduced in flowing hydrogen at 400°C for 4 hours. The resulting samples were designated Fe-Ru (reduced). A small portion of each reduced sample was exposed to air at room temperature for 24 hours; Fe-Ru (exposed). The air exposed samples were heated in air at 400°C for 4 hours and labeled as Fe-Ru (calcined). The samples were investigated by Mössbauer spectroscopy at all steps of the preparation.

Mössbauer Spectroscopy

All Fe-57 Mössbauer spectra were taken on samples at liquid nitrogen temperatures with the source at room temperature. No correction was made for second order Doppler effects. A Ranger Electronic Mössbauer spectrometer was used in conjunction with a Nuclear Data series 2200 Multichannel Analyzer operating in the multiscaling mode. The source was 54.3 mCi of Co-57 in a host lattice of rhodium metal purchased from the Spire Corp., Bedford, MA. A linewidth of 0.28mm/sec was observed for a $K_4Fe(CN)_6 \cdot 3H_2O$ absorber. Data reduction was carried out by a conventional least-squares Lorenzian lineshape program on a Perkin Elmer Interdata 8/32 computer. All isomer shift values were referenced to a National Bureau of Standards metallic iron foil.

RESULTS AND DISCUSSION

Examples of representative Mössbauer spectra obtained for the various

catalyst samples as a function of treatment are shown on Figures 1 and 2.
Spectra for the reduced, exposed and calcined samples of the ferrous chloride-
ruthenium trichloride and ferrous nitrate-ruthenium trichloride are shown.
Similar spectra were obtained for the ferrous sulfate system to compare with
previously reported results [1]. The ferric chloride system is unique in
that the spectra indicate clearly reduced and oxidized products. The six-line
magnetically split spectrum of α-Fe is clearly visible in the reduced sample
with an additional line defined by the Mössbauer parameters: isomer shift
=0.05 mm/sec (referenced to metallic iron) and a linewidth of 0.34 mm/sec.
This center line could represent a fraction of iron dispersed as very small
particles in the superparamagnetic phase [6] or it could be a signal associated
with a portion of the sample converted to ε-iron in the form of a Ru-Fe alloy
[7]. Subsequent temperature dependent spectra and the shape of the peak indicate that the presence of the alloy phase is probably the source of this
signal.

Note that the reduced phase reacts with oxygen at room temperature (the
exposed sample) to produce a product which exhibits an asymmetric Mössbauer
doublet with peak positions at 0.03 and 0.80 mm/sec. If the doublet is decomposed into a singlet at 0.03mm/sec (corresponding to the largest peak) and
a doublet with an isomer shift of + 0.38mm/sec and a quadrupole splitting of
0.77mm/sec the product can be assigned as a mixture of the Ru-Fe alloy and
superparamagnetic α-Fe$_2$O$_3$. The assignment of the superparamagnetic
α-Fe$_2$O$_3$ is indicative that the metallic particle size in the reduced sample
is quite small. At room temperature the critical particle diameter for the
onset of superparamagnetic behavior in α-iron is less than 15°A (hence the
observed magnetic spectrum of the reduced phase) but the critical particle
size for α-Fe$_2$O$_3$ is 135 Å [6,9]. Chemical evidence of the small particle
size and subsequent high surface area of the metallic particles is exhibited
by the highly exothermic reaction which takes place when the sample is exposed to air; in some instances the sample actually flamed. The product after
calcination shows the characteristic six-line magnetic spectrum of bulk
α-Fe$_2$O$_3$ indicating the particle sintering which takes place during high temperature oxidation.

Although it is not possible to completely characterize the particle size
of the microcrystallites in a solid state mixture by the Mössbauer effect,
some further information can be gained by determining the spectra at different
temperatures. The Fe-Ru chloride (exposed) system was investigated at 83°K,
298°K and 373°K. The six line magnetic splitting was barely apparent at
293°K but was well developed at 83°K. By comparison with earlier work [8]
this would imply that the average effective particle size of the α-Fe$_2$O$_3$
was of the order of 70Å or less.

The spectra for the ferric nitrate samples are similar to the chloride
materials except that the quality of the spectra are poor (the recoil free
fraction is low) and some unidentified product remains in both the reduced
and calcined samples. The reduced sample contains a very large (compared to
the chloride case) peak with an isomer shift of 0.00-0.01 mm/sec which can
again best be assigned to the Ru-Fe alloy. Some vestige of that peak remains even in the calcined sample indicating that the alloy may be more re-
sistant to oxidation than the metallic iron. The variation in line width for
the magnetic peaks in the reduced sample and the poor signal to noise ratio
for the superparamagnetic α-Fe$_2$O$_3$ spectrum in the exposed case are both
indicative of very small, anisotropic particles. The fact that oxidation is
essentially complete when the sample is exposed to air at room temperature
also clearly demonstrates the high surface area of the iron metal component
in these samples.

The Fe-Ru sulfate samples are different as was observed earlier [1] in that

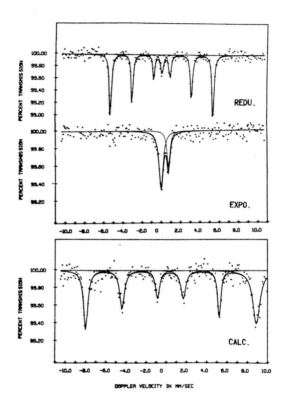

Fig. 1. Fe-57 Mössbauer Spectra of Fe-Ru Bimetallic Catalyst
(Chloride-Chloride) System.

the reduced sample contains some ferrous sulfide and the calcined sample is
not converted totally to $\alpha\text{-}Fe_2O_3$.

Further work is presently underway in our laboratory to substantiate the
particle size studies by x-ray diffraction and scanning electron microscopy.
However, the fact that we have constructed a Mössbauer cell which allows the
samples to be examined essentially in situ greatly enhances the utility of
this tool. The reduced samples are so very air and moisture sensitive that it
is almost impossible to get authentic characterization measurements on the reduced material. Then results clearly illustrate the potential of Fe-57
Mössbauer spectroscopy as a rapid, analytical tool for assessing the solid
state chemistry of materials containing iron in any form.

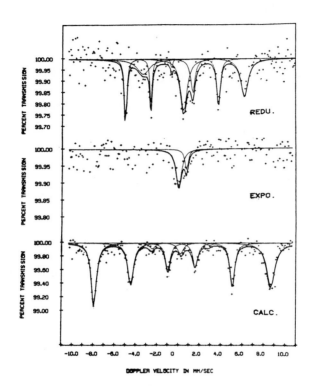

Fig. 2. Fe-57 Mössbauer Spectra of Fe-Ru Bimetallic
(Nitrate-Chloride) System.

ACKNOWLEDGEMENT

The authors would like to thank the National Science Foundation for the partial financial support of this project by Grants No. CHE 76-17434 and CHE-7912999.

REFERENCES

1. M. L. Good, M. D. Patil, J. T. Donner and C. P. Madhusudahan, ACS Sym. Series, Recent Chemical Applications of Mössbauer Spectroscopy, in press.

2. M. L. Good, M. Akbarnejad and J. Donner, ACS Division of Petroleum Chemistry Preprints, Las Vegas, Nevada, September 1980.

3. W. N. Delgass, G. L. Haller, R. Kellerman and J. H. Lunsford, Spectroscopy in Heterogenous Catalysis (Academic Press, New York 1979) p. 132.

4. F. L. Ott, T. Fleisch and W. N. Delgass, J. Catalysis 60, 394 (1979).

5. M. A. Vannice, Y. L. Lam and R. L. Garten, ACS Adv. Chem. Series, 178, 25 (1979).

6. G. B. Raupp and W. N. Delgass, J. Catalysis 58, 337 (1979).

7. J. M. Williams and D. I. C. Pearson, J. de Physique, 37, C6-401 (1976).

8. W. Kündig, H. Bömmel, G. Constabaris and R. H. Lindguist, Phys. Rev. 142, 327 (1966).

9. F. J. Berry, Adv. Inorg. and Radiochem., 21, 255 (1978).

Kaufmann and Shenoy, editors
Nuclear and Electron Resonance Spectroscopies Applied to Materials Science

MÖSSBAUER STUDY OF IRON AND TIN EXCHANGED INTO VICTORIAN BROWN COAL

J.D.CASHION, P.E.CLARK AND P.S.COOK
Department of Physics, Monash University, Clayton, Victoria 3168, Australia.

F.P.LARKINS AND M.MARSHALL
Department of Chemistry, Monash University, Clayton, Victoria 3168, Australia.

ABSTRACT

Mössbauer effect measurements have been taken of iron and
tin introduced into brown coal to act as catalysts for
hydrogenation. Spectra taken with ^{119}Sn show that after
introduction from solution, the tin appears as SnO_2.
During subsequent hydrogenation in an autoclave, some of the
SnO_2 is converted to β-tin and SnS. Ion-exchanged iron is
shown from ^{57}Fe measurements to occupy at least one site
not populated in the native coal. Line broadening in this
site and the other two original sites indicate the existence of variations in the local environment.

INTRODUCTION

The last few years have seen an enormous re-awakening of interest in
increasing the use of coal in many areas, ranging from the traditional one of
burning it more or less unchanged from its mined state to various techniques
of converting it into other fuels - solid, liquid and gaseous. Mössbauer
effect studies of coal have shared in this upsurge, and since the lone pioneering study in 1967 [1] there have been at least seventeen more papers since
1977.

All except two of these papers have dealt with high rank U.S. coals and
^{57}Fe has been used exclusively. With its sensitivity to the chemical state of
the atom being investigated, we might expect many varied bonding arrangements
of iron in a material as inhomogeneous as coal. However the U.S. studies have
shown that all the iron is in mineral form, principally as sulphides, carbonates and silicates, with sulphates and oxides present in weathered coals (see
e.g. [2]), and it has only been in studies on some lower rank Australian brown
coals [3,4] from the La Trobe Valley in Victoria in which organically bonded
iron has been identified.

Victorian brown coal is a low rank (67% C), low sulphur (<1%) coal which
has typically 67% water content as mined. It dries rapidly to approximately
15% water on exposure to the atmosphere, but will regain moisture if dried
below this figure. The organically bonded iron occurs in several similar but
distinct sites [4] and it appears that in the bed-moist state the majority of
its ligand are water molecules.

Coal is rarely considered as a "material" in the materials science context
and its use has continued for centuries with little improvement in our knowledge of its detailed structure. It is only recently that microscopic tools
such as resonance techniques have come to be applied to it to supplement the
only partially interpretable data obtained from the more established techniques such as x-ray diffraction.

In this paper we will describe experiments carried out in conjunction with
catalytic hydrogenation studies and involve the first coal-based ^{119}Sn studies

carried out on coal containing tin introduced from solution as well as [57]Fe studies on ion-exchanged and inherent iron. These studies are aimed at determining the mode of operation of the iron or tin catalyst and, in conjunction with yield efficiency results, to try and interpret on a microscopic basis the reasons for differing yields from different coals or from different catalytic conditions.

TIN EXCHANGED COAL

The main details of the chemical processes involved in this investigation have been described recently by Hatswell et al. [5]. For the tin catalyst runs, Morwell Drum 66 coal was air-dried at room temperature, ground to -60 mesh BS sieve and then gravity floated with CCl_4 to remove some of the silica and pyrite/marcasite (99% total mass remaining). After drying, it was mixed with 0.05M $SnCl_2.2H_2O$ or 0.01M $SnCl_4.5H_2O$ in a 30:1 ratio of solution to dry coal and stirred under nitrogen. The mixture was then filtered and washed to constant pH and dried under nitrogen to 30% water content. The tin concentration after this operation was approximately 9.5 wt% ($SnCl_2$ treatment) or 3.7 wt% ($SnCl_4$ treatment) on a dry basis indicating that not only had most of the Ca, Mg, Na and Fe ions been replaced by Sn, but some of the H ions as well.

Two types of hydrogenation experiments were carried out. In one, 50 g of coal N_2 dried at 105°C were slurried 1:1 with tetralin and heated in a 1 dm^3 autoclave to 375-400°C in 1½ hours under an initial hydrogen pressure of 10 MPa. Final temperature was held for 1 hour before overnight cooling. In the other, 3.5 g of coal ($SnCl_4$ treated and subsequently N_2 dried at 105°C) were heated to 310°C in 40 minutes in a 70 cm^3 autoclave under the same initial hydrogen pressure. After the desired time at 310°C, the autoclave was cooled by air-blast to 200°C in 10 minutes.

Fig. 1(a) shows the [119]Sn spectrum of the washed coal showing that there was no tin present before the ion exchange process. After the ion exchange, the spectrum in Fig. 1(b) was obtained showing a single broadened resonance which is in fact an unresolved quadrupole split doublet. From the parameters (isomer shift = 0.02(2) mm s^{-1} with respect to $CaSnO_3$, quadrupole splitting = 0.57(2) mm s^{-1}) this is undoubtedly due to SnO_2. It is interesting to note that SnO_2 was obtained from both $SnCl_2$ and $SnCl_4$ starting solutions. Carboxylate-tin species are very unstable and presumably hydrolyse during the treatment to form tiny precipitates of SnO_2. Photoelectron spectroscopy also identified the material as SnO_2 whereas x-ray diffraction measurements did not reveal any crystalline SnO_2. Proof of a uniform distribution of tin atoms through the coal was given by electron microprobe analysis.

Fig. 1(c) shows the final CH_2Cl_2 insoluble residue after the conversion had been taken to completion (run #6 in [5]). Further lines are now evident at positive velocities, due principally to SnS (IS = 3.10(3) mm s^{-1}, QS = 0.93(3) mm s^{-1}) and a single line due to β-Sn (IS = 2.48(3) mm s^{-1}). There is another weak line at 1.5(1) mm s^{-1} which has not been identified but which may be a hydride. The existence of β-Sn and SnS was also detected by x-ray diffraction.

The transition during the hydrogenation process can be seen in more detail in Fig. 2 which shows the spectra of two samples which were heated to 310°C in hydrogen without tetralin for (a) 20 min and (b) 60 min. The SnO_2 peak is being reduced in intensity and the changes show that the tin is certainly an active agent during the hydrogenation. These spectra are part of a complete study of the time and temperature dependence of catalytic conversion using tin which will be published elsewhere.

In Fig. 2(a) we can see the commencement of the growth of the β-Sn peak with at least two other peaks which we have not yet positively identified. After one hour in the autoclave (Fig. 2(b)), the SnS doublet is starting to appear.

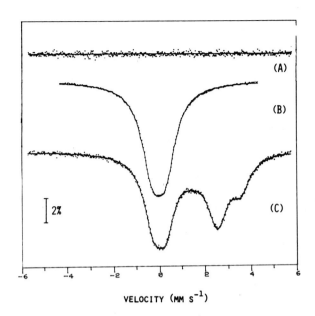

Fig. 1 ^{119}Sn Mössbauer spectrum of (a) coal after flotation, (b) coal after
treatment with $SnCl_2$ solution and (c) residue after hydrogenation at
375-400°C.

The temperature of 310°C was chosen because the coal has lost virtually all
the CO_2 to be given off and hydrogen consumption has become significant. We
believe that the catalytic reaction commences by the hydrogen reducing the
SnO_2 to tin metal, which is liquid at this temperature. This can then disperse
very easily throughout the coal to be well situated to act as a catalyst. The
reaction proceeds with the tin scavenging both oxygen from the coal matrix and
organic sulphur. The conversion efficiency of this method of catalysis is
comparable to that obtained on a test run with a commercial cobalt molybdate
catalyst. It can be seen that the Mössbauer spectra provide a readily interpretable microscopic picture of the changes taking place in the catalyst
during the hydrogenation.

IRON EXCHANGED COAL

In the case of the iron ion exchanged samples the procedure was essentially
the same as that described for the tin case but using 0.05M $FeSO_4.7H_2O$ (run
#3 in [5]) as the solution. The coal already contained iron and Fig. 3(a)
shows the spectrum of this iron after the CCl_4 flotation had removed most of
the FeS_2. Two doublets are visible as marked by the bar diagrams, the lengths
of the bars being proportional to the intensity of the peaks. The outer
doublet with IS = 0.31(4) mm s^{-1} and QS = 0.62(4) mm s^{-1} is due mainly to
pyrite. The inner doublet with IS = 0.20(4) mm s^{-1} and QS = 0.40(4) mm s^{-1} is
due to organically bonded iron, and we note that these parameters differ from

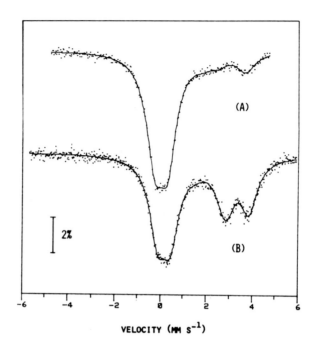

Fig. 2 ^{119}Sn Mössbauer spectrum of coal after hydrogenation at 310°C (a) for 20 min, and (b) after 60 min.

any obtained in the characterization study of Victorian brown coals [4]. We believe that this is due to the different hydration levels of the coals. The linewidths are slightly broader than those obtained from the bed-moist samples in [4], being 0.34 - 0.40 mm s^{-1}. This indicates some variations in the surroundings of the iron atoms in each of the two sites, with the possibility that there is another weak organically bonded component under the pyrite doublet, probably corresponding to the non-magnetically split component in sample #15 of [4].

The spectrum after the ion exchange, Fig. 3(b), surprisingly showed an extra site being populated which was iron-free originally. This site has a larger quadrupole splitting than the other two sites, 0.96(4) mm s^{-1} and so must be in a less symmetrical electronic environment. The isomer shift is also considerably more positive, 0.48(4) mm s^{-1}, although it is difficult to tell if the iron is in a trivalent or low spin divalent state. Further experiments at low temperatures should resolve this ambiguity.

The ratios of the populations of the original two sites also showed a change after the ion exchange, with the more asymmetrical site being favoured. Since this site and the new one have both been populated by replacing Ca^{2+}, Mg^{2+} and Na^+ which probably have lower co-ordination and also lower valence for Na^+, it is reasonable to expect the sites to be less symmetrical. The linewidths of these sites are all larger than for Fig. 3(a) being 0.40 - 0.50 mm s^{-1}, with the new doublet being the broadest. This suggests that there are at least two and possibly many more sites involved in this doublet and also that the changed populations of the original doublets are due, at least in part, to sites which are similar rather than identical to those inherent in the coal.

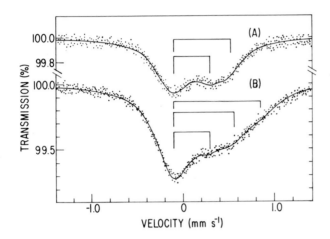

Fig. 3 ^{57}Fe Mössbauer spectrum of coal (a) after flotation, and (b) after treatment with FeSO$_4$ solution showing population of new site.

The hydrogenation efficiency of the iron catalyzed sample was not quite as high as that obtained with the tin and cobalt molybdate catalysts [5]. However unlike the tin case, where it seems that all the tin behaves essentially the same, the iron catalyst provides a variety of sites which can change with the type of coal and possibly with the ion exchange conditions. It is interesting to compare this situation with other current research using both inherent and added pyrite as a catalyst. We hope that the organically bonded iron should be more evenly distributed and better situated for catalytic action. It is in this area that a microscopic probe provides invaluable information for elucidating the reason for different yields, which may not be well correlated with simple bulk measurements such as the total iron content.

In the case of the above organically bonded iron sites, we do not yet have a good knowledge of the environment of the sites. However it is fortunate that this is not necessary for our purposes since we only need to be able to correlate different sites with particular properties in order to be able to enhance the use of the coal in that application.

ACKNOWLEDGEMENTS

This work has been carried out with the assistance of the Victorian Brown Coal Council, the Australian Research Grants Committee and the Australian Institute of Nuclear Science and Engineering.

REFERENCES

1. J.F. Lefelhocz, R.A. Friedel and T.P. Kohman, Geochim. and Cosmochim. Acta 31, 2261 (1967).

2. F.E. Huggins and G.P. Huffman in Analytical Methods for Coal and Coal Products, Vol. III, C. Karr Jr. ed. (Academic Press, New York, 1980). P.A. Montano in Recent Chemical Applications of the Mössbauer Effect, J.G. Stevens, G.K. Shenoy eds. (A.C.S. Adv. in Chem. Series, New York) in press.

3. H.N.S. Schafer, Fuel, 56, 45 (1977).

4. J.D. Cashion, B. Maguire and L.T. Kiss in Recent Chemical Applications of the Mössbauer Effect, J.G. Stevens, G.K. Shenoy eds. (A.C.S. Adv. in Chem. Series, New York) in press.

5. M.R. Hatswell, W.R. Jackson, F.P. Larkins, M. Marshall, D. Rash and D.E. Rogers, Fuel, 59, 442 (1980).

NMR STUDY OF SODIUM ION MOBILITY IN A SINGLE CRYSTAL OF
NA β"-ALUMINA AND WATER EFFECTS IN POLYCRYSTALLINE SAMPLES

CECIL E. HAYES and DAVID C. AILION
Physics Department, University of Utah, Salt Lake City, Utah 84112

ABSTRACT

We measured the temperature dependence of T_1 in a Mg-stabil-
ized single crystal of Na β"-alumina. A comparison with
earlier measurements on Li-stabilized polycrystalline samples
indicates that, below 500K, the sodium ion kinetics are not
directly affected by the stabilizing element. Above 500K
our measured T_1 shows a dependence on sample composition.
Observations of NMR linewidth and signal intensity vs. water
content indicate that, at low water concentration (< 2%), all
sodium is in contact with water in its conduction plane;
at higher water concentrations, the additional water may be
forming complexes with some of the sodium.

INTRODUCTION

The superionic conductors sodium β- and β"-alumina have been under intense
study [1] in recent years because of their possible application in energy storage technology. Nuclear magnetic resonance techniques have been applied to
obtain information about ionic motion and structure. Although Na β"-alumina
exhibits higher conductivity, most studies have been on Na β-alumina, since
single crystals of β-alumina have been more available. A few reports of NMR
studies on polycrystalline Na β"-alumina [2,3,4] and, more recently, on single
crystals [5] have appeared.

In this paper we report NMR relaxation time measurements on a single crystal
of Na β"-alumina and compare them to our earlier results [3] for polycrystalline
samples. We also describe additional studies of the effects of water on powdered samples of polycrystalline Na β"-alumina [4].

NMR BACKGROUND

Placing isolated sodium nuclei (spin 3/2) in an external magnetic field H_o
(in the z-direction) creates four equally spaced energy levels whose splitting
is equal to ℏ times by the Larmor frequency ω_L, which is proportional to H_o.
A strong rf pulse at frequency ω_L can tip the equilibrium magnetization away
from H_o such that it will precess around H_o at ω_L and induce a signal in an
rf pickup coil. The signal strength is proportional to the magnetization component in the x-y plane and will vary as $\sin\phi$, where the precession angle ϕ is
proportional to both the amplitude and duration of the rf pulse. Hence, a 90°
pulse will maximize the pickup signal. A 180° pulse will invert the magnetization but will not induce a signal in the pickup coil.

Because the sodium nucleus also has an electric quadrupole moment, electric
field gradients at the nucleus due to the neighboring charges cause the NMR line
to split into three lines, a central line and two satellites. According to
perturbation theory, the satellite transitions are shifted to first order in
the quadrupolar interaction strength whereas the central transition has zero

first-order shift and is shifted only to second order. The shift is a function of crystal orientation relative to applied magnetic field. Since all orientations are present in a powdered sample, the satellite lines are spread over such a wide range of frequencies that they cannot be detected easily. The central transition gives a second-order quadrupolar broadened powder pattern characterized by a cusp at each end of the distribution. For quadrupolar shifted levels, the pulse length required for maximum signal of the central transition is only 45°, whereas a 90° pulse will invert the magnetization [6].

In this work, the NMR signals were recorded using a method described by Clark [7]. Repetitive sampling pulses at a fixed frequency cause those nuclei which are on resonance to precess. The signal induced in the pickup coil is recorded as the magnetic field slowly sweeps through the line. When studying relaxation times, one or more preliminary pulses are applied to prepare the spin system either by inverting the magnetization or by equalizing the population (saturation). At some variable time later, the sample pulse (either 90° or 45°, as we discussed earlier) is used to measure the recovering signal.

SAMPLES

Na β"-alumina has a rhombohedral unit cell made up of three spinel blocks of aluminum and oxygen separated by loosely packed layers of oxygen and sodium. Each block could ideally accommodate the formula $Na_2 Al_{11} O_{17}$ except for the lack of charge neutrality. In practice the structure is stabilized by substituting Li or Mg in some of the aluminum sites. The single crystal used in this study has a composition $Na_{1.67} Mg_{0.67} Al_{10.33} O_{17}$ [8]. The polycrystalline sample is $Na_{1.72} Li_{0.30} Al_{10.70} O_{17.06}$ [9] and consists of particles of size 70 microns or less containing grains of about 10 microns diameter.

SPIN-LATTICE RELAXATION IN DRY SAMPLES

In our previous paper [3] we studied the nuclear relaxation time T_1 in powdered polycrystalline lithium-stabilized Na β"-alumina. The dotted lines in Fig. 1 summarize our earlier findings. Curve (a) was taken at 24 MHz using the inversion-recovery technique. There is a broad motionally-induced minimum in T_1 at a temperature of about 312 ± 12K. From 330 to 500K, the T_1 data can be characterized by a thermally activated process with an activation energy $E_A = 0.37 \pm 0.04$ eV. At higher temperatures there is bending over of T_1 which suggests a new motional process at elevated temperatures. This interpretation is consistent with the frequency dependence of T_1 observed in this temperature region in curve (b) at 18 MHz. According to NMR theory [10], T_1 should be proportional to ω^2 at temperatures below the T_1 minimum.

For temperatures lower than the T_1 minimum, our T_1 values measured in powder [3] depend on the pulse sequence. Curve

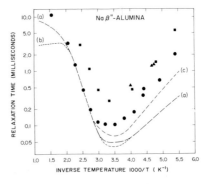

Fig. 1. Spin lattice relaxation times for Na^{23} β"-alumina. Curves are for polycrystalline samples: inversion-recovery (a) at 24 MHz and (b) at 18 MH; saturation-recovery (c) at 24 MHz. Data points are for a single crystal at 24 MHz: inversion-recovery (●) for θ = 90° and (■) for θ = 0°; saturation-recovery (▲) for θ = 0°.

(c) was obtained by attempting to saturate the magnetization with a series of equally spaced 45° pulses before observing the signal with a 45° sampling pulse. These T_1 values and their slope are greater than those of curve (a) obtained from a single inverting pulse followed by the sampling pulse.

The temperature dependence of T_1 in the single crystal for two orientations are given as data points in Fig. 1. The orientations are θ = 0° and 90°, where θ is the angle between $\vec{H}_o$ and the crystal c-axis. Both orientations provide the same high and low temperature slopes and the same temperature at the minimum, since the slopes are determined by the sodium ion kinetics rather than the field orientation. The difference in magnitude of T_1 between θ = 0° and 90° arises because the orientation influences the effectiveness of the electric field gradient fluctuations in causing spin relaxation. Our low temperature T_1 powder data is lower than that of either single crystal, since it was taken from the high field cusp of the powder pattern and corresponds approximately to θ = 42° [11]. A few measurements on the single crystal at θ = 42° agreed in magnitude with the powder data. The excellent agreement over the temperature range 333 – 500K between data for the powdered material and that for the single crystal at θ = 90° indicates that the motional process involved here is not influenced by differences in sample composition. There is also approximate equality between the two low temperature slopes of the single crystal and that of curve (c). Several data points were taken using the saturation-recovery technique for θ = 0° in the single crystal, but they don't differ significantly from T_1 values obtained with the inversion-recovery pulse sequence. It is not clear now why there was a difference in T_1 values for the two pulse techniques in the powdered sample. However, our single crystal data and the saturation-recovery data [curve (c)] give the same activation energy (0.14 ± 0.01 eV) over the range 190K to 286K.

An important question concerns the effect of the stabilizing element, Li^+ or Mg^{++}, on the Na mobility. It should be noted that below 500K our NMR relaxation time studies show little difference in Na ion kinetics between powder samples stabilized by Li^+ and the single crystal stabilized by Mg^{++}. The lithium or magnesium, which substitutes for aluminum in sites far from the conduction plane, probably exerts no direct influence on sodium motion.

At the highest temperature there is evidence that the T_1 for the single crystal (θ = 90°) is also bending over but to a lesser degree than for the powdered sample. T_1 for another powdered sample (not shown) bent over more strongly than curve (a). The fact that we observed different T_1 behavior above 500K for two lithium stabilized samples as well as the magnesium stabilized sample suggests that the stabilizing element is probably not the dominant influence on sodium motion in this temperature range. Ionic conductivity measurements in polycrystalline samples [12] and in single crystals [13] indicate a decrease in activation energy for Na β"-alumina at high temperatures. Boilot et al. [14] have recently found evidence for short-range ordering of the sodium vacancies into a superlattice structure consisting of five sodiums and one vacancy for every six sites. They measured the coherence length for this ordered array of vacancies to be 70 Å at temperatures below 300K but to decrease smoothly to 20 Å above 600K. Furthermore, they suggest that the decrease in activation energy is related to the decrease in coherence length. Our single crystal was close to the ideal composition for such ordering and apparently exhibits a lower than usual ionic conductivity [8]. Thus, there is likely to be less breakdown of this ordering at higher temperatures in the single crystal than in the powder, which contains slightly more sodium, as is consistent with our observations that the slope of T_1 in the former is less than in the latter.

286

EFFECTS OF ABSORBED WATER

In an earlier paper [4] we described the effects of absorbed water on the NMR lineshape and on relaxation times of powdered polycrystalline samples of Na β"-alumina. For completeness, we will summarize the former results while giving the details of our new findings.

When a sample of powdered polycrystalline Na β"-alumina is exposed overnight to 100% relative humidity at room temperature, it absorbs water equal to about 16% of its dry weight. All but 2% water content can be removed at room temperature by overnight evaporation in low humidity or in a vacuum. Elevated temperatures are needed to remove the remaining water. There appears to be a qualitative difference between the effects on the NMR signal of the loosely bound water (2 to 16%) and those of the more tightly bound water (0 to 2%).

For a sample with up to 2% water the sodium NMR signal shows a second-order quadrupolar broadened powder pattern with a linewidth reduced compared to that of a dry sample. As this water is then removed by baking the sample at successively higher temperatures, the quadrupolar interaction strength increases monotonically to the value in a dry sample. The fact that the area under the curve of the powder pattern remains essentially constant for this range of water content indicates that all the sodium ions interact equally with the water which therefore must be distributed in the conduction plane [4].

High water content destroys the second-order quadrupolar broadened powder pattern and produces a strong narrow Na signal (1.2 gauss wide) sitting on a wide low lying base. The narrow line signal strength was calibrated and found to arise from only 28 ± 12% of the available sodium ions in samples having 16% water content. As the absorbed water evaporated from the powder, the area of the narrow line component varied approximately linearly with water content up to at least 16%. The linearity of this relationship indicates that all the absorbed water is in contact with at least some of the Na nuclei. However, the fact that the linewidth (1.2 g) is much broader than that of a simple liquid implies that the sodium ions are not in solution.

The question remains—does the narrow line arise from Na nuclei which now see zero quadrupole interaction or from nuclei which see first-order but not second-order effects? These possibilities can be distinguished by the fact that spins seeing zero electric field gradient (EFG) will have maximum signal after a 90° pulse, whereas first-order shifted spins respond best to a 45° pulse. Figure 2 shows the signal amplitude vs. pulse width for dry and humid samples. The dry sample signal, which is known to be second-order broadened, peaks at 2.25 μsec (45° pulse) and inverts at 4.5 μsec (90° pulse). The humid sample signal peaks at a value intermediate between 45° and 90° and inverts at 9 μsec (180° pulse).

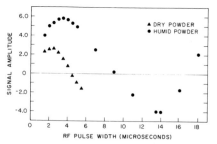

Fig. 2. Signal amplitude versus rf pulse width for humid and dry polycrystalline Na β"-alumina.

A simple analysis indicates that the result is due to about half the spins being in zero EFG and the other half in intermediate sized EFG's (i.e., first-order splitting). Such reductions in EFG may result when sufficient water is available to form a hydration complex around the the sodium.

There is still the question of where the loosely bound water (2 to 16%) is located in the powder particles. Since the Na ions do not all see the same quadrupolar interaction, this water is apparently not uniformly distributed throughout the sample. One possibility would be for H^+ or H_3O^+ ions to exchange with some of the sodium ions in the conduction planes. These sodium ions would be free to migrate out of the crystallite and form a solution of NaOH in a sur- face layer of water. As the water evaporates, the sodium would be forced back into the crystal. We checked for this mechanism by observing the narrow NMR line in a humid powder sample during immersion in anhydrous methanol [15]. If the narrow line had become even more narrow, we would have concluded that the methanol had removed Na^+ ions since the methanol cannot penetrate the sample. In fact, adding methanol eliminated most of the narrow line and partially restored the second-order powder pattern, thereby indicating that the methanol removed water but not Na^+. Either Na^+ was not on the surface or the methanol preferentially stripped H^+ or H_3O^+ ions from the conduction plane and forced surface Na^+ back into the bulk material.

Another possibility is that water penetrates the outer layers of the crystal- lites to greater extent than the interior volume of the crystallite, similar to the findings of Foster and Arbach [16] for water penetrations of Na β"-gallate. We observed some indirect evidence for this mechanism in Na β"-alumina as des- cribed below. Repeated washing of a powdered sample of Na β"-alumina with deion- ized water at room temperature leached out 30 ± 3% of the initial sodium content. The sodium loss was determined by sodium flame analysis of the collected wash water. The lost sodium must have been replaced by H_3O^+ or possibly by H^+ in order to maintain charge neutrality in the crystal. The washed powder was dried first in ambient air and then by heating under vacuum at a series of progres- sively higher temperatures. NMR lineshapes recorded at room temperature for different concentrations of volatile water (Fig. 3) show the same qualitative behavior as those of an unwashed sample following exposure to humidity [4].

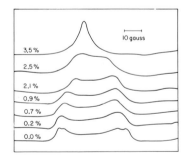

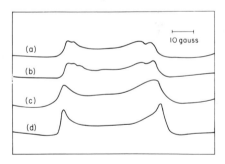

Fig. 3. NMR lineshapes for washed Na β"-alumina powder after successive dryings. Labels give the volatile water content.

Fig. 4. NMR lineshapes for washed and then dried Na β"-alumina powder, recorded (a) at 25°C and (b) at 100°C. Dry unwashed powder was recorded (c) at 25°C and (d) at 100° C.

There are, however, several important differences between the washed and dried powder sample and another sample which was not exposed to water. Curves (a) and (b) of Fig. 4 refer to a washed sample, subsequently dried at 500°C and then recorded at 25°C and 100°C, respectively; whereas curves (c) and (d) refer to a similarly dried sample which had not been washed in water. The peak-to-peak linewidth of the washed sample [Fig. 4(a)] is reduced to 86% of that of the unwashed sample [Fig. 4(c)]. Also the lineshape of Fig. 4(a) shows evidence of two distinct quadrupolar interaction strengths. Most of the sodium ions produce a powder pattern due to the stronger quadrupolar interaction, but a smaller number (perhaps 10%) produce a slightly narrower powder pattern superimposed on the wider one. The second contribution may arise from a surface layer where water penetration and sodium exchange are more complete than in the interior of the crystallite. A comparison of curve (d) with curve (c) shows a pronounced sharpening of the structure at 100°C produced by motional narrowing and/or reduced lifetime broadening. Both effects arise from more rapid ionic motion as the temperature is raised. Comparing curves (a) and (b) gives no similar evidence of increased ionic motion with increased temperature in the washed sample. Thus the H_3O^+ or H^+ ions in the conduction plane have probably significantly altered the sodium ion motion, in agreement with our earlier T_1 observations in powder samples exposed to humidity [4].

CONCLUSIONS

We observed the same relaxation behavior below 500K in a Mg^{++}-stabilized single crystal as in Li^+-stabilized powders, which indicates that the stabilizing element is not affecting the sodium motion. Sodium concentration in the conduction plane is more likely the critical factor at high temperatures. The higher sodium content of the polycrystalline samples may encourage the breakdown of the superlattice ordering as the temperature is raised. From our studies of water effects we conclude that all the Na is in contact with water and each Na experiences the same environment for water content up to 2%, thereby indicating that the water in the conduction plane is distributed uniformly. For water content between 2% and 16%, the extra water gives rise to a considerably narrowed Na line involving up to approximately 28% of the sodium.

We are very grateful to Dr. J. B. Bates for providing us with the single crystal and to Dr. G. R. Miller for providing the polycrystalline samples. This research was supported by the U.S. National Science Foundation under Grant DMR 76-18966.

REFERENCES

1. Fast Ion Transport in Solids, P. Vashishta, J.N. Mundy, and G.K. Shenoy eds. (North Holland, New York 1979).
2. W.C. Bailey, H.S. Story, and W.L. Roth in: Superionic Conductors, G.D. Mahan and W.L. Roth eds. (Plenum Press, New York 1976) p. 370.
3. C.E. Hayes and D.C. Ailion in Ref. 1, p. 297.
4. D.C. Ailion and C.E. Hayes in Ref. 1, p. 301.
5. W.C. Bailey, H.S. Story, A.R. Ochadlick, Jr., and G.C. Farrington in Ref. 1, p. 281.
6. A. Abragam, Principles of Nuclear Magnetism (Oxford University Press, Oxford 1961) p. 36.
7. W.G. Clark, Rev. Sci. Instrum. 35, 316 (1964).
8. J.B. Bates, private communication.
9. G.R. Miller, B.J. McEntire, T.D. Hadnagy, J.R. Rasmussen, R.S. Gorden, and A.V. Virkar in Ref. 1, p. 83.
10. N. Bloembergen, E.M. Purcell, and R.V. Pound, Phys. Rev. 73, 679 (1948).
11. See Ref. 6, pp. 232-241.
12. T. Cole, N. Weber, and T.K. Hunt in Ref. 1, p. 277.
13. G.C. Farrington and J.L. Briant in Ref. 1, p. 395.
14. J.P. Boilot, G. Collin, Ph. Colomban, and R. Comes, preprint.
15. Treatment with methanol was suggested by Dr. T. Cole.
16. L.M. Foster and G.V. Arbach, J. Electrochem. Soc. 124, 165 (1977).

LOW-LYING EXCITATIONS AND LOCAL MOTIONS IN Na β-ALUMINA

MARCO VILLA
Istituto di Fisica "A. Volta" e Gruppo Nazionale di Struttura della Materia del
C.N.R., 27100 Pavia, Italy

JOHN L. BJORKSTAM
Department of Electrical Engineering, University of Washington, Seattle, WA
98195

ABSTRACT

We present an NMR study of sodium motion and distribution in Na β-alumina below 160K. The ^{23}Na lineshape
shows that a rearrangement of sodiums occurs near 100K.
Below this "transition", a fraction of the sodium ions
undergo a local motion that is an efficient mechanism for
^{23}Na spin-lattice relaxation in the interval 50÷80K. This
motion does not have a major influence upon the ^{27}Al relaxation. Below 50K the low-lying excitations typical of
glasses control the spin-lattice relaxation processes of
^{27}Al and, to a certain extent, ^{23}Na.

INTRODUCTION

A still incomplete chapter·of condensed matter physics addresses the problems
of the low-temperature (T) dynamics of disordered materials. For very different
types of "disorder", and over a wide spectrum of substances, certain properties
show large deviations from the behavior of crystals [1]. It would be most helpful to find a "universal" explanation for the rather typical effects of disorder.
Single crystal β-alumina solid electrolytes exhibit typical glasslike behavior
at low temperatures [2-7]. One purpose of this work is to present experimental
evidence that the spin-lattice relaxation times (T_1) of ^{23}Na and ^{27}Al do not display the kind of behavior expected in crystals. A second goal is to show that
near 100K drastic changes in both the dynamics and the arrangement of the diffusing ions occur in β-alumina. Here we will comment upon our data without re-
sorting to models of interpretation that can be very detailed but somewhat
tentative at our level of knowledge.

LOW-T SPIN LATTICE RELAXATION IN DIELECTRIC GLASSES

The low-T anomalies of glasses are caused by the presence of a spectrum of
low energy excitations in addition to the phononic modes described by a Debye-
like density of states. The low lying excitations are often described by two-
level-systems (TLS) with a broad distribution of splitting energies [1,5-7].
However, the nature of these excitations is still virtually unknown. For
quadrupole perturbed nuclear spin systems in the dielectric glasses so far analyzed, it has been demonstrated that these excitations control the spin-lattice
ralaxation process at low-T [8]. A classical theory of relaxation by coupling
with the phonons of a Debye spectrum would predict $T_1 \sim T^{-2}$ and $T_1 \sim T^{-7}$ at temperatures above and below ~ 10 K, respectively [9]. Instead, for quadrupole perturbed spins in dielectric glasses the published results are described by the
empirical law [8]

$$T_1 \propto T^{-\beta}, \ 1 \leq \beta \leq 2 \tag{1}$$

Moreover, T_1 seems frequency independent and it is much shorter than T_1 in the crystalline modifications of the glass. These unusual features are observed, typically, up to ∿100K while many other low-T anomalies vanish above a few degrees K.

^{23}Na NMR RESPONSE OF β-ALUMINAS

The β-aluminas are mixed oxides of aluminum and a monovalent metal with ideal formula $Na_2O \cdot 11Al_2O_3$. The conducting ions (Na^+) lie in mirror planes between compact aluminum oxide spinel blocks. The sodium nucleus has spin I = (3/2) and its NMR response is determined by the interaction between the nuclear quadrupole moment and the electric field gradient (efg) at its position. The NMR response of both ^{23}Na and ^{27}Al shows a strong dependence upon the angle, α, between the normal $\vec{c}$ to the conducting plane and $\vec{H}_o$. In particular, it has been shown [10] that for ^{23}Na at α=0° the rate of the Δm=±1 transition (W_1) is zero. Following a pulse which saturates only the ^{23}Na central line, the magnetization M(t) due to the population difference of the (±1/2) levels is

$$\frac{\Delta M(t)}{M^{eq}} = \frac{M^{eq} - M(t)}{M^{eq}} = \frac{1}{2} \left[\exp(-2W_1 t) + \exp(-2W_2 t) \right] \tag{2}$$

Below 100K, the following phenomena have been observed with ^{23}Na NMR in β-a-lumina: (i) the ^{23}Na satellite lines show a structure [11], (ii) an anomalous temperature dependence of the central line first moment suggests changes of the available ^{23}Na site occupancy probabilities [12], (iii) the intensity of the free induction decay (fid) at α=0° gradually goes to zero with decreasing T, [10], (iv) the central line shows a structure which, "cannot be interpreted in terms of multiple, static, asymmetric sodium sites" [13]. These (probably related) effects still await a proper interpretation.

RESULTS AND DISCUSSION

Union Carbide crystals with 22% excess sodium could be rotated in such a way as to vary the angle α between the magnetic field and crystal $\vec{c}$-axis. For relaxation measurements the fid following a π/2 rf-pulse was integrated with a boxcar, or Fourier transformed. The Bruker SPX spectrometer had a dead time of ≈20μs. By summing the squares of absorption and dispersion components one may obtain the "modulus" of the spectrum, for which the resonance wings are enhanced, but which is free from distortions arising from incorrect phase settings. Figure 1 shows ^{23}Na central line spectra at α=0° and α=45°. Below 100K, the line at α=0° consists of two almost equal peaks which were unresolved in previous studies [11,12]. The high and low frequency peaks are attributed to two sites, to be designated I and II, respectively. By increasing T above ∿100K, the low frequency peak broadens and decreases in intensity while the other increases. Above 150K the α=0° spectrum consists of a single peak which shifts toward low frequency with increasing T. While this last phenomenon is a purely dynamical effect [12], disappearance of the low frequency peak implies a repopulation of site I at the expense of site II. It is this repopulation which causes the "anomalous" ^{23}Na first moment displacment previously observed below 140K [12]. For α=0°, the intensity of the ^{23}Na signal was measured by comparing maximum heights of the ^{23}Na and ^{27}Al fid's. At 12.5K, and 100K the ^{23}Na signal intensity is 94% of the RT signal. Instead, at 80K it is only 77% of the RT value. Figure 1 shows that disappearance of the signal is nearly complete at 80K and $\alpha=45^\circ$. However, at the same angle and temperature, the ^{23}Na signal is easily observable with CW methods. Thus, the disappearance of all or part of the signal is caused by the existence of spin-lattice relaxation times shorter than the dead time.

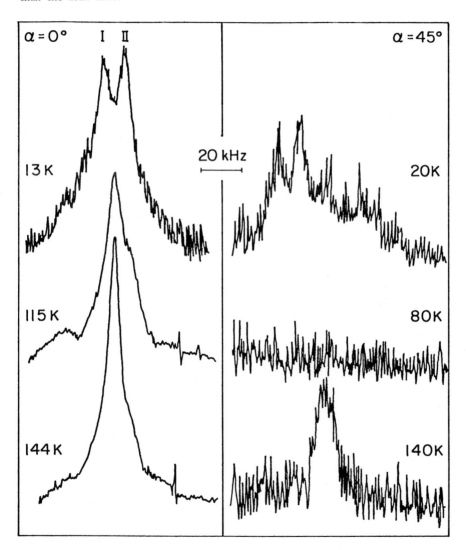

Fig. 1. ^{23}Na central line spectra at 21MHz and different temperatures

The onset of motion responsible for such fast relaxation, as well as differences of the ^{23}Na relaxation in different environments, can be studied by analyzing the recovery $M(t)$ at $\alpha=0°$ in the following way. The slow decay due to the "forbidden" $\Delta m=\pm1$ transitions is measured and its contribution subtracted from $\Delta M(t)$ with the help of Eq. (2). For temperatures above 160K the resulting plot is exponential over more than a decade of amplitude (see Fig. 2a). This fact implies that each spin samples all possible rates of relaxation in a time shorter than $(2W_2)^{-1}$. Somewhere in the 140-160K interval, the relaxation due to the $\Delta m=\pm2$ transitions loses its exponential character. We then define T_{1o}^{-1} and T_{1a}^{-1} as the relaxation rates at the time origin, and when the recovery is $\sim90\%$

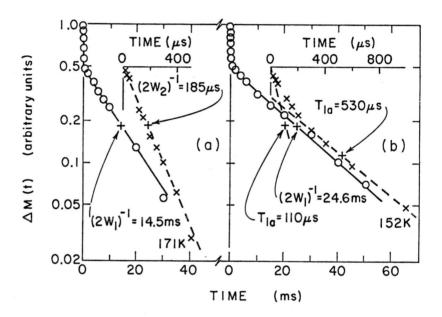

Fig. 2.   Recovery plots of the ^{23}Na magnetization at $\alpha\approx0°$ and 21MHz. The inset shows, in expanded time scales, the recovery due to the $\Delta m=\pm2$ transitions (see text).

completed, respectively (see Fig. 2b). For all the temperature range explored, subtraction of the recovery due to the $\Delta m=\pm1$ transition is easy and meaningful because $(2W_1)^{-1}$ was always found larger than T_{1a} by an order of magnitude or more and the slowest of the relaxation processes always accounts for nearly half the ΔM recovery (see Eq. (2)). Figure 3a shows the behavior with temperature of T_{1o} and T_{1a} for ^{23}Na at $\alpha=0°$. Over the range of experiments, large variations were often noted in the values of T_{1o} dependent upon different choices of the integration time.

We propose the following explanation for the complex behavior of T_{1o}. From 10K to ~60K the relaxation rates of the rapidly decaying spins increase with increasing T. When the signal of these spins begins disappearing in the dead time, T_{1o} increases. After going through a maximum near 60K, the relaxation rates of the rapidly relaxing nuclei steadily decrease with increasing T, thus making the spins again observable and lengthening T_{1o}. It may even be that a decrease in

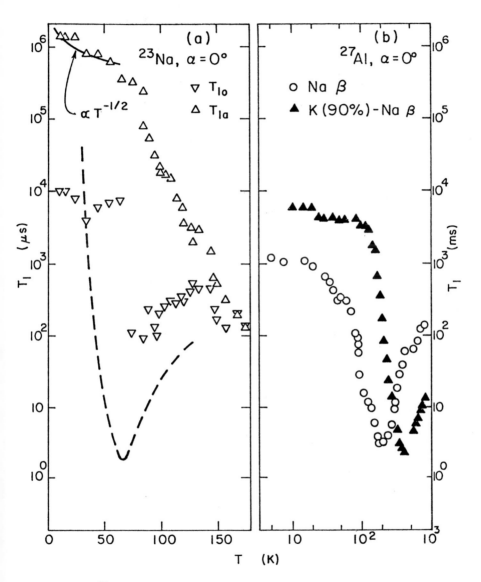

Fig. 3. (a) ^{23}Na relaxation times T_{1o} and T_{1a} in Na β-alumina. The dashed curve represents the guessed behavior of T_1 for the partly invisible fraction of spins.

(b) ^{27}Al T_1 at 21MHz in Na β-alumina (0) and K (90%)-Na β-alumina (▲).

the number of the spins which undergo the relaxationally effective motion contribute to the increase of T_{1o} from 80K to $\sim$140K. Above 140K the long-range diffusive motion becomes the dominant relaxation mechanism and $(2W_2)^{-1}$ goes through a second minimum at $\sim$220K [14]. The dashed curve in Fig. 3a gives an idea of what may be the behavior of T_1 for the rapidly relaxing spins. In drawing it, we arbitrarily assumed $T_1 \propto (1+\omega_L^2\tau^2)\tau^{-1}$ with $\tau^{-1}=\tau_o^{-1} \exp(-E/kT)$. Furthermore we choose a typical attempt frequency $\tau_o^{-1}=10^{12}s^{-1}$. A value of E = 0.05eV was then deduced by requiring that the T_1 minimum occurs at 65K as experimentally observed; i.e., in the middle of the temperature region were the reduction of the signal for $\alpha=0^\circ$ is observed.

The temperature behavior of T_{1a} depends upon the following phenomena: (i) With decreasing T, the fraction of spins which do not undergo the motion responsible for fast relaxation at $\alpha=0^\circ$ exchanges more and more slowly with the rapidly relaxing spins. (ii) Glasslike excitations of β-alumina cause efg fluctuations at the location of the slowly relaxing spins at $\alpha=0^\circ$. This mechanism probably dominates T_{1a} at low T and is responsible for the weak temperature dependence observed below 50K. The solid curve in Fig. 3a represents Eq. (1) with $\beta=0.5$; a value substantially smaller than for dielectric glasses [8].

Figure 3b compares the ^{27}Al T_1 in Na and K(90%)-Na β-alumina in the 5$\div$800 K interval. With increasing T, the following is observed: (i) Below 20 K T_1 is almost temperature and frequency independent. Heat capacity and dielectric constant measurements [3-5] imply a density of low-lying excitations in Na β-alumina which is $\sim$ 2-10 times that in K β-alumina. This is related to the $\sim$6 reduction of low-T ^{27}Al T_1 in Na β-alumina as compared with K β-alumina. It may also be noted that the local motion which reduces the ^{23}Na T_{1o} near 65 K has neglegible effect upon the ^{27}Al T_1. The gradual onset of long range diffusive motion which leads to the higher temperature T_1 minima of Fig. 3b becomes evident near 70 K in Na β-alumina and near 130 K in K-Na β-alumina.

While the overall picture emerging from our NMR results is suggestive, we have yet to identify sites I and II. Furthermore, while the behavior at $\alpha=45^\circ$ suggests a universal local motion, the fact that only a portion of the spins are rapidly relaxed at $\alpha=0^\circ$ implies a dynamical inhomogeneity for this motion.

ACKNOWLEDGMENTS

Research supported by D. O. E. (Grant No. EY-76-S-06-2225)

REFERENCES

1. R.O. Pohl and G.L. Salinger, Ann. N.Y. Acad. Sci. 279, 150 (1976).
2. P.J. Anthony and A.C. Anderson, Phys. Rev. B16, 5178 (1976).
3. D.B. McWhan, C.M. Varma, F.L.S. Hsu, J.P. Remeika, Phys. Rev. B15, 593(1977).
4. P.J. Anthony and A.C. Anderson, Phys. Rev. B16, 3827 (1977).
5. P.J. Anthony and A.C. Anderson, Phys. Rev. B19, 5310 (1978).
6. P. Doussineau, R.G. Leisure, A. Levelut and J.Y. Prieur, J. Phys. (Paris) 61, L65 (1980).
7. S.R. Kurtz and H.J. Stapleton, Phys. Rev. B22, 2195 (1980).
8. J. Szeftel and H. Alloul, J. Non-Cryst. Solids 29, 253 (1978).
9. A. Abragam, The Principles of Nuclear Magnetism, (Clarendon, Oxford)(1961).
10. R.E. Walstedt, R. Dupree, J.P. Remeika and A. Rodriguez, Phys. Rev. B15, 3442 (1977).
11. A. Highe, M. Polak and R. W. Vaughan, in Fast Ion Transport in Solids, Eds. P. Vashishta, J.N. Mundy and G.K. Shenoy, (North-Holland, Amsterdam, 1979) p. 305.
12. J.L. Bjorkstam, P. Ferloni and M. Villa, J. Chem. Phys. (15 Sept. 1980).
13. L.C. West, T. Cole and R.W. Vaughan, J. Chem. Phys. 68, 2710 (1978).
14. J.L. Bjorkstam, M. Villa, S. Aldrovandi, M. Corti and J.S. Frye, These Proceedings.

MIXED ALKALI EFFECTS IN β-ALUMINAS

JOHN L. BJORKSTAM
Department of Electrical Engineering, University of Washington, Seattle, WA
98195

MARCO VILLA, SERGIO ALDROVANDI AND MAURIZIO CORTI
Istituto di Fisica "A. Volta" e Gruppo Nazionale di Strutture della Materia del
C.N.R., 27100 Pavia, Italy

JAMES S. FRYE
Regional NMR Facility, Colorado State University, Ft. Collins, Colorado 80523

ABSTRACT

This paper presents a NMR study of mixed alkali effects in
β-aluminas. It is shown that the BPP theory of relaxation
does not describe the NMR results. However, with an elementary analysis of the data, NMR yields information about the
cation dynamics which substantially agrees with the diffusion
measurements.

The mixed alkali (MA) effect is a phenomenon often observed in vitreous e-
lectrolytes [1]. When one monovalent metal oxide ($M_2^A O$) is progressively substituted by another ($M_2^B O$) while the total metal ion concentration ($n_A + n_B$) is kept
constant, certain physical properties of these glasses are a highly nonlinear
function of the relative concentration $x \equiv n_A / (n_A + n_B)$. Properties related to the
ionic mobility, such as electrical conductivity, ionic diffusion, and dielectric
relaxation and losses exhibit the most pronounced non-linear dependence upon x.
Accordingly, the cation transport properties have been those most extensively
investigated in mixed ionic glasses. While deviations from this kind of behavior have been noted for some glass systems [2,3], MA effects upon conductivity have been observed also in crystalline β-aluminas [4]. Since β-aluminas are
a well characterized class of materials [5] and their cations can be easily and
reversibly exchanged, these electrolytes are a good model system for studying
MA effects, as well as the phenomenon of ion transport in disordered materials.
This paper will summarize a NMR study of mixed cation dynamics in β-aluminas
[6]. We show that for β-aluminas there is an overall order of magnitude agreement between the observed NMR response of ^{7}Li, ^{23}Na and ^{27}Al and that predicted
with simple considerations on the basis of diffusional [7] and vibrational
[8-10] properties. We assume that the frequency, τ_J^{-1}, of the diffusional
jumps follows the law

$$\tau_J^{-1} = \tau_o^{-1} \exp(-E/kT) \tag{1}$$

and will identify the NMR correlation time, τ, with τ_J. Here, τ_o^{-1} is the attempt frequency determined from infrared spectroscopy [8-10]. Quite generally,
it is expected that a line-narrowing process is half completed at the temperature, T_Δ, where τ^{-1} becomes equal to the rigid lattice NMR line width, Δ_{RL}.
Moreover a minimum of the spin lattice relaxation time (T_1) is reached at the
temperature, T_{min}, where $\omega_L \tau = 1$ with ω_L = Larmor frequency. Thus, with the ex-

perimental values of τ_0^{-1}, T_Δ and T_{min} and the conditions $\Delta_{RL}\tau=1$ at T_Δ or $\omega_L\tau=1$ at T_{min} we can calculate activation energies which will be indicated by E^*. These energies are compared with those for tracer diffusion (E_T) or conductivity (E_σ).

A formal interpretation of the NMR results is often carried on within the framework of the Bloembergen-Purcell-Pound (BPP) theory which gives [11]

$$T_1 \propto (1+\omega_L^2\tau^2)\tau^{-1}; \quad \Delta^2=\Delta_{res}^2+\Delta_{RL}^2\frac{2}{\pi}\tan^{-1}(a\Delta\tau) \quad (2)$$

with Δ_{res} the residual linewidth after completion of the narrowing process. In the following, by "linewidth" Δ we mean the peak-to-peak separation of the absorption derivative and we choose a=1 in Eq. (2). Unfortunately, Eq's. (2) do not describe the following phenomena, often observed in solid electrolytes, as well as metal hydrides and bronzes [12]. 1) The Arrhenius plot of T_1 gives different activation energies above and below T_{min} and T_1 is not proportional to ω_L^2 below T_{min}. These phenomena are illustrated by ^{23}Na [13] and ^{27}Al relaxation data for Na β-alumina in Figs.1(a) and 1(c). 2) The attempt frequency estimated through Eq. (2) is smaller than the expected $10^{12} \div 10^{13}s^{-1}$ by orders of magnitude (prefactor anomaly). In Fig. 1(b) we have applied the BPP equation to calculate τ for the low-T line narrowing process of the ^{23}Na central line in β-alumina. However, using the values τ_0^{-1} ($1.7 \cdot 10^{12}s^{-1}$) [8] and E (0.16eV) [7], the position of T_Δ is accurately predicted (see arrow in Fig. 1(b)). 3) T_1 is frequency independent near T_{min}. Figures 1(c) and 1(d) report relaxation times for the ^{27}Al central line when the $\vec{c}$ axis forms an angle (α) of 45° and 0°, respectively, with $\vec{H}_0$.

To a certain extent, these phenomena have been explained by calculating T_1 with conceptually simple models of motion. According to these models we have [12]

$$\text{for } \omega_L\tau \ll 1 \quad T_1^{-1} \propto \tau(\omega_L\tau)^{\beta-1} \quad (E_{NMR}=\beta E)$$

$$\text{for } \omega_L\tau \gg 1 \quad T_1^{-1} \propto \tau(\omega_L\tau)^{-(\gamma+1)} \quad (E_{NMR}=\gamma E) \quad (3)$$

Here, β and γ are model-dependent parameters with values in the 0.25÷1 range and E_{NMR} describes the temperature variations of T_1 in an Arrhenius plot. For example, the solid lines in Fig. 1(a) have been drawn using the results of a two dimensional-continuum diffusion (2D-cd) model with the single activation energy of 0.16eV [6]. This is in contrast to a previous interpretation in which the BPP theory, together with a distribution of activation energies for τ, was used to fit the T_1 vs. T results [13]. Since the conductivity and diffusion are well described by single activation energies [7], the validity of a model with a distribution of E-values must be questioned. The 2D-cd model is interpreted in a way which is consistent in concept [6] with a recent theory of Wolf [14] in which extended regions surrounding excess, charge-compensating, O^{2-} ions differ from regions far from such ions. However, Wolf suggests that the apparently different activation energies above and below T_{min} are a result of different mechanisms while, as evident from Eq. (3), this observation can be accomodated with a single activation energy.

The BPP model of relaxation through hopping presumes that each hop is responsible for the fluctuations which cause relaxation. In the diffusion model many jumps take place in the time τ over which the significant fluctuation occurs. We believe that the T_1 minimum of Fig. 1(a) is a result of long range fluctuations in ^{23}Na environment due to the presence of the excess compensating

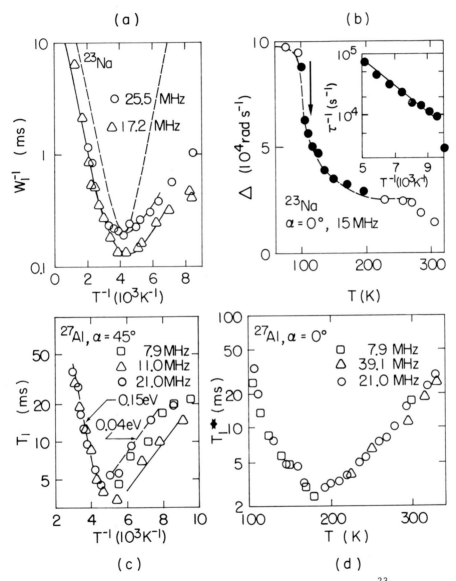

Fig. 1 The NMR response of Na β-alumina: (a) Relaxation data for ^{23}Na fitted
with a diffusion model [6] (solid curves) and with Eq. (2) (dashed curve).
(b) ^{23}Na Δ vs T. An estimate of the correlation times , made by using Eq. (2)
and the data represented by filled circles gave E= 0.044 eV and τ_0^{-1}=1.1(10^6)
s^{-1} (see inset). (c) Frequency dependence of the ^{27}Al T_1 at α=45° (see text).
(d) Frequency dependence of the ^{27}Al T_1^* for α=0°.

O^{2-} ions. Many localized hops may be required for a given conducting ion to escape from the region surrounding a particular compensating O^{2-} ion. While the barriers for individual hops may differ, over the spin lifetime these differences are averaged to give the single value characteristic of long range motion.

The solid line in Fig. 1(c) represents the T_1 values at 11MHz expected from the data at 21MHz on the basis of Eq. (3). However, the values of T_1 at 7.9 MHz in Fig. 1(c), as well as the data of Fig. 1(d), cannot be interpreted with Eq. (3). These deviations from the predictions of Eq. (3), imply the existence of spin-diffusion phenomena. Since features of the ^{27}Al central line depend upon both ω_L and α, one can qualitatively account for the behavior of Figs. 1(c) and 1(d). The presence of significant spin diffusion would imply the existence of inhomogeneity in the local motion dynamics over the spin lifetime.

In comparing Figs. 1(a) and 1(d) it is quite clear that, at a given crystal orientation α, the ^{27}Al T_{min} is lower than that for ^{23}Na. If one sets $\omega\tau_J=1$ at the T_{min} of Fig. 1(d), and assumes the activation energy characteristic of diffusion or conductivity in Na β-alumina, Eq. (1) gives $\tau_o^{-1}=6(\pm4)\times10^{12}s^{-1}$, in order of magnitude agreement with infrared results [7]. The large uncertainty in τ_o^{-1} reflects the unresolved frequency dependence in Fig. 1(d). Thus, while τ should not be identified as the <u>local</u> jump time τ_J responsible for the ^{23}Na T_{min} it does seem to relate directly to the ^{27}Al T_{min}. This reflects the fact that the local motion is effective in "rattling" the spinel cage in which the aluminums reside [6]. It may be further noted that choosing $\omega\tau=1$ at the ^{27}Al T_{min} of Fig. 1(d), and $\Delta_{RL}\tau=1$ at the temperature of ^{23}Na line narrowing in Fig. 1(b), gives the activation energy, 0.17(±0.008)eV, as well as the τ_o^{-1} value above.

While the temperature at which ^{23}Na line narrowing occurs may be related to the Na^+ motion, the detailed dependence of linewidth upon T in Fig. 1(b) is not yet understood. We believe the narrowing results from motional averaging over different electric field gradient (efg) tensors within the unit cell. An adequate theory will require a knowledge of individual efg tensors at the physically and chemically inequivalent ^{23}Na sites, and residence times for these sites.

For K-Na, Rb-Na and Tl-Na β-aluminas, as well as for pure Na β-aluminas, the ^{27}Al T_1 displays a single minimum [6] which we associate with the dominant transport motion. The values of T_{min}, which depend strongly upon crystal composition, are listed in Table I. Setting $\omega_L\tau_J=1$ at T_{min} we have used Eq. (1), together with experimental values of τ_o^{-1} from refs. [8-10], to determine the E* values shown. A comparison of E* with E_T and E_σ for the K-Na and Rb-Na crystals shows a typical MA effect while no significant effect is noted with partial Tl substitution. A similar result was noted for Tl^+ containing glasses [3].

While Δ_{RL} for ^{23}Na is essentially the same in mixed and pure crystals, the RT $\Delta/2\pi$ - value is 5.5±0.3 kHz for Tl-Na and Rb=Na, but only 2.5±0.2 kHz for Na β-alumina. This implies a reduce Na^+ mobility in the mixed crystals. The low ^{23}Na concentration did not allow similar experiments in the K-Na sample.

TABLE I. ACTIVATION ENERGIES FROM ^{27}Al T_{min} AT 21MHz

Composition	$\tau_o^{-1}(s^{-1})$	T_{min}(K)	E*(ev)	E_T(eV)[1]	E_σ(eV)[1]
Na (100%)	1.7(10^{12})	190	0.154	0.165	0.141
Ag (100%)	8.4(10^{11})	190	0.143	0.176	0.173
K(90%)-Na	2.4(10^{12})	420	0.354	0.232	0.294
Tl(60%)-Na	1.1(10^{12})	475	0.368	0.357	0.355
Rb(55%)-Na	2.5(10^{12})	740	0.626	0.311	

[1]These values refer to pure crystals [7].

For the Li$^+$ containing β-aluminas the two resolved minima in the ^{7}Li and ^{27}Al T$_1$ plots (Fig. 2) indicate the presence of motions with widely different time scales. Previously reported studies on flux grown Li(61%)-Na(39%) β-alumina show a ^{7}Li T$_1$ minimum at 500K and 19MHz which shifts to 240K upon heat treatment [15]. The minimum is broader than we have found and did not show the lower temperature minimum evident in Fig. 2.

While much remains to be done for a full understanding of the Li-substituted crystals, we shall call attention to certain features evident in Fig. 2. Since the high-T minimum is the same in both the Li-Na and Li-Ag crystals we attribute this to the ^{7}Li motion. The fact that the low temperature minimum in Li-Na coincides with that due to ^{23}Na motion in Na β-alumina gives further credence to

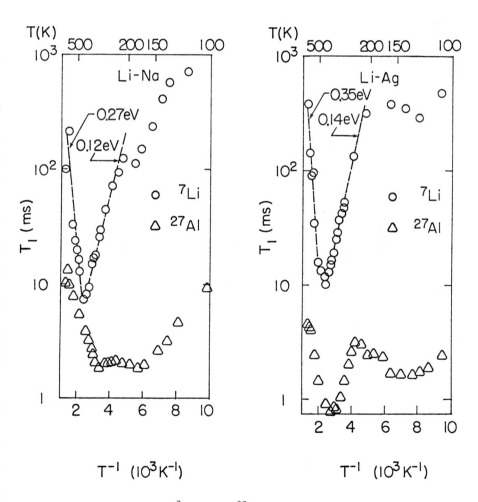

Fig. 2. Relaxation data for ^{7}Li(o) and ^{27}Al(△) in Li-Na and Li-Ag at 21MHz.

300

this assumption. It is worth noting that if one takes $\omega\tau=1$ at the high-T minimum and $\Delta_{RL}\tau=1$ at the temperature where ^{7}Li line-narrowing is half completed one finds $E=0.27$eV, $\tau_0^{-1}=1.2(10^{13})$s^{-1} for the Li-Na crystal and $E=0.38$eV, $\tau_0^{-1}=2.6$ (10^{13})s^{-1} in Li-Ag. These τ_0^{-1} values are in remarkably good agreement with the Raman results of ref [10].

Results of treating the data in another way are given in Table II. Here we use the attempt frequencies of Na$^+$ and Ag$^+$ to extract E* from the T_1 minima at low T in Li-Na and Li-Ag, respectively. Instead, the Li$^+$ attempt frequency [10], is used in connection with the high-T T_1 minima and the ^{7}Li line-narrowing process. For the Li-Na crystal, the fact that the Na$^+$ dynamics changes little in comparison with Na β-alumina is confirmed by analysis of the ^{23}Na lineshape. While this feature is in contrast with suggested inhibition of Na$^+$ motion based upon electrochemical results [16] it is confirmed by internal friction experiments [15]. For the Li-Ag crystal, the NMR data suggests a substantial enhancement of Ag$^+$ mobility. However, this conclusion is based only upon the ^{27}Al NMR evidence and the analogy with a similar phenomenon observed in glasses [3]. The high-T T_{min}'s, the temperatures (T_Δ) for the ^{7}Li line-narrowing, and E_{NMR} for ^{7}Li T_1 when interpreted with Eq. (3), yield very consistent values of the activation energy for Li$^+$ motion in Li-Na and Li-Ag crystals. However, as is the case for Na β-alumina, T_{min} of the mobile ion gives somewhat higher activation energies, due to a shift of ~100K in the T_{min} of ^{7}Li with respect to the corresponding T_{min} of ^{27}Al. An interpretation similar to that given above for Na β-alumina may be suggested.

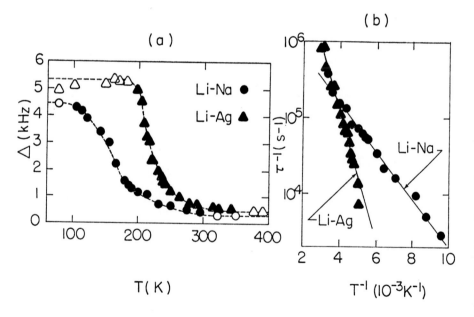

Fig. 3. (a) ^{7}Li Linewidths vs. T in β-aluminas. (b) Correlation times calculated with Eq. (2) and filled symbols of Fig. 3(a). (c) The solid lines represent Eq. (2) with $\tau_0^{-1}=3(10^6)$s^{-1}, $E=0.063$eV for Li-Na and $\tau_0^{-1}=2.9(10^8)$s^{-1}, $E=0.168$eV for Li-Ag.

TABLE II. ACTIVATION ENERGIES FOR Li β-ALUMINAS

Composition	E* from ^{27}Al T_1 (eV)	E* from ^{7}Li T_1 (eV)	E* from ^{7}Li Δ (eV)	E_T(eV)	E_σ(eV)
Li(50%)-Na					
Na$^+$(Low-T_{min})	0.14	0.14		0.250 [1]	0.27 [3]
Li$^+$(High-T_{min})	0.30	0.39	0.28		
Li(85%)-Ag					
Ag$^+$(Low-T_{min})	0.1	0.1			
Li$^+$(High-T_{min})	0.34	0.43	0.37	0.357 [2]	0.355 [2]

[1] Reference [18] (results obtained between 400-700 K)
[2] Activation energies for pure Li β-alumina [7].
[3] Reference [16].

Our NMR activation energies for Li$^+$ motion in the "as exchanged" Li(50%)-Na crystals cover the range 0.27-0.30 eV. This is consistent with recent conductivity results on similar crystals [16] which were carefully annealed. This suggests a neglegible influence of H_2O upon our result and that it is the Li$^+$ jump which control both the Li$^+$ and Na$^+$ long range motion. In contrast, while our NMR activation energies in the Li-Ag crystal are consistent with earlier experiments [2,17], the more recent results 16 give 0.24 eV for diffusion in pure Li β-alumina. We must therefore presume a substantial influence of water contamination upon these results.

In general, it cannot be expected that diffusion and NMR experiments necessarily detect precisely the same aspects of the mass transport phenomenon. While NMR samples motions over microscopic lengths (∿ a few Å), the tracer diffusion coefficient can be controlled by percolation phenomena. This may account for the differences between E* and E_T for Na$^+$ in Table II. The same crystal was used in both experiments. Nevertheless we are gratified with the order of magnitude agreement with other experiments and anticipate that more careful analyses of these and future data will contribute substantially to our understanding of MA properties.

ACKNOWLEDGEMENTS

Research supported by the U.S. Department of Energy (Grant No. EY-76-06-2225). Part of the measurements have been performed at the Colorado State University Regional NMR Facility funded by NSF, Grant No. CHE-78-18581.

REFERENCES

1. D. E. Day, J. Non-Cryst. Solids, 21, 343 (1976).

2. P. J. Hayward, Phys. Chem. Glasses, 17, 54 (1976); 18, 1 (1977).

3. S. Sakka, K. Matusita and K. Kamiya, Phys. Chem. Glasses, 20, 25 (1979).

4. G. V. Chandrashekhar and L. M. Foster, Solid State Commun. 27, 269 (1978).

302

5. P. Vashishta, J. N. Mundy and G. K. Shenoy, Eds. Fast Ion Transport in Solids, (North-Holland, New York (1979)).

6. For experimental, mathematical and interpretative details see: J. L. Bjorkstam, S. Manzini and M. Villa, in ref. [5], and J. L. Bjorkstam and M. Villa, Phys. Rev. B (Dec. 1980) and J. Phys. (Paris) (Feb. 1981).

7. J. H. Kennedy in Solid Electrolytes, Ed. S. Geller (Springer-Verlag, Berlin) p. 105 (1977).

8. D. B. McWhan, S. J. Allen, Jr., J. P. Remeika and P. D. Dernier, Phys. Rev. Lett., 35, 953 (1975).

9. Ph. Colomban and G. Lucazeau, J. Chem. Phys., 72, 1213 (1980).

10. T. Kaneda, J. B. Bates and J. C. Wang, Solid State Commun., 48, 469 (1978).

11. A. Abragam, The Principles of Nuclear Magnetism, (Clarendon Press, Oxford) Chap. X, (1961)

12. J. L. Bjorkstam and M. Villa, Magn. Reson. Rev., 6, 1 (1980).

13. R. E. Walstedt, R. Dupree, J. P. Remeika and A. Rodrigues, Phys. Rev., B15, 3442 (1977).

14. D. Wolf, J. Phys. Chem. Solids, 40, 757 (1979).

15. R. E. Walstedt, R. S. Berg, J. P. Remeika, A. S. Cooper and B. E. Prescott, Ref. 5, p. 355.

16. J.L. Briant and G.C. Farrington, J. Electrochem Soc., In press.

17. W. Roth and G.C. Farrington, Science, 196, 1332 (1977).

18. B. Seegmiller, M.S. Thesis, University of Washington (Seattle), 1977 (Unpublished).

SPIN SUSCEPTIBILITY OF INTERCALATED GRAPHITE AND DOPED POLYACETYLENE

JAMES W. KAUFER
Department of Physics, University of Pennsylvania, Philadelphia,
Pennsylvania 19104

SEIICHIRO IKEHATA
Department of Physics, University of Pennsylvania, Philadelphia,
Pennsylvania 19104 and Department of Physics, University of Tokyo,
Tokyo 113 Japan

ABSTRACT

We have examined the electronic susceptibility of intercalated graphite and doped $(CH)_x$ using ESR [1] . The
dopant in both cases was AsF_5. For metallic intercalated
graphite and for heavily doped $(CH)_x$, the low frequency
Schumacher-Slichter technique was used to determine the
density of states at the Fermi energy. In the case of
doped $(CH)_x$, susceptibility measurements as a function
of temperature allowed separate determination of the
Curie and Pauli contributions.

INTRODUCTION

Intercalated graphite and doped polyacetylene have recently attracted
a great deal of attention. Contributing to the interest is the wide variety of intercalant [2] and dopant [3-7] species available, the very high
conductivities obtained with some dopants, and the structural anisotropy
of these materials indicating the possibility of unusual transport mechanisms.[8,9]
    A large number of donor and acceptor atomic and molecular species
may be intercalated in graphite. Like the parent graphite, these materials exhibit large anisotropies in conductivity. The AsF_5 - graphite
compounds yield particularly high anisotropies, σ_a/σ_c 10^6, with σ_a close
to that of copper.[10] This anisotropy leads to the expectation that
these compounds may be treated as 2-d metals, using a graphite rigid-band
model. [11] By using magnetic resonance techniques, we have measured the density of states at the Fermi energy in stages I and II AsF_5-graphite. Use of
the rigid band model allowed determination of the fractional charge transfer to the intercalant.
    Polyacetylene in its undoped state is the simplest linear conjugated
polymer. It is a direct gap semiconductor with a band gap of ~1.4 eV.
By doping with various donor or acceptor compounds the conductivity may
be varied from 10^{-9} up to $>10^3 (\Omega\text{-cm})^{-1}$.
    Upon doping the conductivity rises, goes through a semiconductor-
metal transition at a few percent doping level, after which the conductivity slowly increases up to the maximum doping level of about fourteen
percent. Although the bulk properties of $(CH)_x$ show a small anisotropy[12],
NMR studies show that the intrinsic anisotropy may be very large.[13]

This 1-d character along with the broken symmetry in the ground state allows for the possibility of describing the transport in the lightly doped polymer in terms of nonlinear solitons[8,9,14] rather than in terms of traditional semiconductor doping. Again, using magnetic resonance we have measured the Curie and Pauli susceptibilities as a function of doping in order to elucidate the nature of the transport in lightly doped $(CH)_x$ and the semiconductor-metal transition.

The susceptibility measurements were carried out using the Schumacher-Slichter technique, which will be discussed in the following paragraphs.

THEORY

The Schumacher-Slichter technique is a method for determining absolute spin susceptibility. [15] The nuclear resonance of a constituent species is used to calibrate the electron resonance. The ESR and NMR signals are detected with the same apparatus; only the static field being changed to observe one resonance or the other. The experiment is performed in a standard CW NMR spectrometer at typical NMR frequencies.

There are several advantages to this method. First, it is a resonance experiment so it looks only at the electron spin paramagnetism, no diamagnetic contribution need be estimated. Second, a carefully calibrated standard is not needed. Third, skin depth losses can be avoided for metallic samples since it is a low (radio) frequency measurement and since the reference NMR is observed under identical conditions.

There are some disadvantages. The method requires that the material have a detectable nuclear signal. If it does not, however, it could be mixed with an inert substance which does. The stoichiometry of the material must be well known in order to determine the number of nuclear spins. In addition, if the material has a large magnetoresistance, the loading of the sample coil may change too much from the electron to the nuclear resonance to make the comparison valid.

In a typical CW NMR experiment one measures a derivative signal, $Y(H)$, which is proportional to $d\chi''/dH$, the derivative of the absorption lineshape. The proportionality between these signals depends on factors such as:

    a) H_1 - rf magnetic field amplitude

    b) H_m - modulation field amplitude

    c) G - gain of detection system

    d) misc effects - filling factor, circuit Q, etc.

The fourth class of effects deals with circuit parameters and sample geometry which will be the same for the ESR and NMR lines. The first three factors may be adjusted from one resonance to the other since, barring saturation or gross over modulation, the signal intensity is linear in them. For convenience we take them as constant in the following

The absorption is related to the static susceptibility, χ_o, in the following manner:

$$\chi_o = \frac{1}{\pi} \int \frac{\chi''(\omega)\,d\omega}{\omega}$$

$$= \frac{\alpha}{\pi} \int\int \frac{Y(H)}{H} d^2H$$

$$= \frac{\alpha\gamma}{\pi\omega_o} \int\int Y(H)\,d^2H$$

Where the last line is true for the case that the resonance is sharply peaked about H_0, as is generally true for the NMR line, or for the case that the lineshape is Lorentzian and is not saturated which is generally true for the ESR line. The substitution $\omega = \gamma H$ has been made; α is the proportionality constant mentioned above. Taking A as the area under the absorption curve, the static susceptibility is:

$$\chi_o = \frac{\alpha \gamma A}{\pi \omega_o}$$

One may now write:

$$\chi_{on} = \frac{\alpha \gamma_n A_n}{\pi \omega_o} \qquad \text{(nuclear resonance)}$$

$$\chi_{oe} = \frac{\alpha \gamma_e A_e}{\pi \omega_o} \qquad \text{(electron resonance)}$$

$$\chi_{oe} = \frac{NJ(J + 1)}{3KT} (\gamma_n \hbar)^2 \frac{\gamma_e A_e}{\gamma_n A_n}$$

Where the last line is obtained by dividing the second by the first and using the expression for the Curie susceptibility of the nuclei. Here N is the concentration of nuclear spins. The final expression gives the absolute electronic susceptibility in terms of the ratio of the areas under the two absorption cures.

A method closely related to this technique is one using a calibrated spin standard. In this case the measurement is carried out in a standard microwave ESR spectrometer, the sample and the standard placed in the same or equivalent positions in the cavity. As both materials are in the cavity simultaneously, the standard must have a resonance far from the material being studied. Ruby (CR^{3+} in Al_2O_3) is often chosen because it has signals far from $g = 2$ and obeys a Curie law. In this case one has:

$$\chi_{oe} = \chi_{or} \frac{\gamma_e A_e}{\gamma_r A_r} \quad ,$$

where the subscript r refers to the ruby.

Among the advantages to this method are: The sample need not have a strong nuclear signal, nor need its stoichiometry be well known. Only small amounts of sample are needed due to the sensitivity of microwave ESR. The disadvantages include: Metallic samples must be carefully treated to avoid skin depth problems. A well calibrated standard must be used, and a determination made that it and the sample being investigated are in equivalent positions in the cavity (ie., same H_1 and H_m). The sample weight must be accurately determined, this may be a problem for small air sensitive samples.

Measuring the temperature dependence of χ and plotting the data as χ vs $1/T$ allows determination of the Curie and Pauli contributions, since χ_c is inversely proportional to T and χ_p is temperature independent. Once χ_p is known the density of states may be determined from the relation:

$$N(E_F) = \chi_p / \mu_B^2 \quad .$$

Here, μ_B is the Bohr magneton and $N(E_F)$ is the density of states at the Fermi energy for both signs of spins.

EXPERIMENT

Graphite

Samples of stage I and II AsF_5-graphite were prepared from highly oriented pyrolitic graphite using methods previously described.[16-18] Dimensions were 8mmx25mmx3mm after intercalation. Excess AsF_5 was pumped off prior to sealing in rectangular Pyrex tubing, with the sample held at -96°C to prevent de-intercalation. The sample coil was wound on the Pyrex ampoule with the coil axis ⊥ graphite c-axis; this allowed penetration of the rf field along the poorly conducting c-axis. Uniform stage I and II was verified by neutron diffraction on companion samples.

The measurements were carried out at 37MHz at room temperature with a simple Q-meter CW NMR spectrometer, which was interfaced to a PDP 11/40 minicomputer to store data and compute areas. The ESR line of both samples was Lorentzian with a peak-to-peak width of ≈0.5G. The ^{19}F nuclear resonance was used in the Schumacher-Slichter technique, and as had been previously observed [16], the NMR line of both samples was motionally narrowed. The resulting long relaxation times ($T_1 \sim T_2 \sim$ 250ms) necessitated the use of a 1.4Hz modulation frequency. The density of states was found to be .060 and .056 states/eV/carbon for stages I and II respectively.

Polyacetylene

The $(CH)_x$ study is actually two separate experiments. The first, an initial study, involved a sample heavily doped by a method similar to that developed by Shirakawa, [19] and described elsewhere.[3-7] The second involved a series of samples from lightly to heavily doped by a process designed to achieve more uniform doping [1]

In the first experiment the sample was doped to saturation, pumped on to remove excess AsF_5, then sealed off. A stoichiometry of $[CH(AsF_5)_{.12}]_x$ was determined by weight uptake. The same spectrometer as for the graphite was used. The ESR line was Lorentzian with a width of ~0.5G. The ^{1}H and ^{19}F signals were used in the Schumacher-Slichter technique; both lines were Gaussian with widths of 4.0 and 2.5G respectively. The density of states was found to be .18 states/eV/carbon.

In the second experiment samples of $[CH(AsF_5)_y]_x$ with .0004 $\leq y \leq$ .138 were prepared. The lightly doped samples, .0004 $\leq y \leq$ .0025 were studied at 10GHz using a ruby spin standard [20]. χ_P and χ_C were determined from $\chi(T)$ 77 $\leq T \leq$ 295°K as a function of dopant level. The heavily doped samples, .0025 $\leq y \leq$ .138 were studied at 10MHz using the ^{1}H and ^{19}F signals. The temperature dependence between 4.2 and 295°K allowed determination of χ_P and χ_C. The Pauli susceptibility is plotted versus dopant level in figure 1.

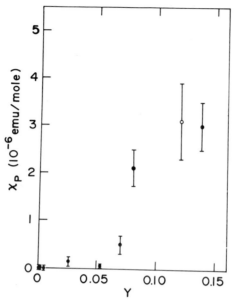

FIGURE 1.

X_P vs. Dopant Level

DISCUSSION

Graphite

Treating AsF_5-graphite in a rigid band picture, where the intercalant accepts electrons from the otherwise unchanged 2-d graphite bands, allows a simple determination of the charge transfer per intercalant molecule. The new Fermi energy is obtained by reading from the theoretical density of states versus energy curve, the energy corresponding to the measured density of states. The charge transfer is then obtained by integrating the density of states from the new Fermi energy to that of pure graphite. Using a 2-d graphite dispersion curve [11] we find the fractional charge transfer per intercalant to be f = .24e and .48e for stages I and II AsF_5-graphite respectively.

A possible oxidation reaction [21] for arsenic pentafluoride intercalated into graphite is:

$$3AsF_5 + 2e^- \rightarrow 2AsF_6^- + AsF_3$$

The charge transfer would be 2e/3 if this reaction went to completion. Recent X-ray photoelectron spectroscopy (XPS) data [22] indicates that this reaction does not go to completion. It is proposed that the equilibrium condition depends on the stage; the higher the stage, the further towards completion the reaction goes, hence the larger the charge transfer (up to a maximum of 2e/3). In this view all three arsenic fluorides would exist as intercalants, charge transfer occuring to the AsF_6^- only, while

the AsF_5 and AsF_3 remain neutral. The estimate given for the charge transfer from this work [22] are f = .27e and .35e for stages I and II AsF_5-
graphite respectively, in reasonable agreement with our data.

In light of this small charge transfer, the high conductivities
in these two compounds must be understood in terms of high mobility.
It has been suggested [23,24] that in the acceptor compounds there are
unusually small electron-phonon scattering rates in the graphite planes
and that this feature correlates with the degree of anisotropy. In
fact, for the donor compounds (charge transfer ~1e), the anisotropies
are much lower, as are the basal plane conductivities.

Polyacetylene

The semiconductor-metal (SM) transition and the transport in the
lightly doped polymer are of considerable interest. There have been
attempts to explain these phenomena in terms of the creation by doping of
small metallic regions in a matrix of semiconducting $(CH)_x$. [25] This
model is based in part on early data which suggested a linear dependence
of χ_p on dopant level. The SM transition would then be due to percolation
between these regions. This view is inconsistent with our data.

Transport at light doping levels may be understood in terms of the
soliton doping mechanism. [8,9,14] The ground state of a $(CH)_x$ chain
has a broken symmetry; it can exist in two phases, one having double
bonds to the right, the other with double bonds to the left. A potential
defect structure in the chain involves one phase on the right side, the
other phase on the left. In fact, theoretical calculations show that this
domain wall or soliton structure will exist in all chains with an odd number
of carbon atoms.[26] The soliton may be neutral with spin one half, or
charged, with spin zero, if a donor or acceptor is nearby. To minimize bond
strain energy the soliton would be spread over several lattice constants.[9]
This delocalization leads to a high mobility for the neutral soliton. Magnetic measurements [13] indicate that the ESR line in undoped $(CH)_x$ is due
to neutral solitons.

The importance of the soliton concept for doping is that less energy
is required to form a charged soliton than to add an electron to the conduction band or a hole to the valence band. [9] Thus in the case of
dilute doping one creates solitons, loosely bound to dopant ions, which
conduct through an activated process and by hopping. [14] Further doping
increases the density of solitons until the SM transition occurs, where a
metallic state is achieved. In this case one expects to see little or no
Pauli susceptibility until or after the SM transition. We find a small
χ_p until well after the SM transition, in agreement with this model. It
should be noted that in light of the sharp dependence of χ_p with y one
can understand data indicating a more linear dependence of χ_p on dopant
level in terms of a non-uniform doping process where the SM transition
is smeared out over a broader range. Spatial fluctuations in the dopant
concentration would lead to a continuous increase of χ_p with y. This
non-uniform doping may be avoided by following a more careful procedure.

CONCLUSIONS

As advances in physics, chemistry and materials science generate increasing numbers and varieties of novel conducting compounds, there will be an increasing need to characterize and understand these materials. In this the electronic susceptibility will be of fundamental importance. The Schumacher-Slichter technique provides an accurate means of obtaining this information, without the use of highly specialized facilities.

AKNOWLEDGMENTS

This work was supported by the National Science Foundation Materials Research Laboratory Program at the University of Pennsylvania. (DMR-79-23647)

REFERENCES

1. The data described in this manuscript have been previously published in the following articles: B. R. Weinberger, J. Kaufer, A. J. Heeger, J. E. Fischer, M. Moran and N. A. W. Holzwarth, Phys. Rev. Lett. 41, 1417 (1978); B. R. Weinberger, J. Kaufer, A. J. Heeger, A. Pron, and A. G. MacDiarmid, Phys. Rev. B. 20, 223 (1979); S. Ikehata, J. Kaufer, T. Woerner, A. Pron, M. A. Druy, A. Sivak, A. J. Heeger and A. G. MacDiarmid, Phys. Rev. Lett. 45, 1123 (1980).

2. See for example, Mater. Sci. Eng. 31 (1977).

3. H. Shirakawa, E. J. Louis, A. G. MacDiarmid, C. K. Chiang and A. J. Heeger, Chem. Commun., 578 (1978).

4. C. K. Chiang, C. R. Fincher, Jr., Y. W. Park, A. J. Heeger, H. Shirakawa, E. J. Louis, S. C. Gau, and A. G. MacDiarmid, Phys. Rev. Lett, 39, 1098 (1977).

5. C. K. Chiang, M. A. Druy, S. C. Gau, A. J. Heeger, H. Shirakawa, E. J. Louis, A. G. MacDiarmid, and Y. W. Park, J. Amer. Chem. Soc. 100, 1013 (1978).

6. C. K. Chiang, Y. W. Park, A. J. Heeger, H. Shirakawa, E. J. Louis and A. G. MacDiarmid, J. Chem. Phys. 69, 5098 (1978).

7. Y. W. Park, M.A. Druy, C. K. Chiang, A. J. Heeger, A. G. MacDiarmid, H. Shirakawa and S. Ikeda, J. Polym. Sci. Polym. Lett. Ed. 17, 628 (1979).

8. M. J. Rice Phys. Lett. 71A, 152 (1979).

9. W. P. Su, J. R. Schrieffer, and A. J. Heeger, Phys. Rev. Lett. 42, 1698 (1979). and Phys. Rev. B 22, 2099 (1980).

10. G. M. T. Foley, C. Zeller, E. R. Falardean, and F. L. Vogel Solid State Commun. 24, 371 (1977).

11. G. S. Painter and D. E. Ellis, Phys. Rev. B 1, 4747 (1970).

12. C. R. Fincher, Jr., M Osaki, M. Tanaka, D. Peebles, L. Lauchlan, A. J. Heeber and A. G. MacDiarmid, Phys. Rev. B 20, 1589 (1979).

13. M. Nechstein, F. Devreaux, R. L. Greene, T. C. Clarke, and G. B. Street, Phys. Rev. Lett. 44, 356 (1980).

14. W. P. Su, S. Kivelson, and J. R. Schrieffer in: Physics in One Dimension, J. Bernasconi, and T. Schneider eds. (Springer,Berlin, Heidleberg and New York 1980).

15. R. T. Schumacher and C. P. Slichter, Phys. Rev. 101, 58 (1956).

16. B. R. Weinberger, J. Kaufer, A. J. Heeger, E. R. Falardeau, and J. E. Fischer, Solid State Commun. 27, 163 (1978).

17. S. K. Khanna, E. R. Falardean, A. J. Heeger, and J. E. Fischer, Solid State Commun. 25, 1059 (1978).

18. E. R. Falardean, L. R. Hanlon, and T. E. Thompson, Inorg. Chem. 17, 301 (1978).

19. H. Shirakawa and S. Ikeda, Polym. J. 2 231 (1971); H. Shirakawa, T. Ito, and S. Ikeda Polym. J. 4 460 (1973); T. Ito, H. Shirakawa, and S. Ikeda, J. Polym. Sci. J. Polym. Chem. Ed. 13, 1943 (1975).

20. National Bureau of Standards Electron Paramagnetic Resonance In-Tensity Standard: SRM-2601.

21. N. Bartlett, R. N. Biagioni, B. W. McQuillan, A. S. Robertson, and A. C. Thompson, J. Chem. Soc. Chem. Commun. 200 (1978); N. Bartlett, B. McQuillan, and A. S. Robertson, Mater. Res. Bull 13, 1259 (1978).

22. M. J. Moran, J. E. Fischer, W. R. Salaneck, J. Chem. Phys. 73, 629 (1980).

23. J. E. Fischer, and T. E. Thompson, Phys. Today 31, No. 7, 36 (1978).

24. L. Pietronero, S. Strassler, H. R. Zeller and M. J. Rice, Physica 99B, 499 (1980).

25. Y. Tomkiewicz, T. D. Schultz, H. B. Broom, T. C. Clarke and G. B. Street, Phys. Rev. Lett. 43, 1532 (1979).

26. W. P. Su, Solid State Commun. (in press).

Published 1981 by Elsevier North Holland, Inc.
Kaufmann and Shenoy, editors
Nuclear and Electron Resonance Spectroscopies Applied to Materials Science

ELECTRONIC STRUCTURE OF IRON SUBSTITUTED LITHIUM INTERCALATED TaS_2

M. Eibschütz, D. W. Murphy and F. J. DiSalvo
Bell Laboratories, Murray Hill, New Jersey   07974

ABSTRACT

We have used the ^{57}Fe Mössbauer effect to study the electronic spin configuration of Fe substituted in the layer compound $Li_xTa_{1-y}Fe_yS_2$ ($0 \leq x \leq 1$; $0 \leq y \leq 0.1$). The Mössbauer effect results show that the Fe is localized and in Fe^{2+} valence state but the electronic configuration depends on the lithium concentration. At room temperature the isomer shift of Fe^{2+} increases from 0.56 mm/s for $x = 0$ (no Li) to 0.77 mm/s for $x = .86$ (at $y = 0.05$). The isomer shift results reflect the changes in the electronic configuration of the host from $Ta^{4+}(5d^1)$ to $Ta^{3+}(5d^2)$ and an increase in ionicity as the lithium content increases from 0 to 1.

Recently, there has been a great deal of interest in the intercalation of lithium into the transition metal layer dichalcogenides because of their potential use in an effective cathode reaction in Li-anode secondary cells [1,2]. The major impetus behind used of the dichalcogenides is that some exhibit high electrochemical activity, high energy density and reversibility as cathodes materials in lithium metal cells [1-3]. One system that displays reversible electrochemical behavior is $Li/2H-TaS_2$ [3,4]. We report here the ^{57}Fe Mössbauer effect (ME) measurement in a different TaS_2 polymorph $1T-Li_xTa_{1-y}Fe_yS_2$ ($0 \leq x \leq 0.86$; $y=0.05$). The Mössbauer isomer shift (IS) gives information about the valence state of Fe and indirectly of Ta as a function of the lithium content. We will show that the iron is localized and in the Fe^{2+} state. The IS increases with lithium concentration reflecting the changes in the electronic configuration of the host from Ta^{4+} ($5d^1$) to Ta^{3+} ($5d^2$) and increasing ionicity.

$1T-Fe_yTa_{1-y}S_2$ is a layer compound having the CdI_2 structure [5-7]. $1T-TaS_2$ supports an incommensurate charge density wave (CDW) below $\sim 600K$ which becomes commensurate at $\sim 200K$ [8,9]. X-ray and electron diffraction studies in $1T-Fe_yTa_{1-y}S_2$ indicate that the Fe is essentially randomly distributed on octahedrally coordinated Ta sites; no evidence of Fe ordering is seen at any $y \leq \frac{1}{3}$. Doping $1T-TaS_2$ with Fe rapidly destroys the commensurate CDW but the incommensurate CDW is present to $y \sim 0.1$, as evidenced by the existence of sharp satellite diffraction peaks. For $y > 0.1$ the peaks become diffuse. The low temperature electrical resistivity increases rapidly with increasing the doping level, y, and with decreasing temperature.

The magnetic susceptibility and Mössbauer isomer shift [7,8] in $1T-Fe_yTa_{1-y}S_2$ ($y \leq 1/3$) demonstrated the presence of a dynamic low spin-high spin (LS-HS) transition of Fe^{2+} with increasing temperature. It was established that the iron was present solely as Fe^{2+}, that its magnetic electrons were localized, and that the LS-HS energy gap was significantly modulated as a function of T by a strictive interaction with the local lattice environment. As a result of this interaction, the mean displacement coordinate of local ion motion is modulated in a distinctive manner as a thermal population of HS levels takes place.

$1T-Fe_yTa_{1-y}S_2$ was prepared from stoichiometric mixtures of Ta, Fe and S sealed in a small quartz tube under vacuum (10 m Torr or less) and reacted initially for several days at 800°C. After this initial reaction, the samples were opened, ground, pressed into pellets at 50,000 psi, and resealed in the quartz tube with enough excess sulfur to give approximately 1-2 atm. vapor pressure at 950°C. These samples were reheated to 950°C for 7 days, so that diffusion of the cations could occur producing essentially homogeneous pellets, and later quenched into cold water after a several day anneal at 750°C. The lithiated compound $Li_xFe_{0.05}Ta_{0.95}S_2$ was prepared by reaction of n-butyl lithium with 1T material. The maximum amount of Li incorporated in the sample studied was $x = 0.86$. X-ray powder diffraction patterns on the lithiated sample appeared to show a single crystallographic phase having the 1T-structure.

312

The ^{57}Fe Mössbauer spectra were obtained in a standard transmission geometry with a conventional constant acceleration spectrometer using a ^{57}Co in Pd source. Powder absorbers for ME measurements were made in a dry argon atmosphere by mixing the material with boron nitride powder. The ME spectra of $Li_xFe_{0.05}Ta_{0.95}S_2$ are shown in Fig. 1. The ME spectrum for x = 0 shows two resonance lines due to the electric field gradient at the iron nucleus. The asymmetry in the line intensities of the spectrum is due to nonrandom orientation of small single-crystal platelets in a powder sample. The spectra were analyzed by fitting to a sum of two Lorentzian curves of independent position, width and dip. At x=0 the hyperfine parameters are: $|1/2\ e^2qQ| = 0.56 \pm 0.01$ mm/s and IS = 0.56 ± 0.01 mm/s. The magnitude of IS at room temperature corresponds to Fe^{2+} in an intermediate state. The Fe atoms fluctuate between low spin (S=0) and high spin (S=2) state faster than the inverse of the spectral splittings involved of 10^{-7} sec. In $1T-Fe_yTa_{1-y}S_2$ (y≤1/3) each Fe atom rapidly changes its electronic state so that the Mössbauer spectrum is time averaged and only one set of resonance lines is seen [7].

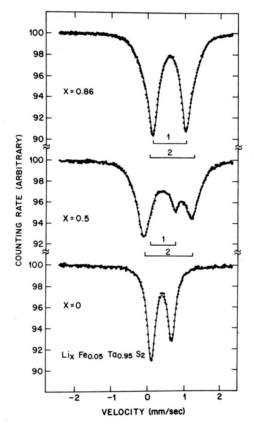

Fig. 1. Room temperature ^{57}Fe Mössbauer absorption spectra of $Li_xFe_{0.05}Ta_{0.95}S_2$ where x=0 to 1. The solid line represents the sum of two or four Lorentzian least squares fits to the data (see text).

The ME spectrum for x = 0.86 is different than the spectrum without lithium. The spectrum shows two sets of quadrupole splitting (QS) resonance lines. The two main resonance lines which correspond to one QS accounts for about 80% of the spectrum area. The hyperfine parameters are given in Table I. The minority contribution to the spectrum is also due to Fe^{2+} in HS state. The IS value of this minority Fe of 0.77 mm/s is in good agreement with the value predicted for Fe^{2+} in HS state.[7] The slightly smaller value predicted for IS of 0.75 mm/s is possibly due to the absence of some Li in the larger particles, since the maximum amount of Li in the sample is x = 0.86 [10]. This value of the high-spin IS is in very good agreement with IS for FeS (0.77 mm/s) and other sulfides with Fe in octahedral sites [11,12].

TABLE I
Room Temperature Mössbauer Effect Parameters for
$Li_xFe_{0.05}Ta_{0.95}S_2$ Compounds

x	IS(1)*a (mm/s)	IS(2)*a (mm/s)	QS(1)a,b (mm/s)	QS(2)a,b (mm/s)
0	0.56	-	0.56	-
0.5	0.61	0.74	0.66	1.34
0.86	0.77	0.75	0.90	1.42

* IS with respect to Fe metal

a The standard deviation for IS and QS is ± 0.01 mm/s
b $QS = |e^2qQ/2|$

The ME spectrum for x = 0.5 is slightly more complex and is different than the spectrum for x = 0 or x = 0.86. The spectrum was analyzed with two sets of quadrupole splitting resonance lines which correspond to two different Fe in two different electronic states. The IS results as a function of x are plotted in Fig. 2. The IS value of 0.61 mm/s indicates that one type of the iron is Fe^{2+} in an intermediate state due to the dynamics LS-LH in the compound. This value is slightly larger than the IS = 0.56 mm/s for x = 0 indicating an increase in ionicity in the compound with lithium concentration. The IS value of 0.74 mm/s indicates that the other iron is Fe^{2+} in the high spin state close to the value for x = 0.86. These IS results reflect a change in the electronic structure of the host from Ta^{4+} ($5d^1$) to Ta^{3+} ($5d^2$). These IS results reflect the change in electronic configuration of Fe and indirectly the valence of Ta as a function of the amount of intercalation.

The appearance of fully HS Fe^{2+} with increasing lithiation may be understood by considering the ionic radii of the cations [13]. The radii increase in the sequence Fe^{2+} (S=O) < Ta^{4+} < Ta^{3+} < Fe^{2+} (S=2). Thus, as the hexagonal unit cell parameters increase on lithiation [14] a higher spin state is favored for iron. Since we observed two types of iron for lithiated $Ta_{1-y}Fe_yS_2$ even though the 1T polymorph has only one crystallographic cation site, these samples must be either two phase or exhibit a superlattice. We were unable to confirm this by x-ray powder diffraction.

In summary using the ME we have been able to show that the iron in $Li_xFe_{0.05}Ta_{0.95}S_2$ is Fe^{2+} at all x but changes from intermediate spin to HS at a certain concentration of lithium. Fe is not reduced by the intercalation of Li but reflects the change in the electronic structure of the host.

314

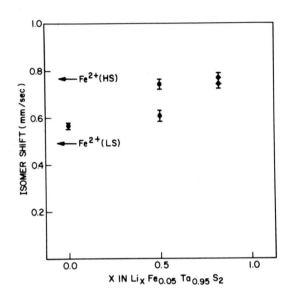

Fig. 2. Isomer shift of the Mössbauer absorption x-ray of ^{57}Fe in $Li_xFe_{0.05}Ta_{0.95}S_2$ (x=0 to 1) expressed relative to metallic iron at 295K. The arrows represent values expected for pure $Fe^{2+}(LS)$ and $Fe^{2+}(HS)$ from Reference 7.

REFERENCES

1. M. S. Whittingham, Science 192, 1126 (1976).

2. D. W. Murphy and F. A. Trumbore, J. Cryst. Growth 39, 185 (1977).

3. M. S. Whithingham Prog. Sol. State Chemistry 12, 1 (1978).

4. S. Basu and W. L. Worrell in: Electrode Materials and Processes for Energy Conversion and Storage, J. D. E. McIntyre, S. Srinivasan and F. G. Will, eds. (The Electrochemical Society, Princeton, New Jersey 1977) p. 861.

5. R. M. Fleming and R. V. Coleman, Phys. Rev. Lett. 34, 1502 (1975).

6. M. Eibschütz and F. J. DiSalvo, Phys. Rev. Lett. 36, 104 (1975).

7. M. Eibschütz, M. E. Lines and F. J. DiSalvo Phys. Rev. B15, 103 (1977).

8. J. A. Wilson, F. J. DiSalvo and S. Mahajan Adv. Phys. 24, 117 (1975).

9. F. J. DiSalvo, J. A. Wilson, B. G. Bagley and J. V. Waszczak Phys. Rev. B12, 464 (1975).

10. M. Eibschütz in: Symposium on Recent Chemical Applications of Mössbauer Spectroscopy, March 1980, J. Am. Chem. Soc. Conf. Proceedings (in press).

11. S. Hafner and M. Kalvius, Z. Kristallogr 123, 443 (1966).

12. M. Eibschütz, E. Hermon and S. Shtrikman, J. Phys. Chem. Solids 28, 1963 (1967).

13. R. D. Shannon and C. T. Prewitt, Acta Cryst. B25, 946 (1969).

14. D. W. Murphy, F. J. DiSalvo, G. W. Hull and J. W. Waszczak, Inorg. Chem. 15, 17 (1976).

Published 1981 by Elsevier North Holland, Inc.
Kaufmann and Shenoy, editors
Nuclear and Electron Resonance Spectroscopies Applied to Materials Science

MÖSSBAUER EFFECT STUDY OF THE ABSENCE OF FE MAGNETISM IN SUPERCONDUCTING $Sc_2Fe_3Si_5$ AND Th_7Fe_3 *

J. D. CASHION,[†] G. K. SHENOY, D. NIARCHOS,
P. J. VICCARO AND CHARLES M. FALCO
Argonne National Laboratory, Solid State Science Division,
Argonne, Illinois 60439

ABSTRACT

There are only a few Fe-based compounds which are superconducting since the depairing of Cooper pairs is unavoidable
in the presence of an Fe moment. In $Sc_2Fe_3Si_5$ and Th_7Fe_3,
Mössbauer measurements on ^{57}Fe show that there is no magnetism at the Fe site in either compound. The isomer shift
values indicate the reason behind this identical magnetic
behavior to be quite different in its electronic origin.

INTRODUCTION

It is a well known fact that a small quantity of a paramagnetic impurity
such as iron in most of the superconductors will degrade its superconducting behavior dramatically. This results from the exchange interaction between the
conduction electron and the paramagnetic spin which breaks the Cooper pairs.
This phenomenon is well described by the Abrikosov-Gorkov theory [1], according
to which the depression in the superconducting transition temperature depends on
the spin and the amount of the paramagnetic impurity, its exchange interaction
strength with the conduction electrons and the density of conduction electrons
at the Fermi level.

In spite of this fact, there are a few iron-containing superconductors.
The natural question regarding these materials is "why are they superconducting?" A very weak exchange coupling can keep the electrons from depairing as
is the case with some of the rare-earth based Chevrel phase compounds [2]. However, this is not the only explanation and one important thing to investigate is
the magnetic state of the iron in such superconductors.

We have used the Mössbauer effect to investigate the magnetism of iron
atoms in various materials. This nuclear spectroscopic method permits one to
deduce information regarding the magnetic moment on iron through the Zeeman
interaction of the nucleus. In this paper we present Mössbauer studies of two
of the iron based superconductors, viz. Th_7Fe_3 and $Sc_2Fe_3Si_5$.

IRON-BASED SUPERCONDUCTORS

In Table I we have listed the compounds containing more than 14 atomic percent of iron which exhibit superconductivity. We have pointed out in the last
section various possible reasons for these iron rich compounds to be superconducting. To achieve either the absence of iron magnetism, or a weak coupling
of the iron moment with the superconducting electrons, the crystal structure
and the electronic structure need to be very unusual. Only under such special
circumstances can one avoid depairing of the Cooper pairs by the magnetic atoms.

The value of T_c for the compounds in Table I is not small. This fact
strongly suggests that the superconductivity in them could be due to d-
electrons. In addition to the Fe atom itself, the d-electrons in these com-

pounds can usually be contributed by one other element constituting these compounds. Thus it is important to find which of the d-electrons form the superconducting pairs.

Table I: SUPERCONDUCTING TRANSITION TEMPERATURES
OF IRON BASED COMPOUNDS

Compound	$T_c(K)$	Reference
Th_7Fe_3	1.86	3
U_6Fe	3.9	4
Zr_2Fe	0.17	5
$Sc_2Fe_3Si_5$	4.6	6
$Y_2Fe_3Si_5$	2.4 - 2.0	6
$Lu_2Fe_3Si_5$	6.1 - 5.8	6
$TlFe_3Te_3$	4.0	7,8

MÖSSBAUER RESULTS AND DISCUSSION

The Mössbauer effect is an excellent microscopic tool for measuring the local magnetization. In a magnetically ordered material, the moment on Fe (local or non-local) produces a field at the nucleus both from the unquenched orbital momentum and from the polarization of the s-electrons by the spin-only part of the d-moment. As a result one realizes a hyperfine magnetic field at the nucleus which interacts with the nuclear moment. Such a nuclear Zeeman effect can easily be measured with Mössbauer spectroscopy. The value of the hyperfine field is roughly proportional to the moment on the atom and permits us a sensitivity of the order of 0.03 μ_B in the measurements.

In paramagnetic materials, one can perform similar experiments by the application of external magnetic fields. The field experienced by the nucleus, $\vec{H}_n$, is given by

$$\vec{H}_n = \vec{H}_{ext} + \vec{H}_i \qquad (1)$$

where $\vec{H}_i$ is the induced field and $\vec{H}_{ext}$ is the applied magnetic field. $\vec{H}_i$ would be zero if the iron atom should carry no magnetic moment and as a result $\vec{H}_n$ will be equal to $\vec{H}_{ext}$. In fact, this is true if the material is diamagnetic, since the induced (negative) moment is too small to produce a measurable effect on the nucleus. On the other hand, if a magnetic moment is present on the atom or induced by a neighboring paramagnetic atom, one will observe a non-zero H_i.

We shall illustrate these ideas with specific examples. The ternary compound $Sc_2Fe_3Si_5$ is a superconductor (Table I) and from the Mössbauer studies of this compound using the 14.4 keV resonance radiation from ^{57}Fe one can investigate the magnetic state of iron. By measuring the nuclear Zeeman effect in this material up to 60 kG, we have demonstrated [9] that H_i is equal to zero (see Fig. 1). Thus Fe has no magnetic moment in this lattice.

The superconducting behavior of Th_7Fe_3 has been known for many years. No microscopic information on the magnetism of Fe in this lattice is available. Our Mössbauer studies performed in 70 kG again show that the field measured through the nuclear Zeeman effect is the same as the applied field (see Fig. 1). Therefore, in this lattice also, Fe has no measurable magnetic moment, the limit being 0.1 μ_B. The quadrupole interaction, $1/2\ e^2qQ$, also determined from this measurement is $+\ 0.49 \pm 0.03$ mm s^{-1} and the asymmetry parameter $\eta = 0.75 \pm 0.10$ [Fig. 2(b)]. The computer fitting of this type of spectrum is described in [9].

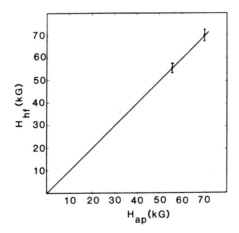

Fig. 1 The observed field at the Fe nucleus in $Sc_2Fe_3Si_5$ and Th_7Fe_3 as a function of external field. The curve shows the absence of any induced field at the Fe site.

Thus in both Th_7Fe_3 and $Sc_2Fe_3Si_5$ the presence of superconductivity at low temperatures is possible because of the absence of magnetism in both these materials. This should be contrasted with the behavior of Chevrel phase compounds in which the rare-earths have large magnetic moments, but they exhibit superconductivity because of a poor exchange coupling of these moments with the superconducting electrons [2].

The absence of an Fe magnetic moment in its compounds is a rather unusual phenomenon. A complex coordination produces a low-spin zero moment situation in some compounds. Thus for example, when Fe is coordinated by 12 or 13 atoms, the Fe goes into a low spin state with a filled band or filled d-orbital configura-

318

tion as in $V_3Fe_5Si_2$ or α-Mn [10]. We believe this to be the situation also in $Sc_2Fe_3Si_5$. In Th_7Fe_3 on the other hand Fe does not have a complex coordination and the low density of projected d-states at the Fe site seems to be responsible for the absence of a magnetic moment. This occurs, for example, in FeTi [11] which has a very similar isomer shift. Thus the electronic states of Fe are different in $Sc_2Fe_3Si_5$ and Th_7Fe_3, but both result in no magnetism.

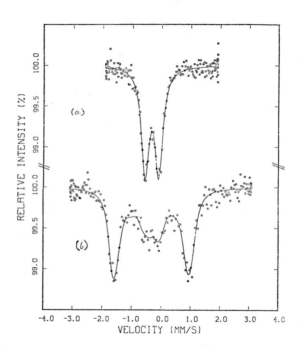

Fig. 2 ^{57}Fe Mössbauer spectra of Th_7Fe_3 taken (a) at room temperature and (b) at 4.2 K in an external magnetic field of 70 kG applied parallel to the gamma ray direction.

One of the useful parameters in Mössbauer spectroscopy which can yield relative information on the electronic structure of an atom in different lattices is the isomer shift. This is basically a measure of the total electron density (primarily from the s electrons) at the nucleus. The average value of the isomer shifts are + 0.18 ± 0.02 mm/s for $Sc_2Fe_3Si_5$ and -0.22 ± 0.02 mm/s for Th_7Fe_3 [Fig. 2(a)], both referred to Fe metal at room temperature. The more positive number for the $Sc_2Fe_3Si_5$ implies a lower electron density which can be ascribed to a large d-shielding (a filled d band) compared to that in Fe metal or Th_7Fe_3. The low density of d-states at Fe in Th_7Fe_3 is produced by a larger density of s electrons compared to Fe metal. Although such an interpretation of isomer shift values for these compounds is qualitatively in agreement with their magnetic behavior, a more detailed understanding requires electronic structure calculations.

Finally, we wish to make a few comments on the nature of the superconducting d-electrons. In $Sc_2Fe_3Si_5$, by comparing with magnetic $Dy_2Fe_3Si_5$ we have demonstrated that the Fe 3d-electrons are probably responsible for the superconductivity [12]. This is also supported by the formation of Fe clusters in this ternary compound, essential for realizing superconducting behavior [13].

In Th_7Fe_3, each of the three types of Th atoms have Th near neighbors whose separation [14] is less than or equal to the separation in Th metal (3.60 Å), itself already a "superconducting distance" since Th metal is superconducting below 1.38 K. Thus it satisfies the cluster criterion and further evidence to support the belief that the superconducting electrons are derived from the 6d electrons of Th is provided by our isomer shift measurement which shows that the Fe 3d electrons have a low density of states.

REFERENCES

*Work supported by the U.S. Department of Energy.

†On leave from the Department of Physics, Monash University, Victoria, Australia.

1. A. A. Abrikosov and L. P. Gor'kov, Sov. Phys. JETP 12, 1243 (1961).

2. B. D. Dunlap, G. K. Shenoy, F. Y. Fradin, C. D. Barnet and C. W. Kimball, J. Mag. Magn. Materials 13, 319 (1979).

3. B. T. Matthias, V. B. Compton and E. Corenzwit, J. Phys. Chem. Solids 19, 130 (1961).

4. B. S. Chandrasekhar and J. K. Hulm, J. Phys. Chem. Solids 7, 259 (1958).

5. E. E. Havinga, H. Damsma and J. M. Kanis, J. Less-Common Metals 27, 281 (1972).

6. Hans F. Braun, Phys. Letters 75A, 386 (1980).

7. W. Hönle, H. G. von Schnering, A. Lipka and K. Yvon, J. Less-Common Metals 71, 135 (1980).

8. M. Potel, R. Chevrel, M. Sergent, M. Decroux, and Ø. Fischer, C. R. Acad. Sci., Ser. C. 288, 429 (1979).

9. J. D. Cashion, G. K. Shenoy, D. Niarchos, P. J. Viccaro and Charles M. Falco, Phys. Letters 79A, 454 (1980).

10. C. W. Kimball, W. C. Phillips, M. V. Nevitt and R. S. Preston, Phys. Rev. 146, 375 (1966).

11. E. V. Mielczarek and W. P. Winfree, Phys. Rev. B11, 1026 (1975).

12. J. D. Cashion, G. K. Shenoy, D. Niarchos, P. J. Viccaro, A. T. Aldred and Charles M. Falco, Proc. 14th Annual Conference on Magnetism and Magnetic Materials, Dallas, Texas, November 11-14, 1980.

13. J. M. Vandenberg and B. T. Matthias, Science 198, 194 (1977).

14. Structure Reports, ed. W. B. Pearson (Oosthoek, Utrecht) Vol. 20, 132 (1956).

Published 1981 by Elsevier North Holland, Inc.
Kaufmann and Shenoy, editors
Nuclear and Electron Resonance Spectroscopies Applied to Materials Science

MICROSCOPIC STUDY OF METAL HYDRIDES USING ELECTRON SPIN RESONANCE[†]

E. L. VENTURINI
Sandia National Laboratories*, Albuquerque, NM 87185

ABSTRACT

Electron spin resonance (ESR) of dilute paramagnetic ions in nonmagnetic metallic hydrides provides microscopic infor- mation about the hydrogen ions in the immediate vicinity of the impurity. By comparing ESR spectra for different host metals and several hydrogen/metal ratios, one can determine material properties including host lattice symmetry, phase boundaries and occupation of available sites by hydrogen. Examples are presented of ESR of dilute Er in group IIIB and IVB metal hydrides, demonstrating the sensitivity and versa- tility of ESR as a spectroscopic technique.

INTRODUCTION

Microscopic information concerning lattice symmetry, phase boundaries and the location, net charge and site energy of hydrogen ions in metals and metal hydrides is of fundamental importance in understanding these materials and optimizing their macroscopic properties for specific applications. Several spectroscopic techniques have been applied to this problem including nuclear magnetic resonance [1], Mössbauer resonance [2], inelastic neutron scattering [3] and photoemission [4]. In this paper several features of electron spin resonance (ESR) of dilute paramagnetic ions in nonmagnetic metallic hydrides are discussed to demonstrate the sensitivity and versatility of this method in answering the questions posed above.

In a typical ESR experiment one measures the derivative of the absorbed microwave power at a fixed frequency versus the applied magnetic field strength. The position of the maximum absorbed power is termed the resonance field, and is usually specified by the ESR g-factor $g = 0.7145 \times$(frequency in MHz)/(resonance field in Oe). All samples were prepared by arc-melting 0.1 atomic % Er into purified Sc, Y or Zr, followed by surface cleaning and then hydriding in a modified Sieverts' apparatus. The polycrystalline samples were then ground into a fine powder, and their derivative ESR absorption spectra recorded at 9.8 GHz and helium temperatures. The combination of dilute Er and low temperature assures narrow absorption linewidths, facilitating the detailed spectral analy- sis presented in the next three sections.

HOST LATTICE SYMMETRY

The ideal metal dihydride MH_2 has the face-centered-cubic (fcc) fluorite structure, where the metal atom is surrounded by eight nearest-neighbor (nn) hydrogen ions forming a simple cube as shown by the solid circles in Fig. 1. [5,6,7] These hydrogen positions are termed tetrahedral (T) sites. The six next-nearest-neighbor (nnn) octahedral (O) sites shown as open circles are vacant in the ideal MH_2 lattice, and fully occupied in the trihydride MH_3. When a paramagnetic Er ion is randomly substituted for a host metal atom, its ESR spectrum reflects the local crystal field arising primarily from the nn and nnn hydrogen ions. Our spectra show clearly the effect of small distortions in the nn cube or the partial occupation of nnn O-sites by hydrogen.

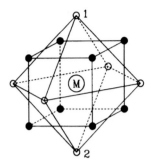

Ⓜ Metal Ion
at Cube Center

● Tetrahedral (T)
Hydrogen Ions

○ Octahedral (O)
Hydrogen Ions

Fig. 1. Metal ion site in the face-centered-cubic hydride lattice.

Fig. 2 shows three distinct derivative ESR absorption spectra taken at 9.82 GHz and 2 K with $ScH_{1.93}$, $ZrH_{2.00}$ and $YH_{2.94}$ metal hydride powders, contrasting Er ions in cubic, axial and biaxial site symmetry, respectively. The powder samples have ESR spectra which are averages over all orientations of the magnetic field with respect to the crystal axes. The simplest spectrum is the top trace for Er in $ScH_{1.93}$, where the host lattice is fcc, and the Er is surrounded by eight nn T-site hydrogen ions with all O-sites vacant (Fig. 1).[7] The cubic site symmetry results in an isotropic g-factor g_{cubic} = 6.778(3), corresponding to a Γ_7 doublet ground state for trivalent Er in a cubic crystal field.[8] This requires that the eight T-site hydrogen ions be negatively charged. The top trace has small resolved hyperfine ESR signals on either side of the main absorption due to the 23% of natural Er isotopes with a nuclear spin of 7/2.

The middle spectrum in Fig. 2 is for Er in $ZrH_{2.00}$, and is characteristic of a magnetic ion with axial site symmetry, in agreement with x-ray data which show the host lattice to be face-centered-tetragonal (fct) for hydrogen/metal ratios between 1.73 and 2.00.[5] In this structure the metal atoms are surrounded by eight nn T-site hydrogen ions with nnn O-sites vacant, but the simple cube in Fig. 1 is "squashed" to a c/a lattice constant ratio of 0.89. The powder ESR spectrum for ions in axial sites has a positive peak and a nearly "normal" derivative absorption labelled $g_{\parallel}$ and $g_{\perp}$, respectively, where $g_{\parallel}$ is the g-factor when the magnetic field is along the high symmetry axis, and $g_{\perp}$ is the g-factor with the field normal to this axis (Fig. 2). The ESR g-factors for Er in $ZrH_{2.00}$ are $g_{\parallel}$ = 7.55(2) and $g_{\perp}$ = 6.418(5).

The bottom trace in Fig. 2 shows an Er ESR spectrum in hexagonal $YH_{2.94}$. The metal ion site in this material is biaxial [7], giving rise to three distinct ESR g-factors, g_1 = 10.8(1), g_2 = 5.02(2) and g_3 = 3.74(1). The Er ion is surrounded by eight nn T-site and six nnn O-site hydrogen ions. The simple cube plus octahedron shown in Fig. 1 are severely distorted as evidenced by the large splitting of the g-factors from the isotropic cubic value of 6.8.

The striking differences in the Er spectra for the three metal hydride hosts shown in Fig. 2 demonstrate the sensitivity of ESR to host lattice symmetry. This contrasts sharply with recent Mössbauer measurements showing that, within the experimental resolution, Er in $ZrH_{1.86}$ has approximately cubic site symmetry [9], and the Er site symmetry in YH_3 can be treated as axial [10].

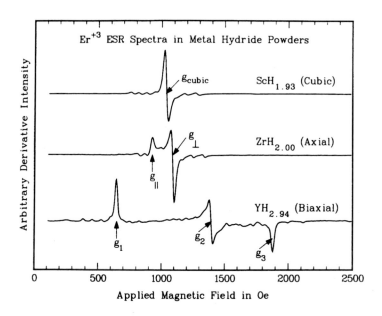

Fig. 2. Er ESR spectra in three metal hydride hosts taken at 9.82 GHz and 2 K.

SITE OCCUPATION BY HYDROGEN

The ESR spectrum in Fig. 3 for Er in $ScH_{1.993}$ contains two new signals not seen in the relatively low gain top trace of Fig. 2 for Er in $ScH_{1.93}$.[11] The cubic signal occurs at the same g-factor in both spectra, indicating Er ions on identical sites, while the the two new signals reflect Er ions in two distinct axial sites with $g_{\parallel}^{(1)} = 9.1(1)$, $g_{\perp}^{(1)} = 5.45(2)$, $g_{\parallel}^{(2)} = 11.0(3)$ and $g_{\perp}^{(2)} = 4.07(2)$. X-ray data show only the fcc host lattice for ScH_x when $1.9 < x < 2.0$.[7] The intensity of the axial signals in Fig. 3 increases rapidly as x approaches 2, leading to the conclusion that they are due to occupation of nnn O-sites by hydrogen. [11,12] The resonance in Fig. 3 with superscript "1" is attributed to Er ions in an axial site having eight nn T-site hydrogen ions plus one nnn O-site hydrogen, indicated by the site numbered "1" in Fig. 1. Axial signal "2" in Fig. 3 arises from Er ions surrounded by eight T-site hydrogens plus two nnn O-site hydrogens on opposite sides of the magnetic probe, shown in Fig. 1 by sites numbered "1" and "2".[12]

Using a numerical least-squares fit to the three components of the Er ESR powder spectrum in Fig. 3, we obtain the integrated intensities of the three signals, and hence the relative number of Er ions in each of the three sites. [11] The relative numbers are a direct measure of the nnn O-site occupation by hydrogen ions. The change in occupation versus hydrogen/metal ratio is determined by analyzing ESR spectra from a series of samples. The distribution of hydrogen ions among available sites is a function of temperature and the site energies. A lattice-gas model has been developed which explains the ESR data, and yields site energy differences for hydrogen both adjacent to the Er impurities and in the bulk host hydride.[12,13]

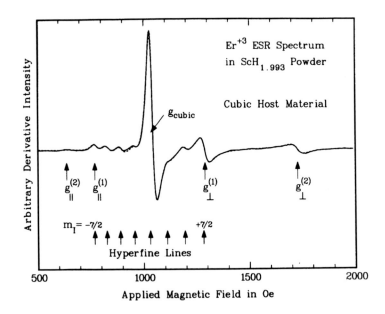

Fig. 3. ESR spectrum showing three distinct Er ion sites in $ScH_{1.993}$.

Using this model, we find that ScH_x has few bulk O-sites occupied by hydrogen for $x < 2$, while YH_x has substantial numbers of hydrogen ions on O-sites for x as low as 1.9.[13] In addition there are appreciable changes (10%) in site energy differences with isotope as shown by ESR spectra for hydrides and deuterides of Sc and Y.[14] These examples demonstrate the sensitivity of ESR to the location and partial site occupation by hydrogen in metal hydrides.

It is important to note that O-site filling by hydrogen produces distinct ESR signals (Fig. 3) in contrast to nuclear magnetic resonance which measures hydrogen location from changes in the linewidth of a single absorption.[15] However, ESR determines directly only the hydrogen distribution adjacent to the magnetic probe, and information about the bulk host hydride requires a model such as the lattice-gas calculation.

PHASE BOUNDARY DETERMINATION

A final intriguing application of ESR in metal hydrides is the determination of phase boundaries by studying the spectra of mixed phase material. YH_x has an fcc structure for $1.9 < x < 2.0$, and is hexagonal for $2.7 < x < 3.0$.[16] Fig. 4 shows that a sample with $x = 2.64$ has an ESR spectrum which is a superposition of the top and bottom traces in Fig. 2, i. e., a mixture of fcc and hexagonal phases. Concentrating first on the top spectrum in Fig. 4 recorded at 2 K, a detailed numerical fit to the cubic and biaxial signals gives a ratio of 10 ± 2 for Er ions in the hexagonal phase relative to those in the cubic phase.

Using a simple lever rule for mixed phases, one expects a fraction α of the hexagonal phase and $1-\alpha$ of the cubic phase satisfying $\alpha C_{hex} + (1-\alpha)C_{cub} = x$,

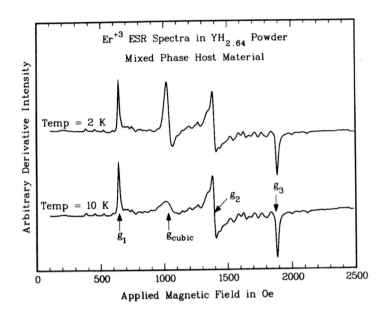

Fig. 4. Spectra comparing ESR signal intensities in $YH_{2.64}$:Er at 2 and 10 K.

where x is the hydrogen/metal ratio for the mixed phase sample, and the C's represent the lowest single phase composition for the hexagonal hydride (2.7) and the highest for the cubic hydride (2.0). For x = 2.64 the lever rule gives α = 0.92. Hence the ratio of hexagonal to cubic material is α/(1-α) = 10.7 in good agreement with the experimental ESR value of 10±2. The analysis above can be reversed, and by using ESR intensities for several mixed phase samples with 2.0 < x < 2.7, we can determine the single phase boundaries independent of the conventional pressure-composition-temperature measurements.

Fig. 4 also demonstrates the ability of ESR to distinguish metallic and insulating samples. Comparing the top trace at 2 K with the bottom trace for the same material at 10 K, the hexagonal phase signal intensity remains essentially unchanged, while the cubic phase absorption linewidth broadens dramatically with a corresponding decrease in intensity. In fact the cubic ESR signal linewidth increases linearly with rising temperature as expected for Korringa relaxation of the Er spin via conduction electron scattering.[17] The fcc phase is metallic, and the slope of the ESR linewidth versus temperature, 5.4(1) Oe/K, is proportional to the square of the conduction electron density of states at the Fermi level at the Er ion site. The hexagonal phase is insulating, and Fig. 4 indicates little change in its ESR signal linewidth from 2 to 10 K.

CONCLUSIONS

In this brief paper we have highlighted several important features of ESR spectroscopy of dilute paramagnetic Er ions in three nonmagnetic metallic hydrides. This technique provides detailed microscopic information on the host lattice symmetry, the location, net charge and site energies of hydrogen isotopes, and the phase boundaries present in these materials. ESR is shown to be both a sensitive and versatile method for exploring these properties, and compares favorably with the various other resonance probes.

REFERENCES

[†]This work sponsored by the U. S. Department of Energy under Contract No. DE-AC04-76-DP00789.

[*]A U. S. Department of Energy facility.

1. R. M. Cotts in: Hydrogen in Metals I, Basic Properties, G. Alefeld and J. Völkl eds. (Springer-Verlag, Berlin, 1978) Ch. 9.

2. F. E. Wagner and G. Wörtmann in: ibid., Ch. 6.

3. T. Springer in: ibid., Ch. 4; K. Sköld in: ibid., Ch. 10.

4. J. H. Weaver et al., Phys. Rev. B19, 4855 (1979).

5. R. L. Beck and W. M. Mueller in: Metal Hydrides, W. M. Mueller, J. P. Blackledge and G. G. Libowitz eds. (Academic Press, New York, 1968) Ch. 7.

6. W. M. Mueller in: ibid., Chs. 8 and 9.

7. J. P. Blackledge in: ibid., Ch. 10.

8. K. R. Lea et al., J. Phys. Chem. Solids 23, 1381 (1962).

9. G. K. Shenoy et al., J. Physique 40, Suppl. C-2, 180 (1979).

10. B. Suits et al., J. Magn. Magn. Mater. 5, 344 (1977).

11. E. L. Venturini, J. Appl. Phys. 50, 2053 (1979).

12. E. L. Venturini and P. M. Richards, Solid State Commun. 32, 1185 (1979).

13. E. L. Venturini and P. M. Richards, Phys. Lett. 76A, 344 (1980).

14. E. L. Venturini, J. Less-Common Met. 74 45 (1980).

15. D. L. Anderson et al., Phys. Rev. B21, 2625 (1980).

16. L. N. Yannopoulos et al., J. Phys. Chem. 69, 2510 (1965).

17. J. Korringa, Physica 16, 601 (1950).

Published 1981 by Elsevier North Holland, Inc.
Kaufmann and Shenoy, editors
Nuclear and Electron Resonance Spectroscopies Applied to Materials Science

MÖSSBAUER EFFECT INVESTIGATION OF HYDROGEN RICH TERNARY HYDRIDES OF YFe_2, $DyFe_2$ and $ErFe_2$*

P. J. VICCARO, D. NIARCHOS, G. K. SHENOY AND B. D. DUNLAP
Argonne National Laboratory, Argonne, Illinois 60439, USA

ABSTRACT

The hydrogen rich ternary hydride phases of RFe_2H_x (R = Y, Dy, Er) with x ⩾ 4 have been investigated using ^{57}Fe, ^{161}Dy and ^{166}Er Mössbauer spectroscopy and bulk magnetization. In all cases, the presence of hydrogen results in a dramatic reduction in the moment on Fe as well as the magnetic transition temperature. The data also imply a simultaneous weakening of the R-Fe magnetic exchange.

INTRODUCTION

Many rare earth based RFe_2 Laves phase intermetallic compounds are known to readily absorb hydrogen at ambient temperature [1,2]. This absorption results in the formation of ternary hydride phases with well defined hydrogen concentrations. For a given phase, the magnetic and electronic properties can be quite different from those of the respective host intermetallic compound especially in hydrides with large hydrogen concentrations. The origin of the difference is associated with the presence of hydrogen in the metallic lattice and its perturbation of the surrounding metal atoms.

In an attempt to determine the effects of hydrogen on intermetallics of this type, we have conducted bulk magnetization and ^{57}Fe, ^{161}Dy and ^{166}Er Mössbauer effect experiments on RFe_2H_x hydrides with R = Y, Dy, Er. We present here results for the hydrogen rich ternary hydride phases with x ⩾ 4. The Mössbauer effect is useful in monitoring the microscopic environment of the metal atom in question and the changes in this environment due to the incorporation of hydrogen. In particular, the isomer shift, which is related to the total electron density at the nucleus of the isotope in question, gives information concerning the local electronic structures and densities. The magnetic hyperfine field (H_n) on the other hand, can be related to the local magnetic moment. The coupling of these moments through the magnetic exchange interaction gives rise to the spontaneous or bulk magnetic moment measured with bulk magnetization. The results from these measurements for the hydrogen rich ternary phases will be compared to those for lower order hydride phases and the parent intermetallic.

EXPERIMENTAL

The preparation of the parent intermetallics and their hydrides has been previously described [1,3]. Hydrogen concentrations were maintained in a given phase through surface poisoning with SO_2 gas. Subsequent determinations of the actual hydrogen concentrations in the samples studied were made through volumetric analyses of evolved hydrogen.

For the Mössbauer experiments, sources consisting of ^{57}Co in Rh,

---

*Work supported by the U.S. Department of Energy.

$^{161}Gd_{0.5}Dy_{0.5}F_3$, and $Y_{0.7}{}^{166}Ho_{0.3}H_2$ were utilized for the ^{57}Fe (14.4 keV), ^{161}Dy (25.6 keV) and ^{166}Er (80.6 keV) resonances respectively. Magnetization data were collected with a magnetometer based on a Faraday balance in external magnetic fields up to 15 kG. X-ray diffraction films (Debye-Scherrer camera) were made using CuK_α radiation.

RESULTS AND DISCUSSION

Well defined hydride phases (RFe_2H_x) are formed with $x = 2$, 3.5 and 4 for the R atoms considered here [1-4]. In addition we have recently observed a new phase with the composition $RFe_2H_{4.5}$. The structure relative to the metal atoms of all hydrides with $x = 2$ and $x = 3.5$ retain the cubic C15 Laves structure of the parent intermetallic with an expansion of approximately 5% and 8%, respectively, in the lattice constant. A distortion of the cubic lattice to a rhombohedral one is observed for $ErFe_2H_{4.1}$ [4]. For $DyFe_2H_{4.5}$ and $YFe_2H_{4.5}$, however, the lattice remains cubic with an approximately 9% expansion of the unit cell.

The fact that, in all but the case of $ErFe_2H_{4.1}$, the incorporation of hydrogen affects the size but not the symmetry of the unit cell facilitates a comparison of the magnetic properties of the hydrides to the respective parent intermetallics. In the latter compounds, these properties are almost completely determined by the Fe-Fe and Fe-R magnetic interaction with the former dominating [5]. Both $ErFe_2$ and $DyFe_2$ are ferrimagnetic with antiparallel R-Fe spin coupling via the RKKY interaction [5]. The orientation of the net moment is different in the two compounds due to different R anisotropies. For YFe_2 the R-Fe interaction is absent since Y is diamagnetic, and the compound is ferromagnetic. In all three intermetallics both the R and Fe each occupy a single site in the lattice. More than one inequivalent magnetic site can be observed for Fe depending on the orientation of total magnetic moment (see Fig. 1).

The bulk magnetization data for the hydrides $YFe_2H_{4.5}$, $DyFe_2H_{4.5}$ and $ErFe_2H_{4.1}$ are dramatically different from those of the respective intermetallic compounds in which either ferromagnetic (YFe_2) or ferrimagnetic ($DyFe_2$, $ErFe_2$) behavior is observed [5]. The Er based ternary, for example, shows paramagnetic behavior down to 5 K. From previous Mössbauer measurements [4] it is known that a magnetic transition occurs only below 4 K compared to 587 K for $ErFe_2$. For the Y based hydride, paramagnetic behavior is observed down to 5 K associated either with the $YFe_2H_{4.5}$ phase or a small amount (< 7%) of an impurity phase detected in the sample. The ferromagnetic transition temperature for YFe_2 on the other hand is 542 K [5]. In the case of $DyFe_2H_{4.5}$ a magnetic transition at 20 ± 3 K is seen in the data which again is reduced from the value of 635 K for $DyFe_2$. In all three ternary hydrides, the strong magnetic character of the parent intermetallic is significantly weakened by the presence of hydrogen. For the hydrides with smaller hydrogen concentrations, a gradual weakening is also observed as hydrogen concentration increases [3,4,6]. From the above data, it appears that an abrupt change in the magnetic properties occurs in going from $RFe_2H_{3.5}$ to RFe_2H_x with $x \geq 4$.

Similarly, dramatic changes in the ^{57}Fe hyperfine field are observed in $YFeH_{4.5}$, $ErFe_2H_4$ and $DyFe_2H_{4.5}$ compared to the respective RFe_2 parent intermetallics at 4.2 K (see Fig. 1). In particular, a nearly complete collapse of the hyperfine magnetic field appears to occur at these concentrations. External magnetic field experiments confirm that the collapse reflects a real reduction of the Fe moment from about 1.6 μ_B in unhydrided compounds less than 0.2 μ_B on hydrogenization $x \geq 4$. Since the magnetic interaction is driven by the Fe-Fe exchange, it is not surprising that the strongly reduced T_c values observed in the bulk magnetzation data are encountered for these phases in which the Fe moment is also found to be small.

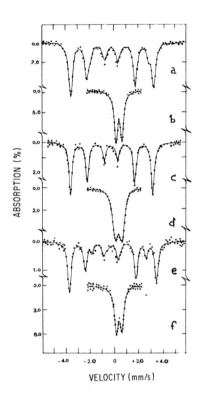

Fig. 1 ^{57}Fe Mössbauer spectra at 4.2 K for (a) YFe$_2$, (b) YFe$_2$H$_{4.5}$, (c) DyFe$_2$, (d) DyFe$_2$H$_{4.5}$, (e) ErFe$_2$ and (d) ErFe$_2$H$_{4.5}$. The spectra for YFe$_2$ and ErFe$_2$ show two magnetic Fe sites in a 1:3 population ratio expected for the magnetic structure of these compounds. DyFe$_2$ shows one magnetic site. The hydrides all show the absence of a magnetic hyperfine field and are characterized by a single quadrupole doublet.

Furthermore, the applied field data indicate that as in the parent inter-metallic compound, Fe occupies only one type of site in the respective ternary hydride phase even in ErFe$_2$H$_{4.1}$ where a distortion of the lattice occurs. The ^{57}Fe Mössbauer spectrum in each case is characterized by a single quad-rupole interaction giving rise to the doublet structure shown in Fig. 1. These results show that the two distinct Fe sites which appear in the dihydride phase RFe$_2$H$_2$ [4,6] are again equivalent in the higher order phases with x > 4, presumably due to a filling by hydrogen available of interstitial holes.

The ^{161}Dy Mössbauer data for the ternary hydride DyFe$_2$H$_{4.5}$ at 4.2 K show the presence of two Dy sites with a population ratio of approximately 1:1. The hyperfine fields were determined to be 5680 ± 80 kG and 5190 ± 80 kG. As men-tioned, this ternary hydride retains the cubic C15 structure of DyFe$_2$ in which Dy occupies one type of crystallographic site. The presence of two Dy sites in the hydride phase must then be associated with differences in the local con-figuration of surrounding hydrogen.

Since the data in $DyFe_2H_{4.5}$ were taken below the magnetic transition temperature ($\sim$ 20 K), both magnetic exchange and crystalline electric field interactions are present at the Dy sites. In the case of $DyFe_2$ and $DyFe_2H_2$ [3] where the magnetic transition temperatures are high due to the presence of an Fe-Fe exchange, the magnetic exchange dominates. This produces a nearly pure $|J_z| = 15/2$ ground state derived from the J = 15/2 free ion term of Dy^{3+} which, in turn, results in a hyperfine field of 6200 kG corresponding to a local moment of 10 μ_B on Dy.

Crystal field effects, on the other hand, would produce a ground state which is an admixture of several $|J_z>$ states and consequently would result in a smaller local moment and hyperfine field on Dy. Therefore the observation of reduced values of H_n at both Dy sites in $DyFe_2H_{4.5}$ imply a weaker exchange interaction relative to the crystal field occurs in the hydride phase. A similar effect was seen in $ErFe_2H_{4.1}$ [4]. In both cases, a magnetic exchange weaker than observed in the lower order hydride phases and the respective parent intermetallic compounds is understandable in view of the severe reduction of the Fe moment caused by hydrogen in the higher order hydride phases.

REFERENCES

1. H. A. Kierstead, P. J. Viccaro, G. K. Shenoy and B. D. Dunlap, J. Less-Common Metals 66, 219 (1979).

2. H. A. Kierstead, J. Less-Common Metals 70, 199 (1980).

3. P. J. Viccaro, J. M. Friedt, D. G. Niarchos, B. D. Dunlap, G. K. Shenoy, A. T. Aldred and D. G. Westlake, J. Appl. Phys. 50, 2051 (1979).

4. P. J. Viccaro, G. K. Shenoy, B. D. Dunlap, D. G. Westlake and J. F. Miller, J. de Phys. (Paris) 40, C2-198 (1979).

5. K. H. J. Buschow, Rep. Prog. Phys. 40, 1179 (1979).

6. P. J. Viccaro, D. Niarchos, G. K. Shenoy, B. D. Dunlap and A. T. Aldred, to be published.

Published 1981 by Elsevier North Holland, Inc.
Kaufmann and Shenoy, editors
Nuclear and Electron Resonance Spectroscopies Applied to Materials Science

THE ANOMALOUS SPECTRAL DENSITY FUNCTION FOR DIFFUSIVE MOTION OF HYDROGEN
IN LaNi$_5$H$_7$*

H. CHANG and I. J. LOWE
Physics Department, University of Pittsburgh, Pittsburgh, PA 15260

R. J. KARLICEK, JR.[†]
Chemistry Department, University of Pittsburgh, Pittsburgh, PA 15260

ABSTRACT

Measurements of T_1 (B_o = 0.94T), and T_{1r} as a function
of temperature have been carried out on a LaNi$_5$H$_7$ sample
at 4 different rotating magnetic field values. The T_1
and T_{1r} data are consistent with earlier data by Karlicek
and Lowe [3,4], in which an asymmetry in the slopes of the
log T_{1r} vs. T^{-1} plot was found. The new data has been
analyzed assuming a spectral density function $J(\omega,\tau_c)$ of
form $J(\omega,\tau_c) = A(\tau_c)B(\Omega)F(\omega\tau_c)$, with $\tau_c = \tau_{c\infty} \exp(E_a/kT)$.
This assumption leads to a spectral density function that
fits <u>all</u> our data well, with E_a = 39 KJ/gm-Atom H, and
$J(\omega) \sim \omega^{-1.35}$ in the high frequency limit. This E_a agrees
well with the E_a obtained from diffusion constant measurements.

INTRODUCTION

Hydrogen motion in the intermetallic hydride β-LaNi$_5$H$_x$ ($x \sim 6$ to 7) has
been studied in recent years by both nuclear magnetic resonance [1-5] and
neutron scattering techniques [6]. The data of Karlicek and Lowe [3,4] could
be interpreted as suggesting the existence of two activation energies that
influenced hydrogen motion. They measured the diffusion constant of hydrogen
in LaNi$_5$H$_{6.5}$ (331°K to 375°K) using an alternating pulsed field gradient (APFG)
technique [5], yielding an activation energy E_a of 40 KJ/gm-Atom H in the
log D vs. T^{-1} plot. They also measured NMR relaxation times T_1 in the
laboratory frame (B_o = 1.4T) and T_{1r} in the rotating frame (B_1 = 1.0 mT)
from 77°K to 375°K. The log T_{1r} vs. T^{-1} plot was found to be asymmetrical
with E_a's of 39KJ/gm-Atom H on the short correlation time side of T_{1r} minimum
and 20KJ/gm-Atom H on the long correlation time side. A similar plot for
T_1, after the electronic contribution was removed, gave E_a of 20 KJ/gm-Atom H
in the long correlation time region. The observed T_1 and T_{1r} minima, occur-

*Supported by NSF Grant DMR 78-15441
†Now at Bell Telephone Laboratories, Murray Hill, N.J.

332

ring at $\omega_o \tau_c \approx 1$ and $2\omega_1 \tau_c \approx 1$ respectively, scaled according to E_a of
39 KJ/gm-Atom H.

   In order to investigate the asymmetrical non-BPP-type behavior of T_1
and T_{1r} in β-LaNi$_5$ hydride, we have carried out further measurements of
T_1 (B_o = 0.94T) and T_{1r} for four different rotating magnetic field values
from 120°K to 320°K. This data has been analyzed, using a frequency temperature
superposition technique [7], to fit one empirical spectral density function
of the form

$$J(\omega, \tau_c) = A(\tau_c) B(\Omega) F(\omega \tau_c) \qquad (1)$$

with
$$\tau_c = \tau_{c\infty} \exp(E_a/kT), \qquad (2)$$

and one activation energy. We report here the results of such an analysis.

NMR RELAXATION TIMES MEASUREMENTS

   β-LaNi$_5$ hydride used in this experiment was prepared in a sealed quartz
tube using the technique described by Karlicek [8]. Hydrogen concentration
was determined to be 7.0 H/La by the drop in pressure due to absorption of

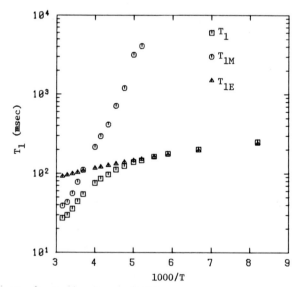

Fig. 1 T_1 (msec) vs. 1000/T
T_{1M} and T_{1E} curves are obtained from a
computer fit of experimental data, T_1,
to Debye function plus Koringa relation.

hydrogen by the sample. Grinding and transfer operations were conducted under an Ar atmosphere to prevent Ni contamination during hydrogenation; the sample underwent only one absorption.

A 40 MHz home-made pulsed NMR spectrometer [9] was used for our measurements. T_{1r} measurements were carried out at four rotating field values, 10.6G, 15.9G, 21.0G, and 26.7G. The data of the T_1 and T_{1r} measurements are plotted vs. 1/T in Fig. 1 and Fig. 2 respectively. Both the T_1 curve and the T_{1r} curve with B_1 = 10.6G are consistent with earlier data, which has only one T_{1r} curve at 10.6G.

T_1 was found to have two contributing mechanisms, electronic (T_{1E}) with a Koringa constant K = 29.1 ± 0.8 sec°K and motional (T_{1M}) with a slope of 21.0 ± 0.7 KJ/gm-Atom H in the long correlation time region. T_{1r} vs. 1/T curves are asymmetrical with slopes of 36 KJ/gm-Atom H on short correlation time side and 20 KJ/gm-Atom H on long correlation time side of T_{1r} minima. Field dependence was found to be $\omega_1^{-1.35}$ at low temperatures. T_{1r} minima are broad and the field dependence of T_{1r} starts before T_{1r} minima on the short correlation time side. In the temperature range from 180°K to 220°K the free induction decay displays two lines, one is motionally narrowed and has a shorter T_{1r} than the other [4,8]. T_{1r} in the long correlation time region was found to be non-exponential. Existence of two components can

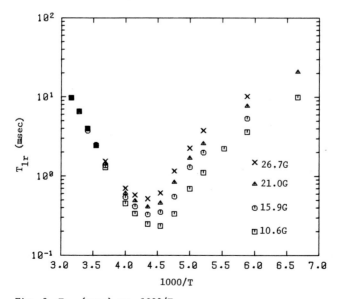

Fig. 2 T_{1r} (msec) vs. 1000/T
Experimental data at the four rotating magnetic field values, 10.6G, 15.9G, 21.0G and 26.7G.

explain the non-exponentiality in this temperature range, but not for lower temperatures, where only rigid lattice line shape was observed. T_{1r} for both components obeys about the same field dependence. Data for the rigid-lattice component were used for analysis.

EMPIRICAL SPECTRAL DENSITY FUNCTION

The well-known equations relating dipolar relaxation rates to spectral density functions for like spins are as follow [10],

$$\frac{1}{T_{1M}} = \frac{3}{2} \gamma^4 \hbar^2 I(I+1) \sum_k [J_{jk}^{(1)} (\omega_0) + J_{jk}^{(2)} (2\omega_0)] , \qquad (3)$$

$$\frac{1}{T_{1r}} = \gamma^4 \hbar^2 I(I+1) \sum_k [\frac{3}{8} J_{jk}^{(0)} (2\omega_1) + \frac{15}{4} J_{jk}^{(1)} (\omega_0) + \frac{3}{8} J_{jk}^{(2)} (2\omega_0)] . \qquad (4)$$

As long as T_{1r} is much shorter than T_{1M}, which is the case for β-LaNi$_5$ hydride over the temperature range of measurements, the contribution from the last two terms in Eq. (4) is negligible for a large enough B_0 field, and shall be dropped initially and treated as a small correction later on. The spectral density function we obtained from T_{1r}^{-1} data is $\sum_k J_{jk}^{(0)} (2\omega_1)$.

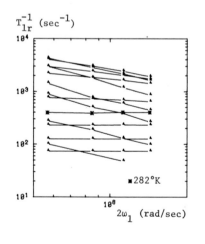

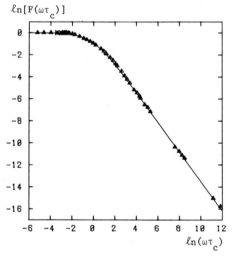

Fig. 3 T_{1r}^{-1} (sec^{-1}) vs. $2\omega_1$ (rad/sec)
Each set of four points connected by straight lines corresponds to data at a temperature. (*-282°K).

Fig. 4 $\ln[F(\omega\tau_c)]$ vs. $\ln(\omega\tau_c)$
Master curve obtained by translating sets of points in Fig. 3 relative to the set corresponding to 282°K. The smooth curve shows the function

$$\frac{1}{(1+\omega\tau_c)^{1.35}} .$$

For each temperature, the values of T_{1r}^{-1} were plotted against $2\omega_1$ on a log-log scale (Fig. 3). Each set of four points, corresponding to four rotating field values, could be translated both horizontally and vertically so as to lie on a master curve, which eventually contained all the points (Fig. 4). This procedure [7] is equivalent to fitting T_{1r}^{-1} data to a spectral density function of the form listed in Eqs. (1) and (2). $F(\omega\tau_c)$ describes the master curve, A is an amplitude function depending only on τ_c, B is a function needed to give the correct dimensionality, and Ω is a temperature-independent frequency, characteristic of the material. The shift factors are $\ell n(\tau_c/\tau_c^*)$ horizontally and $\ell n[A(\tau_c)/A(\tau_c^*)]$ vertically, where τ_c^* is the value of τ_c at the temperature relative to which sets of points were moved. Fig. 5 and Fig. 6 show $\ell n\ \tau_c$ vs. $1/T$ and $\ell n\ A$ vs. $\ell n\ \tau_c$ plots respectively, from which the value of E_a and the function $A(\tau_c)$ were determined.

$F(\omega\tau_c)$ has the following properties,

$$F(\omega\tau_c) = 1, \quad \omega\tau_c \ll 1; \quad \text{and} \quad F(\omega\tau_c) = (\omega\tau_c)^{-1.35}, \quad \omega\tau_c \gg 1.$$

We find that $F(\omega\tau_c)$ can be fitted reasonably well by $\dfrac{1}{(1 + \omega\tau_c)^{1.35}}$.

Thus the empirical spectral density function for β-LaNi$_5$ hydride can be

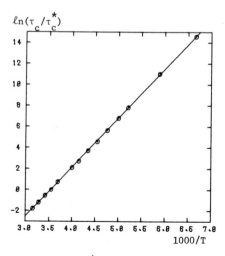

Fig. 5 $\ell n(\tau_c/\tau_c^*)$ vs. 1000/T

Horizontal shifting factors plotted against 1/T to get activation energy. The slope of the straight line is E_a = 39 KJ/gm-Atom H.

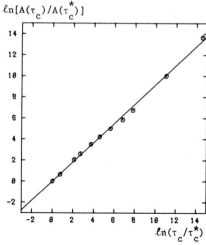

Fig. 6 $\ell n[A(\tau_c)/A(\tau_c^*)]$ vs. $\ell n(\tau_c/\tau_c^*)$

Vertical shifting factors plotted against horizontal shifting factors. The slope of the straight line is 0.92.

written as

$$J(\omega,\tau_c) = B(\Omega) \frac{\tau_c^{1-m}}{(1 + \omega\tau_c)^{2-n}} , \quad 0<m,n<1, \tag{5}$$

with one activation energy E_a. The parameters obtained from β-LaNi$_5$ H$_{7.0}$ data are listed in Table I. The apparent activation energies or slopes in T_{1r} vs. $1/T$ plot now become $(1-m)E_a = 36$ KJ/gm-Atom H on the short correlation time side and $(1-n+m)E_a = 17$ KJ/gm-Atom H on the other side as predicted by the spectral density function of Eq. (5).

The correction of T_{1r} due to T_{1M} was made by making the assumption of random isotropic motion which leads to $J^{(0)}/6 = J^{(1)} = J^{(2)}/4$ [11]. Defining a reduced spectral density function $J(\omega)$ from $J_{jk}^{(0)}(\omega) = \frac{4}{5} r_{jk}^{-6} J(\omega)$, Eqs. (3) and (4) can be rewritten as

$$\frac{1}{T_{1M}} = \frac{1}{3} M_2 \{J(\omega_0) + 4J(2\omega_0)\} , \tag{6}$$

$$\frac{1}{T_{1r}} = \frac{1}{2} M_2 \{J(2\omega_1) + \frac{5}{3} J(\omega_0) + \frac{2}{3} J(2\omega_0)\} , \tag{7}$$

where $M_2 = \frac{3}{5} \gamma^4 \hbar^2 I(I+1) \sum_k r_{jk}^{-6}$ is the powder average second moment.
Using the $\omega^{-1.35}$ dependence for $J(\omega)$ in the high frequency limit to account for $J(2\omega_0)$ terms, we obtained

$$\frac{1}{2} M_2 J(\omega_0) = .58 (\frac{1}{T_{1M}}) , \tag{8}$$

$$\frac{1}{2} M_2 J(2\omega_1) = \frac{1}{T_{1r}} - 1.12 (\frac{1}{T_{1M}}) . \tag{9}$$

The correction term in Eq. (9) is small at low temperatures as expected. The values of $J(\omega_0)$ were found to fall on the master curve well when being shifted according to the same E_a for T_{1r}.

Table I.

E_a		39 ± 3	KJ/gm-Atom H	Fig. 5
m		0.08 ± 0.02		Fig. 6
n		0.65 ± 0.02		Fig. 4
$\tau_c^*(282°K)$		5.7×10^{-8}	sec	
$\tau_{c\infty}$		3.7×10^{-15}	sec	
T_1 slope $\omega\tau_c \gg 1$		21.0 ± 0.7	KJ/gm-Atom H	Fig. 1
K		29.1 ± 0.8	sec · °K	
T_{1r} slope $\omega\tau_c \ll 1$		36 ± 1	KJ/gm-Atom H	Fig. 2
slope $\omega\tau_c \gg 1$		20 ± 2	KJ/gm-Atom H	

DISCUSSION

The presence of unequal slopes on the two sides of the T_{1r} minimum had previously been interpreted in terms of two separate activation energies governing two types of hydrogen motion [4]. This explanation is supported by the existence of two separate hydrogen lines which undergo motional narrowing at different temperatures, and two inequivalent hydrogen sites in β-LaNi$_5$ hydride as shown in the neutron scattering results for β-LaNi$_5$ deuteride [6]. This explanation, however, must be discarded because it leads to an ω^2 dependence of the spectral density function in the high frequency limit, and fails to justify the observed deviation from ω^2 dependence. Using scaling theory, an empirical spectral density function (5) is obtained by temperature-frequency superposition technique that gives one activation energy E_a, which is in good agreement with E_a obtained from diffusion constant measurements, and that yields asymmetrical T_{1r} vs. $1/T$ curves with non-quadratic frequency dependence. What remains to be explained is the origin of such a peculiar spectral density function.

The non-BPP type of relaxation behavior observed for β-LaNi$_5$ hydride is by no means unique. Similar type behavior has been found in NMR relaxation times measurements for other metal hydrides [7], β-aluminum [12], hydrogen-bonded liquids like glycerol [13], etc., and also in dielectric relaxation measurements for a wide range of materials [14]. Distribution of activation energies or correlation times [12] has been introduced to modify single BPP spectral density function to fit experimental data. However, no physical model can justify any particular distribution function, and no distributions can successfully account for the entire frequency behavior. In explaining their T_1 data for glycerol, Fiorito and Meister [13] proposed a model for translational relaxation, in which each spin makes many jumps before it can effectively relax other spins. They have derived a post-peak $\omega^{3/2}$ dependence of T_1 in this diffusive fluid-like limit, while in the jump limit the BPP spectrum is recovered. Ngai et al [14,15] has carried out studies in finding a mechanism of dielectric relaxation to account for what they recognized as universality of dielectric response. This theory is associated with the existence of correlated states and their infrared divergent response to the transitions induced by time varing Hamiltonian. It has been extended to NMR relaxation, mechanical relaxation, and 1/f noise. However, the existence of such correlated states in β-LaNi$_5$ hydride needs to be justified.

CONCLUSION

We conclude that the NMR relaxation mechanism of β-LaNi$_5$ hydride can not be described by simple BPP theory. There is a mixing of short correlation time behavior into the long correlation time region such that the frequency dependence of the spectral density function is less quadratic in the high frequency limit. The activation energy for this empirical spectral density function agrees well with that from diffusion constant measurements. We are now looking for a model of hydrogen motion in β-LaNi$_5$ hydride that can give rise to this spectral density function.

338

REFERENCES

1. T. K. Halstead, J. Solid State Chem., 11 (1974), 114.

2. T. K. Halstead, N. A. Abood and K. H. J. Buschow, Solid State Commun., 19 (1976), 425.

3. R. F. Karlicek, Jr. and I. J. Lowe, Solid State Commun., 31 (1979), 163.

4. R. F. Karlicek, Jr. and I. J. Lowe, J. Less Common Metals, 73 (1980), 219.

5. I. J. Lowe and R. F. Karlicek, Jr., J. Mag. Resonance, 37 (1980), 75.

6. P. Fisher, A. Furrer, G. Busch and L. Schlapbach, Helv. Phys. Acta, 50 (1976), 421.

7. T. C. Jones, T. K. Halstead and K. H. J. Buschow, J. Less Common Metals, 73 (1980), 209.

8. R. F. Karlicek, Jr., Thesis, University of Pittsburgh.

9. I. J. Lowe and R. F. Karlicek, Jr., J. Mag. Resonance, 32 (1978), 199.

10. A. Abragam, The Principles of Nuclear Magnetism, Oxford University Press, London, 1961.

11. N. Bloembergen, E. M. Purcell and R. V. Pound, Phys. Rev., 72 (1948), 679.

12. R. E. Walstead, R. Dupree, J. P. Remeika and A. Rodriguez, Phys. Rev. B, 15 (1977), 3442.

13. R. B. Fiorito and R. Meister, J. Chem. Phys. 56 (1972), 4605.

14. K. L. Ngai, Comments Solid State Phys., 9 (1980), 127, 141.

15. K. L. Ngai and C. T. White, Phys. Rev. B, 20 (1979), 2475.

Semiconductors

Published 1981 by Elsevier North Holland, Inc.
Kaufmann and Shenoy, editors
Nuclear and Electron Resonance Spectroscopies Applied to Materials Science

PROTON NMR SPIN-LATTICE RELAXATION TIME CHARACTERIZATION OF a-Si(H) STRUCTURE

M. E. LOWRY, R. G. BARNES, D. R. TORGESON, AND F. R. JEFFREY
Ames Laboratory-USDOE* and Department of Physics, Iowa State
University, Ames, Iowa 50011

ABSTRACT

NMR data are presented for reactively sputtered amorphous
silicon-hydrogen alloys (a-Si(H)). Measured differences in
two of the samples are attributed to two distinct morphologies: a mixed phase (monohydride and dihydride) and a purely monohydride composition. Features of the mixed phase
morphology have been modeled. Room temperature, 35 MHz
spin-lattice relaxation times are presented for a series of
monohydride samples prepared with systematically varied
sputtering parameters. A correlation of proton T_1 with the
density of ESR states tentatively is suggested.

*Operated for the U.S. Department of Energy by Iowa State
University under contract No. W-7405-Eng-82. This research
was supported by the Director for Energy Research, Office
of Basic Energy Sciences, WPAS-KC-02-02-02.

INTRODUCTION

Electron Spin Resonance (ESR) spectroscopy has been a fundamental tool in
probing the structure of amorphous silicon-hydrogen alloys [a-Si(H)] for several years [1]. However, Nuclear Magnetic Resonance (NMR) is relatively new to
the field [2]. One of the more important triumphs of ESR spectroscopy in defining the structure of a-Si(H) has been its role in unmasking the function of the
incorporated hydrogen, which act to satisfy "dangling bonds" or defects in the
ideal continuously connected, tetrahedrally coordinated network of pure a-Si [3].
This discovery led to the now commonly held belief that glow discharge (GD)
a-Si(H) and the reactively sputtered (in the presence of H) material were essentially the same. However, there seems to be at least one difference between
the two materials. Infrared transmission spectroscopy has identified two main
classes of bonding type for the incorporated hydrogen: monohydride and dihydride [4,5]. Using the sputtering process one is able to reliably control the
type of bonding [5]. Verification of the existence of dihydride and monohydride configurations by NMR techniques, addressed by at least one group of
workers [2], forms a major portion of this communication.

THEORY AND METHOD

Our NMR measurements have concentrated on the spin-lattice relaxation time T_1
of the protons in various a-Si(H) samples. $(T_1)^{-1}$ represents the rate of transfer of absorbed rf energy from the proton spin system to the "lattice". This
decay of energy content of the spin system can be represented by the Bloch
equation [6]:

$$dM_z(\tau)/d\tau = [M_0 - M_z(\tau)]/T_1 \qquad (1)$$

here, $M_z(\tau)$ represents the component of the nuclear magnetization parallel to
the static field H_0 and M_0 represents the value of M_z at equilibrium. The

particular method used to measure T_1 will, of course, determine the boundary conditions to be satisfied by Eq. (1). We have utilized the inversion recovery (IR) method which uses a 180, τ, 90 NMR pulse sequence [7] and dictates the initial condition that $M_z(\tau=0)=-M_0$. This leads to the solution $M_z= [1-2\exp(-\tau/T_1)]$. The spin-lattice relaxation times are then extracted from the slopes of $\ln[M_0-M_z(\tau)]$ vs. τ data sets. This analysis has been accomplished using an on-line computer employing a linear regression program.

In samples which have several different proton-lattice couplings, the simple analysis given above is inadequate. For the case of two protonic environments, using the IR method, the longitudinal magnetization becomes

$$M_z = M_{0a}[1-2\exp(-\tau/T_{1a})] + M_{0b}[1-2\exp(-\tau/T_{1b})] \tag{2}$$

where the subscripts a and b refer to the two kinds of spin-lattice couplings.

The rf absorption spectra were obtained using the method of Fourier transformation [7] of an off-resonance free induction decay (FID), instead of the more familiar continuous wave NMR technique. The pulse method was employed because the large value of T_1 observed makes accumulation of direct (cw) rf absorption spectra very time consuming. The Fourier transformations have been accomplished using a computer on-line with the NMR pulse spectrometer and a standard Fast Fourier Transform program. These data will be referred to by the acronym "FFTFID".

It is important to know how many protons are in each sample as a measure of the hydrogen concentration. To this end, proton spin counts were made by measuring the equilibrium magnetization M_0 for each sample. Comparing M_0 for each a-Si(H) sample with M_0 for a standard sample of water yields the approximate number of protons in each of the a-Si(H) samples.

DATA

The most significant results are perhaps the spin-lattice relaxation times from sample C-38 at room temperature compared with the corresponding data from sample C-41. Sample C-38 has been characterized by infrared transmission spectroscopy [4] as containing both monohydride and dihydride bond configurations (i.e., mixed phase). Sample C-41 indicates only monohydride bonds in the ir spectra. Figure 1 illustrates a typical semi-log plot of $M_0-M_z(\tau)$ vs. τ, for C-38. The non-linearity of the plot indicates more than one relaxation rate is present in the sample. Equation (2) yields a reasonable fit. The parameters of the fit were obtained by an asymptotic linear regression analysis. Linearity was invoked by considering Eq. (2) in the following asymptotic regions of time:

$$\tau \ll T_{1b} \qquad \ln(B_1-M_z) \approx \alpha - \tau/T_{1a} \quad , \tag{3a}$$

$$\tau \gg T_{1a} \qquad \ln(B_2-M_z) \approx \beta - \tau/T_{1b} \quad . \tag{3b}$$

Where $B_1=M_{0a}-M_{0b}$, $B_2=M_{0a}+M_{0b}$; α, β are constants. A "window" of time is chosen and the data is fit linearly using either Eq. (?a) or (3b). As the "window" width is diminished the approximations become better and the parameters then asymptotically approach their proper values. T_{1a} must be significantly less than T_{1b}, for the method to succeed. By repeating this fitting procedure on several data sets for sample C-38, we have determined that 69±11% of the protrons are in short T_1 configurations (with $T_{1a}=1.4\pm0.5$ sec). The remaining protons are in bonding configurations with a long relaxation time ($T_{1b}=8.8\pm1.0$ sec). In contrast, sample C-41 is characterized by a single (to within 5%) proton spin-lattice relaxation rate $T_1=35\pm5$ sec.

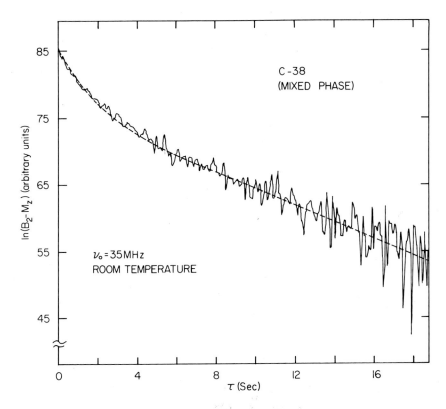

Fig. 1. The solid curve is a logarithmic plot of the total baseline (B) minus the magnetization (M_z) vs. time for an inversion recovery T_1 experiment for sample C-38. The non-linear nature of this plot is typical of a sample having more than one T_1. The dashed line is a fit assuming only two distinct T_1's (as in Eq. 2).

To further characterize these two samples (C-38 and C-41), FFTFID data have also been accumulated. The spectrum for C-38 was fit with a wide Gaussian shape function (FWHM=24.9 kHz, 87% of the intensity) and a narrow Gaussian (FWHM=4.47 kHz, 13%). The FFTFID spectrum of C-41 has been fit with a wide Gaussian shape function (FWHM=26.0 kHz, 74%) and a narrow Lorentzian (FWHM= 4.43 kHz, 26%). These FFTFID spectra were both taken at 300°K and 35 MHz. The resonances exhibited no shifts, within experimental error. Using the spin counting method described above, these "sample-wide average" proton densities were obtained: $(5.3\pm1.8)\times10^{21}$ protons/cm^3 and $(3.2\pm1.2)\times10^{21}$ protons/cm^3 for C-38 and C-41 respectively. In addition to the NMR studies, we have measured the integrated ESR spectra for C-38 and C-41 samples at low microwave power (to avoid saturation). Sample C-38 shows 80 times the concentration of unpaired electron spins compared with sample C-41.

In addition to characterizing a-Si(H) alloys by bonding type using NMR and ESR techniques, we have begun to study the proton resonance of strictly monohydride materials as a function of one sputtering parameter. Nominally, we have sought to vary only the hydrogen gas flow rate which affects the partial pressure of H in the sputtering chamber (hence the plasma chemistry of the deposition process itself), and observe the effect this has on the NMR characterization of the protonic environment. Table 1 presents the preliminary results of these experiments. The room temperature T_1 values indicate a maximum around 13 c.c./min H_2 flow rate. Presently, we have ESR data on sample C-41. ESR data from the remaining samples will be of importance in determining the proton relaxation mechanism.

TABLE I.
Room temperature proton T_1's and "sample-wide average" proton densities (from NMR measurements) are presented for reactively sputtered a-Si(H) of various H_2 sputtering flow rates. A single T_1 is evident for each sample; infrared data indicates solely monohydride composition for each.

SAMPLE	T_1 (sec, ±15%)	H DENSITY ($10^{21}/cm^3$, ±5%*)	H_2 FLOW RATE (c.c./min, ±5%)
C-53	8.1	4.9	36
C-70	22.3	3.9	24
C-41	35.0	3.2	25
C-51	44.8	3.1	13
C-52	14.7	1.8	5.5
C-45	2.8	1.0	3.0

*Relative uncertainty. Relative densities are more precise than absolute densities.

ANALYSIS AND DISCUSSION

It is possible to relate the second moment [6], M_2, of a component of a FFTFID spectrum with the spatial density of protons that give rise to that component. The van Vleck formula [6] for the second moment provides that $M_2 \sim \sum_{ij} (r_{ij})^{-6}$.
We may approximate this sum for an amorphous material by relating the nth nearest neighbor distance, R_n, to the density of protons ρ_H. It is easily seen that $4\pi R_n^3/3 = (n+1)/\rho_H$, where the approximation is better for higher n values. This leads to $M_2(kHz^2) = 7.28 \times 10^{-44} \rho_H^2 (cm^{-3})$ (5). Using Eq. (5) and the "sample wide" average spin density for either C-38 or C-41 the predicted second moments are at least two orders of magnitude less than those calculated directly from the FFTFID spectra. This implies that substantial inhomogeneity or "clustering" exists in the protonic spatial distribution, as has been suggested [2], for both the mixed phase sample (C-38) and the monohydride (C-41). Application of Eq. (5) to the Gaussian component of the C-38 spectrum yields proton densities of 39 protons per $10^3(\text{Å})^3$ and 7 protons per $10^3(\text{Å})^3$ for the wide and narrow components, respectively. Analysis of FFTFID data for C-41 indicate the Gaussian component originates from protons in clusters of average densities of 40 per $10^3(\text{Å})^3$. Care must be taken with the Lorentzian component, since M_2 is not defined for a Lorentzian function. We have used the sample spin count, the relative areas of the two components, and the volume of the sample to calculate the average proton density of 0.8 protons per $10^3(\text{Å})^3$ for the narrow component of the C-41 FFTFID spectrum. Preliminary temperature measurements from 77 to 300K indicate no appreciable narrowing of the narrow component of the C-41 FFTFID with increasing temperature. Motional narrowing is apparently not the cause of the narrow line shape. Calculations by Myles, et al. [8] yield line shapes of NMR resonances for systems of strongly interacting, randomly

distributed spins. In the dilute limit these lineshapes do approximate a
Lorentzian shape. The C-41 FFTFID also yields "extra" intensity in the wings
[11].

With this interpretation, consistent models for the protonic structure in both
samples (C-38 and C-41) emerge. For the monohydride (C-41) some protons are
randomly distributed (26%) while the rest (74%) reside in some type of clusters
(probably the internal surfaces of microvoids). The two component spin-lattice
relaxation observed in C-38 and the FFTFID spectrum [9] peculiar to this mixed
phase material may be explained by a model in which columns of monohydride
material have grown, via a process of nucleation during sputtering, amidst a
matrix of $(SiH_2)_n$ chains [10]. The protons in the dihydride matrix are the
source of the wide Gaussian FFTFID component and the shorter relaxation time
($T_{1a}=1.4$ sec). The protons in the monohydride islands account for the narrow
resonance and the longer relaxation time ($T_{1b}=8.8$ sec). T_{1b} is, however, considerably shorter than the observed relaxation time for the isolated monohydride material. This apparent discrepancy can be resolved if one considers
spin diffusion [6]. The spin energy of the monohydride protons may either
relax to local centers or diffuse to the faster relaxing centers in the dihydride phase. Relaxation rates are additive. Hence, we find $T_{1diff}=11.8$ sec.
We show in Ref. [11] that this spin diffusion limiting of the cross-relaxation
of the monohydride protons to the faster relaxing dihydride protons allows us
to estimate the average diameter of the monohydride columns to be 140 Å. This
value agrees well with electron microscopy observations [10] that have been
reported for GD a-Si(H).

Knights, et al. [12] reported an increase of the ESR intensity with increasing H
content after a certain level of H incorporation has been reached. Knight's
work, our scaling of proton T_1's with ESR spin density in samples C-38 and C-41,
and our T_1 data vs. hydrogen concentration given in Table I lead to a plausible
model in which the proton T_1's may be correlated with ESR-active electron
concentration. Whether the electrons are actually the source of the protonic
relaxation, or the presence of the electronic states is a result of protonic
"disorder modes," as suggested by Carlos and Taylor [13], remains to be determined through more extensive temperature and frequency studies of the relaxation
processes. We note that our preliminary temperature, T_1 data do not contradict
the model of Carlos and Taylor.

CONCLUSIONS

We have presented NMR evidence for the existence of two phases of a-Si(H) in the
same sample. By correlating the resonance data with ir data we were led to con-
struct a model in which a dihydride phase surrounds columnar regions of the
monohydride phase. Further, we were able to determine the size of these
columns, in good agreement with other observations [10].

We have also suggested the possibility of a positive correlation of the protonic
relaxation rate with the ESR spin concentration. However, validation of this
proposed correlation must await completion of our ESR studies of these same
samples. A more complete understanding of the proton relaxation mechanisms in
our several a-Si(H) samples will await completion of a detailed study of the
temperature and frequency dependence of the proton spin-lattice relaxation times.
In view of the well established spin-dependent recombination phonomenon [14], it
is probable that the observed ESR electrons are associated with the short photo-
carrier lifetime problem in a-Si(H) solar cells. Thus, any probe (such as proton relaxation) which can shed light on these electronic states deserves
thorough investigation.

346

ACKNOWLEDGMENTS

It is a pleasure to thank Dr. M. L. S. Garcia for her much needed assistance with many aspects of this work. One of the authors (M.E.L.) thanks Dr. F. Borsa for several beneficial conversations regarding interpretation of relaxation processes in these materials.

REFERENCES:

1. M. H. Brodsky and R. S. Title, Phys. Rev. Lett. 23, 581 (1969).

2. J. A. Reimer, R. W. Vaughan, and J. C. Knights, Phys. Rev. Lett. 44, 193 (1980).

3. E. A. Davis, in Topics in Applied Physics: Amorphous Semiconductors (Springer-Verlag, 1979), p. 48.

4. G. Lucovsky, R. J. Nemanich, and J. C. Knights, Phys. Rev. B 19, 2064 (1979).

5. F. R. Jeffrey, H. R. Shanks, and G. C. Danielson, J. Appl. Phys. 50, 7034 (1979).

6. A. Abragam, The Principles of Nuclear Magnetism (Oxford, 1961).

7. T. C. Farrar and E. D. Becker, Pulse and Fourier Transform NMR (Academic Press, 1971).

8. C. W. Myles, C. Ebner, and P. A. Fedders, Phys. Rev. B 14, 1 (1976).

9. For the monohydride columns surrounded by dihydride regions in the mixed phase sample, it may have been appropriate to use three components for the fit, i.e., a narrow Lorentzian and wide Gaussain for the monohydride and a wider Gaussian for the dihydride phase. We believed this to be too many parameters to be justified by the S/N of these spectra.

10. J. C. Knights and R. A. Lujan, Appl. Phys. Lett. 35, 244 (1979).

11. M. E. Lowry, F. R. Jeffrey, R. G. Barnes, and D. R. Torgeson, to be published in Sol. State Commun.

12. J. C. Knights, G. Lucovsky, and R. J. Nemanich, J. Non-Cryst. Sol. 32, 393 (1979).

13. W. E. Carlos and P. C. Taylor, Phys. Rev. Lett. 45, 358 (1980).

14. I. Solomon, in pp. 196-207 of Ref. 3.

Published 1981 by Elsevier North Holland, Inc.
Kaufmann and Shenoy, editors
Nuclear and Electron Resonance Spectroscopies Applied to Materials Science

PROTON NMR STUDIES OF AMORPHOUS PLASMA-DEPOSITED FILMS

JEFFREY A. REIMER[*] and ROBERT W. VAUGHAN[**]
Division of Chemistry and Chemical Engineering, California Institute of
Technology, Pasadena, California 91125

JOHN C. KNIGHTS
Xerox Palo Alto Research Center, 3333 Coyote Hill Road, Palo Alto, California
94304

ABSTRACT

Proton magnetic resonance data are presented for the hydrogen alloys of plasma-deposited amorphous boron, silicon,
carbon, silicon carbide, and silicon nitride. Linewidth
and lineshape analysis leads to the conclusion that
hydrogen nuclei are clustered in a-Si/C:H, a-C:H, and
a-Si/NiH. The a-Si/C:H and a-C:H data show that hydrogen
exists in two phases. Modeling of linewidths in a-Si/C:H
indicates that the two phases are heavily hydrogenated
carbon clusters imbedded in a weakly hydrogenated silicon
lattice. Evidence is also presented for the presence of
motionally narrowed hydrogen spectra in a-B:H, a-Si/N:H
and a-C:H. "B NMR spectra in a-B:H show no evidence of
motional narrowing. It is suggested that the hydrogen
nuclei giving rise to the motionally narrowed spectra
are associated with disorder modes.

INTRODUCTION

Hydrogenated amorphous thin films have attracted a great deal of attention
recently because of their desirable electronic and/or optical properties.
Plasma-deposited silicon nitride films have been utilized successfully as
passivation layers in high reliability silicon-integrated circuits (1). The
remarkable properties of hydrogenated amorphous silicon (a-Si:H) vis-à-vis
pure amorphous silicon has lead to the development of p,N and Schottky
barrier junctions as well as moderately efficient solar cells (2). The role
of hydrogen in these films is not well understood although it is generally
accepted that hydrogen passivates dangling bond intrinsic defects and thus
decreases the density of states in the gap (2).
Recent proton magnetic resonance results have posed some new questions on
the role hydrogen plays within plasma-deposited a-Si:H. Lineshape results
(3) have shown a-Si:H is characterized by two domains which differ in the
local density of hydrogen. Spin-lattice relaxation measurements (4) were
recently explained by postulating the existence of hydrogen-containing
disorder modes that may contribute to the electronic states in the gap. It
is the purpose of this paper to show proton magnetic resonance spectra of
amorphous boron (a-B:H), carbon (a-C:H), silicon carbide (a-Si/C:H), silicon
nitride (a-Si/N:H). Lineshape and linewidth analysis of these data furnish
new information on the spatial homogeneity of these films as well as yield
new evidence for hydrogen-containing disorder modes in a-B:H, a-Si/N:H,
and a-C:H.

EXPERIMENTAL

The samples were prepared by an rf-diode deposition system described previous-

ly (5). All samples were deposited onto ~ 2-inch diameter aluminum foil substrates in thicknesses of 1-100µ resulting in sample masses of 5-100 milligrams
after removal of the substrates with dilute acid etch. Analysis of the
a-Si/N:H and a-Si/C:H samples was done by Rutherford backscattering. Proton
magnetic resonance spectra were taken with a Fourier transform solid state
spectrometer (6) operating at 56.4 Mhz (1.3 Tesla). The NMR spectra were
obtained by Fourier transforming 1,000 averages of the free induction decay
(FID) following a 90° pulse. Spin-lattice relaxation times (T_1) were determined by the inversion-recovery method (7). The "B spectrum
was obtained on a standard pulse spectrometer operating at 86.7 Mhz (6.3
Tesla). Only the $\frac{1}{2} \to -\frac{1}{2}$ transistion is shown.
To obtain the data shown in Tables 1 and 2 all proton FID spectra were least
squares fitted assuming the broad component to be Gaussian and the narrow
component Lorentizian which gave values for the linewidths and areas of the
two components and the errors were estimated from the reproducibility of the
fits resulting from a variety of initial starting parameters.

RESULTS

Figure 1 shows a comparison of spectra at room temperature and 125 K for the
samples described in Table 1. The observed linewidths reflect contributions

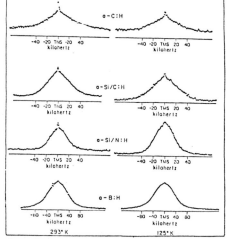

Figure 1: Proton free-induction decay. spectra for, from top to bottom,
amorphous carbon, silicon carbide, silicon nitride, and boron
(cathode). At left are spectra at room temperature, at right, 125K.

from (i) homonuclear dipolar interactions, (ii) heteronuclear dipolar interactions, (iii) chemical shift interactions. When multiple pulse cycles (8-12)
were applied (which removes the homonuclear dipolar broadening), the lines
narrowed significantly. Thus, for a-C:H, a Si/C:H and a-Si/N:H, the homonuclear dipolar interaction is the dominant broadening mechanism in the
proton spectra. In a-B:H, the heteronuclear dipolar interaction due to nearby ^{10}B and ^{11}B nuclei was estimated (15) and found to be as large or larger
than the homonuclear dipolar interaction.
The spectra for all the samples may be decomposed into at least two components.
Multicomponent lineshapes in these films may arise from the following
situations: (i) motion in a domain or at a defect site which decouples spin-

spin communication (15) with the rest of the dipolar reservoir; (ii) the existence of spatially isolated ($\geq$8 A (14)) dipolar reservoirs with differing densities of spins such that there is no spin-spin communication between them; or (iii), both (i) and (ii). The first case has been observed in partially crystalline polyethylene (20) and case (ii) in amorphous silicon hydrogen films (3). Effects on the lineshape due to motion may often be removed by reducing the temperature; hence, cases (i) and (ii) may be readily distinguish-ed. Inspection of Figure 1 and Table 2 shows that at low temperature, where motional narrowing is not expected, both a-C:H and a-Si/C:H spectra still exhibit two-component behavior. In a-Si/N:H, a-B:H, and a-C:H a small fraction of the protons have lineshapes that do appear to be motionally narrowed at room temperature. However, this fraction accounts for no more than 3% of the total proton spins. Hence, we conclude that a-Si/C:H is best described by. case (ii), a-B:H and a-Si/N:H by case (i), and a-C:H by case (iii). The low-temperature spectra for a-B:H and a-Si/N:H show no evidence for case (ii); how-ever, we shall present evidence suggesting that the hydrogen nuclei in a-Si/N:H are clustered. In a-B:H we estimate (15) the heteronuclear dipolar interaction to be 70-120 khz; hence, no structural information is expected from proton lineshapes.

TABLE 1

SAMPLE		HYDROGEN DISTRIBUTION				T_1 (Seconds)	LINE WIDTH (FWHM, kHz)	
		%	Atom %	%	Atom %		Broad	Narrow
a-$C_{65}H_{35}$	293°K	93.5	32.7±.4	6.5	2.3±.3	0.4±.1	78.4±2.6	1.6±.7
	130°K	91.1	32.1±.4	8.9	3.1±.4		76.8±2.8	6.6±.2
a-$B_{75}H_{25}$ (cathode)	293°K	98.9	24.8±.1	1.1	0.3±.1	4.5±.3	118±3	1.8±.1
	124°K	98.7	24.8±.1	1.3	0.3±.1		121±3	11.7±.6
a-$B_{79}H_{21}^{(a)}$ (anode)	293°K	98.5	20.7±.1	1.5	0.3±.1	5.1±.3	83.7±1.4	3.6±.7
a-$Si_{31}N_{46}H_{21}$	293°K	97.0	22.3±.2	3.0	0.7±.2	4.6±.4	35.1±.3	2.2±.1
	124°K	97.8	22.5±.2	2.2	0.6±.2		55.6±.2	6.0±.2
a-$Si_{20}C_{57}H_{23}$	293°K	94.5	21.5±.3	5.5	1.2±.3	0.2±.05	69.0±.3	7.4±.3
	126°K	95.3	21.6±.3	4.7	1.1±.3		78.4±.4	7.0±.3
a-$Si_{26}C_{36}H_{38}^{(a)}$	293°K	98.1	37.5±.3	1.9	0.7±.3		45.6±.4	1.9±.2

(a) Spectra not shown in Figure 1.

Further insight on the distribution of hydrogen can be derived from analysis of the low-temperature data shown in Figure 1 and Table 1 according to models for local proton configurations. The homonuclear and heteronuclear dipolar interactions in polycrystalline or amorphous solids give rise to Gaussian lineshapes whose second moments are given by, respectively (15)

$$M_2^{II} = 3/5 \, \gamma_I^4 h^2 I \, (I + 1) \sum_j 1/r_{ij}^6 \qquad (1)$$

$$M_2^{IS} = 4/15 \, \gamma_I^2 \gamma_S^2 h^2 S \, (S + 1) \sum_j 1/r_{ij}^6$$

where λ_I is the nuclear gyromagnetic ratio for a nucleus of spin I and r_{ij} is the internuclear distance between spins i and j. For a-C:H, a-Si/C:H, and a-Si/N:H two types of proton configurations to which Equation 1 simply applies have been considered. The first is one of local clustering into configuration such as CH_3, $(CH_2)_n$, $(SiH_2CH_2)_n$, $SiNH_2$, and a hydrogenated $[111]$ c-carbon

surface. The corresponding Si-H clustering has been calculated previously (3).
The results of these calculations and an estimate (16) of the linewidth (FWHM)
for isolated CH_2 are as follows: CH_2, 33.1 khz (SiH_2, 13.5 khz); CH_3, 46.9 khz
(SiH_3, 19.2 khz); $(CH_2)_n$, 44.1 khz $(SiH_2)_n$, 17.1 khz); $(SiH_2CH_2)_n$, 36.1 khz;
and $C[111]$, 57.4 khz ($Si[111]$, 23.5 khz) and $SiNH_2$, 37.8 khz. Given the
sensitivity of the calculation to small changes in internuclear distances
and the errors involved in the fitting procedures, we conclude that any of the
local configurations on carbon could give rise to the broad component in both
a-C:H and a-Si/C:H spectra. Since the width of the broad component is >45 khz
in both of the a-Si/C:H samples and the broadest component expected for protons
clustered on silicon is 23.5 khz and on $(SiH_2CH_2)_n$ is 36.1 khz, we conclude that
for silicon carbide the broad line is due to predominantly "polyacetylene" type
environments. This is especially true for the higher-carbon-content sample
where the observed broad component linewidth is identical to that of amorphous
carbon. This implies that the spatial inhomogeneity in the protons is reflective
of carbon clusters which are heavily hydrogenated imbedded in a weakly hydrogenated
amorphous silicon lattice.

TABLE 2

SAMPLE	ATOM % NARROW	FWHM NARROW (kHz)	HOMOGENEOUS[a] WIDTH (kHz)
a-C:H[b]	3.1	6.6	1.7
a-Si$_{.26}$C$_{.36}$H$_{.38}$	0.7	1.9	0.4
a-Si$_{.20}$C$_{.57}$H$_{.23}$	1.2	7.0	0.6
a-Si$_{.31}$N$_{.46}$H$_{.23}$	23.0[c]	55.6	12.9

(a) Calculated with use of Eq. (1).

(b) The low temperature values were chosen to eliminate motion effects.

(c) Total hydrogen content.

The second situation to which Equation 1 may be applied simply is one where the
spins are distributed uniformly on a cubic lattice whose spacing is equal to the
inverse cube root of the spin density. The results of applying this model
for Equation 1 to the narrow component of the a-Si/C:H and a-C:H lineshapes as
well as all the protons in a-Si/N:H, are clustered. Local clustering of hydrogen
on nitrogen atoms is sufficient to explain the observed linewidth in a-Si/N:H.
However, in a-Si/C:H and a-C:H the clustering in the narrow component is not
dominated by local multiply bonded species such as SiH_x or CH_x since those species yield much larger linewidths. We conclude that the narrow component is
composed primarily of clustered monohydride species (CH or SiH).
There are several implications of these results from a-Si/C:H and a-C:H for
models of microstructure and optimcal properties. Microstructure in a-C:H
films prepared from ethylene, acetylene, and styrene (21-23) has been observed
by SEM, TEM, and low-angle x-ray scattering. In those studies it was concluded
that there are two phases, specifically, spheres imbedded in a polymer binder.
It may be possible to identify the broad component in the FID spectra of a-C:H
with those protons in the polymer binder and the narrow component with monohydride clustering on internal spheres. However, more detailed experiments are
needed to confirm these suggestions.
The observation of carbon clusters in a-Si/C:H is consistent with recent
photoluminescence results (24) on films prepared from SiH_4/C_2H_4 mixtures. In
those data the visible photo-luminescence consists of two bands which change

in intensity with changing composition. The relatively low hydrogen content of the silicon-rich regions of these films (<1.5 atom %, from Table 2) represents a hydrogen content not accessible by plasma deposition of SiH_4 to form a-Si:H. Thus, the optical properties of plasma deposited a-Si/C:H may prove useful in the study of a-Si:H. Furthermore, the heavy hydrogen clustering for the carbon-rich regions may have implications in film growth and/or the chemistry of the plasma above the films.

The room temperature spectra of a-C:H, a-Si/N:H and a-B:H all show a small narrow component that increases in width as the temperature is decreased. Since these components do not decrease in intensity with extended evacuation of the samples, they are not due to adsorbed gases. We conclude that they are motionally narrowed (8) hydrogen nuclei which are indigenous to the films. Estimates of the dipolar broadening, including the effects of motion (8,17), in rotating local bonding configurations such as methyl groups yield values much larger than those observed. The central transition of the "B spectra for a-B:H is a typical dipolar-broadened, second-order quadrapular powder pattern and shows no evidence for motionally narrowed "B nuclei. We suggest that the motionally narrowed proton lines correspond to hydrogen nuclei rapidly moving in a disorder mode. The existence of hydrogen-containing disorder modes has recently been proposed (4) in a-Si:H in order to explain the temperature and field dependence of proton spin-lattice relaxation (T_1) as well as other relaxation measurements (25). Their presence in a-C:H, a-Si/N:H, and a-B:H indicates that such modes may be ubiquitous to hydrogen-containing amorphous materials. The relationship between these modes and defects affecting optical properties warrants further investigation.

In conclusion, we have used magnetic resonance results to show that for a series of plasma-deposited inorganic thin films, hydrogen is spatially inhomogeneous on two levels. The first consists of hydrogen clustering which we have observed in a-C:H, a-Si/C:H, and a-Si/N:H. The second, observed in a-C:H and a-Si/C:H, we conclude to be a two-phase compositional inhomogeneity. Furthermore, we have shown evidence that in a-Si/C:H the inhomogeneity is due to heavily hydrogenated a-Si lattice. Finally, we have observed a small fraction of motionally narrowed hydrogen nuclei attributed to disorder modes within the lattices of a-B:H, a-C:H, and a-Si/N:H. These results imply that film micro-structure, compositional inhomogeneity, and rapidly moving hydrogen nuclei may be important in understanding both defect structures and optical properties in plasma-deposited inorganic thin films.

The authors thank S. I. Chan and T. M. Duncan for helpful comments on the manuscript and T. Sigmon for assistance with the Rutherford backscattering analysis. The authors are grateful to R. A. Lujan for his expert assistance in sample preparation. This work was supported by the National Science Foundation under grant no. DMR-77-21394.

*IBM T. J. Watson Research Center, Yorktown Heights, NY 10598

**deceased

REFERENCES

1. A. K. Sinha, Solid State Tech. Apr., **133** (1980).

2. H. Fritzche, C. C. Tsai, P. Persans, Solid State Tech. Jan., **55** (1978).

3. J. A. Reimer, R. W. Vaughan, J. C. Knights, Phys. Rev. Lett., **44**, 193 (1980).

4. W. E. Carlos and P. C. Taylor, Phys. Rev. Lett., **45**, 358 (1980).

5. R. A. Street, J. C. Knights, and D. K. Biegelsen, Phys. Rev. **B18**, 1880 (1978).

6. R. W. Vaughan, D. D. Elleman, L. M. Stacey, W.-K. Rhim, and J. W. Lee, Rev. Sci. Inst. $\underline{43}$, 1356 (1972).

7. T. C. Farrar and E. D. Becker, Pulse and Fourier Transform NMR (Academic Press, New York, 1971).

8. C. P. Slichter, Principles of Magnetic Resonance (Springer-Verlag, Berlin, 1978).

9. M. Mehring in: NMR Basic Principles and Progress, P. Diehl, E. Fluck, R. Kosfeld eds. (Springer-Verlag, Berlin, 1976).

10. U. Haeberlen in: Advances in Magnetic Resonance, Supplement 1, J. S. Waugh ed. (Academic Press, New York, 1976).

11. W.-K. Rhim, D. D. Elleman, L. B. Schreiber, R. W. Vaughan, J. Chem. Phys. $\underline{60}$, 4595 (1974).

12. D. P. Burum, W.-K. Rhim, J. Chem. Phys. $\underline{71}$, 944 (1979).

13. A. Guivarich, J. Richard, M. LeConteller, E. Ligeon, J. Fontenille, J. Appl. Phys. $\underline{51}$, 2167 (1980).

14. Reference (15) shows the probability of mutual spin flips between neighbors to be $\sim\sqrt{m_2}/30$.

15. A. Abragam, The Principles of Nuclear Magnetism (Oxford, London, 1961).

16. Both CH_2 and SiH_2 give rise to Pake doublets. In an amorphous matrix such doublets would be expected to broaden considerably and the values of 13.6 and 33.1 khz for SiH_2 and CH_2, respectively, are an approximate measure of the linewidth.

17. H. S. Gutowsky, G. E. Pake, J. Chem. Phys. $\underline{18}$, 162 (1950).

18. W. E. Carlos, U. Stron, P. C. Taylor, Phys. Rev. Lett. $\underline{45}$, 358 (1980).

19. J. A. Reimer, R. W. Vaughan, and J. C. Knights, manuscript in preparation.

20. R. G. Pembleton, R. C. Wilson, and B. C. Gerstein, J. Chem. Phys. $\underline{66}$, 5133 (1977).

21. M. R. Havens, K. G. Mayhan, W. J. James, J. Appl. Poly. Science $\underline{22}$, 2793 (1975).

22. M. R. Havens, K. G. Mayhan, W. J. James, ibid. $\underline{22}$, 2799 (1978).

23. A. Moshonov, Y. Avny, J. Appl. Poly. Science $\underline{25}$, 771 (1980).

24. D. Engemann, R. Fischer, J. Knecht, Appl. Phys. Lett. $\underline{32}$, 567 (1978).

25. J. A. Reimer, R. W. Vaughan, and J. C. Knights, submitted, Phys. Rev. B.

353

LASER AND THERMAL ANNEALING OF Co-IMPLANTED Si STUDIED BY MOSSBAUER SPECTROSCOPY

G. LANGOUCHE, M. de POTTER, M. VAN ROSSUM, J. DE BRUYN, I. DEZSI[+], R. COUSSEMENT
Instituut voor Kern- en Stralingsfysika, Leuven University, Belgium
+ on leave from Central Research Institute for Physics, Budapest, Hungary

ABSTRACT

Mössbauer spectroscopy was used to study the lattice
location of Fe in Si. Strikingly different spectra were
recorded depending on the implantation dose and
implantation temperature. Drastic changes were also
observed in the spectra upon thermal treatment or laser
irradiation of the samples. Implantation profiles of
several of these sources were also measured. Laser
irradiation and thermal annealing above 400° C results
in surface segregation of the implanted ^{57}Co activity.

Ever since the first days of the Mössbauer effect, it was believed that
Mössbauer spectroscopy would yield interesting information on the microscopic
lattice surrounding impurity atoms introduced in semiconductors. One of the
most likely impurity atoms for such a study was ^{57}Fe, the best studied Mössbauer
isotope. A series of electrical, optical and diffusion experiments [1] on iron-
doped silicon had already suggested that two species of iron might be present in
these samples, one of which was interstitially located.
Already in 1961 a Mössbauer experiment was reported [2] in which ^{57}Fe was
introduced into Si and Ge by diffusion of the ^{57}Co parent isotope. The large
variations in the Mössbauer spectra [3] immediatly revealed that the surrounding
of the Fe atom in Si depends very much on the diffusion temperature and the
quenching speeds used in the preparation of the samples. It was soon realized
that the extremely low solubility of Co and Fe in Si, S = 2 x 10^{16} atoms/cm^3
at 1300 °C, and their very high diffusion coefficient, D = 10^{-5} cm^2/s at
1300 °C [4], make the formation of precipitates in this system extremely
likely [5].
In more recent years implantation techniques were used [6-9] to introduce
^{57}Fe in various semiconductors. With this technique, precipitates are not
readily expected, but the damage introduced during the implantation process
can easily affect the final lattice site. The first implantation experiment
[6], which made use of ^{57}Fe recoil implanted into Si, could not easily be
interpreted. It was not clear whether the two resonances observed in the
Mössbauer spectra were due to the presence of two lattice sites, or due to a
quadrupole interaction at a single non-symmetric lattice site.
Very recently new Mössbauer studies were undertaken to solve the puzzle of
Fe impurities in Si and Ge. It was observed [10] that strikingly different
Mössbauer spectra were recorded depending on the implantation dose. Above
an implantation dose of 10^{14} atoms/cm^2, where the sample is known to be
amorphized, it was demonstrated [11-12] that the symmetric doublet in the
Mössbauer spectrum is due to a quadrupole interaction. Below this implantation
dose, an additional single line fraction is present in the Mössbauer spectra,
and this line is attributed to Fe atoms that come to rest at a symmetric
undamaged lattice site, probably outside the amorphous zone that surrounds the
implantation track [13].

354

We want to report here some recent results on the thermal behaviour of Fe-implanted Si. A detailed description of these thermal and laser-annealing studies, combined with profile measurements, will be given elsewhere [14]. Laser annealing of a ^{57}Co-implanted Si sample qith a pulsed ruby laser (0.9 J, 20 ns pulse) results in a complete precipitation of the radioactive Co atoms at the surface of the sample. The Mössbauer spectrum (Fig. 1) of this

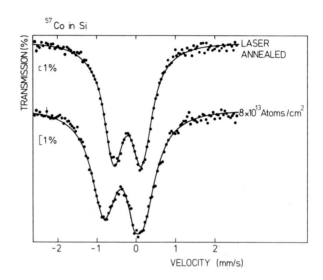

Fig. 1. Mössbauer spectra of ^{57}Co in Si, as-implanted and laser-annealed.

surface layer is a quadrupole doublet with splitting Δ = 0.69(4) mm/s and isomer shift δ = - 0.20(2) mm/s with respect to a potassium ferrocyanide absorber. This result was recently confirmed by Rutherford backscattering [15] measurements on laser annealed Fe-implanted Si samples, and it was shown that for large implantation doses (10^{16} atoms/cm^2) the precipitate has a peculiar cell structure.

Thermal annealing of ^{57}Co-implanted Si results in a gradual collapse of the Mössbauer spectrum as a function of annealing temperature (Fig. 2). The profile measurements demonstrated that for longer annealing times at elevated temperatures (such as 3 hours at 600 °C) or when the target is heated continuously at 400 °C or more during the implantation process, a precipitated layer is also found at the surface. Moreover, the similarity of the Mössbauer spectrum after laser annealing and after thermal annealing to 400 °C strongly suggests that already at this annealing temperature, when the recrystallization of the amorphous layer starts to occur, the Co atoms start forming a segregated phase.

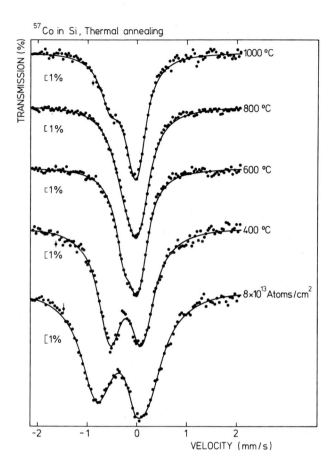

Fig. 2. Mössbauer spectra of ^{57}Co implanted in Si and thermally annealed at different temperatures.

In conclusion we can state that, although the interpretation of the spectra was not clear originally, Mössbauer spectroscopy has been very helpful to understand the way in which an iron impurity is incorporated in a Si lattice. In particular, it was clearly demonstrated that the surrounding of the Fe impurity depends strongly on the implantation dose and the thermal treatment of the sample.

We acknowledge the help of R. Vanautgaerden for the implantations, and of Prof. J. D'Olieslager for the use of his laser set-up.

356

REFERENCES

1. C.B. Collins, R.O. Carlson, Phys. Rev. 108, 1409 (1957).

2. M. de Coster, H. Pollak, S. Amelincks in Proceedings of the second international conference on the Mössbauer effect (Sacaly 1961), D.M.J. Compton, A.H. Schoen Eds. (J. Wiley and Sons, New York 1962) p. 289.

3. P.C. Norem, G.K. Wertheim, J. Phys. Chem. Solids 23, 1111 (1962).

4. A.G. Milness, Deep impurities in semiconductors (J. Wiley and Sons, New York 1973).

5. W. Bergholz, W. Schröter, Phys. Stat. Sol. a49, 489 (1978).

6. G. Latshaw, Ph. D. Thesis, Stanford University (1971).

7. B. Sawicka, J. Sawicki, J. Stanek, J. de Physique 37, C6-893 (1976).

8. G. Weyer, G. Grebe, A. Kettschau, B.I. Deutsch, A. Nylandstedt Larsen, O. Holck, J. de Physique 37, C6-893 (1976).

9. G. Langouche, I. Dézsi, J. De bruyn, M. Van Rossum, R. Coussement, J. de Physique 40, C2-547 (1979).

10. G. Langouche, I. Dézsi, M. Van Rossum, J. De bruyn, R. Coussement, Phys. Stat. Sol. b93, K107 (1979).

11. B.D. Sawicka, J.A. Sawicki, Phys. Lett. A64, 311 (1977).

12. G. Langouche, I. Dézsi, M. Van Rossum, J. De bruyn, R. Coussement, Phys. Stat. Sol. b89, K17 (1978).

13. L.W. Howe, M.H. Rainville, H.K. Haugen, D.A. Thompson, Nucl. Instr. Meth. 170, 419 (1980).

14. M. de Potter, G. Langouche, J. De bruyn, M. Van Rossum, R. Coussement, I. Dézsi, Hyp. Int. (to be published).

15. C.W. White, S.R. Wilson, B.R. Appleton, J. Narayan in Laser and electron-beam processing of materials, C.W. White, P.S. Peercy Eds. (Academic Press, New York 1980) p. 124.

16. I. Dézsi, R. Coussement, G. Langouche, B. Molnar, D.L. Nagy, M. de Potter, J. de Physique 41, C1-425 (1980).

357

SILICIDE FORMATION AT Fe-Si INTERFACES STUDIED BY MÖSSBAUER SPECTROSCOPY AND
RUTHERFORD BACKSCATTERING

R. L. COHEN, L. C. FELDMAN, K. W. WEST AND P. J. SILVERMAN
Bell Laboratories, Murray Hill, N. J.   07974

EXTENDED ABSTRACT

We have deposited the Mössbauer isotope Fe^{57} on (111) Si surfaces at room
temperature and measured the Mössbauer spectrum using conversion electron
Mössbauer spectroscopy for enhanced surface sensitivity. Samples were examined
as a function of substrate preparation and thermal anneal. Rutherford backscattering was used to monitor sample preparation. The Fe^{57} layers have been
coated with natural iron (2.2% Fe^{57}) or silver to prevent them from oxidizing
when exposed to air. The known distinctive Mössbauer spectra of Fe, FeSi,
$FeSi_2$, Fe in Si, and Si in Fe allows the identification of these phases in the
samples.
        Fig. 1 shows the interface structure observed in these experiments immediately
after the room-temperature deposition: a thin layer of FeSi is formed at the
interface between the Fe^{57} and the silicon. The unreacted Fe^{57} is in the form
of an alpha-Fe layer with some silicon dissolved in the iron. Spectra demonstrating this are shown in figs. 2, 3, and 4. These conversion-electron
Mössbauer spectra have been obtained with no annealing.
        Fig. 2 shows that when 7.2Å of Fe is deposited, the in situ reaction converts
all of the iron to a form which has a doublet spectrum corresponding to the
literature values for the line positions of FeSi. All other compounds in the
Fe-Si system have significantly different line positions. Fig. 3 shows that
when more Fe is deposited on Si, FeSi is also found, but an additional broad

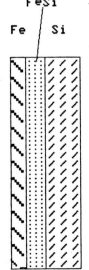

spectrum is observed. This corresponds to the spectrum which
would be observed for alpha-Fe containing large amounts of Si.
The presence of the Si in Fe phase shows that the FeSi is forming by diffusion of the Si through the silicide layer and into
the iron. When a thick layer of Fe (62Å, fig. 4) is deposited,
almost all is in the form of alpha-Fe, and the lines from FeSi
are relatively weak. Heating at temperature from 200-650°C
enhances the formation of the silicide, and the Mössbauer
spectrum becomes that of bulk FeSi. After annealing, no $FeSi_2$
or Fe-Si alloys are observed. Some samples show evidence of an
interfacial barrier which prevents silicide formation and
stabilizes the Fe layer against anneals of 2 hours at 650°C.
        The results show that conversion electron Mössbauer
spectroscopy can be an effective probe for identification of Fe
compounds formed on the 1-2 monolayer scale.

Fe̱Si

Fe | Si

Fig. 1

358

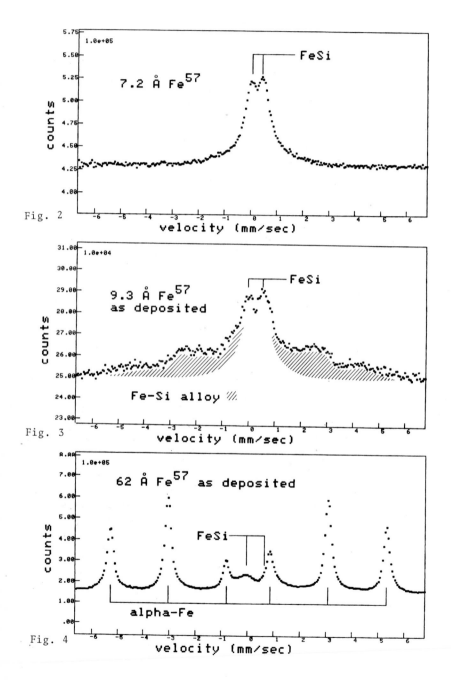

Fig. 2

Fig. 3

Fig. 4

LASER ANNEALING OF Te-IMPLANTED III-V SEMICONDUCTORS
STUDIED BY MÖSSBAUER SPECTROSCOPY

M. VAN ROSSUM, I. DEZSI[+], G. LANGOUCHE[x], J. DE BRUYN AND R. COUSSEMENT
Instituut voor Kern- en Stralingsfysika, Leuven University, Belgium
[+]On leave from Central Research Institute for Physics, Budapest, Hungary

ABSTRACT

    The laser annealing behaviour of Te-implanted GaAs, GaSb,
GaP and InP has been studied by Mössbauer Spectroscopy.
The spectra of the as-implanted samples are characterized
by a quadrupole split multiplet, showing that the Te ions
come to rest at non-substitutional lattice sites. The laser annealing is shown to shift the implanted impurities
towards substitutional positions. The efficiency of the
laser annealing procedure is very similar for all lattices
under study.

INTRODUCTION

    The usefulness of ion implanted III-V semiconductors as device materials has
been widely recognized. However, ion implantation of the dopants causes severe
radiation damage, which has te be removed by some annealing procedure. Unfortunately, several III-V compounds (especially GaAs) dissociate during thermal
annealing and therefore require the use of special encapsulation techniques.
Therefore, the applications of the laser annealing is of particular interest
for these materials.
    The possibility of using Mössbauer Spectroscopy for the study of implanta-
tions in III-V compounds has already been demonstrated for the case of ^{119}Sn
[1-2]. In this contribution, we present the first Mössbauer Effects study on
the laser annealing mechanism in compound semiconductors. ^{129}Te^m was choosen
as the radioactive impurity, since Te is known as a shallow n-type dopant in
III-V compounds; the Mössbauer experiments were performed on the daughter nucleus ^{129}I.

EXPERIMENTAL

    The ^{129}Te^m ions (obtained from neutron irradiation of stable ^{128}Te) were im-
planted in GaAs, GaP, GaSb and InP slices cut perpendicular to the <111> direction. The accelerating voltage was set at 85 kV and the usual implanted dose
varied around 5×10^{14} Te at/cm^2. The implantations were performed at room temperature. The Mössbauer absorber was a CuI powder sample, containing 10 mg
^{129}I/cm^2. Both source and absorber were kept at 4.2 K during the Mössbauer
measurements; the source was moved by a conventional electromechanical velocity
transducer. A Xe-filled proportional counter was employed for the detection of
the 27.7 keV Mössbauer gamma-rays.
    The laser annealings were performed with a Q-switched ruby laser (λ=694 nm)
delivering pulses of 20 ns; the standard energy density of the pulses was about
1.5 J/cm^2. The laser irradiations were carried out in air and the implanted
area was scanned by manual adjustment.

_____

[x] Bevoegd verklaard navorser N.F.W.O.

RESULTS

Figure 1 displays the Mössbauer spectrum of the as-implanted GaAs sample. Only those $^{129}Te^m$ which decay to ^{129}I contribute to the resonance, which therefore reflects the chemical bond of I in the lattice. It can easily be fitted with a single quadrupole split component, although the sidewing of the quadrupole pattern does not clearly appear because of insufficient statistics. The spectra of the other implanted materials are qualitatively similar to this one and only the values of the hyperfine interaction parameters differ to some extent; these have been tabulated in table I.

Laser annealing caused extended damage ripples at the surface of the samples. Nevertheless, only a small fraction (<1%) of the implanted activity was lost during the annealing, thus showing that no substantial material evaporation did occur. The Mössbauer spectra of laser-treated samples exhibit an unsplit resonance on the negative velocity side (figure 2 and table II); the centroid of this line is situated at a slightly lower negative velocity than the quadrupole split resonance. Figure 2 shows the spectrum of GaAs, which is qualitatively representative of the three other materials as well.

DISCUSSION

Laser annealing studies of ion-implanted III-V compounds have mostly been restricted to the case of GaAs [3]. The electrical activity of the Te dopants, which is usually low after implantation, has been shown to improve more significantly after laser annealing than after thermal treatment [4].

Our data indicate that the Te atoms occupy similar sites after implantation in the four lattices under study. Differences in isomer shift and quadrupole splitting values may be attributed to the first or second neighbour element. Since the existence of a quadrupole interaction implies a non-cubic surrounding, it cannot correspond with a regular substitutional position, but an "off-center" quasi-substitutional position cannot be excluded. The landing of implanted Te ions at a non-substitutional site with lower than cubic symmetry seems to be a common feature of all diamond-type semiconductors, since it has been observed in Si and Ge as well [5].

In these hosts, it was proved that the quadrupole interaction is not necessarily related to the bulk amorphization; there is not enough experimental evidence at this moment to make a similar statement about the III-V compounds.

The nature of the non-substitutional landing site in Si and Ge has been investigated in detail, and the possibility of a vacancy associated Te has been considered [6]. In GaAs, the formation of V_{Ga} Te complexes has been assumed to account for the low electrical activity of the implanted layers after thermal annealing [7], but little is known about the as-implanted site. Our measurements suggest that this as-implanted site is similar in type-IV and type III-V semiconductors.

After laser annealing, the disappearance of the quadrupole interaction indicates that the Te impurities have shifted towards cubic sites, making a substitutional location very probable. This is in agreement with channeling measurements on laser annealed Te implanted GaAs, which measured a high substitutional (>80 %) Te fraction [8]. The residual defects may be partly responsible for the line broadening which is observed in many spectra; in any case, there is no direct evidence for a remaining non-substitutional component. It may be observed that the isomer shift ratio of the as-implanted and laser annealed sites is nearly constant (table II), showing a striking similarity between the rearrangement processes in the different lattices.

Our laser annealing results on III-V compounds are similar to the ones obtained on Si and Ge [5-9]. Here too, a predominant single line component after annealing was attributed to a high substitutional Te function. However, the electronic density at this site (as derived from isomer shift values) is significantly higher in type IV-semiconductors. This may be due to the partly ionic character of the bond between Ga/In and Te in III-V compounds.

CONCLUSION

We have demonstrated the ability of the Mössbauer Effect to probe the behaviour of Te implanted in III-V semiconductors. The impurities are found to occupy non-substitutional (possibly vacancy-associated) lattices sites after implantation. The laser annealing is a very efficient technique for shifting these atoms towards substitutional positions. Moreover, the Mössbauer results show the same general behaviour for Te implanted in fourfold coordinated semiconductors.

TABLE I

hosts	Δ	η	δ	Γ
GaAs	-3.15 (3)	0.3	-0.92 (1)	1.67 (4)
GaSb	-3.75 (4)	0.3	-1.05 (1)	2.48 (6)
GaP	-3.12 (3)	0.3	-0.86 (1)	1.90 (5)
InP	-2.59 (3)	0.3	-0.88 (1)	1.70 (4)

Δ = quadrupole splitting
δ = isomer shift vs. CuI
Γ = linewidth, all values in mm/s
η = asymmetry of the electric field gradient

TABLE II

hosts	δ	Γ
GaAs	-0.67 (1)	1.84 (4)
GaSb	-0.70 (1)	2.54 (5)
GaP	-0.59 (1)	2.28 (4)
InP	-0.53 (1)	1.48 (3)

all values in mm/s

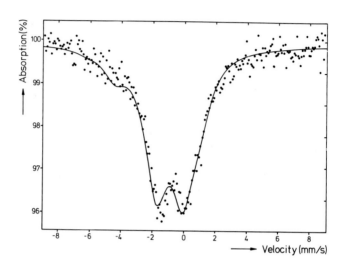

Fig. 1.   Mössbauer spectrum of ^{129}I after implantation of ^{129}Te^m in GaAs.

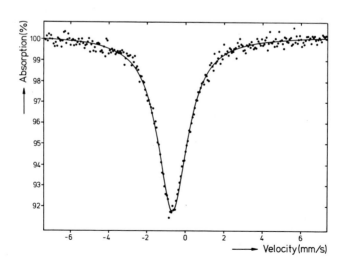

Fig. 2.   Same sample as figure 1 after laser annealing.

ACKNOWLEDGMENTS

The authors wish to thank Dr. J. D'Olieslager for his kind assistance during the laser annealings. This work has been supported by the I.I.K.W. and the S.C.K.-K.U.Leuven association.

REFERENCES

1. G. Weyer, J.W.Petersen, S. Damgaard, H.L. Nielsen and J. Heinemeier, Phys. Rev. Lett. 44, 155 (1980).

2. G. Weyer, S. Damgaard, J.W. Petersen and J. Heinemeier, J. Phys. C13, L181 (1980).

3. J.A. Golovchenko and T.N.C. Venkatesan, Appl. Phys. Lett. 32, 147 (1978).

4. P.A. Pianetta, C.A. Stolte and J.L. Hansen, to be published in Appl. Phys. Lett.

5. J. De bruyn, R. Coussement, I. Dézsi, G. Langouche and M. Van Rossum, contribution to the "Vth International Conference on Hyperfine Interactions", Berlin, July 1980.

6. I. Dézsi et al., to be published.

7. B.K. Shiu, Y.S. Park and J.E. Ehret, in: Ion Implantation in Semiconductors 1976, F. Chernow, J.A. Borders and D.K. Brice eds. (Plenum Press, New York 1977) pp. 107.

8. B.Y. Sealy, S.S. Kular, K.G. Stephens, D. Sadana and G.R. Booker, in: Proceedings of the International Conference on ion beam modification of materials, J. Gyulai, T. Lohner and E. Pasztor eds. (Budapest 1978) pp. 751.

9. J. De bruyn, G. Langouche, M. Van Rossum, M. de Potter and R. Coussement, Phys. Lett. 73A, 356 (1979).

Published 1981 by Elsevier North Holland, Inc.
Kaufmann and Shenoy, editors
Nuclear and Electron Resonance Spectroscopies Applied to Materials Science

NMR IN LIQUID SEMICONDUCTING Se_xTe_{1-x}

R. DUPREE AND D.J. KIRBY
Department of Physics, University of Warwick, Coventry CV4 7AL, U.K.

J.A. GARDNER
University of Warwick and Department of Physics, Oregon State University,
Corvallis, Oregon 97331, U.S.A.

ABSTRACT

For $x > 0.2$, liquid Se_xTe_{1-x} alloys are semiconducting
near the liquidus and have strongly temperature-dependent
magnetic susceptibilities. We have studied the electronic
paramagnetic centers by measuring their influence on the
nuclear magnetic resonance shift and relaxation times of
^{77}Se and ^{125}Te in liquid $Se_{0.5}Te_{0.5}$. We find that the
electronic centers are highly localized, their average hyperfine contact interaction with ^{125}Te is three times that
with ^{77}Se, and the electronic spin correlation time is of
order 1 psec at 350°C and decreases with rising T.

INTRODUCTION

Molten chalcogenides display semiconducting-like electronic transport
properties [1] and strongly temperature-dependent magnetic susceptibilities.
In liquid Se and Se-Te alloys up to approximately 80 atomic % Te, the susceptibility near the liquidus rises rapidly with temperature due to thermally-
activated paramagnetic centers. [2] We report here a study of these centers
in liquid $Se_{0.5}Te_{0.5}$ using the nuclear magnetic resonance of ^{77}Se and ^{125}Te as
experimental probes. The resonance frequency shift and nuclear relaxation
rates caused by interaction between the nuclear magnetic moment and magnetic
fields due to these centers provide information about the electronic structure
and dynamics of the centers. In the present case interpretation of the data is
simplified because both nuclei have spin ½, so there is no quadrupole moment,
and dipole interactions with nearby nuclei are negligible because of diffusional narrowing at all temperatures studied. The nuclear resonance is shifted by
interaction with the diamagnetic moment of nearby electrons (the chemical
shift) and with the paramagnetic moment of the thermally-induced centers.
Since the chemical shift is not strongly temperature-dependent and in this
case contributes negligibly to nuclear relaxation we confine our discussion to
the paramagnetic shift K.

EXPERIMENTAL DETAILS AND RESULTS

Measurements of the resonance shifts, transverse (T_2) and longitudinal
(T_1) relaxation times were made by standard pulse resonance methods. Samples
were sealed in quartz ampules about which a Pt or Ag resonance coil was wound.
The sample and coil were heated by a water-cooled molybdenum-wound furnace
mounted in the room-temperature insert of a superconducting magnet. RF pulses
having peak power of approximately one kW were transmitted to the coil, and the
nuclear response following the perturbation was measured by detecting the RF
voltage induced in the coil by the rotating nuclei. By measuring the time and

frequency dependence of the signal following a single ($\approx$ 90^0) pulse, both the resonance shift and free induction decay time (T_2^*) could be determined. T_1 was measured by the usual 180^0 - 90^0 pulse sequence, and T_2 by the 90^0 - 180^0 spin echo method. [3] Above approximately 500^{0}C, T_1, T_2 and T_2^* were equal within experimental uncertainty, so careful relaxation time measurements of T_1 and T_2 were only performed at lower temperatures. The resonance shifts and relaxation times for ^{125}Te are shown vs. temperature in Fig. 1 and for ^{77}Se in Fig. 2. In both figures the resonance shifts are defined by

$$K(T) = \frac{f(T) - f_0}{f_0} \qquad (1)$$

where $f(T)$ is the resonance frequency at temperature T and f_0 is the resonant frequency in the limit of small T. Since the chemical shift is temperature-independent, by this definition K is only the paramagnetic shift. All data in Fig. 1 were taken with an external field of approximately 44 kG, where the ^{125}Te resonance frequency is $\approx$ 59 MHz. In Fig. 2 the shift data shown were taken at 44 kG (where the ^{77}Se resonance frequency is $\approx$ 36MHz), but essentially identical shifts were found at 72 kG. The relaxation times in Fig. 2 were measured at 72 kG. At 44 kG, the relaxation times are somewhat smaller at low T but at high T are approximately the same as those shown in Fig. 2.

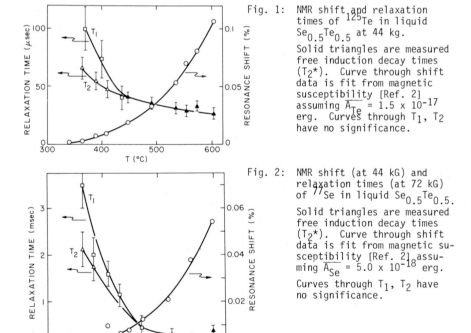

Fig. 1: NMR shift and relaxation times of ^{125}Te in liquid Se$_{0.5}$Te$_{0.5}$ at 44 kg. Solid triangles are measured free induction decay times (T_2^*). Curve through shift data is fit from magnetic susceptibility [Ref. 2] assuming $\overline{A_{Te}}$ = 1.5 x 10^{-17} erg. Curves through T_1, T_2 have no significance.

Fig. 2: NMR shift (at 44 kG) and relaxation times (at 72 kG) of ^{77}Se in liquid Se$_{0.5}$Te$_{0.5}$. Solid triangles are measured free induction decay times (T_2^*). Curve through shift data is fit from magnetic susceptibility [Ref. 2] assuming $\overline{A_{Se}}$ = 5.0 x 10^{-18} erg. Curves through T_1, T_2 have no significance.

In these liquids, K arises from the contact interaction between the nuclear magnetic moment and the electronic spin magnetic moment density at the nucleus. Although the electronic spin is strongly localized, the short lifetime of the paramagnetic centers and the rapid diffusion of the nuclei compared to the NMR precession period mean that the interaction is averaged over all sites. We shall assume that the paramagnetic centers are randomly distributed over the Se and Te atoms since there is good reason to believe the bonding in these alloys is random [2]. In this case, it is straightforward to show that the shift K_n of nuclear type n is related to the total molar paramagnetic susceptibility χ_p by

$$K_n = \frac{1}{N_o \gamma_e \gamma_n h^2} \overline{A_n} \chi_p \tag{2}$$

where N_o is Avogadro's number and γ_e, γ_n are respectively the electronic and nuclear gyromagnetic ratios. $\overline{A_n}$ is the total hyperfine coupling constant per paramagnetic center. Comparison of the resonance shifts in Figs. 1 and 2 with the known susceptibility [2] of $Se_x Te_{1-x}$ yields excellent agreement with this relationship if $\overline{A_{Se}}$ is taken to be 5.0×10^{-18} erg and $\overline{A_{Te}}$ is 1.5×10^{-17} erg. The lines fit to K in Figs. 1 and 2 are computed from χ_p using these coupling constants.

If the nuclear relaxation is also dominated by the electronic centers, the relaxation rates are given by [3]

$$T_1^{-1} = \frac{S(S+1)c}{3h^2} \overline{A_n}^2 \frac{2\tau}{1 + \omega_s^2 \tau^2} \tag{3}$$

$$T_2^{-1} = \frac{S(S+1)c}{3h^2} \overline{A_n}^2 \tau [1 + \frac{1}{1 + \omega_s^2 \tau^2}] \tag{4}$$

where

$$c = \frac{\chi}{N_o \mu_B^2 S(S+1)/3 \, kT} \tag{5}$$

is the concentration of the paramagnetic centers and S is their spin; in this case $S = \frac{1}{2}$. τ is the correlation time for spin fluctuations and ω_s is the electronic Larmor frequency. When $\omega_s \tau \gtrsim 1$, $T_1 > T_2$, and τ can be calculated from the ratio of the measured relaxation times,

$$\tau = \frac{1}{\omega_s} [2(\frac{T_1}{T_2} - 1)]^{\frac{1}{2}} . \tag{6}$$

At the lowest temperatures shown in Figs. 1 and 2, T_1 and T_2 are sufficiently different that τ and $\overline{A_n}^2$ can be obtained with reasonable accuracy. At $360^o C$, we find that $\tau_{Te} \approx 1.3$ psec, $(\overline{A_{Te}}^2)^{\frac{1}{2}} \approx 0.7 \times 10^{-17}$ erg, $\tau_{Se} \approx 0.9$ psec, and $(\overline{A_{Se}}^2)^{\frac{1}{2}} \approx 2 \times 10^{-18}$ erg. Using these values for $\overline{A_n}^2$, we can compute τ_n as a function of T, and it decreases rapidly with rising T. Above $500^o C$, $\omega_s \tau_n \ll 1$ for both nuclear types. This is consistent with our observation that $T_1 = T_2$ above $500^o C$. The low temperature relaxation times measured for [77]Se at 44 kG are also in good agreement with the smaller values expected because of the lower value of ω_s at the lower field.

DISCUSSION

It is instructive to compare the NMR results in this alloy with those measured for pure liquid Se by Warren and Dupree [4]. The ^{77}Se hyperfine coupling constant in the alloy is larger than the value of 2.9×10^{-18} erg they find for pure Se. This indicates that the s-p ratio at a Se paramagnetic center is somewhat larger in the alloy, but in both cases, the center has predominantly p-character. The flucuation time τ_{Se} is somewhat smaller in the alloy. In both cases, the ratio of $(\overline{A_{Se}^2})^{\frac{1}{2}}$ to $\overline{A}_{Se}$ is roughly one half, indicating that the major hyperfine interaction is at the nucleus of the paramagnetic center. If the electronic spin density were significantly delocalized, the sum of A_{Se}^2 over the nuclei of the center would be very much smaller than $(\overline{A_{Se}})^2$. These observations strongly support the identification of the paramagnetic centers as uncharged dangling bonds. Other possible types of paramagnetic center are expected to have more delocalized spin. [5] The decreased flucuation time is consistent with the higher concentration (at a given temperature) of paramagnetic centers and weaker bonding in the alloy.

Comparison of the ^{77}Se and ^{125}Te data in the alloy is also instructive. As expected, the heavier Te atom has a larger hyperfine coupling energy, but the ratio of $(\overline{A_{Te}^2})^{\frac{1}{2}}$ to $\overline{A}_{Te}$ is rather similar to that of ^{77}Se, indicating that the electronic structure of a center at a Te atom is similar to that of one at Se. This fact and the approximate equality of τ_{Se} and τ_{Te} add support to our initial assumption of random bonding in alloy.

ACKNOWLEDGEMENTS

This research was supported in part by the UK Scientific Research Council. One of us (JAG) wishes to thank the SRC for a fellowship during the time this work was done.

REFERENCES

1. M. Cutler, Liquid Semiconductors, (Academic Press, New York 1977).

2. J.A. Gardner and M. Cutler, Phys. Rev. B 20, 529 (1979).

3. A. Abragam, Principles of Nuclear Magnetism (Oxford University Press, London, 1961).

4. W.W. Warren Jr. and R. Dupree, Phys. Rev. B 22, 2257 (1980).

5. D. Vanderbilt and J.D. Joannopoulos, Phys. Rev. Lett. 38, 656 (1979).

Published 1981 by Elsevier North Holland, Inc.
Kaufmann and Shenoy, editors
Nuclear and Electron Resonance Spectroscopies Applied to Materials Science

MOTIONAL CORRELATION TIME OF DILUTE ^{111}Cd IMPURITIES IN Se-RICH LIQUID Se-Te ALLOYS

D. K. GASKILL, J. A. GARDNER, K. S. KRANE, and K. KRUSCH
Department of Physics, Oregon State University, Corvallis, Oregon 97331

R. L. RASERA
Department of Physics, Oregon State University and University of Maryland
Baltimore County, Catonsville, Maryland 21228

ABSTRACT

The motional correlation time, τ_c, in liquid Se and Se-rich
Se-Te alloys has been investigated between 500 and 900°C
using time differential perturbed angular correlations of
γ-rays from dilute ^{111}Cd impurities. In all alloys we find
$\tau_c \propto \exp(E_0/kT)$ at low T where $E_0 = 0.36$ eV. τ_c deviates
from this relation at high T. At low T, τ_c is tentatively
identified as the lifetime of a Cd to host molecule bond,
and at high T as the average lifetime of bonds in the host
molecule.

INTRODUCTION

The Perturbed Angular Correlation (PAC) experimental technique is not really
a resonance method, but because it is a measurement of interactions between
nuclei and their environment, it is informative to include it in this symposium. Physical information from PAC experiments is quite similar to that
obtained from Mössbauer, NMR, and ESR hyperfine measurements.

PAC utilizes a radioactive isotope which decays via a 2-step gamma-ray cascade. The experimental work reported in this paper utilizes the isotope
^{111}In, which decays to ^{111}Cd* by electron capture. The ^{111}Cd* subsequently
decays by a 171 keV gamma ray (γ_1) to an intermediate ^{111}Cd state with lifetime $\tau_N = 121$ nsec, which then decays to the ^{111}Cd ground state by emission
of a 245 keV gamma (γ_2). In the absence of magnetic fields or electric field
gradients at the Cd nucleus, γ_2 would be emitted with an anisotropic probability proportional to

$$W(\theta) = [1 + A_2 P_2(\cos\theta) + A_4 P_4(\cos\theta)] \qquad (1)$$

where θ is the angle of emission of γ_2 with respect to the direction of γ_1,
and the nuclear angular correlation constants are given by $A_2 = -0.180$, $A_4 = 0.002$, for this cascade. In general, however, a Cd nucleus in condensed
matter will be subject to magnetic and/or electric fields due to nearby elec-
trons and ions. Thus the nucleus will be subjected to torque during the time
between emission of γ_1 and γ_2, so in general the angular distribution of Eq. 1
must be modified to include the effects of reorientation of the nuclear ensemble by interaction with these fields. For a polytropic sample (i.e., one
with no preferred overall orientation such as a liquid or powder) then

$$W(\theta,t) = [1 + A_2 G_2(t) P_2(\cos\theta) + A_4 G_4(t) P_4(\cos\theta)] \qquad (2)$$

The functions $G_2(t)$ and $G_4(t)$ are normalized to unity at $t = 0$ and are governed by the hyperfine interactions. The purpose of a PAC experiment is to measure G_2 and G_4 and relate them to the electronic properties of the material being investigated. Because in our case A_4 is quite small, the last term in Eq. 2 contributes negligibly to $W(\theta,t)$, and we shall drop it. It is easy to show [1] that $G_2(t)$ can be determined experimentally by measuring $W(\theta,t)$ at $\theta = 90^0$ and 180^0. Then

$$A_2 G_2(t) = 2 \frac{W(180^0,t) - W(90^0,t)}{W(180^0,t) + 2W(90^0,t)} \quad . \qquad (3)$$

Normally, it is possible to measure $W(\theta,t)$ between $t = 0$ and several times τ_N. Details of the experimental procedure for measuring $W(\theta,t)$ are given elsewhere. [2,3]

USE OF PAC TO STUDY ELECTRONIC PROPERTIES OF LIQUIDS

Both magnetic fields interacting with the nuclear magnetic moment and electric field gradients interacting with the nuclear quadrupolar moment can perturb the nucleus significantly in its intermediate state, but the latter is dominant in this experimental work, so we shall ignore magnetic fields in this discussion. We shall also discuss only the situation for which electric field gradients in the liquid fluctuate rapidly with respect to τ_N. For such a situation, it has been shown [4] that G_2 should decrease exponentially with time as

$$G_2(t) = e^{-\lambda t} \qquad (4)$$

where

$$\lambda = 2.487 \ \tau_c \langle \nu_Q^2 \rangle \quad . \qquad (5)$$

Here $\langle \nu_Q^2 \rangle$ is the average square quadrupolar interaction frequency and τ_c is the correlation time of electric field gradient fluctuations. The fluctuations can be caused either by physical tumbling of the molecule to which the nucleus is bound or by electronic processes which cause the chemical bonding to change rapidly. In either case τ_c is a quantity of physical interest, but in the present experiments we have investigated a class of liquids in which τ_c is expected to be dominated by chemical fluctuations. $\langle \nu_Q^2 \rangle$ can be roughly estimated for these liquids and should not depend strongly on temperature. Thus our PAC measurements provide a semi-quantitative value for τ_c and a relatively accurate measure of its temperature dependence. This information on chemical kinetics is of considerable interest and is not often measureable by other means.

EXPERIMENTAL RESULTS AND DISCUSSION

The work reported here on Se-rich liquid Se-Te alloys is an extension (to materials expected to have more rapid chemical bonding fluctuations) of PAC measurements on liquid Se. [2,3] For pure Se, τ_c varies from approximately 20 psec at 900^0C to 200 psec at 500^0C. Over most of the range τ_c varies with temperature as

$$\tau_c \propto e^{E_0/kT} \qquad (6)$$

where $E_0 = 0.36$ eV, but the data deviate significantly from this relation above 800°C. We find qualitatively similar behavior in the Se-Te alloys. The log of experimentally determined values of λ are shown vs T^{-1} in Fig. 1. In this figure, the right-hand ordinate shows the value of τ_c obtained from λ using Eq. 5 and $\langle \nu_Q^2 \rangle^{\frac{1}{2}} = 125$ MHz. [2] In Fig. 2 we show τ_c for pure Se on a compressed scale and include low temperature correlation times obtained recently by Warren and Dupree [5] from ^{77}Se NMR line-broadening chemical-shift anisotropy in pure liquid Se.

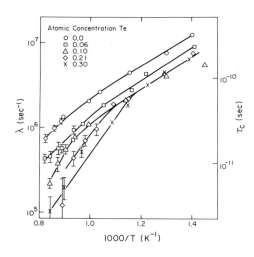

Fig. 1: Log λ vs T^{-1} for Cd in liquid Se-Te alloys. The right hand ordinate shows τ_c if $\langle \nu_Q^2 \rangle^{\frac{1}{2}} = 125$ MHz. The straight lines through the points at low T have slope $E_0 = 0.36$ eV. The curves through the points at high temp are only a guide to the eye.

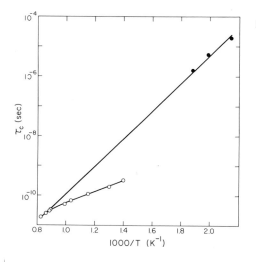

Fig. 2: Log τ_c vs T^{-1} in liquid Se. The solid points are intrinsic correlation times measured by NMR [Ref. 5]. Open points are for Cd impurities measured by PAC, assuming $\langle \nu_Q^2 \rangle^{\frac{1}{2}} = 125$ MHz.

Liquid Se is a weakly-bonded polymeric liquid in which the polymer size decreases with temperature due to thermal dissociation. At the highest temperatures investigated in this work, the average Se polymer size is estimated to be of order 10^2 atoms. [5,6] The Se-rich Se-Te alloys are qualitatively similar with somewhat weaker average bonding [7] so the polymer size and correlation times are expected to be somewhat smaller than for pure Se. Since electric field gradients at ^{129}I and ^{125}Te tracers are somewhat smaller in Te than in Se [8], $\langle \nu_Q^2 \rangle$ may also be slightly reduced. The correlation times measured in PAC experiments will be the minimum of the polymer rotation time, the Se-Se (or in the alloys an average Se-Se, Se-Te, and Te-Te) bond lifetime, or the lifetime of the bond between the Cd tracer and the polymer molecule. In these liquids there is probably no physical difference between the first two correlation times, since the molecules almost certainly "rotate" by breaking and reforming bonds between adjacent molecules. There is of course a very important difference between this "intrinsic" correlation time and the "tracer" correlation time due to Cd bond fluctuations, and it is important to identify which of the two is the measured quantity.

The existence of two different temperature regions for λ which is evident in Fig. 1 suggests that in one region, τ_c is intrinsic and is tracer-related in the other. This suggestion is further supported by the NMR measurements of τ_c shown in Fig. 2. The NMR correlation times are intrinsic and are consistent with an extrapolation to low temperature of the pure Se high-T region but not the low-T PAC data. The implication is that the high-T PAC correlation times are intrinsic and that the low-T correlation times are extrinsic in the sense that τ_c is dominated by Cd-polymer bond fluctuations. In principle this identification can be proven definitively by PAC measurements with other isotopes, and we are presently exploring this possibility.

It is clear from Fig. 1 that addition of small amounts of Te to Se causes an overall reduction of λ. Since the activation energy in the low-T "extrinsic" region is unchanged by adding Te, it is possible that at low T the change may be due to reduction of $\langle \nu_Q^2 \rangle$, and that τ_c is not affected by alloying. This possibility can be checked by PAC measurements at sufficiently low temperature that $\tau_c \gg \tau_N$, $\langle \nu_Q^2 \rangle^{-\frac{1}{2}}$ where $G_2(t)$ becomes independent of τ_c, and $\langle \nu_Q^2 \rangle$ can be determined directly. Although we are unlikely to achieve this condition in the liquid state, the corresponding quenched glass should be an adequate substitute. Unfortunately, our preliminary measurements at lower temperatures have been inconclusive because of experimental difficulties in sample preparation.

The PAC data in the interesting high-T region are relatively inaccurate because the decay of $G_2(t)$ is very small within the experimentally accessible time range. Nonetheless it is clear that the relative decrease of λ with addition of Te is greater than in the low-T region. Although part of this decrease may be ascribed to change of $\langle \nu_Q^2 \rangle$, τ_c must also be decreased by addition of Te. This is in accord with naive expectations, since adding Te is known to weaken the bonding. The "intrinsic" τ_c temperature dependence cannot be determined with confidence from the present data. If τ_c is activated in this region, the activation energy is of order 0.9 eV for the higher Te-content alloys, approximately the same as the slope of the line which connects the low and high-T "intrinsic" pure Se correlation times in Fig. 2

More accurate data in this region are required before one can draw any reliable conclusions about the temperature dependence of τ_c in this region however.

ACKNOWLEDGMENTS

We appreciate several fruitful discussions with Professor Melvin Cutler.

This work was supported in part by the Office of Naval Research.

REFERENCES

1. H. Frauenfelder and R.M. Steffen in Alpha- Beta- and Gamma-Ray Spectroscopy, edited by K. Siegbahn (North Holland, Amsterdam, 1965) Chap. 14A.

2. R.L. Rasera and J.A. Gardner, Phys. Rev. B 18, 6856 (1978).

3. D.K. Gaskill, J.A. Gardner, K.S. Krane, K. Krusch, and R.L. Rasera, Proceedings of the 5th International Conference on Hyperfine Interactions, Berlin, 1980, in press.

4. A. Abragam and R.V. Pound, Phys. Rev. 92, 943 (1953).

5. W.W. Warren, Jr. and R. Dupree, Phys. Rev. B 22, 2257 (1980).

6. A. Eisenberg and A.V. Tobolsky, J. Polymer Sci. 46, 19 (1960).

7. M. Cutler, Liquid Semiconductors (Academic Press, New York, 1977).

8. C.S. Kim and P. Boolchand, Phys. Rev. B 19, 3187 (1979).

Metals and Alloys

Published 1981 by Elsevier North Holland, Inc.
Kaufmann and Shenoy, editors
Nuclear and Electron Resonance Spectroscopies Applied to Materials Science

A MÖSSBAUER STUDY OF MICROSTRUCTURAL AND CHEMICAL CHANGES IN FE-9NI STEEL
DURING TWO-PHASE TEMPERING

B. FULTZ AND J. W. MORRIS, JR.
Materials and Molecular Research Division, Lawrence Berkeley Laboratory and the
Department of Materials Science and Mineral Engineering, University of
California, Berkeley, California 94720, USA

ABSTRACT

   Two-phase tempering of martensitic Fe-9Ni steel serves
to enhance the low temperature toughness and forms austenite
precipitates in this material. Hyperfine field effects in
Fe-Ni alloys were systematized so that tempering induced
chemical composition changes in the martensite could be
quantified by Mössbauer spectrometry. The kinetics of segregation of alloy elements from the martensite into the
fresh austenite can be determined simultaneously with the
amount of austenite which has formed.

INTRODUCTION

   Fe-9Ni steel with the commercial "quenched and tempered" heat treatment has
seen increasing application as a structural material in cryogenic temperature
environments. The notch toughness of this material at low temperatures is
dramatically improved after tempering the martensitic quenched material for one
hour or so at $600°C$. Transmission electron microscopy [1,2] has shown that
during this tempering treatment crystals of austenite, γ phase, form between
the crystals of tempered martensite, α_t phase.
   Through study of the tempering-induced microstructural changes we hope to
eventually clarify the mechanism of the low temperature mechanical property
improvements in Fe-9Ni steel. Any such fundamental study must quantify the
amount of precipitated γ phase and also determine the alloy chemistry of the γ
and α_t phases because the chemical composition of the γ phase will influence
its formation and stability against the martensite transformation. During
tempering the characteristic diffusion distances of Mn, Cr, and Si in the α
phase are only a few thousand angstroms or less, so we expect the kinetics of
chemical segregation to play an important role in the γ phase precipitation
process. A method for studying both the amount of γ phase which has precipitated and the extent of alloy element segregation to the γ or α phases is necessary in our research program, but the small dimensional scale ($<1\mu m$) and the
small fraction of precipitated γ phase (a few percent) makes this precipitation
process and associated chemical segregation effects amenable to study by only a
few experimental techniques.
   By simultaneously quantifying the extent of this precipitation reaction and
the associated kinetics of alloy element segregation, Mössbauer spectrometry is
serving as a unique and valuable tool in materials science. The method of
phase analysis by Mössbauer spectrometry uses the fact that Fe^{57} nuclei in the
paramagnetic γ phase see no hyperfine field and exhibit a single absorption
peak near zero Doppler shift energy, whereas peaks from Fe^{57} nuclei in the
ferromagnetic α phase are well removed from zero Doppler shift energy because
of the large hyperfine magnetic field and the nuclear Zeeman effect. We refer
to the literature for a description of this method of phase analysis [3], but
in the work reported here we will describe how Mössbauer spectrometry proves

378

efficient in simultaneously determining chemical changes in the α phase which accompany the precipitation of the γ phase during tempering. Such chemical information is obtained from perturbations in the hyperfine magnetic fields at Fe^{57} nuclei which are situated next to alloy element atoms. We have systematized the perturbations in Fe^{57} hyperfine magnetic fields in quenched, α phase Fe-Ni binary alloys and in Fe-Ni-X* ternary alloys so that this information could be used to quantitatively determine the chemical composition of the α phase in these materials. We then tempered these quenched alloys to form some γ phase and followed the alloy element segregation process by determining the chemical changes in the α phase. These chemical changes in the α phase were small, so a difference spectrum procedure was employed to compare the Mössbauer spectra before and after tempering.

EXPERIMENTAL

In the study of alloy element effects on room temperature Mössbauer spectra we used a constant acceleration transmission geometry spectrometer with an Austin Science Associates electromechanical drive, a Kr and CO_2 filled proportional counter, and a microcomputer setup for 1024 channel multichannel scaling. In many experiments small peak shifts were measured, so we closely monitored the performance of the Doppler drive unit by collecting a spectrum from the same pure iron foil after alternate runs. It was found that drive stabilities of better than 0.1 data channels at the $I = \pm^1/_2 \rightarrow I = \pm^3/_2$ absorption peaks occurred for periods of up to two weeks. However, two spectra taken at different times for which the $I = +^1/_2 \rightarrow I = +^3/_2$ peak was shifted by 0.4 data channels, for example, could be compared on the same Doppler shift energy axis by adding the data channels together in pairs. In this example 25% of the n+1st data channels were added to the nth data channels to shift one spectrum downwards by 0.2 data channels, and 25% of the n-1st data channels were added to the nth data channels to shift the other spectrum upwards by 0.2 data channels. The data of Fig. 3 and Fig. 4 were processed in this way.

The 100 mCi Co^{57} in Pd radiation source caused the proportional counter to show some evidence of saturation behavior. The counter recovery time was comparable to the time spent scanning the absorption peaks, and a problem could be seen for 25 µm natural iron absorbers which exhibited a large dip in count rate at the Mössbauer peaks. In these specimens an asymmetrical distortion of the absorption peaks could be readily found by comparing peak shapes of the two spectra taken with opposite accelerations of the Doppler drive. To suppress this distortion, the single channel analyzer window was set much wider than the detector energy resolution for 14.4 keV γ-rays, and very thin (5µm) specimens were used. The very thin specimens also ensured that absorber thickness distortion effects could be neglected.

Alloys were prepared by melting 99.99% purity components in a vacuum furnace under He backpressure. The furnace cooling rate was sufficient to ensure that the high temperature γ phase transformed to α phase martensitically without the chemical segregation effects associated with α phase precipitation. Negligible weight losses were found after melting, and no surface reactions were observed for any of the melted alloys, so chemical compositions were determined and controlled by the weights of the component elements. After cold rolling to 50 µm, the foils were prepared by heating 50-100°C into the γ phase region of the Fe-Ni phase diagram for 10 minutes in an inert atmosphere and water quenched. The quenched foils were then chemically polished to their final thickness. Before tempering, the same foil used for collecting the quenched α phase spectrum was sealed in an evacuated pyrex ampule with a low Ar backpressure. No surface oxidation was observed after tempering.

*"X" was one of the elements Mn, Cr, Si, or C which are alloy additions of <1% concentration in commercial Fe-6Ni and Fe-9Ni steel.

MÖSSBAUER SPECTROMETRY RESULTS AND INTERPRETATION

Ni Analysis. In α phase Fe-Ni alloys, the dependence of the isomer shift on the concentration of Ni is small (See Fig. 1). At lower Ni concentrations we have found an isomer shift of 0.0035 mm/sec/%Ni, larger than the 0.0023 mm/sec/ %Ni determined by Vincze and Campbell [4], but a shift of 0.032 ± 0.003 mm/sec was found for our Fe-12.0 at.% Ni alloy. We believe that we have observed a very small electric quadrupole splitting in this alloy which is twice as large as that reported for pure α-phase iron [5].

Much larger and easier to measure changes in Mössbauer spectra with changes in Ni concentration arise from hyperfine magnetic field effects. An Fe-9% Ni alloy is not a dilute solution of Ni in Fe, and there is a high probability than an Fe^{57} nucleus will experience perturbations in its hyperfine magnetic field from several adjacent Ni atoms. The perturbation from each Ni atom is not large so no structure is resolvable in Mössbauer peaks of Fe-Ni alloys. Nevertheless, for quantitative analyses of Ni composition it is sufficient to consider the change in the numerically calculated first moments and the change in the full width at half maximum (FWHM) of the $I = \pm^1/_2 \rightarrow I = \pm^3/_2$ absorption peaks as a function of Ni composition. From Fig. 1 we see that when an α phase Fe-9Ni alloy experiences a small decrease in Ni concentration the decrease in the FWHM of the $I = \pm^1/_2 \rightarrow I = \pm^3/_2$ peaks, the decrease in isomer shift, and the decrease in the mean hyperfine magnetic field will all be proportional to the Ni concentration change. We have shown that when the shift in the first moment and the increase in the FWHM of a family of normalized lorentzian functions are both proportional to composition, and in the case when the changes in all the nth moments of a family of peaked curves scale by a factor proportional to the composition raised to the nth power, the height from valley to peak of a difference spectrum of two such peaked functions generated for slightly different compositions will be closely proportional to the composition difference.

The $I = \pm^1/_2 \rightarrow I = \pm^3/_2$ peaks from Fe-<10 at.% Ni alloys fulfill the criteria for the proportionality of difference spectrum heights to small Ni composition changes. An Fe-8.8 at.% Ni alloy foil was tempered for 1 hr at 600°C to produce 1½% γ phase (see the small peak around channel 124 in Fig. 2). The difference of spectra taken from this material before and after tempering (also shown in Fig. 2) gives the same shape around the $I = \pm^1/_2 \rightarrow I = \pm^3/_2$ peak positions as for the normalized Fe-6.0 at.% Ni and Fe-8.8 at.% Ni alloys shown in Fig. 3, but with only about 1/5 the relative intensity. The proportionality between difference spectrum height and Ni concentration change indicates that the Ni content of the α-phase was reduced by 1/5 x (8.8-6.0)% ≈1/2% after tempering. A Ni content of 33% in the 1½% γ-phase is hence implied. This is larger than 21% Ni in the γ phase as predicted from the Fe-Ni phase diagram but shows the predicted trend. In Fe-Ni-X ternary alloys such differential evidence of Ni depletion in the α phase after tempering could be found on the high Doppler shift energy side of the $I = \pm^1/_2 \rightarrow I = \pm^3/_2$ absorption peaks (see Fig. 5), but the difference on the low Doppler shift energy side was dominated by the hyperfine magnetic field effects of the X element. These X element hyperfine field effects were first found in a commercial Fe-Ni alloy steel by Kim and Schwartz [6] who suggested that they arose from the small number of Fe^{57} nuclei near C atoms.

Mn, Cr, Si, and C Analysis. Early experiments by Wertheim et al. [7] and succeeding studies [4,8,9] have examined the hyperfine magnetic field effects induced at Fe^{57} nuclei by moderately dilute concentrations of alloy elements in α phase Fe. There are many probable local environments for an Fe^{57} nucleus in a non-dilute solid solution, so there is often a problem with the uniqueness of the empirical parameters which specify hyperfine magnetic field changes for Fe^{57} nuclei which are situated near X element atoms. Less ambiguous chemical information may be obtained for dilute solutions of X in Fe. Assuming that

380

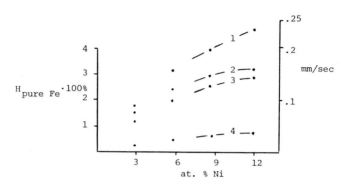

Fig. 1

1 Δ FWHM of $I=-\frac{1}{2} \rightarrow I=-\frac{3}{2}$ peak (mm/sec)
2 Δ FWHM of $I=+\frac{1}{2} \rightarrow I=+\frac{3}{2}$ peak (mm/sec)
3 $<H> - <H_{pure\ Fe}>$ $(H_{Fe} \cdot 100\ \%)$
4 Isomer shift with respect to pure Fe (mm/sec)

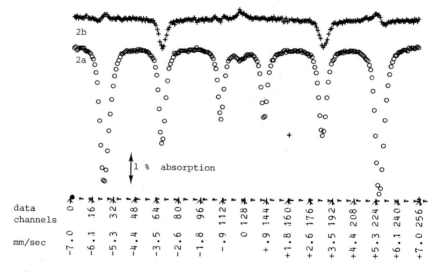

Fig. 2a o Fe-8.8 at.% Ni tempered 1 hr at 600°C
     2b + difference of quenched Fe-8.8 at.% Ni spectrum (not shown) minus
           spectrum of 2a (same absorption scale)

the martensitic quenched material is a random solid solution, the Mössbauer
spectrum of an Fe-dilute X alloy should show a main peak from Fe^{57} nuclei whose
hyperfine fields are largely unaffected by X atoms and additional satellite
absorption peaks from Fe^{57} nuclei which see hyperfine magnetic field perturba-
tations from one (rarely two) X atom neighbors. The ratio of satellite intensity to total peak intensity should be determined as in [7] by the binomial

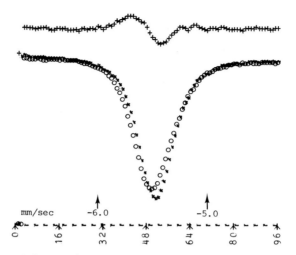

mm/sec   -6.0   -5.0

Fig. 3 o quenched Fe-8.8 at.% Ni
        * quenched Fe-6.0 at.% Ni
        + difference spectrum of above (* - o)   (same absorption scale)

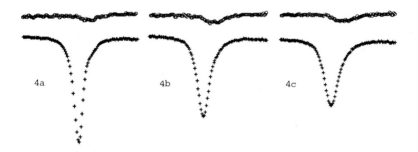

Fig. 4   Normalized spectra:
     4a   + quenched Fe-0.74 at.% Mn binary alloy spectrum
          o difference spectrum of + spectrum minus pure Fe spectrum
     4b   + quenched Fe-3.0 at.% Ni-0.76 at.% Mn spectrum
          o difference spectrum of + spectrum minus Fe-3.0 at.% Ni spectrum
     4c   + quenched Fe-6.0 at.% Ni-0.76 at.% Mn spectrum
          o difference spectrum of + spectrum minus Fe-6.0 at.% Ni spectrum

probability distribution but in dilute alloys this ratio will be nearly equal
to the concentration of X times the total number of nearest neighbor sites
around an Fe^{57} nucleus at which an X element atom will produce a strong hyperfine magnetic field perturbation.
    The satellite intensity for our Fe-0.77 at.% Mn alloy (Fig. 4) is revealed
by the difference spectrum method described below. The 6+% satellite to total

peak integrated intensity ratio found with this difference spectrum indicates that there are 9±2 sites around each Mn atom at which an Fe^{57} nucleus will experience a substantial perturbation in its hyperfine magnetic field. This would appear to correspond to the first nearest neighbor shell in the bcc structure which contains 8 atoms.

   In our analysis of Fe-Ni-X materials, we assumed that their Mössbauer peaks also consisted of a main peak, which had the same shape as the peak from an Fe-Ni binary alloy, and an additional satellite peak from Fe^{57} which experienced a large perturbation in their hyperfine magnetic fields due to the presence of an adjacent X element atom. The difference spectrum procedure used to extract the satellite peak intensity did not start with peaks of equal integrated intensity; the area enclosed by an Fe-Ni peak was always less than the area of the corresponding peak of the Fe-Ni-X spectrum by an amount equal to the satellite peak area. In addition to this normalization problem, and additional parameter for the difference spectrum method was necessary in the case of Fe-Ni-X ternary alloys and Fe-X binary alloys as well. The main absorption peak from these materials is due to Fe^{57} nuclei which are out of range of the strong local electron spin disturbances of the alloy element atoms, but this peak is known to exhibit a hyperfine magnetic field change of a small amount proportional to the X element concentration. Thus in preparing the difference spectra, it was necessary to vary both the shift of the main peak and the peak area normalization.

   We expect that there will be little intensity due to the satellite peak on the high Doppler shift energy side of the absorption peaks of Fe-Ni-X alloys because Mn, Cr, and Si are known to induce hyperfine magnetic field perturbations at neighboring Fe^{57} nuclei which are large and negative in α-phase Fe. For a given main peak shift, the tails on the high Doppler shift energy side of Fe-Ni-X and Fe-Ni Mössbauer peaks were least squares fit, and the peak normalization which gave the best fit was used for generating the difference spectrum. In this way, other values of peak shifts were then used to generate difference spectra, and the best value of peak shift was chosen on the basis of giving a smooth difference spectrum baseline in the region of the tip of the main part of the absorption peak. With our assumption of the Fe-Ni-X peaks being composed of an Fe-Ni component and a satellite component, a properly normalized and shifted Fe-Ni-X peak should match well the Fe-Ni peak on the high Doppler shift energy side, and then smoothly drop below the Fe-Ni peak intensity as the peaks are examined from high to low Doppler shift energy. Dips or sudden rises in the baseline of the difference spectra would hence not be expected and were grounds for rejecting most difference spectra. The least squares fitting procedure gave results for the satellite peak intensity which were within ~2% of the satellite intensities found in the difference spectra generated with the authors' discretion; this discrepancy was, in part, attributed to variations in the asymmetric peak distortion described in the experimental section.

   Difference spectra of quenched Fe-Ni-0.75 at.% Mn and Fe-Ni alloys are shown in Fig. 4. The satellite peak intensity was consistently larger than predicted for hyperfine field perturbations from Mn atoms only in the first nearest neighbor shell. Based on these integrated intensities there appear to be 13 ± 2 nearest neighbor sites around each Mn atom at which an Fe^{57} nucleus will experience a strong perturbation in its hyperfine magnetic field in Fe-3.0 at.% Ni and Fe-6.0 at.% Ni alloys. No such difference in hyperfine field effects were found for Cr in Fe, Fe-3.0 at.% Ni or Fe-6.0 at.% Ni hosts for which there are 14 ± 2 strongly perturbing sites around an Fe^{57} nucleus. We believe that Cr and Mn neighbor satellite peaks extracted from the broadened peaks of Fe-9Ni and Fe-12Ni hosts showed an intensity increase which indicates 20 ± 5 strongly perturbing neighbor sites. A quantitative basis has now been provided for determination of small Mn and Cr concentration changes in α phase Fe-Ni-X alloys

383

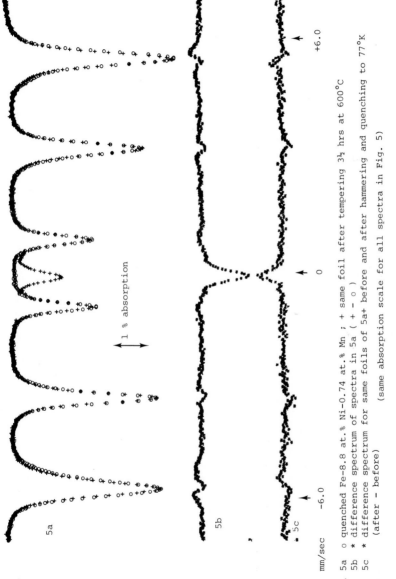

1 % absorption

mm/sec   −6.0   0   +6.0

Fig. 5a o quenched Fe-8.8 at.% Ni-0.74 at.% Mn ; + same foil after tempering 3½ hrs at 600°C
5b * difference spectrum of spectra in 5a (+ − o)
5c * difference spectrum for same foils of 5a+ before and after hammering and quenching to 77°K
(after − before)   (same absorption scale for all spectra in Fig. 5)

by examination of spectrum intensity differences on the low Doppler shift energy sides of the Mössbauer absorption peaks.

Figure 5 shows transmission Mössbauer spectra of an Fe-8.8 at.% Ni-0.76 at.% Mn foil which were taken before and after $3\frac{1}{2}$ hrs. tempering at 600°C to produce 6% γ phase. The difference spectrum around the $I = \pm^1/_2 \rightarrow I = \pm^3/_2$ (and $I = \pm^1/_2 \rightarrow I = \pm^1/_2$) absorption peaks shows a change on the high Doppler shift energy side which indicates a reduction of ~0.6 at.% Ni. After roughly correcting the intensity of the satellite peaks on the low Doppler shift energy side for this Ni reduction, we find a decrease in Mn neighbor satellite intensity which indicates a loss of ~0.2 at.% Mn from the α phase. The metastable γ phase was then transformed martensitically to α phase by quenching the foils in liquid nitrogen and lightly hammering them at room temperature. The difference spectrum of Fig. 5c shows a transformation of half of the γ phase and a simultaneous gain in average Ni content of the α phase. No change in Mn content of the α phase is evident, however. Very similar tempering effects were found in Mössbauer spectra of Fe-9Ni-C, Fe-9Ni-Si, and Fe-9Ni-Cr alloys. The Si, Cr, and especially the C segregation processes appear to occur more readily than the Mn segregation.

CONCLUSIONS

After hyperfine magnetic field effects are systematized so that chemical composition changes in the α phase can be quantified by Mössbauer spectrometry, a difference spectrum method can be used to detect small chemical composition changes in α phase Fe-Ni-X alloys. This procedure is being used to determine the kinetics of alloy element segregation from martensite into the austenite precipitated during a tempering treatment.

ACKNOWLEDGEMENT

This work was supported by the Division of Materials Sciences, Office of Basic Energy Sciences, U.S. Department of Energy under contract No. W-7405-Eng-48.

REFERENCES

1. T. Ooka et al., Jour. Japan Inst. Metals 30, 442 (1966).

2. C. K. Syn, B. Fultz, and J. W. Morris, Jr., Met. Trans. A 9, 1635 (1978).

3. L. H. Schwartz in: Applications of Mössbauer Spectroscopy Vol. I, chapter 2. R. Cohen ed. (Acad. Press, N.Y. 1976).

4. I. Vincze and I. A. Campbell, Jour. Phys. F 3, 647 (1973).

5. J. J. Spijkerman et al., Phys. Rev. Lett. 26, 323 (1971).

6. K. J. Kim and L. H. Schwartz, Mater. Sci. Eng. 33, 5 (1978).

7. G. K. Wertheim et al., Phys. Rev. Lett. 12, 24 (1964).

8. M. B. Stearns, J. Appl. Phys. 36, 913 (1965).

9. W. E. Sauer and R. J. Reynik, J. Appl. Phys. 42, 1604 (1971).

TDPAC-STUDIES OF ELECTRIC FIELD GRADIENTS IN AMORPHOUS
METALLIC SYSTEMS

P. HEUBES, D. KORN, G. SCHATZ and G. ZIBOLD
Fakultät für Physik, Universität Konstanz, D-7750 Konstanz,
Fed. Rep. Germany

ABSTRACT

The time differential perturbed γ-γ angular
correlation technique (TDPAC) is applied to
the amorphous metallic systems Ga, Bi, $In_{50}Au_{50}$
and $In_{80}Ag_{20}$. The electric field gradient
tensor probed by ^{111}Cd nuclei shows a broad
probability distribution with a relative width
of 0.4 - 0.5 for all systems, as suggested by
a continuous random structural model.

INTRODUCTION

Outstanding features of amorphous metals are their local atomic
arrangement, lacking long-range periodicity, and their electronic
structure. In recent years several structural models have been
proposed which are roughly characterized in terms of "microcry-
stalline" and "continuous random" models [1,2]. In case of amorphous metals and metallic alloys,the radial distribution analysis
from diffraction studies favours the dense random packing hypothesis although an unambiguous determination by diffraction techniques is principally not possible [1]. Recently additional information on the local atomic arrangement has been derived from
the electric field gradient (EFG) acting onto probe atoms [3,4].
This quantity depends on the symmetry of the charges surrounding
the probes and therefore provides direct experimental information
on the angular distribution of local atomic arrangements. The
time differential perturbed angular correlation technique (TDPAC)
complements other EFG sensitive methods (e.g. Mössbauer spectroscopy), since one can work with a low number of probes ($> 10^{10}$ probe
nuclei per sample). This special feature turns out to be important
in the case of metallic systems since most of these samples are
only available in the amorphous state by quench condensation onto
cooled substrates keeping the deposited films below a critical
thickness. Mössbauer effect experiments, however, require an absorber thickness not below a certain thickness (typically 1 µm)
in order to obtain a reasonable absorption effect. This restric-
tion is not present in TDPAC investigations. Another feature related to the exclusive EFG sensitivity of the TDPAC technique is
avoiding possible complications from other interactions (e.g.
isomer shifts). Since samples with structural disorder usually
exhibit a broad distribution of interaction frequencies,an un-
equivocal separation of different contributions is a very difficult task. The conclusions drawn from TDPAC results, however, are
not affected by this ambiguity.

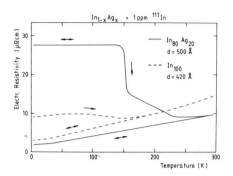

<u>Fig. 1</u>

Electrical resistivity of quench-condensed In-Ag alloys as a
function of temperature.

In the present work we have extended previous investigations of
amorphous Ga [4] to Bi metal and In-Au and In-Ag metallic alloys.
For the TDPAC measurements the 172-247 keV γ-γ cascade of ^{111}Cd
after ^{111}In decay was used with the intermediate state $I^{\pi} = 5/2^{+}$
($T_{1/2}$ = 84 ns). In the presence of a non-vanishing EFG this state
splits up into three quadrupole energy levels. Therefore three
transition frequencies   contribute   to the time differential
perturbed angular correlation (e.g. fig. 2b), from which the
splitting parameters V_{zz} and $\eta = (V_{xx}-V_{yy})/V_{zz}$ can be derived.
Here V_{ij} are the components of the EFG tensor at the
probing nucleus (for details see ref. [5]). The amorphous film
samples were prepared by quench condensation on a sapphire substrate at 9 K in vacuum of 10^{-8} mbar. Less than 1 ppm radioactive
^{111}In atoms had been added to the bulk material before evaporation
reaching total activities of about 5 μCi in the film samples.
The amorphous state was studied immediately after condensation
revealing in all cases a broad and asymmetric distribution of
quadrupole transition frequencies caused by non-unique EFGs.
Subsequently the crystalline states were restored by annealing
the film samples above the phase transformation temperatures and
verified by monitoring the known behaviour of electrical resistivity. With the exception of Bi then all samples exhibited well-
defined quadrupole interactions caused by unique EFGs in accordance with regular lattice locations of the probe nuclei in non-
cubic crystals. The derived components V_{ij} of the EFG tensor agree
well with those previously observed in the corresponding bulk
materials. Fig. 1 and fig. 2 show typical results obtained from
In-Ag films (500 Å) with the composition $In_{80}Ag_{20}$. The amorphous
to crystalline transformation (T_{cryst} = 150K) is accompanied by
a striking change of the electrical resistivity (fig. 1) as well
as of the TDPAC spectra R(t) (fig. 2). The crystallization process
clearly decomposes the sample into pure In metal and the intermetallic compound In_2Ag which both have a tetragonal structure
($V_{zz} \neq 0$, $\eta = 0$).

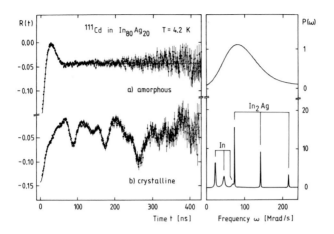

Fig. 2

Left: TDPAC spectra of [111]Cd after ^{111}In decay in (a) amorphous and (b) crystalline $In_{80}Ag_{20}$.

Right: Results of least squares fits to the data displaying the quadrupole transition frequencies which contribute to the TDPAC spectra.

A quantitative analysis of the quadrupole frequency distribution observed in the amorphous state has been performed in ref. [4] for Ga. Within this framework the experimental data are compared with probability distributions $P(V_{zz}, \eta)$ of the EFG derived from various computer simulations of the amorphous structure, applying the point-charge model and the hard-sphere approach. We have performed this procedure for all investigated systems with the result that all experimental data are consistent with the hard sphere random packing model. The EFG distribution $P(V_{zz}, \eta)$ predicted by this structural model [4] can be approximated by the following analytical relationship:

$$P(V_{zz}, \eta) = (2/\pi)^{1/2} \eta/\Delta_{zz} \exp\{-1/2[(V_{zz}-V^o_{zz})/\Delta_{zz}]^2\} \qquad (1)$$

The main features of this distribution are the linear dependence on the asymmetry parameter η and the Gaussian broadening of the component V_{zz} around V^o_{zz}. Making use of this relationship the TDPAC spectra $R(t)$ have been analysed within a least squares fit procedure underlying the following perturbation factor:

$$G_{kk}(t) = \int_{-\infty}^{+\infty} \int_{0}^{1} P(V_{zz}, \eta)\, G_{kk}(t, V_{zz}, \eta)\, dV_{zz}\, d\eta \qquad (2)$$

The solid lines in fig. 2 represent the least squares fit results for $In_{80}Ag_{20}$. In table 1 the results for Ga, Bi, $In_{50}Au_{50}$, and $In_{80}Ag_{20}$ are summarized. The amorphous state of each metallic system is characterized by its distinct mean value V_{zz}^{o} whereas the relative broadening $\Delta_{zz}/V_{zz}^{o} = 0.4 - 0.5$ is a common feature of all samples in accordance with the EFG distribution predicted by the hard sphere random packing model.

In order to examine the sensitivity of the experimental EFG data on the spatial distribution of atoms within the first few coordination shells, additional experiments have been performed on In-Ag alloys with low Ag concentration (< 10 at. %). These alloys cannot be forced into the amorphous state (dashed line in fig. 1). The TDPAC results obtained immediately after condensation clearly contradict the perturbation factor (2) and exhibit the unique EFGs observed also after annealing including minor distributions. These distributions originate probably from the very small grain sizes of the quench-condensed samples.

This work was supported by the Deutsche Forschungsgemeinschaft.

Table 1

Sample properties and electric field gradient parameters measured with ^{111}Cd probes after ^{111}In decay in amorphous Ga, Bi, $In_{50}Au_{50}$ and $In_{80}Ag_{20}$.

Sample	Ga	Bi	$In_{50}Au_{50}$	$In_{80}Ag_{20}$
d [Å]	2000	1000	800	500
T_m [K]	4.2	4.2	77	4.2
T_{cryst} [K]	17	16	200	150
V_{zz}^{o} [10^{17} Vcm^{-2}]	2.7(1)	5.1(2)	4.7(3)	3.1(1)
Δ_{zz} [10^{17} Vcm^{-2}]	1.1(1)	2.5(2)	2.4(2)	1.3(1)
Δ_{zz}/V_{zz}^{o}	0.41(3)	0.49(5)	0.52(6)	0.42(3)

d = film thickness; T_m = temperature during TDPAC measurements; T_{cryst} = crystallization temperature.

REFERENCES

1. J.L. Finney, in: The structure of non-crystalline materials, ed. P.H. Gaskell (Taylor and Francis, London, 1977) p. 35 and references therein

2. D.E. Polk and B.C. Giessen, in: Metallic Glasses (American Society for Metals, Metals Park, Ohio, 1976) p. 1

3. J. Bolz and F. Pobell, Z. Physik $\underline{B20}$ (1975) 95

4. P. Heubes, D. Korn, G. Schatz and G. Zibold, Phys. Lett. $\underline{74A}$ (1979) 267

5. E.N. Kaufmann and R.J. Vianden, Rev. Mod. Phys. $\underline{51}$ (1979) 161

Published 1981 by Elsevier North Holland, Inc.
Kaufmann and Shenoy, editors
Nuclear and Electron Resonance Spectroscopies Applied to Materials Science

LASER-PULSE MELTING OF HAFNIUM-IMPLANTED NICKEL STUDIED WITH TDPAC AND RBS/CHANNELING TECHNIQUES

L. BUENE*, E. N. KAUFMANN**, M. L. McDONALD
Bell Laboratories, Murray Hill, New Jersey 07974

J. KOTHAUS, R. VIANDEN, K. FREITAG
Institut für Strahlen-und Kernphysik, University of Bonn, West Germany

C. W. DRAPER
Western Electric Engineering Research Center, Princeton, New Jersey 08540

ABSTRACT

The perturbed angular correlation technique has been applied to study the local environment of tantalum in nickel after ion implantation of hafnium and after laser-pulse melting. The magnetic hyperfine interaction at the daughter nucleus tantalum in nickel is used to determine the uniqueness of the tantalum lattice site. Several hafnium concentrations were employed and auxiliary measurements using ion-beam channeling, Rutherford backscattering and Auger electron spectroscopy were performed.

INTRODUCTION

Recently it has been demonstrated that impurities implanted into various metal substrates, in particular nickel, can be incorporated in the substrate lattice to concentrations exceeding the equilibrium solubility of the system by laser-pulse melting of a surface layer followed by rapid resolidification. The resolidification process is epitaxial and the impurity concentration redistributes according to an effective distribution coefficient applying to the moving solid-liquid interface [1]. The question arises as to how, at the microscopic level, the solute atoms incorporate into the solid as the interface moves past them. For example, one might imagine that as the impurities incorporate into the lattice beyond their normal solubility, defects must also be incorporated during the solidification process in order to minimize total strain energy. Such a question can be examined using the hyperfine interaction which sees the solid from the viewpoint of the impurity atom.

The nickel lattice is an excellent choice for this experiment for several reasons. It is magnetic thus providing a convenient hyperfine interaction. It has been studied with respect to laser pulse melting more extensively than any other pure metal [1]. It shows no high temperature allotropic transition thus insuring the simplest possible metallurgical picture. In addition, it is a sufficiently light element to permit Rutherford backscattering experiments as well.

The appropriateness of a hafnium probe arises from the following considerations. The binary phase diagram for the hafnium-nickel system shows no liquid phase miscibility gap and shows a finite (0.5%) but *limited* solid solubility [2]. Thus one expects to easily exceed the equilibrium solubility but at the same time not generate second phase precipitates. The relative atomic radius of hafnium to that of nickel is quite large and therefore might be expected to enhance any defect (for example, vacancy) trapping at the resolidification interface. Hafnium has an isotope, *viz.*, 181-Hf, which is highly appropriate for perturbed angular correlation measurements. The parent halflife is 45 days and the daughter nucleus,

---

* Supported in part by Royal Norwegian Council for Scientific and Industrial Research. Resident visitor at Bell Laboratories. Present address: Veritas, 1322 Høvik, Norway.

** Partially supported while at Bonn through the Heinrich-Hertz Stiftung.

392

181-Ta, displays an excited state at 482keV with appropriate halflife and magnetic moment for this study. In addition, previous work on hafnium-implanted nickel and hafnium-alloyed nickel has been reported. The lattice location for hafnium implanted in nickel has been reported by Kaufmann, et al [3]. Several early references concerning the work on radioactive hafnium implanted in nickel followed by PAC are cited in Reference [3]. More recently a PAC study has been performed by the group of Gerdau [4] which confirms the small but finite solubility of hafnium in nickel through normal metallurgical preparation methods and also at higher concentrations has demonstrated the formation of intermetallic compounds.

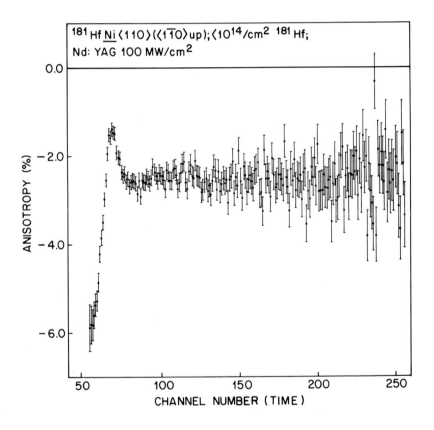

Fig. 1. Spin precession pattern for the 482keV state in 181-Ta in Ni single crystal after implantation of the 181-Hf parent and laser pulse melting. (See Figure 10 of Ref. [5] for the pattern observed before melting).

EXPERIMENTAL DETAILS

Nickel single crystals were cut from a single boule and oriented such that the <110> axis was normal to the plane of the samples. The samples were lapped and electropolished to a mirror-like finish. 181-Hf radioactivity was implanted at the University of Bonn isotope separator. In some cases, inactive 180-Hf was implanted at Bell Laboratories prior to the radioactive implantation in order to increase the concentration of hafnium in the alloy. Subsequent angular correlation measurements were performed in Bonn and at Bell Laboratories. Laser-pulse melting was performed at the Engineering Research Center of Western Electric in Princeton. A 35μm focussed spot from a Nd:YAG Q-switched laser pulsed at 11 kHz was raster-scanned across the sample. We estimate the fluence of radioactive hafnium at less than 10^{14}/cm^2. Typical spin precession spectra for this low-fluence as-implanted case are shown elsewhere in this volume [5] and will not be repeated here. Analysis of the spectra obtained in the low-fluence case, including least-squares fitting with the possibility of two different unique frequencies, indicates preliminarily that there exists a dominant site occupied to about 70% and a second site which is probably a defect associated site due to the radiation damage from ion implantation itself.

Figure 1 shows the spectrum obtained after laser-pulse melting of a low-fluence sample. It is immediately clear that the large amplitude unique oscillations seen previously (Figure 10 of Reference [5]) have disappeared. One sees in Figure 1 the rapid loss of anisotropy characteristic of a nonunique environment. Figures 2a and b present the results for as-implanted and laser-melted cases where a preimplant of approximately 1×10^{16}/cm^2 was applied. Here the unique but damped oscillations in the as-implanted case are further reduced in amplitude with increased damping after laser melting. The solid lines in Figure 2 are least-squares fits to the data. An analogous pair of spectra are shown in Figures 3a and b for the higher preimplantation concentration of hafnium of 4×10^{16}/cm^2. In the as-implanted case, one sees virtually no oscillation in the spectra whereas, after laser melting, some oscillatory behavior is recovered. In all cases the anisotropy at the time zero of these spectra is larger than that found in the oscillation at times greater than zero. Some hafnium therefore resides in sites where strong nonunique interactions destroy the anisotropy in short times. These spectra also show that some hafnium is incorporated substitutionally in the regrowth process.

Auxiliary Rutherford backscattering measurements were performed and three spectra are shown in Figure 4. The top spectrum corresponds to an as-implanted case at a fluence comparable to the sample referred to in Figure 3. One sees that aside from a deep tail, the hafnium peak is nonsubstitutional since its amplitude is independent of whether the ion beam is channeling along <110> directions or is incident in a random direction. One also sees from the host scattering that a peak near the surface implies an essentially amorphized layer. In the middle spectrum of Figure 4, one sees the regrowth of the nickel because the large surface peak is no longer present and one also sees the redistribution in depth of the hafnium such that the deep hafnium is now substitutional. This redistribution is consistent with the spin precession results. A fraction is found on reasonably good lattice sites while the remainder resides in a surface layer which is presumably highly nonunique. One also sees from Figure 4 that laser-pulse melting in air (middle spectrum) as compared to helium (lower spectrum) alters the character of the surface layer. It appears that in the absence of oxygen, the surface peak displays some substitutionality and is significantly smaller. Thus we conclude that hafnium at the surface, when melting occurs in air, is incorporated in an oxide layer which accounts for a highly nonunique environment. This conclusion is corroborated by Auger electron spectroscopy which shows both a reduction in oxygen *and* hafnium concentration at the surface when melting occurs in helium. The oxidation of laser treated metals has been pointed out in Reference [6].

Our preliminary analysis shows that, whereas in as-implanted samples one can decompose the spectra into two frequency components, after laser pulse melting, only a single component with a unique frequency can be discerned superimposed on a component with a rapid loss of anisotropy (presumably hafnium in the oxide layer). Thus no direct evidence for a unique defect association in the regrowth incorporation process has been obtained.

Thermal annealing experiments have also been carried out on low-fluence samples. Although for lack of space the data cannot be displayed here, the results show that after annealing at temperatures above approximately 600°C most of the hafnium atoms leave the unique site and yield a strongly damped spectrum. This contradicts the expectation based on the binary-phase diagram wherein annealing at high temperature should simply cause substitutional diffusion of the hafnium impurity so long as the concentration is below the solubility limit. We are presently investigating the possibility that the hafnium can oxidize *in situ* in nickel or that the hafnium indeed migrates to the surface and oxidizes.

394

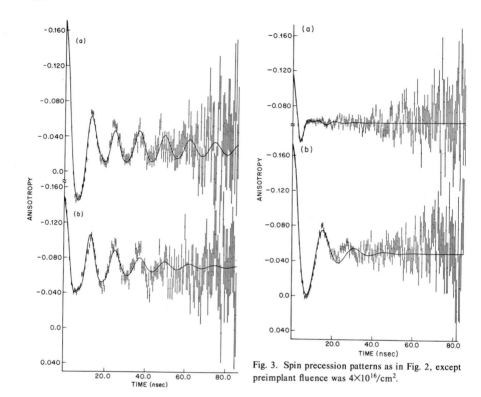

Fig. 3. Spin precession patterns as in Fig. 2, except preimplant fluence was $4\times10^{16}/cm^2$.

Fig. 2. Spin precession pattern for 181-Hf implanted Ni which was preimplanted with $1\times10^{16}/cm^2$ 180-Hf a) as-implanted and b) after laser-pulse melting.

It is interesting to note that in a low-fluence sample, laser-pulse melting leaves all hafnium in the nonunique surface layer whereas at higher hafnium concentrations, only a portion (40%) is segregated to the surface. This implies a saturation effect is operative. To check this, a low-fluence sample which had been laser-pulse melted was subsequently over-implanted with inactive hafnium to a fluence of $1\times10^{16}/cm^2$ and remelted. The resulting spin precession spectrum is shown in Figure 5. It is clear that a fraction of the hafnium now occupies unique lattice sites. Thus reincorporation into the lattice through recoil implantation followed by melting with an excess of hafnium atoms present is possible. Details of these results and further work will be published elsewhere [7]. One important test still to be made is laser-pulse melting using a single large-diameter irradiated region rather than a raster-scanned sequence in which the same point is remelted multiple times. The multiple melting and thus repetitive resegregation of the hafnium (oxide) provides a greater opportunity for collection at the surface than would a single pulse. The actual frequencies and frequency damping found in the data shown here indicate that at supersaturation for hafnium in nickel, the lattice is under a good deal of strain such that the field experienced by the hafnium is distributed around a mean value. In addition, the hyperfine field at greater hafnium concentrations is found to be smaller than at the lower concentrations due to the dilution effect of adding nonmagnetic impurities [4].

One of us (ENK) is grateful for the hospitality shown him during the performance of some of this work at the University of Bonn and for support from the Heinrich-Hertz Stiftung. We thank W. F. Flood and D. C. Jacobson for sample preparation assistance.

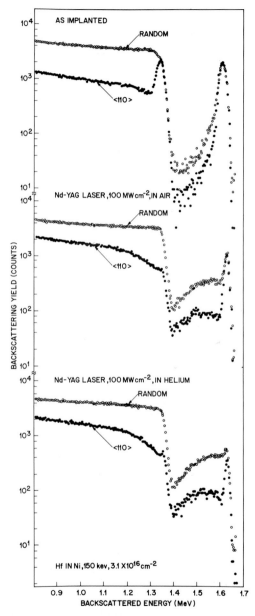

Fig. 4. Rutherford backscattering/channeling energy spectra observed using a 1.8MeV ^{4}He$^+$ beam, on a <110> Ni crystal as-implanted with 3.1×10^{16}/cm^2 180-Hf (top), after laser pulse melting in air (middle), and after melting in a helium atmosphere (bottom).

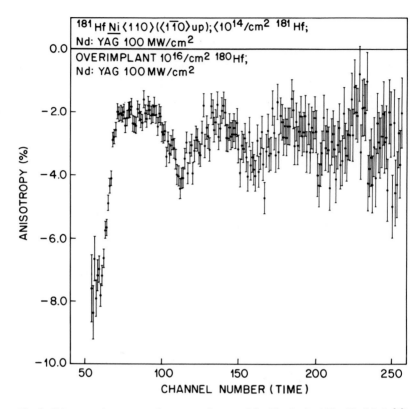

Fig. 5. Spin precession pattern of same sample as used for Fig. 1. (and Fig. 10 of Ref. [5]) after overimplantation with $10^{16}/cm^2$ 180-Hf and laser pulse remelting.

REFERENCES

1. L. Buene, J. M. Poate, D. C. Jacobson, C. W. Draper and J. K. Hirvonen, Appl. Phys. Letters, 385 (1980) and in Mat. Res. Soc. Proceedings (1980) *1*, in press.

2. W. G. Moffatt, *The Handbook of Binary Phase Diagrams* (General Electric Co., Schenectady, 1978) (12/79 update).

3. E. N. Kaufmann, J. M. Poate and W. M. Augustyniak, Phys. Rev. *B7*, 951 (1973).

4. E. Gerdau, H. Winkler, W. Gebert, B. Giese and J. Braunsfurth, Hyperfine Interact. *1*, 459 (1976): E. Gerdau, H. Winkler, B. Giese and J. Braunsfurth, Hyperfine Interact. *1*, 469 (1976).

5. E. N. Kaufmann, this volume (Figure 10).

6. S. M. Metev et al., J. Phys. D*13*, L75 (1980).

7. J. Kothaus et al., to be published.

THE ELECTRIC FIELD GRADIENT IN NONCUBIC METALS AND ALLOYS

W. WITTHUHN, U. DE, W. ENGEL, S. HOTH, R. KEITEL, W. KLINGER, and R. SEEBÖCK
Physikalisches Institut der Universität Erlangen-Nürnberg, 852 Erlangen, Germany

ABSTRACT

The electric field gradient (efg) present in noncubic solids
causes an energy-splitting of the nuclear levels via the
quadrupole hyperfine interaction. During the last few years
the perturbed angular correlation method has proved a unique
experimental tool for investigating this interaction especially in metals. The basic principles of the method are
discussed. Recent experimental results are given for pure
metals and highly diluted systems as well as for alloys and
intermetallics. The last section deals with theoretical aspects of the temperature dependence of the efg in pure metals.

INTRODUCTION

The knowledge of the charge distribution in a solid is of interest for the understanding of its basic properties and of great importance for many applications
in material science. In noncubic solids this charge distribution creates an
electric field gradient (efg), which causes an energy-splitting of the nuclear
levels due to the quadrupole hyperfine interaction (QI) between the nuclear
quadrupole moment Q and the efg at the site of the nucleus. This interaction can
be studied by a variety of experimental methods. In metals, however, they are
limited in general to a few favourable cases or are restricted to low temperatures. Stimulated by the work of Raghavan [1] the time-differential perturbed
angular correlation (distribution) methods (TDPAC TDPAD) have been demonstrated
during the last years to be a unique experimental tool for investigating this
interaction. Here no limitations due to skin effect exist as in classical NMR
or NQR measurements and no low temperatures in the mK range are needed as required for specific heat and nuclear orientation experiments. Thus, the full
temperature range from mK up to the melting points of the metals or alloys
can be covered. In this respect the method is superior to the Mössbauer-effect
where the signal depends on a Debye-Waller-factor and decreases with increasing
temperatures. A recent comprehensive review of the electric field gradients in
metals has been given by Kaufmann and Vianden [2] . Here a detailed comparison
of the experimental methods as well as a survey of the available experimental
data and present status of the theoretical understanding of the origin of the
efg can be found.

METHOD

The perturbed angular correlation methods have been described highly detailed
in Ref. [3] , here only the principles will be mentioned. The detection of the
quadrupole interaction is based on the observation of the anisotropic γ-radia-
tion patterns emitted from an isomeric nuclear state with a nonrandom nuclear
spin orientation. The useful lifetimes of the excited levels range from a few
ns up to some μs, in favourable cases up to a few ms, the limits being given
by the interaction strength or the time resolution of the apparatus and the
statistics or measuring time.

In the "angular correlation" method (TDPAC) the nuclear orientation is produced by means of a γ-γ-cascade, where the detection of the emission direction of the first radiation populating the isomeric state of interest selects an ensemble of nuclei with definite spin orientation. The intensity distribution of the second radiation is then anisotropic in space and correlated to the direction of the first one. It should be noted that the correlation is a direct consequence of the conservation of angular momentum and thus is governed by the angular momenta involved only (level spins and γ-ray multipolarities). The influence of the quadrupole interaction results in a perturbation of this correlation in time.

The "angular distribution" method (TDPAD) can be treated in an analogous way. Here the isomeric levels are populated by a nuclear reaction. The angular momentum is transferred by the reaction preferentially in a plane perpendicular to the ion-beam direction. Thus the spins of the excited nuclei are oriented again. The distribution of the emitted deexcitation γ-radiation is then anisotropic with respect to the beam axis.

The observed γ-ray intensity is given by

$$N(\Theta,t) = N_o \, e^{-t/\tau} \, W(\Theta,t) \, , \tag{1}$$

where τ is the meanlife of the isomeric state and Θ gives the angle between the two radiations (or between the γ-ray and the beam-axis). The angular correlation $W(\Theta,t)$ can be written (for details see Ref. [3]):

$$W(\Theta,t) = \sum_{k,even} A_{kk} \, G_{kk}(t) \, P_k(\cos\Theta) \, . \tag{2}$$

In the case of an axially symmetric efg with random orientation (polycrystalline source) the perturbation factor $G_{kk}(t)$ is given by

$$G_{kk}(t) = \sum_n s_{kn} \, \cos \, (n\omega_o t) \tag{3}$$

where the frequency ω_o contains the information on the quadrupole coupling constant $\nu_Q = (e^2 qQ)/h$:

$$\omega_o = \frac{3\pi}{I(2I-1)} \, \nu_Q \tag{4}$$

(half-integer nuclear spin I is assumed).

Conventionally the the measured intensities of multi-detector systems are combined properly to an intensity ratio R(t), which is independend of the exponential decay and is directly proportional to the product $A_{kk} G_{kk}(t)$.

RESULTS

The results of our present series of quadrupole interaction studies in metallic environment are summarized in the table I. The main emphasis laid on the temperature dependence of the efg, in the binary alloys the concentration dependence was investigated also. Fig. 1 shows two typical TDPAD-time spectra for [120]Sb in tin at different target temperatures. The solid lines are fits of the theoretical patterns for a polycrystalline sample and nuclear spin I = 3, with

Host lattice	Mg	Zn			Cd			In	
Probe atom	^{111}Cd	^{67}Zn	^{67,69,71}Ge	^{111}Cd	^{111}Cd	^{117}In	^{113,116}Sn	^{111}Cd	^{117}In
Measured dependence	T	T	T	T	T		T	T	T
Method	C	D	D	C	C	C	D	C	C
Ref.		4,5	5,6		4	4	4,7	4	4

Host lattice	Sn		Tl	Bi	CdMg	InTl	InCd	InPb	InBi
Probe atom	^{111}Cd	^{120}Sb	^{111}Cd	^{111}Cd	←		^{111}Cd		→
Measured dependence	T	T	T	(T)	←		T and X		→
Method	C	D	C	C	←		C		→
Ref.	4	8			9	10	ł0	10	

T = temperature; x = concentration; C = TDPAC; D = TDPAD

Table I SUMMARY OF THE INVESTIGATED METALS AND ALLOYS

the amplitude and the basic frequency ω_0 being the only free parameters. The figure shows excellent agreement between the experimental data and the predicted modulation pattern. Furthermore the increase of the interaction frequency with decreasing temperature can be seen clearly. This temperature dependence is depicted in Fig. 2. It follows closely the relation [4] :

$$eq(T) / eq(O) = 1 - B \, T^{3/2} \, . \tag{5}$$

The different impurity atoms in tin measured up to now show a different relative temperature variation of the efg, characterized by the slope parameter B (cf. Eq. 5); some of the available data are shown in Fig.2 also. Recently a first attempt was undertaken, to relate this variation of B to charge-, mass-, and size-differences of host and impurity atoms in Tl metal [11] . Assuming a Debye-Waller factor to be responsible for the temperature dependence of the efg in the pure lattice, the authors show systematic trends by introducing an effective Debye temperature, which in turn depends on mass ratios and effective force constants. Following this approach we find rough agreement in the tin case also [8] . At the present status of the theory no definite conclusions can be drawn. Systematic trends, however, seem to emerge from the experimental data. They indicate a realistic chance for the future, to get information on local vibrational modes at impurity sites from quadrupole hyperfine experiments.

The application of the TDPAC method to binary alloys is still in an explorative phase and only a few experiments have been published up to now. Fig. 3 shows the variation of the efg at the ^{111}Cd probe nucleus in InCd alloys as a function of temperature and concentration. At low Cd concentrations (up to approximately 5 at % Cd) the alloy forms the tetragonal (In)-phase. Here the efg follows the $T^{3/2}$ relation as in pure metals. The solid lines of the three most In-rich alloys in Fig. 3 are fits of Eq. 5 to the data. At higher Cd-concentrations, however, the efg changes rapidly (the solid lines of these alloys in Fig. 3 are drawn only to guide the eye) if the boundary to the cubic ß-phase around 300 K is approached.

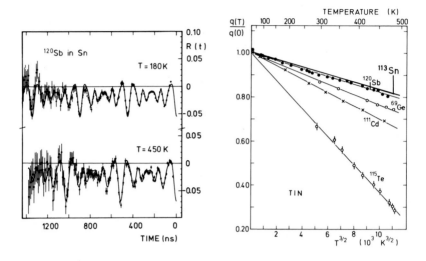

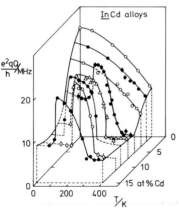

Fig. 1   TDPAD time spectra for ^{120}Sb in tin showing the modulation of the γ-ray intensity at different temperatures.

Fig. 2   Temperature dependence of the efg in tin measured with different probe-atoms. For Refs. of the experimental data see Ref. 2.

The values of the efg measured in this cubic phase are small compared to those in the tetragonal phase. They also show only a very weak temperature dependence, which is a well-known (but not yet explained) characteristic of the efg in cubic local disturbed systems [12] . Thus, the cubic-noncubic phase transition clearly shows up in the measurements of the efg. A second series of discontinuities appear around 120 K (see Fig. 3). The efg-values measured below this transition temperature show again the typical characteristics of a cubic system. This leads to the conclusion that the cubic ß-phase is formed again. This phase transition has not been noted before.

A second interesting problem is the dependence of the efg on the axial ratio c/a. This ratio can be varied via pressure or temperature. The relative change of c/a with pressure, however, is rather small and the variation of temperature also changes the vibrational state of the lattice, therefore masking the c/a depen-

Fig. 3   The efg for ^{111}Cd in In-rich InCd alloys as function of temperature and Cd-concentration.

dence. A convenient way of changing c/a continuously over a wide range is achieved
by changing the composition of binary alloys. In the present investigations on
tetragonal In-rich alloys and hexagonal CdMg alloys the c/a ratios were varied
by about 20%. Over this range the efg follows linearly the c/a ratio (for details
see Refs. 9,10), as expected from simple point-charge calculations. This result
supports for the first time in binary alloys the universal correlation between
elctronic and ionic field gradients observed in pure metals and highly diluted
systems [13] .

The $T^{3/2}$ relation (Eq. 5) of the quadrupole coupling constant [4] has stimulated
experimental and theoretical work. The current status of the theoretical understanding (reviewed in Ref. 2) of this temperature dependence is still unsatis-
fying. We have investigated the efg in pure nontransition metals starting from
the total Hamiltonian of the many body system consisting of ions and conduction
electrons, using the adiabatic approximation for the Schrödinger equation. Here
the ions move in a weak anharmonic effective potential, which originates from the
direct Coulomb interaction of the ions and the energy of the conduction electrons,
the latter representing an indirect interaction of the ions via the electron gas.
The total efg in lowest order is identical with the one derived from a simple
screened potential (for details see Ref. 14). In order to calculate its temperature dependence the total efg was expanded in a power series of the ionic displacement differences up to second order. Thus automatically displacement correlations are taken into account. Explicit expressions for the temperature of the
efg are obtained for low and high temperatures. In the low temperature limit the
influence of the anharmonicity of the potential are neglected and the Debye
approximation was used. This leads to the expression [14] :

$$eq = a + bT^4 \qquad (T \ll \Theta) \qquad (6)$$

Where a and b are temperature independent constants. This result differs from
previous calculations based on the mean square displacements only. In the high
temperature limit the influence of the anharmonic parts of the effective potential were taken into account. One obtains:

$$eq = a' + b'T + c'T^2 \qquad (T > \Theta) \qquad (7)$$

These expressions were fitted to the experimental data for In [15] . The results
are shown in Figs. 4 and 5.

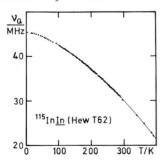

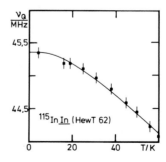

Fig. 4   Temperature dependence of the efg in pure In metal. The solid
line is a fit of Eq. 7 to the data (Θ of In: 108 K).

Fig. 5   Low-temperature part of the data of Fig.4. In the fit the expression
given by Eq. 6 was used.

REFERENCES

1. P. Raghavan and R. S. Raghavan, Phys. Rev. Lett. 27, 724 (1971)
   R. S. Raghavan and P. Rahhavan, Phys. Lett. 36A, 313 (1971).

2. E. N. Kaufmann and R. J. Vianden, Rev. Mod. Phys. 51, 161 (1979)

3. W. D. Hamilton, editor, The Electromagnetic Interaction in Nuclear Spectroscopy (North-Holland, Amsterdam 1975)

4. J. Christiansen et al., Z. Physik B24, 177 (1976)

5. G. Hempel et al., Phys. Lett. 55A, 51 (1975)

6. R. Böhm et al., Hyperfine Interactions 4, 763 (1978)

7. R. Keitel et al., Proc. Intern. Conf. on Hyperfine Interactions,
   Berlin 1980, to be published in Hyperfine Interactions

8. R. Seeböck, Diplomarbeit, Erlangen and to be published

9. U. De et al., see Ref. 7

10. S. Hoth et al., see Ref. 7

11. P. Heubes et al., Hyperfine Interactions 7, 93 (1979)

12. A. Weidinger et al., Phys. Lett. 65A, 247 (1978)

13. R. S. Raghavan et al., Phys. Rev. Lett. 34, 1280 (1975)

14. W. Engel, Thesis, Erlangen (1979)
    W. Engel et al., see Ref. 7

15. R. R. Hewitt and T. T. Taylor, Phys. Rev. 125, 524 (1962)

Published 1981 by Elsevier North Holland, Inc.
Kaufmann and Shenoy, editors
Nuclear and Electron Resonance Spectroscopies Applied to Materials Science

INTERSTITIAL-ATOM TRAPPING AND DEUTERIUM LOCALIZATION AT ^{57}Co
IMPURITIES IN Cu

C.T. MA, B. MAKOWSKI, M. MARCUSO* AND P. BOOLCHAND
Department of Physics, University of Cincinnati, Cincinnati, Ohio 45221

ABSTRACT

  A high purity Cu target foil doped with ^{57}Co and cooled
to 80 K has been deuteron irradiated (E = 2 MeV to .5
MeV) using a Van de Graaff accelerator. In-situ Mössbauer
spectra of the target were taken as a function of deuterium dose and post irradiation annealing. Two new ^{57}Co
sites are observed in addition to the original substitutional site. The new site '1', populated at low temperature following the irradiations is identified with a ^{57}Co
impurity atom having Cu-interstitials trapped in its immediate vicinity. The new site '2', populated on annealing
the target above 360 K is believed to represent a ^{57}Co
impurity having deuterium localized in its vicinity.

INTRODUCTION

  Microscopic trapping of elementary defects and defect clusters at solutes in
cubic metals has been demonstrated by resonance techniques. In recent years,
particular interest has also focussed on the interaction of light interstitial
atoms (μ, p, d) with defects and solutes in metals. As an example, the localization of μ^+ at Mn solutes [1] and vacancies [2] in Al host has been recently
demonstrated. In this paper, we describe a new experimental method to study
the interaction of p and d with defects and solutes in metals. The method
utilizes investigating p or d irradiated metal targets using nuclear gamma
resonance (NGR or the Mössbauer effect). We have applied this method to the
case of d irradiated Cu and in this paper show that valuable new information
on localization of d and trapping of Cu-interstitial clusters at Co solutes in
Cu can be obtained.

EXPERIMENTAL

  We have developed [3] an experimental facility to irradiate target foils
cooled to 10 K or 80 K with p or d using a 2-MeV Van de Graaff accelerator.
The facility includes a cold-finger assembly forming part of a Janis super-
vary-temp liquid helium dewar which is mated to the accelerator beam line.
Provision for a heater and thermometry on the cold finger is made to vary the
temperature of the target.
  High-purity Cu target foil (Marz grade from Materials Research Corporation)
of 61μ thickness was doped with ^{57}Co to prepare a 0.5 mCurie Mössbauer source.
The source exhibited a linewidth of .22(1) mm/s for the inner two lines of Fe
taken with a 12.5μ thick foil. The target, cooled to 80 K, was subjected to
d irradiation at a beam current of 10 μamps and beam energies ranging from 2
MeV to 0.5 MeV. After a total of 5 hours of irradiations during which the
beam energy was successively reduced starting from 2 MeV, a total integrated

---

* Present address: Argonne National Lab, Argonne, Illinois 60439

dose of 3.7 x 10^{17} corrected for secondary electron e.aission, was achieved on the target.

NGR spectra of the target were taken with an absorber mounted on a constant-acceleration drive. The drive was physically located outside the dewar and at right angles to the accelerator beam line. The target was mounted on the cold finger so that it made an angle of 45° with the beam line and the absorber motion. With this set up spectra of the target could be taken as a function of (a) d irradiation dose and (b) post irradiation annealing in the temperature range 110 K<T<773 K. In the annealing experiments, the target was heated to a desired temperature on the cold finger for 30 min. before cooling back to 80 K for the NGR experiments. For annealing above 330 K, the Cu target was taken out of the dewar and heated to the desired temperature in an evacuated quartz tubing using a resistive furnace. For annealing above 330 K, the NGR experiments were performed with the target and absorber held at room temperature.

RESULTS AND DISCUSSION

Site '1'. Immediately following the d irradiations, clear evidence for at least one new ^{57}Co site in addition to the original substitutional site was observed. In Fig. 1, comparing spectrum B to spectrum A shows the appearance of an irradiation induced satellite line on the higher energy wing of the original single line. In analysing the spectra, the satellite line henceforth labeled as site '1' was fit to a quadrupole doublet. The area ratio $A_1/(A_1 + A_0)$ where A_0 and A_1 stand for the areas under the original line and satellite line, was found to monotonically increase with d dose at first and then to saturate at a value of 0.24 at an integrated dose of 3.7 x 10^{17}. We believe that this saturation value is determined by the temperature of the target during the irradiations. Mössbauer parameters of site '1' from a best fit of the data yield: Δ_1 = + 0.14(2) mm/s and δ_1 = 0.30(2) mm/s relative to the substitutional ^{57}Co site. The positive sign of δ_1 implies that site '1' is characterized by a higher transition energy or a smaller electron density $|\psi(0)|^2$ than the substitutional ^{57}Co site in Cu.

Identification of site '1' is made possible by the post irradiation annealing experiments, the results of which are summarized in Fig. 2. The precipiteous drop in the area ratio A_1/A_T between 270 K and 300 K is rather striking. Stage III recovery of resistivity in pure Cu is known to occur at 285 K in irradiation experiments [4]. This recovery stage is associated with vacancy migration. On this basis one concludes that site '1' represents a ^{57}Co atom having Cu interstitial atoms trapped in its immediate vicinity. We believe that the microscopic configuration of site '1' in analogy to the case of neutron irradiated Al (^{57}Co) [5] is most likely multi-interstitial Cu atoms rather than a mono-interstitial Cu atom trapped at Co. If mono-interstitial Cu atoms were trapped at Co in our experiment, it is likely that caging of Cu-Co dumbells in analogy to Co-Al dumbells [6] would occur, thus making such a defect at best difficult to observe at 80 K. We would further like to point out that the incomplete annealing of site '1' above stage III closely parallels the incomplete resistivity recovery in pure Cu. The latter has been attributed to the existence of large interstitial clusters [7], thus further supporting the proposed identification of site '1'. There is some evidence that above stage III, the microscopic configuration of Cu interstitials trapped at Co may actually be changing. The value of Δ_1 appears to nearly vanish in this temperature range.

It is indeed noteworthy that Co solutes appear to become interstitial trappers in Cu, since the atomic radius of Co (1.25 A) is known to be nearly the same as that of Cu (1.28 A). Second, the reduced electron charge density $\psi|(0)|^2$ at site '1', relative to substitutional Co in Cu is a result in sharp

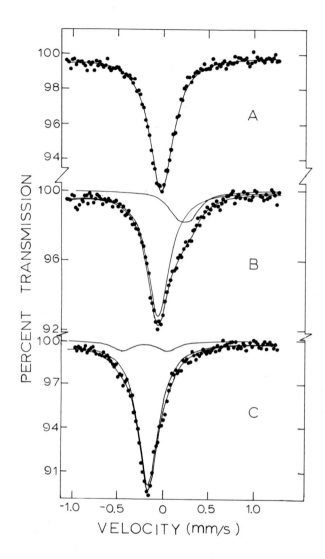

Fig. 1. Mossbauer spectra of Cu(^{57}Co) target taken with a FeCO$_3$ absorber under indicated conditions. A) Target at 80 K before irradiation B) Target at 80 K after 3 x 10^{17} d dose; C) Target annealed at 393 K for 30 min. and studied at 300 K. The second-order Doppler shift between spectrum C and B shows that positive velocity corresponds to a higher transition energy in the source.

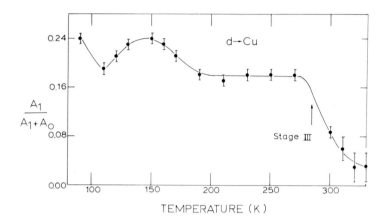

Fig. 2. Results of post-irradiation annealing of d irradiated $\underline{Cu}(^{57}Co)$. The area ratio $A_1/(A_1 + A_0)$ (where A_1 and A_0 denote the area of the site '1' and substitutional-site resonance) is plotted as a function of annealing temperature.

contrast to the one found for Co solutes in Al. Low temperature neutron irradiated $Al(^{57}Co)$ targets exhibit Co sites having multi-interstitial Al atoms trapped in their vicinity. At these defect-associated Co sites, the electron density $|\psi(o)|^2$ is found [5] to be larger than at substitutional ^{57}Co sites in Al. We have studied d irradiated $Al(^{57}Co)$ targets in our experimental set up and in fact confirm that the sign of the satellite line isomer-shift in Al host is opposite to that in a Cu host. Finally, the present result is particularly intriguing since an attempt to observe interstitial trapping following low temperature electron and neutron irradiation of $\underline{Cu}(^{57}Co)$ have been unsuccessful [8]

Site '2'. Annealing the Cu target above 360 K (Fig. 3) revealed the population of a second site, henceforth labeled '2'. This site was found to grow with temperature in the range 360 K<T<400 K and to persist at a fairly constant value $A_2/A_T = 0.10$ all the way up to 573 K. On annealing the source at 673 K, site '2' was found to have completely annealed out. At this point, the narrow linewidth of the target was recovered. We find that site '2' is characterized by $\Delta_2 = 0.62(2)$ mm/s and $\delta_2 = -0.04(2)$ mm/s. There is no detectable evidence of line broadening for the quadrupole doublet of site '2'.

In pure Cu, complete resistance recovery following low temperature 'e' or p irradiation [9] is known to occur at about 650 K, a temperature close to the annealing temperature of site 2. At these elevated temperatures vacancy clusters are the principle defects in Cu [7]. Such clusters are usually trapped by oversized solutes. We therefore consider it unlikely that site '2' represents a vacancy associated Co atom.

We would like to suggest the possibility that site '2' represents the localization of interstitial d at Co solute. The growth of this site and its thermal annealing behavior can be understood as follows: implanted d becomes trapped at vacancies in the implanted depth (<5μ) and the d-vacancy complex is frozen at low temperatures. Above 360 K, the said complex either becomes mobile or it dissociates making feasable for the d to diffuse and seek Co solutes in the

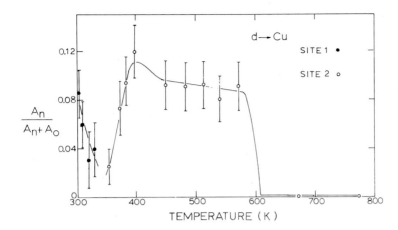

Fig. 3. Results of post-irradiation annealing on d-irradiated $\underline{Cu}(^{57}Co)$. The area ratio $A_n/(A_o + A_n)$ is plotted as a function of annealing temperature. A_n and A_o denote the areas of the site-'n' resonance and the substitutional-site resonance.

bulk (61μ) of the Cu target. The diffusion of d in pure Cu has been measured [10] and from these data one obtains the residence time τ_r of d at octahedral interstitial sites. At 300 K, we obtain $\tau_r \sim 10^{-8}$ sec in pure Cu. In the vicinity of a Co solute however, the d residence time, we presume, will be enhanced because of the local strain field. The nature of the NGR spectrum is determined by the ratio τ_c/τ_{int} where τ_c represents the correlation time and τ_{int} the interaction time [11]. τ_c is related to τ_r through a statistical factor 'ν' which represents the mean number of jumps d must make before it leaves the octahedral interstitial site shell nearest to the Co solute. τ_{int} is related to the nuclear hyperfine interaction E_{int} ($\tau_{int} = h/E_{int}$) which in our case is the quadrupole splitting Δ_2. We estimate $\tau_{int} \sim$ 10^{-8} sec from the measured Δ_2 and conclude that the d diffusion occurs in the quasi-static limit where $\tau_c/\tau_{int} \gg 1$. In such a limit one would in fact observe a static interaction caused by interstitial d in the vicinity of Co as a consequence of slow diffusion. We have recently investigated p irradiated $\underline{Cu}(^{57}Co)$ and do infact observe a high temperature site that completely anneals out at 605 K. The hyperfine interaction at this high temperature site is found to be comparable to the one of site '2'.

The observed quadrupole splitting Δ_2 translates into an EFG of 1.64 A^{-3} at ^{57}Fe on account of interstitial d in a Cu host. The EFG produced by a μ^+ at a near neighbor Cu atom in a Cu host has been calculated [12] from first principles to be 0.26 A^{-3} which agrees well with the experimental value of 0.27 A^{-3}. The substantially larger value of the EFG at an Fe atom near interstitial d in Cu may have its origin in a core enhancement effect. It is clear that a first-principles calculation of the EFG in question will help confirm the microscopic configuration of site '2'.

408

CONCLUSIONS

In-situ Mössbauer spectra of Cu targets doped with ^{57}Co and subjected to d irradiations at low temperatures exhibit two new ^{57}Co sites in addition to the original substitutional ^{57}Co site. The new site '1' populated at low temperatures following the irradiations is identified from post-annealing measurements with a ^{57}Co impurity having Cu-interstitials trapped in its vicinity. The new site '2' populated on annealing the target above 360 K is believed to represent a ^{57}Co impurity having d localized in an adjacent octahedral interstitial site.

ACKNOWLEDGEMENTS

This work was made possible by an award from the University Research Council. We are particularly grateful to Professor J. Anno for making available the use of the Van de Graaff accelerator and to Howard Boeing for technical assistance in the irradiations. We have benefited from valuable discussions with Professors J. Moteff, P. Jena and S.S. Hanna during the course of this work.

REFERENCES

1. O. Hartmann et al., Phys. Rev. Letts $\underline{44}$, 337 (1980).

2. J.A. Brown et al., Phys. Rev. Letts. $\underline{43}$, 1513 (1979).

3. Marc Marcuso, Ph.D. Thesis (unpublished), University of Cincinnati $\underline{1979}$.

4. R.W. Balluffi, J. Nucl. Mat. $\underline{69}$ and $\underline{70}$, 240 (1978).

5. W. Mansel and G. Vogl, J. Phys. F $\underline{7}$, 253 (1977).

6. G. Vogl, W. Mansel and P.H. Dederichs, Phys. Rev. Letts. $\underline{36}$, 1497 (1976).

7. P. Ehrhart and V. Schlagheck, J. Phys. F $\underline{4}$, 1589 (1974).

8. G. Vogl et al., Hyperfine Interactions $\underline{4}$, 681 (1978).

9. O. Echt et al., Phys. Letts. 67A, 427 (1978).

10. L. Katz, M. Guinan and R.J. Borg, Phys. Rev. B $\underline{4}$, 330 (1971).

11. F.E. Wagner and G. Wortmann in: Hydrogen in metals I, G. Alefeld and J. Volkl eds. (Springer Verlag, 1978) pp. 132-167.

12. P. Jena, S.G. Das and K.S. Singhwi, Phys. Rev. Letts. $\underline{40}$, 264 (1978).

Published 1981 by Elsevier North Holland, Inc.
Kaufmann and Shenoy, editors
Nuclear and Electron Resonance Spectroscopies Applied to Materials Science

MUON SPIN DEPOLARIZATION IN METALS WITH DILUTE MAGNETIC IMPURITIES

J. A. BROWN, R. H. HEFFNER, R. L. HUTSON, S. KOHN,* M. LEON, C. E. OLSEN AND
M. E. SCHILLACI
Los Alamos National Scientific Laboratory, Los Alamos, New Mexico 87545
S. A. DODDS, T. L. ESTLE, D. A. VANDERWATER
Rice University, Houston, Texas 77001
P. M. RICHARDS
Sandia National Laboratories, Albuquerque, New Mexico 87185
O. D. McMASTERS
Iowa State University, Ames, Iowa

ABSTRACT

Muon diffusion can be studied in metals without nuclear
moments by the addition of paramagnetic impurities. We describe
the theory and report measurements of muon depolarization in
gold and silver doped with gadolinium and erbium. As well as
diffusion rates, strength of the magnetic interaction and spin
lattice relaxation rate of the paramagnetic ion are inferred from
the data.

INTRODUCTION

One of the most active fields of study using the technique of muon spin
rotation (μSR) is the diffusion of light interstitial particles in a metallic
host lattice. A comparison of muon diffusion with the diffusion of the heavier
hydrogen isotopes offers the possibility to study different mechanisms of partic-
ule motion; the light mass of the muon, for instance, may allow a coherent or
band-like motion which is negligible for the heavier particles.

Information on muon diffusion is derived from the temperature dependence of
the depolarization rate. Depolarization results from interaction of the muon
spin with magnetic moments of the host material, which means the nuclear spins
in the case of nonmagnetic metals treated here. The situation is completely
analogous to that of motional narrowing [1] in NMR. At low temperatures the
muon stays at a particular site for a time long enough to precess in the local
field characteristic of that site. Since the local field varies from site to
site because of random orientation of the host nuclear spins, there results a
spread of precessional frequencies of muons stopped at random which gives rise
to the depolarization. At higher temperatures the muon moves from site to site
and samples many different local fields. The depolarization rate then approaches
zero as only the average local field is seen. The precise manner in which the
depolarization rate decreases with temperature can be used to infer the muon
hopping rate.

There are, however, two limitations inherent in such studies: (a) rapid muon
motion cannot be studied once the depolarization becomes vanishingly small and
(b) many host materials such as Ag and Au have negligible nuclear moments. In
the first case, an upper limit to the observable muon diffusion rate is reached
at low temperatures; in copper no depolarization is observed at temperatures
greater than about 240K [2]. This restricts the comparison with hydrogen
diffusion which is measured at much higher temperatures. In the latter case, no

*Present address: University of California Lawrence Berkeley Laboratory.

information on muon diffusion can be obtained.

These two problems can be circumvented by the introduction of small amounts of paramagnetic impurities in the otherwise nonmagnetic host. Once again the situation is similar to the NMR of fast-moving ions [3]. Although the number of impurities is small, their interaction is very strong compared with the relatiely weak local fields of nuclear moments. If a muon stays an appreciable time in the neighborhood of a paramagnetic spin, it experiences an enormous change in pre-cessional frequency and its polarization is lost from the main signal. This depolarization mechanism is effective only if a muon can move to the vicinity of an impurity since the interaction is short range. (The small fraction of muons which are stopped and remain next to a paramagnetic ion will have their precessional frequencies so greatly increased that they will not contribute to the polarization signal except for times ($\leq 10^{-8}$ sec) shorter than normally measured.) Thus it can provide a rather direct measure of the diffusion rate.

We have performed measurements [4, 5] on Au and Ag doped with Er and Gd which clearly show that muon diffusion can be detected by the influence of paramagnetic impurities. As well as determining muon hopping rates, we are also able to infer magnetic properties such as the electron spin-muon interaction strength and the paramagnetic spin-lattice relaxation rate. These quantities are of interest to basic studies of impurities in metals and of magnetism.

A physical description of the theory of muon depolarization by paramagnetic impurities is given in Sec. II followed by a review of the experimental results. The emphasis is on qualitative features rather than on details which can be found elsewhere [4, 5, 6].

II. QUALITATIVE THEORY

The main qualitative feature of the theory [6] is that the depolarization rate Λ goes through a maximum vs. temperature. This may be understood as follows. At low temperatures the rate is given by the time it takes a muon to diffususe to the vicinity of an impurity

$$\Lambda \text{ (low temperature)} = c\tau_{hop}^{-1} \tag{1}$$

where τ_{hop} is the time spent at a given lattice site and c is the probability that the site is in the neighborhood of an impurity, i.e. on the average 1/c hops are required to find an impurity. Eq. (1) holds as long as the time spent at the impurity is greater than $1/\Lambda_o$, the depolarization time for a muon at rest at that site. In that case only one encounter is required and the time spent at the site is much less than the time required to get to it (we neglect extreme trapping and assume a small impurity concentration so that c<<1). Note that in this limit the depolarization rate depends only on the hopping rate and is independent of details of the interaction.

As temperature is increased τ_{hop} becomes shorter than $1/\Lambda_o$ so that complete depolarization is not possible in a single visit. Then the depolarization rate no longer increases because the increased rate of encountering an impurity is more than offset by the decreased amount of depolarization per encounter. The number of encounters required is of the order of $1/\delta\phi$ where $\delta\phi$ is the magnitude of phase change per encounter and is given by [7]

$$\delta\phi \sim \Delta\omega_e^2 \, \tau_{hop} \, \tau_c \tag{2}$$

(for $\delta\phi<<1$) according to theories of relaxation, where $\Delta\omega_e$ is the frequency shift produced by the impurity and τ_c is the time during which the muon sees a constant paramagnetic spin orientation. $\tau_c = \tau_{hop}$ if the spin is static, but $\tau_c \approx \tau_{SL}$ if the spin-lattice relaxation time τ_{SL} which characterizes fluctuations of the paramagnetic moment is much less than τ_{hop}. The depolarization time is $c^{-1}\tau_{hop}/\delta\phi$, the average time to produce $1/\delta\phi$ encounters, so the rate is

$$\Lambda \text{ (high temperature)} = c \, \Delta\omega_e^2 \tau_c \qquad (3)$$

A simple interpretation of the above is that the relaxation rate experienced by the muon at the impurity site is $\Lambda_o = \Delta\omega_c^2 \tau_c$ so that in the case of rapid hopping where many impurities are seen Λ is just the average relaxation rate of the medium. In contrast to the low temperature case (Eq.(1)), the rate here depends on both the interaction strength $\Delta\omega_e$ and spin-lattice relaxation (for $\tau_{SL} \lesssim \tau_{hop}$).

The depolarization rate goes through a maximum as a function of temperature at a point where $\tau_{hop}^{-1} \sim \Delta\omega_e^2 \tau_c$, as obtained by equating the low temperature, single encounter, rate of Eq. (1) to the high temperature, many encounters, rate of Eq. (3). By comparison motional narrowing of the depolarization rate caused by host nuclear spins occurs in the region $\Delta\omega_n^2 \tau_{hop}^2 \sim 1$. If the impurity spin lattice relaxation rate is negligible ($\tau_c = \tau_{hop}$), this means that the characteristic hopping times which can be measured by paramagnetic impurities and host nuclear spins are in the ratio $|\Delta\omega_n/\Delta\omega_e|$ which is of the order of 10^{-3}-10^{-4}. Thus, as stated in the introduction, diffusion at higher temperatures can be probed by the introduction of paramagnetic ions.

The above treatment has implicity assumed nearest neighbor interactions only and a simple hopping model in which each lattice site is visited only once. If dipolar interactions for which $\Delta\omega^2 \propto \hbar^{-6}$ are used [6,8] the low temperature rate is enhanced by a factor of the order of $(\Lambda_o \tau_{hop})^{\frac{1}{4}}$ where Λ_o is the depolarization rate for a muon fixed at the nearest neighbor distance, but the high temperature rate is basically unaffected. More detailed treatment of hopping which allows for repeated visits to the same impurity [5] can increase the depolarization rate by about a factor of 3.

In summary, we have given a physical description of the depolarization rate due to paramagnetic impurities. It goes through a maximum vs. temperature. On the low-temperature side of the maximum T_2 is strictly proportional to the hopping time if the interactions are nearest-neighbor like, while on the high temperature side the interaction strength and spin lattice relaxation are involved. Hence a study of T_2 through both regions can be used to extract information about both muon diffusion and magnetic spin interactions and dynamics. Since the local field of a paramagnetic ion is orders of magnitude greater than that of a nuclear spin, the depolarization rate by paramagnetic impurities can probe much faster muon hopping than can the interaction with host nuclei.

III. EXPERIMENTAL RESULTS

Considerable data have been obtained on μSR in Au and Ag doped with various amounts of Gd and Er as paramagnetic impurities. We focus here on those aspects which give confirmation of the qualitative features discussed in the previous section. More extensive discussion and data presentation may be found elsewhere [4,5].

Figures 1 and 2 show, respectively, the depolarization rate vs. temperature T for Au:Gd and Au:Er. In both cases the concentration was about 300 ppm. Experiments at 100 ppm verified the predicted linear dependence on c. The

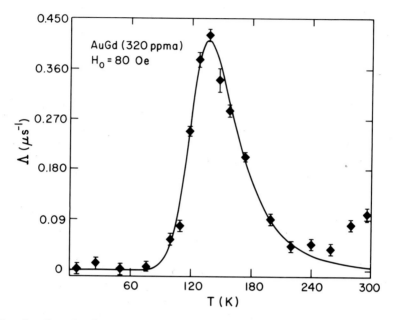

Fig. 1. Muon depolarization rate in Au:Gd

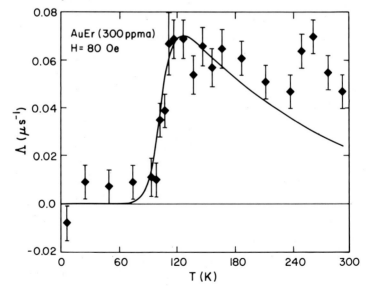

Fig. 2. Muon depolarization rate in Au:Er

following features are noteworthy in relation to the discussion of the previous section. The rate goes through a maximum, and on the low temperature side the dependences are the same for Gd and Er, but there is a big difference on the high temperature side. This is consistent with the prediction that the rate depends only on τ_{hop} on the low temperature side but that it can depend on the spin-lattice relaxation time τ_{SL} on the high temperature side. A simple interpretation is that at high temperatures $\tau_{SL} \gtrsim \tau_{hop}$ for Au:Gd so that $\tau_c \approx \tau_{hop}$ is observed characteristic of activated hopping describing the effective correlation time τ_c. For Au:Er the situation is reversed: $\tau_{SL} < \tau_{hop}$ so that τ_c is proportional to τ_{SL} which varies as T^{-1} for spin-lattice relaxation via coupling to conduction electrons (Korringa rate) [9]. Since Gd is an S-state ion, it is reasonable that it have a much longer τ_{SL} than Er.

The above considerations are borne out by comparison to the quantitative theory [6] shown by the solid curves. The same τ_{hop} is used for Gd and Er and has an activation energy of 0.14 eV. τ_{SL} for Gd is taken to be 7.0 x 10^{-9}/T(K) sec consistent with ESR data [10], and $\tau_{SL} = 10^{-10}$/T(K)sec for Er. A nearest neighbor interaction whose strength is about 5(15) times the nearest neighbor dipole field is needed to fit the Gd(Er) data. This suggests an effective muon-paramagnetic spin interaction perhaps via intermediate coupling with s and d electrons surrounding the local moment of the f electrons.

The theory in its present form predicts only a single maximum whereas the Au:Er data show clear evidence of two maxima and the Au:Gd data give indication of a second peak above room temperature. The origin of this anomaly is not yet clear.

For Ag:Er and Ag:Gd (not shown) Λ increases rapidly above 200K but has not reached a peak at the highest (room) temperature. The activation energy is deduced to be 0.18 eV, but nothing can be said about the other parameters in the absence of a peak.

The diffusion seen here is quite different from that reported [2] in Cu at lower temperatures, which has a very low (0.05eV) activation energy and prefactor characteristic of tunneling. The hopping rate in Cu extrapolated to room temperature is much smaller than observed in this work for Ag and Au. This is consistent with our measurements on Cu:Mn (500 ppm) which show no impurity effects up to room temperature. Clearly it would be of interest to study impurity-doped Cu at higher temperatures to see if a higher activation energy, classical-like process is present similar to that in Ag and Au.

IV CONCLUSIONS

We have shown that muon diffusion can be detected in metals such as Au and Ag which have negligible nuclear moments by the introduction of paramagnetic impurities. Predicted features of the temperature dependence of the depolarization rate have been verified and numbers obtained for the diffusion rate, impurity-muon interaction, and paramagnetic spin-lattice relaxation rate. Further studies are planned at high (above room) temperature and as a function of applied field in order to investigate the two peak structures in materials other than Au:Er and to elucidate the nature of the non-dipolar interaction. The present work, however, is sufficient to show the usefulness of paramagnetic impurities for extending the range of materials in which muon diffusion can be measured and in learning new features of magnetic interactions and spin dynamics.

414

ACKNOWLEDGMENTS

C. Y. Huang is acknowledged for his aid in obtaining gold samples.
This work is supported in part by the U.S. Department of Energy under Contracts
W-7405-ENG-36, DE-AC04-76-DP00789, W-7405-ENG-82 and by the NSF.

REFERENCES

1. Abragam, A., The Principles of Nuclear Magnetism (Oxford University Press
   New York, 1961) Ch 10.

2. V. G. Grebinnik, et al, Sov. Phys. JETP $\underline{41}$, 777 (1976).

3. R. D. Hogg, S. P. Vernon and V. Jaccarino, Phys. Rev. Lett. $\underline{39}$, 481 (1977).

4. **R. E. Heffner**, in Proceedings of 2nd Topical Conference on Muon Spin
   Rotation, Vancouver, B.C., 1980, to be published in Hyperfine Interactions.

5. M. E. Schillaci, et al, as in ref. 4.

6. P. M. Richards, Phys. Rev. $\underline{B18}$, 6358 (1978).

7. Slichter, C. P., Principles of Magnetic Resonance (Harper & Row, New York,
   1963) Ch 5 Sec 7.; Ch 8 of ref. 1.

8. pp. 379-386 of ref. 1.

9. J. Korringa, Physica $\underline{16}$, 601 (1950).

10. C. Rettori, D. Davidov, and E. M. Kim, Phys. Rev. $\underline{B8}$, 5335 (1973).

Published 1981 by Elsevier North Holland, Inc.
Kaufmann and Shenoy, editors
Nuclear and Electron Resonance Spectroscopies Applied to Materials Science

415

STUDY OF INTERNAL INDIUM OXIDATION IN SILVER BY TDPAC

A. F. PASQUEVICH, F. H. SÁNCHEZ*, A. G. BIBILONI*,
C. P. MASSOLO[†] and A. LÓPEZ-GARCÍA*[#]
Departamento de Fisica, Facultad de Ciencias Exactas,
Universidad Nacional de La Plata, C.C. 67, 1900 La Plata,
Argentina

ABSTRACT

Internal oxidation of indium in a silver matrix is
studied by time-differential perturbed angular
correlations. The oxidizer treatments were performed
in air at 1 atm. Our results are consistent with the
formation of different In-O complexes.

The time-differential perturbed angular correlation (TDPAC)
technique permits the study, through the hyperfine-electric quadrupole interaction, of defects and impurities effects on probe
nuclei in cubic metals. Here we report first results on internal
oxidation of indium in silver obtained by this technique.

The indium activity was obtained by means of the
$^{109}Ag(\alpha,2n)^{111}In$ nuclear reaction on 99.99 at % purity and 500
μm thick silver foil targets. The irradiated foils were encap-
sulated in quartz tubes at about 10^{-3} Torr and melted in order to
procure an homogeneous distribution of the radioactive impurities
and to get rid of interactions associated with radiation damage.
Finally, the samples were rolled to a thickness of about 600 μm.
The radioactive impurity concentration in all cases remained below
5 ppm. Before each sequence of oxidation treatments the absence
of any quadrupole interaction was verified. See Fig. 1a for a
typical result. Treatments done in air and in O_2 atmosphere led
to equivalent results. Hereafter we present only those results
obtained in air at p = 1 atm.

For the TDPAC measurements, all done on the γ-γ cascade 171
keV - 243 keV in ^{111}Cd, a conventional two NaI(Tl)-detector setup
with standard electronics was used. The movable detector changed
its position every 1800s and from the measured N(90°,t), N(130°,t)
coincidence counting rates the asymmetry ratios

$$A(t) = 2 \frac{N(180°,t) - N(90°,t)}{N(180°,t) + 2N(90°,t)} \simeq A_{22}^{exp}G_{22}(t) \qquad (1)$$

were calculated after corrections for random coincidences. A
typical result is shown in fig. 1b.

---

*Member of CONICET, Argentina.
[†]Member of CIC Prov. Buenos Aires, Argentina.
[#]Present address: Rutgers University, Nuclear Physics
 Laboratory, New Brunswick, N. J. 08903, U.S.A.

416

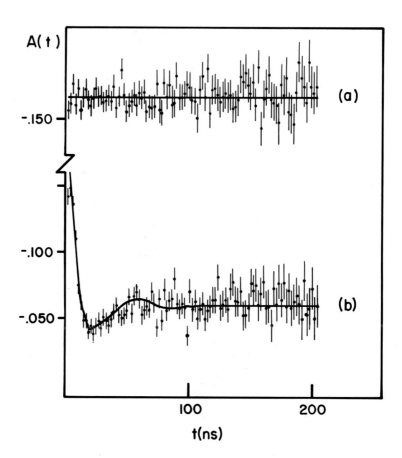

Fig. 1. Typical TDPAC spectra at room temperature. (a) after a treatment of 90 min at 450°C and 0.5 atm N_2. (b) after a similar treatment on O_2. The full line shows the fitted curve to the data.

The calculated A(t)-value has been fitted to the function:

$$A(t) = A_{22}^{exp} \{ f_0 + \sum_{n=0}^{3} s_{2n}(\eta) \ f \ \cos(\omega_n t) \ \exp(-\delta\omega_n t) \} \qquad (2)$$

and

$$f_o + f = 1$$

folded with the measured time resolution curve ($2\tau = 3.44$ ns).
Here f_0 is the fraction of nuclei at unperturbed lattice sites
and f the fraction of nuclei subject to the interaction characterized by the quadrupole frequency $\omega_n = F_n(\eta)\omega_Q$ where

$$\omega_Q = \frac{eQV_{zz}}{40\ h} \quad .$$

The coefficients F_n and s_{2n} are functions of the asymmetry parameter   given in ref. [1]. The exponential function accounts for
a Lorentzian frequency distribution of relative width δ around
ω_n

    In spite of a slight dispersion in the values of the fitted
parameters, the interaction associated to the oxidation was
characterized by the following values:

$$\bar{\omega}_Q = 15.7\ \text{Mhz}, \qquad \bar{\eta} = 0.43, \qquad \bar{\delta}_Q = 5.7\ \text{Mhz}$$

$$\sigma_\omega = 3.1\ \text{Mhz}, \qquad \sigma_\eta = 0.16 \qquad \sigma_\delta = 1.9\ \text{Mhz}$$

where $\bar{\omega}_Q$, $\bar{\eta}$ and $\bar{\delta}$ are mean values and σ is the standard deviation
for the complete set of measurements.

    The evolution of the relative population f of the oxidized In
probe was studied as a function of the treatment temperature at
280, 300, 375, 450, 525, 600 and 675°C. All treatments lasted 90
minutes and then room temperature was reached within 2 minutes.
The measurements were carried on at room temperature. The results
(see fig. 2) show that a drastic change in the relative fraction
f_0 took place around 300°C. A final value of 0° ($f = 100\%$) was
reached at 375°C.

    The fact that this value remained unchanged up to 675°C, and
furthermore after two additional treatments at a pressure as low
as 10^{-3} Torr performed at 750 and 850°C, provides an indication
of the high thermal stability of the formed complexes. The only
way in which we were able to remove the interaction was melting
the probe in vacuum ($p < 10^{-3}$ Torr).

    Other sequences of measurements were carried out keeping the
sample at temperatures above 20°C. We verified that once the
radioactive impurities were completely oxidized, the characteristic parameters of the interaction did not depend on the temperature, at least in the range between 20 and 600°C.

    The rate of the oxidation process at 450°C was also investi-
gated. Measurements following cumulative oxidation treatments of
4 minutes up to a total of 46 minutes, after an initial one of 2
minutes, were performed at room temperature. Our results show
that after 6 minutes treatment, 70 ± 20% of the probes were oxidized. The saturation value of 80 ± 20% was reached after 14

minutes. This saturation value, however, seems strongly dependent on the shape of the sample. In effect, measurements performed on onrolled samples led to a saturation value of about
65%.

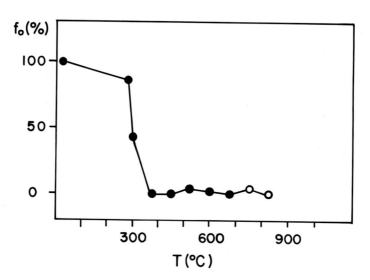

Fig. 2. Fraction of nuclei at unperturbed sites after annealing
treatments of 90 minutes at increasing temperatures. ● in air
at 1 atm; 0 in air at 10^{-3} Torr.

In the interpretation of our results we must take into account
the following facts: the extremely low dilution of In, the high
concentration of oxygen in the silver matrix at the temperatures
used and pressures due to the high solubility of oxygen in this
metal, and also that the diffusion velocity of In in silver is
negligible compared to that of oxygen at the involved temperatures [2]. In consequence we can expect that the In oxidation
occurs *in situ* and therefore each complex includes only one In
atom. Then the high distribution of the interaction parameters
suggests the presence of several different configurations In-O_x,
each one characterized by definite values of ω_Q and η. The
asymmetry parameter $\eta \neq 0$ indicates the existence of configurations with more than one oxygen. Concerning the number of
different configurations involved, it may be pointed out that
those produced by one or more oxygens placed at the center of
any of the eight next neighboring indium fcc cells, would be
favored by size considerations based on the rigid sphere model
and compatible with a bound state In-O, and could probably not
be sufficient to give account of the observed distribution. So,
other interstitial next neighboring sites should be occupied.

Also, lattice distortion effects due to the oxygen pressure cannot be excluded.

Though our results are not conclusive, they may show a fine structure that can be associated to some evolution of the relative fractions of nuclei involved in the different configurations. Further efforts should be made in this direction. The possibility that the impurity oxides cluster or precipitate out of the solution as In_2O_3 can be excluded. The value of 15.7 Mhz for the frequency and the behavior with the temperature of the measurement that we found are quite different from those reported by Salomon [3] in this compound. Our interpretation is in agreement with Ehrlich's [4], whose results (i.e. the increase of the residual resistivity ration RRR) can be explained if clustering of impurity oxides in silver take place only under oxygen pressures as low as 10^{-5} Torr.

We wish to acknowledge the cyclotron staff of the Comisión Nacional de Energía Atómica for the irradiation facilities. This work has been partially supported by CONICET, SECYT and CIC Prov. de Buenos Aires, Argentina and Gesellschaft für Kernforschung MbH, Karlsruhe, West Germany.

REFERENCES

[1] Mendoza-Zelis, L. A., Bibiloni, A. G., Carachoche, M. C., Lopez-Garcia, A., Martinez, J. A., Mercader, R. C., and Pasquevich, A. F., Hyperfine Interactions 3, 315 (1977).

[2] Lazarus, D., in Solid State Physics, vol. 10, Seitz, F. and Turnbull, D. (eds.), (Academic Press, 1969).

[3] Salomon, M., Nucl. Phys. 54, 171 (1964).

[4] Ehrlich, A. H., J. of Mat. Sci. 9, 1064 (1974).

Published 1981 by Elsevier North Holland, Inc.
Kaufmann and Shenoy, editors
Nuclear and Electron Resonance Spectroscopies Applied to Materials Science

DISLOCATION MOTION IN METALS INVESTIGATED BY MEANS OF
PULSED NUCLEAR MAGNETIC RESONANCE

H. TAMLER, H. J. HACKELÖER, and O. KANERT
Institute of Physics, University of Dortmund, 46 Dortmund-50, W. Germany

W. H. M. ALSEM and J. Th. M. DE HOSSON
Dept. of Applied Physics, Materials Science Centre, University of Groningen,
Nijenborgh 18, 9747 AG Groningen, The Netherlands

ABSTRACT

We report the first use of nuclear magnetic resonance to investigate dislocation motion in metals. The spin-lattice relaxation rate in the rotating frame, $T_{1\rho}^{-1}$, of ^{27}Al in polycrystalline, ultrapure Aluminium foils has been measured as
a function of plastic-deformation rate $\dot{\varepsilon}$ for two different
temperatures (77K and 300K). For $\dot{\varepsilon} = 0$, the relaxation rate
is determined by conduction electrons. For a finite deformation rate $\dot{\varepsilon}$, an additional contribution to the relaxation
rate arising from fluctuations in the nuclear quadrupole interaction due to dislocation motion is observed. From the
motion-induced part of the relaxation rate the mean jump
distance of a mobile dislocation is calculated which is determined by the density of lattice defects acting as obstacles for moving dislocations.

INTRODUCTION

Dislocations play a key role in the explanation of the phenomena of slip in
crystals, the mechanism of the process of plastic deformation which is so important in materials science. Strengthening of crystals is brought about by
obstructing the movement of dislocations. A realistic description of the dislocation motion is essential for an understanding of plastic deformation.
    A few years ago, we have shown that nuclear magnetic resonance (NMR) is a
useful tool for studying dislocation motion in non-metallic materials like
alkali halide crystals (1, 2, 3). In the work, the nuclear spin relaxation rate
in the rotating frame, $T_{1\rho}^{-1}$, has been instantaneously measured while a sample
is plastically deforming with a constant deformation rate $\dot{\varepsilon}$. In particular, results have been presented as a function of the deformation rate $\dot{\varepsilon}$, the direction of deformation with respect to the crystal axes, temperature, and concentration of impurity atoms. Parallel to the experimental work, the theoretical
basis for evaluation of the nuclear spin relaxation data has been established
(4). It turned out that from such NMR experiments two microscopic informations
of the dislocation motion can be obtained in principle: (i) The mean jump distance and (ii) the mean time of stay between two consecutive jumps of a mobile
dislocation.
    Now, for the first time dislocations at various velocities in metals, es-
pecially in ultrapure polycrystalline aluminium are studied by means of pulsed
NMR technique. In this paper, first results of the investigation are presented.

422

THEORETICAL BACKGROUND

As discussed for example by Argon (5), the mechanical behaviour of crystalline
material under the influence of a plastic-deformation rate $\dot{\varepsilon}$ is governed by the
Orowan equation (6)

$$\dot{\varepsilon} = \Phi.b.\rho_m.v = \Phi.b.\rho_m.L/\tau_w \; . \tag{1}$$

The physical model underlying Eq.(1) assumes a thermally activated jerky motion
of mobile dislocations of density ρ_m between obstacles in the material. Φ denotes a geometrical factor ($\Phi \simeq 0.5$), while b symbolizes the magnitude of the
Burgers vector. The mean dislocation velocity v may be expressed in terms of
the mean jump width L and the mean waiting time τ_w between two consecutive
jumps.
    While deforming a sample with a constant deformation rate $\dot{\varepsilon}$ the spin-lattice
relaxation rate in a weak rotating field ("locking field") H_1, $1/T_{1\rho}$, of the
nuclei in the material may be decomposed into a "background" relaxation rate
$(1/T_{1\rho})_0$ and the contribution $(1/T_{1\rho})_D$ which is governed by $\dot{\varepsilon}$, i.e. by Eq. (1):

$$\frac{1}{T_{1\rho}} = (\frac{1}{T_{1\rho}})_0 + (\frac{1}{T_{1\rho}})_D \; . \tag{2}$$

In metals $(1/T_{1\rho})_0$ is due to fluctuations in the conduction electron-nucleus
interaction leading to the Korringa relation $(T_{1\rho})_0.T = c$ where the magnitude
of the constant c depends slightly on the strength of the locking field H_1 (7).
For ^{27}Al in aluminium, the value of c is about 1.8 sec $\cdot$ K.
    On the other hand, the dislocation-motion induced part of the relaxation
rate, $(1/T_{1\rho})_D$ is given by (4)

$$(\frac{1}{T_{1\rho}})_D = \frac{\delta_Q.<v^2>}{H_1^2 + H_L^2} \cdot \frac{g_Q(L)}{\Phi \cdot b} \cdot \frac{1}{L} \cdot \dot{\varepsilon} \tag{3}$$

where

$$\delta_Q = \frac{3}{80} \cdot (\frac{eQ}{\gamma h})^2 \cdot \frac{2I+3}{I^2(2I-1)} \; , \tag{3a}$$

is a quadrupole coupling constant and $<v^2>$ denotes the mean-squared electric
field gradient due to the stress field of a dislocation unit length (4).
(Q: quadrupole moment, γ: gyromagnetic ratio, I: spin of the resonant nucleus),
H_L is the mean local field at a resonant nucleus determined by the local dipolar
field H_D and the local quadrupole field H_Q: $H_L^2 = H_D^2 + H_Q^2$ (4). The quadrupolar
geometry factor g_Q in Eq. (3) which depends on the mean step width L was calculated by Hut et al. (1).
    It has to be noted that Eq. (3) is valid only in the "strong-collision re-
gion" where the dislocation motion is slow enough to allow the spins to establish a common spin temperature between successive dislocation jumps. In prac-
tice, the condition is fulfilled for strain rates $\dot{\varepsilon}$ up to about 10 s^{-1} depending
slightly on $H_1^2 + H_L^2$ (2).
    Eq. (3) shows, that for a given strain rate $\dot{\varepsilon}$ the motion-induced part of the
nuclear spin-relaxation rate is proportional to the inverse mean jump distance
$1/L$. Therefore, the relation may be used to determine L.

EXPERIMENTAL DETAILS

For the investigation, ultrapure (99.999%) polycrystalline (size of the grains about 150 μ) aluminium samples were used. To avoid skin effect distortions of the NMR signal the samples consisted of thin rectangular foils with a length of 27 mm, a width of 12 mm, and a thickness of about 50 μ. The sample under investigation was plastically deformed by a servo-hydraulic tensile machine (ZONIC Technical Lab. Inc.) of which the exciter head moves a driving rod with a constant velocity. While the specimen was deforming ^{27}Al NMR measurements were carried out with a BRUKER pulse spectrometer SXP4-100 operating at 15.7 MHz corresponding to a magnetic field of 1.4 T. For the purpose, the NMR head of the spectrometer and the frame in which the rod moves formed a unit which was inserted between the pole pieces of the electromagnet of the spectrometer. The unit could be temperature-controlled between 77 K and 550 K. The spectrometer was triggered by the electronic control of the tensile machine. The trigger starts the nuclear spin relaxation experiment at a definite time during deformation. For the measurement of the nuclear spin relaxation time $T_{1\rho}$ the nuclear magnetization was spin-locked by applying a $\pi/2$ pulse followed by a locking pulse of strength H_1 shifted in rf-phase by $\pi/2$ from the first pulse (8). Immediately before and after the plastic deformation the magnitude of the background relaxation time $(T_{1\rho})_0$ was measured. By means of Eq(2), from the experimental data the motion-induced part of the relaxation time, $(T_{1\rho})_D$, could be determined. All measurements were carried out at plastic deformation ε between 5% and 30%.

To be sure that the increase in the nuclear spin relaxation rate during the plastic deformation is due to dislocation motion the relaxation rate was measured while moving the sample with a constant velocity without any deformation. No change in the relaxation rate was observed within the experimental error.

During the deformation experiment, the stress and the resultant strain are measured separately and simultaneously. The different NMR signals as well as the mechanical data were stored in fast transient recorders (DATALAB DL 905) and then transcribed on magnetic tape. Further processing of the data was carried out by an on-line VARIAN 620 L-computer.

RESULTS AND DISCUSSION

According to Eq. (3), in the strong-collision region a plot of dislocation-induced contribution to $T_{1\rho}$ vs square of the locking field H_1^2 will yield a straight line which may be extrapolated to find the x intercept at $H_1^2 = -H_L^2$. Figure 1 shows such a plot for $(T_{1\rho})_D$ at a constant strain rate $\dot{\epsilon}$ for T = 77 K. The total local field thus obtained is H_L = 4.5 G. On the other hand, for ^{27}Al in polycrystalline aluminium Van Vleck's formula (9) leads to a value for the dipolar local field H_D of about 2.9 G. From both the data, the quadrupolar local field H_Q caused by the stress field of all dislocations in the sample (mobile as well as unmobile forest dislocations) is obtained to be $H_Q = (H_L^2 - H_D^2)^{\frac{1}{2}} \simeq 3.4$ G.

In Fig. 2 the motion-induced part of the relaxation rate, $(1/T_{1\rho})_D$ measured with a locking field H_1 of 10 G is plotted as a function of $\dot{\epsilon}$ for two different temperatures (77 K and 298 K). For comparison results of earlier ^{23}Na NMR measurements on NaCl single crystals (2) are inserted in the figure. The slope of the curve up to $\dot{\epsilon} \simeq 10$ s^{-1} is found to be proportional to the strain rate $\dot{\epsilon}$ as predicted by Eq. (3). As discussed above, from the magnitude of the slope which has a value of about 30 the mean jump width L can be determined providing the other parameters in Eq. (3) are known. The magnitude of the Burgers vector in aluminium is $2.86 \cdot 10^{-8}$ cm. For polycrystalline materials the factor Φ has a value of 0.33 (10). Using the nuclear data of ^{27}Al the quadrupole coupling constant δ_Q was calculated to be $\delta_Q = 1.13 \cdot 10^{-24}$ G^{2}dyne^{-1} cm^4. Taking into account the data of the gradient-elastic constants in aluminium as published by Buttet (11) for the mean-squared electric field gradient $\langle V^2 \rangle$ a value of approximately

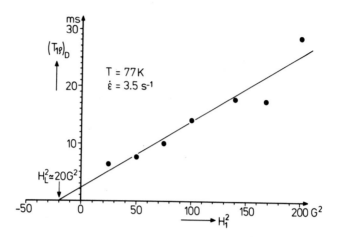

Fig. 1. Plot of the dislocation-induced contribution to the relaxation time
$T_{1\rho}$ vs square of the locking field H_1 of ^{27}Al in aluminium for a constant strain rate $\dot{\varepsilon}$ at 77 K. The plot provides verification of Eq. (3)
with a local field H_L of 4.5 G.

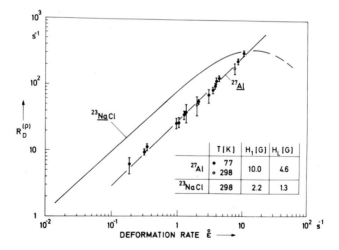

Fig. 2. Dislocation-induced part of the ^{27}Al relaxation rate $1/T_{1\rho}$ in aluminium
as a function of deformation rate $\dot{\varepsilon}$ for two different temperatures.
Locking-field H_1 = 10 G. For comparison, analogous data of ^{23}Na in NaCl
single crystals are inserted in the figure. The straight line confirms
a dislocation induced relaxation mechanism as described by Eq. (3).

$3.7 \cdot 10^{14}$dyne cm^{-2} was obtained. Substituting these data in Eq. (3) and using the theoretical values for the geometry factor g_Q presented in (1) as a function of L an iterative calculation yields a mean step width L of about $1.1 \cdot 10^{-5}$ cm together with $g_Q = 0.85$.

The value of L corresponds roughly with the mean distance between dislocations in deformed aluminium which is of the order of 10^{-5} cm. The agreement would confirm the ideas of dislocation motion in ultrapure f.c.c. metals impeded by forest dislocations. In addition, the slope of the curve in Fig. 2 should be independent of temperature since it is known that for f.c.c. metals the dislocation activation mechanism remains the same between 77 K and 300 K; this is confirmed by the experimental data in Fig. 2. It has to emphasized, however, that the interpretation has necessarily a speculative component at this stage. Further experiments have to be carried out in order to obtain more detailed insight in dislocation dynamics.

ACKNOWLEDGMENTS

The work is part of the research program of the Foundation for Fundamental Research on Matter (F. O. M. - Utrecht/The Netherlands). Financial support by the "Deutsche Forschungsgemeinschaft" (W. Germany) and The Netherlands Organization of the Advancement of Pure Research (Z. W. O. - The Hague) is gratefully acknowledged.

REFERENCES

1. G. Hut, A. W. Sleeswyk, H. J. Hackelöer, H. Selbach, and O. Kanert, Phys. Rev. B 14, 921 (1976).

2. H. J. Hackelöer, H. Selbach, O. Kanert, A. W. Sleeswyk, and G. Hut, Phys. Stat. Sol. (b) 80, 235 (1977).

3. W. H. M. Alsem, A. W. Sleeswyk, H. J. Hackelöer, R. Münter, H. Tamler, and O. Kanert, J. de Physique C6 - 146 (1980).

4. D. Wolf and O. Kanert, Phys. Rev. B 16, 4776 (1977).

5. A. S. Argon, Phil. Mag. 25, 1053 (1972).

6. J. J. Gilman, Micromechanics of Flow in Solids (McGraw-Hill, New York, 1969), p. 157ff.

7. D. Wolf, Spin Temperature and Nuclear Spin Relaxation in Matter, Clarendon Press, Oxford 1979.

8. T. C. Farrer and E. D. Becker, Pulse Fourier Transform NMR, (Academic Press, New York, 1971) p. 91ff.

9. A. Abragam, The Principle of Nuclear Magnetism, (Clarendon, Oxford 1961), Chap. IV.

10. P. Haasen, Physikalische Metallkunde, (Springer-Verlag, Berlin 1974), p. 280ff.

11. J. Buttet, Journ. Phys. F.: Metal Phys. 3, 918 (1973).

Published 1981 by Elsevier North Holland, Inc.
Kaufmann and Shenoy, editors
Nuclear and Electron Resonance Spectroscopies Applied to Materials Science

RADIATION DAMAGE STUDIES OF USn$_3$*

T. K. MCGUIRE and R. H. HERBER
Department of Chemistry, Rutgers University,
New Brunswick, N. J. 08903 USA

ABSTRACT

Uranium stannide, the most stable intermetallic compound
in the Uranium/Sn system, has the Cu$_3$Au(LI$_2$) structure, and
is paramagnetic to liquid He temperature. The magnetic susceptibility data are consistent with 5f band states rather
than a localized electronic structure. Samples of this compound have been prepared both with depleted and ^{235}U enriched,
as well as ^{119}Sn enriched starting material. Detailed ^{119m}Sn
Mössbauer effect studies have been carried out over the temperature range 78$\leq$T$\leq$300 K on ^{60}Co gamma ray irradiated, thermal neutron irradiated (depleted), and thermal neutron irradiated (enriched) samples, using the V 10 position of the
Brookhaven HFR facility for the neutron irradiations
(thermal/fast $\geq$3.6x10^2). The Mössbauer hyperfine parameters
are: IS(78) = 2.45 mm/sec, Q.S.(78) 1.38 mm/sec for the unirradiated samples, consistent with the metallic nature of Sn
in the USn$_3$ structure. No anomalous charge states have been
identified in the Mössbauer spectra of the gamma or neutron
irradiated samples. The free-atom lattice temperature is
~250 K for the unirradiated sample, and shows only minor
changes with ^{60}Co-γ radiation to a total dose of 1000 Megarad.
Thermal neutron radiation (nvt ~5x10^{16}) of the depleted sample results in a lowering of the lattice temperature to
~240 K. Two hours heating at 800°C are sufficient to anneal
the radiation damage as judged by lattice dynamical data. A
sample enriched to 99.94% ^{235}U and irradiated to a fluence
of ~5x10^{16} slow neutrons/cm^2 showed an increase in the free-
atom lattice temperature to ~275 K. A two hour anneal at
800°C restored this parameter to its original value. Resistivity data for these samples and their relationship to spin
fluctuation temperatures will also be discussed.

---

*Supported by the National Science Foundation under Grant DMR 7808615A01

INTRODUCTION

Uranium stannide, USn_3, is the most stable of several known phases in the uranium-tin system. It has the Cu Au(LI_2) structure with space group $Pm3m$ [1] with the tin atom occupying a non-cubic symmetry site. A number of detailed investigations have focussed on the thermodynamic and magnetic properties of this compound in order to elucidate the various contributions to the low temperature specific heat. Murasik et al. [2] have shown USn_3 to be paramagnetic to 4.2 K and Misuik et al. [3] have observed Curie-Weiss behavior down to about 20 K where the susceptibility becomes temperature dependent. Buschow and van Daal [4] have reported resistivity data for this compound, which have been interpreted by them and by Brodsky [5,6] in terms of a spin fluctuation temperature which is defined as the temperature at which the susceptibility becomes temperature dependent ie: the temperature at which local spin fluctuations become more important than thermal relaxation processes. Such a temperature can be evaluated from $\rho(T)$ data and is approximately 11.9 K.

A number of thermodynamic studies of USn_3 have been reported. The free energy of formation is given by $\Delta G = -23000 + 0.60$ T cal mol^{-1} [7] and the heat of sublimation is -78800 cal mol^{-1} [8]. The specific heat has been measured over the temperature interval $0.6 \leq T \leq 430$ K by Bader et al. [9]. These authors assume an equation of the form

$$C = \gamma T + \alpha T^3$$

and find the electronic specific heat coefficient, γ, to be 171 mJ mol^{-1} K^{-2}, in good agreement with the value of 169 mJ $mole^{-1}$ K^{-2} reported by van Maaren et al. [10] and the value of 164 mJ $mole^{-1}$ K^{-2} cited by Brodsky [5]. The high value of γ has been ascribed to a narrow 5f (or hybrid 5f-6d) band at the Fermi surface. van Maaren et al. [10] have also reported a lattice specific heat constant (α) of 10.4 mJ mol^{-1} K^{-1} and a lattice (Debye) temperature of 195 K, in good agreement with the value of 196 K reported in ref. [9].

The present investigation was undertaken to elucidate the effect of radiation damage - especially that due to fission fragment recoil - on the solid state properties of USn_3. In an earlier study [11,12] it was shown that nuclear gamma ray resonance (Mössbauer Effect) techniques are a powerful tool in the study of radiation damage effects in solids, and could be used to advantage in tracing the annealing history of such damaged solids. In particular, using the 23.8 keV gamma radiation in ^{119}Sn, it is possible to probe the effect of radiation damage on the electron density around the tin atom, the lattice temperature of the solid, the distortion of the nearest neighbor symmetry around the metal atom, and a number of related properties of the solid, by measuring the magnitude and temperature dependence of the hyperfine interaction parameters over a temperature range well below the radiation damage annealing temperature. The results of such a study on USn_3 are reported herein.

EXPERIMENTAL

(a) Sample preparation: USn_3 was prepared in various isotopic configurations by arc melting uranium and tin metal under an argon atmosphere. The sample button resulting from the initial stoichiometric mixture of the metals was remelted three times, and annealed at $800^{\circ}C$, to give a homogeneous solid as judged by X-ray powder pattern data. The observed unit cell dimension (a_0) is 4.54 Å. In the present study, the pyrophoric nature of USn_3 has not been confirmed [13], but the prepared samples were stored in an inert atmosphere or under vacuum to avoid reaction with O_2 and/or atmospheric moisture. Three series of samples were prepared for study:

USn_3 depleted U, natural abundance Sn

$U^{119}Sn_3$ depleted U, 85% enriched ^{119}Sn

$^{235}U^{119}Sn_3$ 99.94% ^{235}U, 85% enriched ^{119}Sn

(b) Irradiations: Gamma irradiations using the 1.17 and 1.332 MeV radiations of ^{60}Co were carried out at the Isomedix facility at room temperature on USn_3 [14] sealed under vacuum in quartz ampoules. Total dose rates of 200, 920 and 1000 Mrad were used in these experiments. Neutron irradiations on $U^{119}Sn$ and $^{235}U^{119}Sn$ were carried out at the Brookhaven Hi Flux reactor [15] in position V 10. The thermal flux at this position is 1.8×10^{14} ncm^{-2} sec^{-1} and the fast (>1 MeV) neutron flux is 5×10^{11} ncm^{-2} sec^{-1}. A background sample of USn_3 was also irradiated under these conditions. These samples were irradiated in high purity quartz ampoules.

(c) Mössbauer Spectroscopy: Variable temperature ^{119}Sn Mössbauer effect spectroscopy was effected as described earlier [11,16], with sample temperatures maintained to $\pm$ .5 degrees for the times needed to accumulate the data at each temperature point (normally 2 to 20 hours depending on the resonance effect magnitude). In order to minimize reactor loading by ^{235}U (vide infra) to 2 mgms or less, the Mössbauer experiments were all carried out on ^{119}Sn enriched samples having a total weight of 8-10 mgms. Uranium dilution was effected as needed. The samples were examined as powders of finely crushed particles sealed in Al foil. Spectrometer calibration was effected using the room temperature magnetic hyperfine spectrum of metallic iron. The zero point of the velocity scale was determined from room temperature spectra of $BaSnO_3$, and all isomer shifts are reported with respect to the centroid of such spectra determined before and after the sample spectral runs. Data reduction was effected using a least squares matrix inversion program [17] suitably modified to run on the Rutgers IBM 370/168 computer.

(d) Resistivity Measurements: Resistivities of the samples were determined using a standard 4-probe DC method while the sample temperature was continuously varied over the range $4.2 \leq T \leq 300$ K. Readings were taken with the current flow in both directions to minimize systematic errors of measurement.

RESULTS AND DISCUSSION

The ^{119}Sn Mössbauer spectrum of USn_3 at liquid nitrogen temperature consists of a doublet with well resolved resonance maxima. A typical spectrum is shown in Fig. 1, and the data extracted from the temperature dependent resonance experiments are summarized in Table I. The isomer shift (I.S.) of 2.45 mm sec^{-1} is similar to that observed for metallic (β) tin at room temperature (2.553 ± 0.006 mm sec^{-1}) and is characteristic of the metal formally in the zero oxidation state. The quadrupole splitting (Q.S.) of 1.38 mm sec^{-1} reflects the distortion from cubic symmetry of the nearest neighbor environment around the tin atom. The lattice temperature of 250 K was calculated from the relationship

$$\theta_M = \frac{E_\gamma}{C} \left[\frac{-3}{M_{eff} \ K_B \ d \ln A/dT} \right]^{\frac{1}{2}} \tag{1}$$

$$= 13.381 [d \ln A/dT]^{-\frac{1}{2}}$$

in which E_γ is the ^{119}Sn Mössbauer gamma ray energy, K_B is Boltzmann's constant and $d \ln A/dT$ is the slope of the (normalized) logarithmic temperature

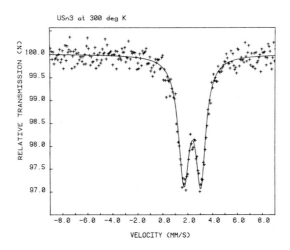

Fig. 1. ^{119}Sn Mössbauer Effect spectrum of USn$_3$ at 300 K. The isomer shift scale is with respect to the centroid of a room temperature BaSnO$_3$ spectrum acquired using the same source as that used in the sample runs.

dependence of the area under the resonance curve. In applying (1) to the USn$_3$ data of the present study, the effective vibrating mass, M$_{eff}$, has been replaced by the "bare" atom mass of ^{119}Sn (thus ignoring any covalency effects in the chemical bonding of tin to its nearest neighbor atoms).

As is reflected in the data summarized in Table I, there is essentially no change in IS, QS or θ_M for USn$_3$ on exposure of up to 1000 M rad of ^{60}Co gamma radiation. No evidence for the existence of anomalous tin atom charge states in these irradiated samples was evident from the ^{119}Sn Mössbauer data. More-over, the resistance ratio [R(T)/R(275)] appeared identical for the gamma irradiated samples and the virgin samples in the region T>T$_{SF}$. These data are summarized graphically in Fig. 2. The spin fluctuation temperature for the 1000 M rad irradiated sample, however, was observed to be ~10 K, a value some-what reduced from that of 11.9 K reported for an unirradiated specimen [4]. This observation can be understood as arising from the onset of local moment formation due to the breakup of the long range correlation implied by the nar-row 5-f band model. This local disorder is not readily detected in the isomer shift parameter extracted from the Mössbauer experiments since no major change in the 5-S electron density around the tin atom is expected to result from the presence of the radiation damage-caused lattice defects.

A sample of U^{119}Sn$_3$ containing only depleted ^{238}U was irradiated (~5 min) to a total nvt of ~5x10^{16} at a reactor position where the slow to fast neutron flux ratio was at least 3.6x10^2. As noted from Table I the Mössbauer spectra for this sample showed the same values of the hyperfine interaction parameters, but the temperature dependence of the recoil-free fraction (as reflected in the area measurements) increased by ~10% with a concomittant decrease in the lattice temperature to 240 K. On annealing at 800°C for 2 hours the original lattice temperature was re-established in this sample. In view of the results obtained

TABLE I
Summary of ^{119}Sn Mössbauer Effect Data

Radiation Treatment		I.S.(78K)[a] mm sec^{-1}	Q.S.(78K)[a] mm sec^{-1}	$\dfrac{d \ln A}{dT}$[b] K^{-1} x10^3	θ_M[c] K
USn$_3$ and U^{119}Sn$_3$					
(a)	none	2.45	1.38	-2.84	250
(b)	^{60}Co gamma-200 Mrad	2.45	1.38	-2.77	254
	-920 Mrad	2.45	1.38	-2.64	260
	-1000 Mrad	2.45	1.38	-2.84	251
(c)	~5x10^{16} ncm^{-2}	2.45	1.38	-3.12	240
	anneal 800°C 2 hrs	2.45	1.38	-2.75	255
^{235}U^{119}Sn$_3$					
(d)	none	2.44	1.37	-2.84	250
(e)	~3.6x10^{16} ncm^{-2}	2.40	1.45	-2.35	276
	anneal 800°C 1 hr	2.40	1.38	-2.80	253
	anneal 800°C 2 hrs	2.43	1.35	-2.73	256

[a]Estimated errors ±0.01 mm sec^{-1}
[b]Estimated errors ±0.09 K^{-1}
[c]Estimated errors ±8 K

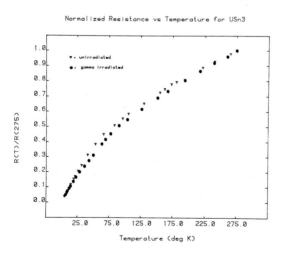

Fig. 2. Resistance Ratio determined by the four-probe technique for virgin
(triangles) and ^{60}Co gamma irradiated (filled circles) USn$_3$.

432

with the ^{60}Co irradiated samples, it is probable that the observed results in this experiment reflect the "background damage" due to the fast neutron flux on ^{238}U, but this point is not definitively established on the basis of the presently available data.

A sample of ^{235}U^{119}Sn$_3$ containing ~2 mgm of the fissile uranium nuclide, was irradiated under the same conditions as the previous sample. The data extracted from the Mössbauer experiments are again included in Table I. As expected, the unirradiated sample gave results completely consistent with those observed for U^{119}Sn$_3$. The neutron irradiated sample, however, gave rise to an asymmetric Mössbauer spectrum with considerably broadened resonance lines, see Fig. 3. Although the increased line widths which are observed suggest the presence of two (or more) distinct tin sites in this sample after neutron irradiation, the individual components could not be resolved from the data. The asymmetric spectra were computer fitted to a two Lorentzian pattern in order to determine the temperature dependence of the total area under the resonance curve. The corresponding d ln A/dT data are shown in Table I, from which it is seen that the apparent temperature dependence has <u>decreased</u> leading to an <u>increase</u> in the calculated lattice temperature, θ_M. A one hour anneal at 800°C is sufficient to restore θ_M to the value observed for the unirradiated sample, but a two hour anneal at 800°C is required to restore the intensity symmetry of the two components of the quadrupole split spectrum.

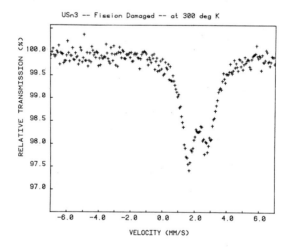

Fig. 3. ^{119}Sn Mössbauer Effect spectrum at 300 K of ^{235}USn$_3$ exposed to a total thermal neutron fluence of ~3.6x10^{16} cm^2. The intensity asymmetry is ascribed to the presence of anomalous tin sites in the sample related to fission particle damage. Two hour annealing at 800°C restores the original symmetric resonance spectrum (See Fig. 1).

Neutron irradiation of a second $^{235}U^{119}Sn$ sample to a total nvt of 3.8×10^{17} cm^{-2} (approximately a factor of 10 higher than in the previous sample) gave rise to a sample which showed macroscopic evidence of radiation damage, possibly due to self heating by the fission fragment tracks in the solid. In contrast to unirradiated USn$_3$ which is brittle and hard, the neutron irradiated sample was malleable and could not be crushed to a powder as before. The ^{119}Sn Mössbauer spectrum showed only a very broad resonance line, inconsistent with the parameters of the virgin matrix. There is some evidence of decomposition and/or phase segregation in this sample related to exceptionally pronounced thermal effects presumably due to ^{235}U fission. Experiments with $^{235}U^{119}Sn_3$ exposed to the same <u>total</u> nvt, but at considerably lower neutron fluxes (to reduce sample heating effects) are currently underway. The present data, although incomplete in the sense of providing a detailed view of fission fragment radiation damage in USn$_3$, clearly demonstrate the applicability of nuclear gamma resonance spectroscopy in the elucidation of these interesting phenomena.

ACKNOWLEDGMENTS

This research was supported in part by the National Science Foundation under Grant DMR 78086 15A01 and a grant from the Center for Computer and Information Services, Rutgers University. This support is herewith gratefully acknowledged. The authors are also indebted to Prof. M. Croft for generously making available the sample preparation and resistance measurement instrumentation used in this study and for numerous fruitful discussions.

434

REFERENCES AND NOTES

[1]   R.E. Rundle and A.S. Wilson, Acta. Cryst. $\underline{2}$, 148 (1949).

[2]   A. Murasik, J. Lecrejewicz, S. Ligenza and A. Zygmundt, Phys. Stat. Solidii $\underline{23}$, K147 (1975).

[3]   A. Misuik, J. Mulak and A. Czopnik, Bull. L'Acad. Pol. Sci. Ser. Chimiques $\underline{20}$, 459 (1972).

[4]   K.H.J. Buschow and H.J. van Daal, A.I.P. Conference Proc. $\underline{5}$, 1464 (1971).

[5]   M.B. Brodsky in "Experimental Studies of Narrow Band Effects in the Actinides" R. Park, Editor, Plenum Press, N.Y., 1977.

[6]   M.B. Brodsky, Phys. Rev. $\underline{9}$, 1381 (1974).

[7]   C.B. Alcock and P. Grievson, J. Inst. Met. $\underline{93}$, 304 (1962).

[8]   A.N. Ivanov and N.F. Praudyuk, Thermo. Nucl. Mat. Symposium, IAEA, Vienna, 735 (1962); Chem. Abstr. $\underline{58}$, 9707 (1963).

[9]   S.D. Bader, G.S. Knapp and H.V. Culbert, Magnetism and Magnetic Materials, AIP Conference $\underline{24}$, 222 (1974).

[10]   M.H. van Maaren, H.J. van Daal, K.H.J. Bushow and C.J. Schinkel, Solid State Comm. $\underline{14}$, 145 (1974).

[11]   R.H. Herber and R. Kalish, Phys. Rev. $\underline{B16}$, 1789 (1977).

[12]   R.H. Herber and R. Kalish, Hyperfine Interactions $\underline{4}$, 666 (1978).

[13]   V.U.S. Rao and R. Vijayaraghavan, J. Phys. Chem. Solids, $\underline{29}$, 123 (1968).

[14]   The authors are indebted to G.R. Dietz of Isomedix Inc. for generously facilitating these exposures, and gratefully acknowledge his assistance in this matter.

[15]   The authors are indebted to D.C. Rorer of the Reactor Division, BNL, for facilitating these irradiations.

[16]   R.H. Herber and Y. Maeda, Inorg. Chem. $\underline{00}$, 000 (1980) and references therein.

[17]   G.K. Shenoy, private communication.

Published 1981 by Elsevier North Holland, Inc.
Kaufmann and Shenoy, editors
Nuclear and Electron Resonance Spectroscopies Applied to Materials Science

GEOMETRICAL STRUCTURE OF LATTICE DEFECT-IMPURITY CONFIGURATIONS DETERMINED

BY TDPAC

M. DEICHER, O. ECHT, E. RECKNAGEL AND TH. WICHERT
Fakultät für Physik, Universität Konstanz, D-7750 Konstanz,
Fed. Rep. Germany

ABSTRACT

The perturbed angular correlation technique (TDPAC) was
applied to determine the orientation of the electric field
gradient tensor induced by lattice defects at the probe
[111]In. The experimental results obtained for self-interstitials, vacancies and defect clusters in Cu, Ag and Au are
shown and their microscopic interpretation is discussed.

1. INTRODUCTION

In this contribution we want to illustrate that the mere tagging of a
defect trapped at a radioactive impurity atom by its characteristic quadrupole
coupling constant $\nu_Q = eQ \cdot V_{zz}/h$ and the asymmetry parameter $\eta = (V_{xx} - V_{yy})/V_{zz}$
does not make the best use of the electric field gradient (efg) in its capa-
city as a tensor (eQ is the nuclear quadrupole moment of the radioactive probe
and V_{ii} are the elements of the diagonalized tensor). In most cases of defect
studies, using nuclear methods like Mössbauer effect or the TDPAC technique
(time differential observation of perturbed $\gamma\gamma$ angular correlation) it has
been omitted to determine the orientation of the efg tensor with respect to
the crystallographic lattice. The orientation and the asymmetry parameter are
just values which can be estimated in the framework of a simple "point charge"
model, at least in case of trapped defects being not too complicated. In such a
model it is assumed that the geometrical behaviour of the efg is mainly governed
by the geometrical arrangement of the metal ions, i.e. by the geometry of the
lattice. Then the efg is easily calculated by the sum of all efg's originating from
each lattice site, so that a vacant lattice site does not give a contribution.
Obviously, the absolute value of the efg is not obtainable in this way, but the
measured orientation of a defect induced efg should become interpretable in this
simple model. The structural information about the probe atom-defect configuration
can be compared with that obtained by other methods like the diffuse scattering
technique, mechanical relaxation measurements |1| or channeling measurements |2|.

2. ORIENTATION OF THE EFG IN LOCALLY DISTORTED CUBIC METALS

The efg is mathematically represented by a symmetric and traceless 3×3
tensor. Fig.1 shows its three parameters of interest in a schematic way:
"Axis", "symmetry" and "orientation". The "orientation" contains the
three rotation angles to obtain a diagonalized matrix with its three com-
ponents V_{xx}, V_{yy} and V_{zz} using the convention $|V_{xx}| \leq |V_{yy}| \leq |V_{zz}|$. When the efg
possesses "symmetry", i.e. $V_{xx} = V_{yy}$ or $\eta = 0$, the orientation of the ten-
sor is presented by the one of the "axis", i.e. the component V_{zz}. In case of
a symmetric efg and for an isomeric state with spin $I = 5/2$, as it holds for the
radioactive probe ^{111}In/^{111}Cd, the time dependence of the observed $\gamma\gamma$ angular

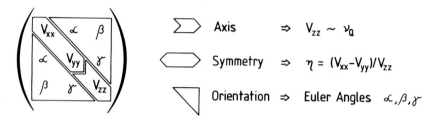

Fig.1: Meaning of the different elements of the efg tensor.

correlation pattern reads

$$G(\vartheta,\varphi,\theta,t) = \sum_{n=o}^{3} S_n(\vartheta,\varphi,\theta) \cos(nc \, \nu_Q t) \qquad (1)$$

with $c = 3\pi/10$. The angles ϑ,φ are spherical coordinates which describe the orientation of the efg with respect to a cartesian coordinate system defined by the direct-
ion of the two $\dot{\gamma}$ quanta emitted with a relative angle θ to each other |3|: The dir-
ection of the z-axis is perpendicular to the plane formed by the two γ rays and the
direction of the x-axis coincides with that of the first γ quantum. In a TDPAC experiment the directions of the two γ quanta are defined by those of two γ-detectors
enclosing the angle θ. For definite directions of the detectors with respect to
the crystallographic lattice of a single crystal the angles ϑ and φ and thereby
the efg orientation can be identified with the lattice directions <h k l>. In
fig.2 the calculated amplitudes S_n are given assuming the detectors to be oriented
along a <110> and < 1$\bar{1}$0 > lattice direction (i.e. $\theta = 90^o$): E.g., the entries $\vartheta = 90^o$
and $\varphi = 45^o$ in the diagram correspond to an efg pointing into one of the six
directions which are equivalent to the <110> direction. It can be seen that the
selected detector geometry is particularly sensitive to orientations about the
<100> direction. As shown by eq.(1) the interaction frequency ν_Q is not affected
by the parameters ϑ,φ and θ. In this way it is possible to predict the amplitudes
S_n for the different efg orientations for any preselected detector orientation
with respect to the crystal. In the following we shall present data of self-
interstitials, vacancies and clusters bound to the probe ^{111}In/^{111}Cd where in
cases of axial symmetry, i.e. $\eta = 0$, the orientation of the symmetry axis, i.e.
of the V_{zz} component, was investigated by the TDPAC. This information was used
to get insight into the microscopic arrangement of the probe-defect configuration
in the lattice.

3. RESULTS AND DISCUSSIONS

3.1 Self-interstitials

Recently it was shown that in the range of the recovery stage I in Cu, which is
around 40 K, two ^{111}In-defect configurations are formed, characterized by $\nu_{Q5} = 46.8$
MHz and $\eta = 0$ and $\nu_{Q6} = 19$ MHz and $\eta = 1.0$ |4|. Due to the low temperature one has
only to take into consideration trapped self-interstitials for the discussion of
these configurations. On the basis of other experiments the defect ν_{Q5} was attributed to a trapped mono-interstitial. For such configurations theoretical calcu-
lations exist |5| which predict for an oversized impurity atom as trapping center
the formation of a "pure" dumbbell at the nearest neighbour site (see fig.3a) and
for an undersized atom the formation of a "mixed" dumbbell (see fig.3b). In the
framework of the point charge model and neglecting possible lattice relaxations one

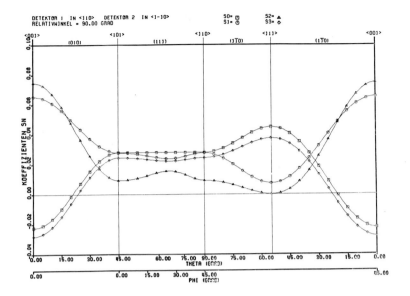

Fig.2: Calculated coefficients S_n for detectors pointing into a <110> and a <1$\bar{1}$0> lattice direction as function of the angles ϑ and φ which define the efg orientation. The angle between the two detectors is $\Theta = 90°$.

Self-Interstitial (●)   Trapped at an Impurity Atom (●)

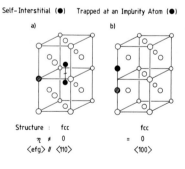

Trapped Monovavancy (□)

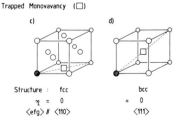

Fig.3: Calculated configurations of a trapped self-interstitial at an oversized impurity (a) and at an undersized impurity (b) in fcc metals [5]. Next neighbour positions of a trapped monovacancy in a fcc lattice (c) and a bcc lattice (d).

438

obtains for the pure dumbbell η ≠ 0 and an efg pointing approximately into a <110>
lattice direction, and for the mixed dumbbell η = 0 and an efg in a <100> direction.
On the left side of fig.4 the Fourier transforms of the experimental time spectra are
shown for different detector orientations; the second orientation corresponds to
the diagram given in fig.2. Already a first glance at the spectra shows the dependence of the amplitudes S_n on the detector geometry. On the right side of fig.4
the experimental values deduced for v_{Q5} are compared with the predictions assuming efg orientations into <100>, <111> and <110> directions. (In this graph the
given S_n values are multiplied by 0.10, the relative fraction of the configuration v_{Q5}.) It is evident, that from the experiment a <100> orientation of the
efg has to be deduced. A comparison with fig.3a and b shows that the mixed dumbbell explains the experimental result. However, this finding is in contrast to
the theoretical predictions, because the In-atom is an oversized impurity in the
Cu lattice.

3.2 Vacancies

In table I the different ^{111}In-defect configurations are collected which have
been found up to now in the recovery stage III of Cu, Ag, Au and Mo. All data were
measured in Konstanz. Except for Mo |6|, the orientation measurements were performed

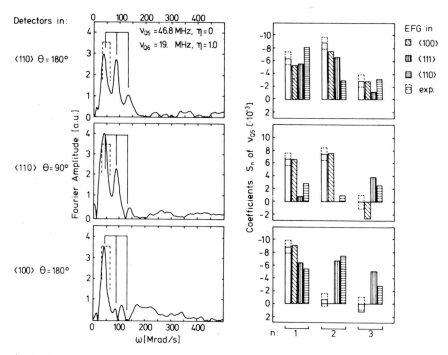

Fig.4: Fourier-transforms of measured time spectra after H⁺-irradiation at
52 K in a Cu single crystal recorded for different detector-crystal orientations and different angles Θ between the detectors (left side). Comparison of
the measured coefficients S_n with theoretically calculated S_n assuming the three
efg orientations <100>, <111> and <110> (right side).

Table I: DEFECTS TRAPPED IN STAGE III USING THE PROBE [111]In

Lattice	No.	ν_Q[MHz]	η	\<efg\> in
Cu(fcc)	1	116.	0	\<110\>
	2	181.	0	not\<110\>
Ag(fcc)	1	82.	0	\<110\>
Au(fcc)	1	102.	0.45	—
	2	101.	0	not\<110\>
	3	91.	0	\<110\>
	4	40.	0	\<111\>
Mo(bcc)	1	125.	0	\<111\>
	2	155.	1.0	—
	3	98.	0	\<100\>

recently, whereas the several defect configurations are known from earlier experiments (Cu |7|, Ag |8|, Au |9|, Mo |10|). On the basis of the published experiments one should assume that ν_{Q1} in Cu and Ag, ν_{Q3} in Au and ν_{Q1} in Mo characterize trapped monovacancies. In fig.3c and d the microscopic situation is shown for a trapped monovacancy in a fcc- and bcc-lattice, respectively |5|. The predicted efg orientations are \<110\> and \<111\> with η = 0. Turning back to table I we recognize two points: The defect configurations mentioned above as candidates for a trapped monovacancy show indeed the expected efg orientation. Secondly, in case that several defect configurations are observed in stage III, the other ones do not fulfill the postulated geometrical conditions for a trapped monovacancy. Both points show that the information obtained by TDPAC experiments is interpretable in a consistent way and that the In-monovacancy configurations in fcc- and bcc-metals exhibit common geometrical features.

3.3 Defect clusters

Concerning defect clusters information from TDPAC experiments is also available: In Ni|11|, Cu |12| and Au |9| [111]In atoms trapped at defect clusters have been observed. In all cases the observed defect configurations show a high thermal stability, i.e. they are stable up to recovery stage V, which is interpreted by the dissolution of clusters. Fig.5 shows the results of TDPAC experiments to determine the orientation of the efg in Cu. For all time spectra three fits were performed assuming that the efg points into \<100\>, \<111\> or \<110\> marked by the dashed, solid and dotted line respectively. From the second and third spectrum it is visible that the efg points into a \<111\> lattice direction. The same results were observed in Ni and Au. As shown in more detail in ref.|13| one can assume that the ^{111}In atoms are within a planar loop which consists of a stacking fault on a {111} plane as shown in fig.6: The stacking sequence of the fcc lattice ABCAB is altered into AB.AB with the result that a local hcp structure originates with the lattice constants c = $(4/3)^{1/2} \cdot a_{fcc}$ in the ideal case and $a_{hcp} = (1/2)^{1/2} \cdot a_{fcc}$. In such a loop an ^{111}In/[111]Cd probe experiences an axially symmetric efg pointing into a \<111\> direction in accordance with the experimental results.

The authors wish to thank R. Minde for experimental help and are grateful to the Bundesminister für Forschung und Technologie for financial support.

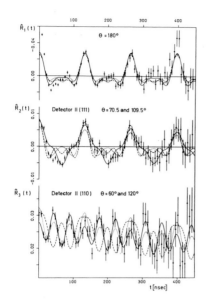

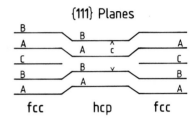

Stacking Faults in fcc Metals
efg ∥ ⟨111⟩, $\eta = 0$

{111} Planes

Fig.5: Influence of the detector orientation on the R(t) spectrum in the case of ν_{Q3} = 52 MHz in a copper single crystal. Fits assuming that the axis of the efg points into <100> (dashed line), <111> (solid line) or <110> (dotted line). The two lower spectra show clearly, that the efg direction coincides with a <111> direction.

Fig.6: Alteration of the stacking sequence ABCAB of the <111> planes in a fcc lattice by one missing plane resulting in a local hcp structure.

REFERENCES

1. Properties of Atomic Defects in Metals, edited by N.L. Peterson and R.W. Siegel, J.Nucl.Mat. 69 & 70, 1978.

2. M.L. Swanson, L.M. Howe and A.F. Quenneville, J.Nucl.Mat. 69 & 70, 372(1978)

3. K. Alder, Helv. Phys. Acta 26, 761 (1953).

4. M. Deicher, R. Minde, E. Recknagel and Th. Wichert
   in: Proceedings of the Fifth International Conference on Hyperfine Interactions, Berlin 1980, edited by H. Haas and G. Kaindl, to be published in
   Hyperfine Interactions.

5. P.H. Dederichs, C. Lehmann, H.R. Schober, A. Scholz, R. Zeller,
   J.Nucl.Mat. 69 & 70, 176 (1978).

6. A. Weidinger, R. Wessner, Th. Wichert and E. Recknagel,
   Phys.Lett.72A, 369 (1979).

7. Th. Wichert, M. Deicher, O. Echt, E. Recknagel, Phys.Rev.Lett. 41, 1659 (1978)

8. M. Deicher, O. Echt, E. Recknagel and Th. Wichert in ref.4.

9. M. Deicher, E. Recknagel and Th. Wichert, Rad.Effects, to be published.

10. A. Weidinger, R. Wessner, E. Recknagel and Th. Wichert
    in: Proceedings of the International Conference on Ion Beam Modification
    of Materials, Albany, USA, 1980, to be published in Nucl.Instr.Methods.

11. F. Pleiter, Hyperfine Interactions 5, 109 (1977)

12. O. Echt, E. Recknagel, A. Weidinger, Th. Wichert, Z.Physik B32, 59 (1978)

13. M. Deicher, O. Echt, E. Recknagel and Th. Wichert, in ref.4.

Published 1981 by Elsevier North Holland, Inc.
Kaufmann and Shenoy, editors
Nuclear and Electron Resonance Spectroscopies Applied to Materials Science

INVESTIGATION OF LATTICE DEFECTS IN HCP METALS

R. KEITEL, W. ENGEL, S. HOTH, W. KLINGER, R. SEEBÖCK, and W. WITTHUHN
Physikalisches Institut der Universität Erlangen-Nürnberg,
8520 Erlangen, Germany

ABSTRACT

With the perturbed angular correlation and distribution methods
we studied the trapping and annealing of lattice defects in the
hcp metals Cd and Zn. The defects were produced (i) by proton
irradiation, (ii) by quenching, and (iii) by heavy ions recoiling after a nuclear reaction. As probe ions we used ^{111}Cd
(i,ii) and ^{67}Ge, ^{69}Ge, ^{71}Ge, ^{113}Sn, and ^{116}Sn (iii).
The In-impurities in Zn and Cd are trapping centers for vacancies. At low temperatures (100 to 200 K) simple defect configurations are observed. In the in-beam experiments a sharp annealing step due to detrapping of vacancy type defects is found.

INTRODUCTION

Irradiation of metals and alloys with fast particles results in a change of material properties as, for example, volume, electrical resistivity, brittleness etc.
Understanding and control of these effects is of great technological importance,
e.g. for the construction of nuclear power plants.
The changes in macroscopic properties are caused by microscopic crystal lattice
defects, such as vacancies, interstitials and their agglomerates. The defects are
produced by collisions of the irradiating particles with lattice atoms. In the
vicinity of a defect the electric charge distribution is changed, and thus the
defect manifests itself in a specific electric field gradient (efg) at the sites
of neighbouring atoms.
Because of the microscopic nature of the defects, nuclear methods are well suited
for their investigation. In these methods an atomic nucleus serves as a probe
which monitors its surroundings via the hyperfine interaction on a short range
scale of several atomic distances.
In the present contribution a summary is given of the methods used and the results obtained in our investigations of radiation induced lattice defects in the
hcp metals zinc and cadmium.

EXPERIMENTAL METHODS

All experiments were carried out using the methods of time differential observation of the perturbed angular γ-ray correlation (TDPAC) and distribution (TDPAD).
For details see ref. /1/. In both methods a radioactive nucleus in an excited
state with a half life of the order of a few nanoseconds up to milliseconds
serves as the microscopic probe. The measured quantity is the intensity of the γ-
radiation depopulating this isomeric state. In the ensemble of the observed probe
nuclei the nuclear spins have to be aligned in order to obtain an anisotropic γ-
ray angular distribution. This is achieved in the case of TDPAC by observing the
γ-quantum which depopulates the isomeric state in coincidence with the populating
one (observation of a $\gamma-\gamma$-cascade).
In the case of TDPAD the alignment is produced by populating the isomeric state
via a nuclear reaction with a pulsed particle beam.

Due to the electric quadrupole interaction of the quadrupole moment Q of the isomeric state with the electric field gradient $V_{zz} = \partial^2 V/\partial z^2$ (V is the electrostatic potential), the γ-ray anisotropy is time dependent. This results in a time
dependent modulation of the exponential γ-decay intensity which is measured time
differentially with NaI(Tl) or Ge(Li) detectors. Experimental details are given in
/2/. By simultaneous measurement with 4 detectors (TDPAC) or 2 detectors (TDPAD)
the exponential decay intensity can be eliminated and the so called perturbation
factor $G_{22}(t)$ can be deduced. In the simple case, where all probe nuclei experience the same axially symmetric efg in a polycrystalline sample, the theoretical
expression for $G_{22}(t)$ is given by

$$G_{22}(t) = \sum_n s_{2n} \cdot \cos(n\omega_0 t). \qquad (1)$$

For half integer nuclear spin I, the frequency ω_0 contains the information on the
efg strength by

$$\omega_0 = 6/(4I(2I-1)\hbar) \; eQV_{zz}. \qquad (2)$$

The quantity $|eQV_{zz}/h|$ is often called the quadrupole coupling constant. In the
case of a non-axially symmetric efg the perturbation factor (1) is in general
nonperiodic and depends on the asymmetry parameter $\eta = (V_{xx} - V_{yy})/V_{zz}$. If different fractions $f^{(i)}$ of the probe nuclei are subject to different efgs, the
perturbation factor has to be written as

$$G_{22}(t) = \sum_i f^{(i)} \cdot G_{22}^{(i)}(t). \qquad (3)$$

Typical examples for experimentally obtained perturbation factors are shown in
fig. 1a.
Like all hyperfine interaction methods (e.g. NQR, MS) both, TDPAC and TDPAD are
highly locally sensitive due to the r^{-3} dependence of the electric quadrupole interaction. The experiments can be performed with a very low probe (= impurity)
concentration of the order of 10^{-8}. The defect induced efg is defect-specific
with respect to magnitude and symmetry. As the probe atoms are often impurities,
trapping of migrating defects is possible. At the present state of solid state
theory, however, it is not possible to calculate the efg caused by a given defect
configuration.
In the TDPAC method the radioactive probe atoms are introduced into the sample
by diffusion or ion-implantation. The production of defects by irradiation and
the annealing programme are performed in the same way as in "classical" methods
such as resistivity annealing.
In the TDPAD method the probe atoms are produced within the sample via a nuclear
reaction. Recoiling through the crystal lattice they generate an extended defect
cascade in which they come to rest. Subsequently the annealing of these correlated, i.e. probe produced, defects can be observed within a time interval determined by the nuclear half life. Thus the TDPAD method is microscopic in space as
well as in time. Probe nuclei with suitable half lives are available in most elements.

EXPERIMENTAL RESULTS AND DISCUSSION

Trapping of radiation induced defects at the probe ions and subsequent detrapping
were studied in the hcp metals cadmium and zinc.
Using the TDPAC-method, uncorrelated defects were observed after proton irradiation and quenching. The ^{111}In daughter-nucleus ^{111}Cd($5/2^+$ state, $T_{1/2}$ = 84 ns)
served as probe.

445

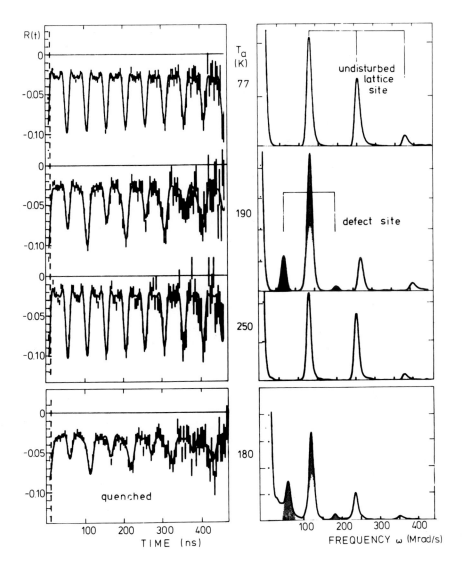

Fig. 1a
TDPAC time spectra after different
annealing temperatures. The upper
three spectra are obtained after
proton irradiation, the bottom one
from a quenched sample

Fig. 1b
Frequency spectra obtained from a
weighted Fourier analysis of the
corresponding time spectra of fig. 1a
(arbitrary units).

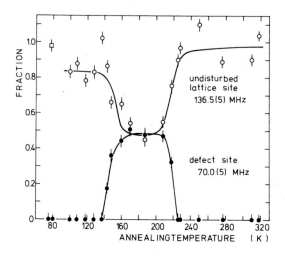

Fig. 2
Relative fractions of [111]In
ions with (●) and without
(o) trapped defect in zinc
as a function of annealing
temperature

Table I   TDPAC results

system	temperature of trapping	probes with defects	$\|eQV_{zz}/h\|$ with defect	η_D	E_B (eV)	probable defect configuration
[111]InCd	105...140 K	15 %	103(3) MHz	0.99(1)	0.003	vacancy type /3/ (mono- or di-)
[111]InZn	140...220 K	50 %	70.0(5) MHz	⩽ 0.15	0.007	vacancy type (mono- or di-)

Table II   TDPAD results

system	observed time range	efg with defect	detrapping temperature	E_B (eV)	ref.
[67]GeZn	800 ns	$\gg V_{zz}^{\circ}$	420(10) K	0.06(5)	/4/
[69]GeZn	4 µs	not resolved	"	0.07(6)	/4/
[71]GeZn	500 ns		"	0.07(3)	/4/
[113]SnCd	900 ns	0.85 V_{zz}°	400(10) K	0.09	this
[116]SnCd	1.6 µs	$\eta = 0.84(5)$	"		work

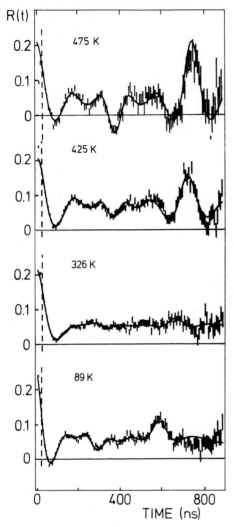

Fig. 3
TDPAD modulation spectra taken at different temperatures for the system ^{113}SnCd

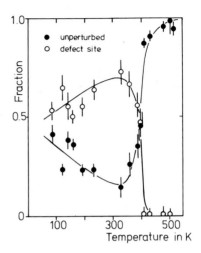

Fig. 4
Relative fractions of ^{113}Sn probe ions in cadmium with (o) and without (●) trapped defect as a function of target temperature

With the TDPAD-method correlated defects were observed after (α,xn)-reactions using isomeric probe levels in

^{67}Ge (9/2$^+$, $T_{1/2}$= 70 ns),
^{69}Ge (9/2$^+$, $T_{1/2}$= 2.9 μs),
^{71}Ge (5/2$^-$, $T_{1/2}$= 73 ns),
^{113}Sn (11/2$^-$, $T_{1/2}$= 82 ns), and
^{116}Sn (5$^-$, $T_{1/2}$= 370 ns).

The results of these experiments are summarized in tables I and II. In the following the most recent experiments on the systems ^{111}CdZn and ^{113}SnCd will be discussed more in detail.

In the TDPAC-experiment on ^{111}CdZn the ^{111}In mother activity was diffused into natural zinc foils of high purity (999,999 %). The irradiation was performed at 77 K with 6 MeV protons up to a defect concentration of approximately 100 ppm. The TDPAC time spectra in fig. 1a were measured at 77 K after isochronal annealing of the samples (holding time 10 min) at the indicated temperatures. The bottom spectrum was obtained after quenching the sample from about 300 °C into an ice-salt cold-mixture of -20 °C. The solid lines are least squares fits of function (3) to the data. Fig. 1b shows the corresponding frequency spectra obtained by a weighted fourier analysis. The striking change in the 190 K spectrum is due to a well defined defect configuration, trapped in the nearest neighbourhood of the ^{111}In ion. This defect-probe complex is formed at 140 K and dissolves at 220 K (see fig. 2). This allows the deduction of a defect-probe binding energy E_B = 0.007 eV. Identical spectra obtained in the quench experiment prove that the defect is of vacancy type. It is characterized by an efg with a quadrupole coupling constant of $|eQV_{zz}/h|$ = 70.0(5) MHz and nearly axial symmetry with $0 \leq \eta_D \leq 0.15$. Using the simple point charge model for symmetry considerations, a di-vacancy with both vacancies in the basal plane or a monovacancy out of the basal plane are the most likely configurations.

TDPAD-spectra of the system ^{113}SnCd, measured at different target temperatures, are shown in fig. 3. At 475 K the efg of the polycrystalline cadmium lattice is observed. With decreasing temperature the modulation of the spectra becomes less pronounced due to a broad efg distribution caused by different defect configurations of the complex defect cascade. Below 230 K a well defined modulation is observed due to a defect, trapped by about 60 % of the probe ions (see fig. 4). At 90 K the resulting efg is 15 % smaller than the efg V_{zz}^0 at an unperturbed lattice site. The asymmetry parameter is η_D = 0.84(5). From symmetry considerations within the point charge model a monovacancy in the basal plane as well as a divacancy with one vacancy in the basal plane yields a consistent explanation of the data. This probe-defect complex dissolves within a small temperature range of 20 degrees at 400 K (fig. 4). From this an upper limit of E_B = 0.09 eV for the defect-probe binding energy E_B was estimated.

This work was financially supported by the Bundesministerium für Forschung und Technologie.

REFERENCES

1. H. Frauenfelder and R.M. Steffen in: Alpha-, Beta- and Gamma-Ray Spectroscopy, K. Siegbahn ed. (North-Holland, Amsterdam 1965)

2. J. Christiansen, P. Heubes, R. Keitel, W. Klinger, W. Loeffler, W. Sandner, W. Witthuhn,
   Z. Phys. B24, 177 (1976)

3. W. Witthuhn, A. Weidinger, W. Sandner, H. Metzner, W. Klinger, and R. Böhm, Z. Phys. B33, 155 (1979)

4. G. Hempel, H. Ingwersen, W. Klinger, W. Loeffler, W. Sandner, and W. Witthuhn, Phys. Lett. 55A, 51 (1975)
   R. Böhm, J. Christiansen, W. Klinger, R. Keitel, W. Loeffler, W. Sandner, W. Witthuhn,
   Hyp. Int. 4, 763 (1977)

Published 1981 by Elsevier North Holland, Inc.
Kaufmann and Shenoy, editors
Nuclear and Electron Resonance Spectroscopies Applied to Materials Science

NUCLEAR MAGNETIC RESONANCE ON RARE-EARTH NUCLEI IN
RE-Fe$_2$ INTERMETALLIC COMPOUNDS

Y. BERTHIER
Laboratoire de Spectrométrie Physique, BP 53X,
Grenoble Cédex, France

R.A.B. DEVINE
Physics Department, P.O. Box 248046, University of Miami,
Coral Gables, Florida   33124

R.A. BUTERA
Department of Chemistry, University of Pittsburgh,
Pittsburgh, Pennsylvania   15260

ABSTRACT

NMR results in RE-Fe$_2$ intermetallic compounds at low
temperatures are reported. Using a spin-echo spectrometer
in the U.H.F. range, hyperfine field parameters have been
measured on ^{167}Er, ^{163}Dy, ^{169}Tm and ^{159}Tb at 1.4K. These
compounds have high Curie temperatures and the magnetic
moment of the rare-earth, $g_J \mu_B \langle J_z \rangle$ is nearly saturated.
The value of the iron moment is also known from Mossbauer
measurements and accurate data for the hyperfine field at
the rare-earth site obtained from NMR enables us to extract
the internal field due to conduction electrons.

INTRODUCTION

Nuclear magnetic resonance (NMR) is a technique which enables us
to proble macroscopically into the local environments in which nuclei
find themselves. We have applied the technique to the study of rare-
earth/metallic intermetallic compounds in the ordered phase and it is
primarily these measurements we shall discuss and with particular
reference to the RE-Fe$_2$ Laves phase compounds. Let us first consider
the resonance phenomena in these compounds and look at the magnetic
field seen by a rare-earth nucleus. In an ordered magnetic compound
such as the RE-Fe$_2$ a RE nucleus sees a combination of magnetic fields.
Let us call the total field seen by a rare-earth nucleus, H_T. A RE
ion has 4f electrons in an unfilled shell and as seen by the nucleus,
these electrons will create a magnetic field. If the 4f shell is
half filled (Gd^{3+}) a "spin only" type field is observed [1] typically
of the order of -300 kOe. The negative sign arises because this
field is created by a core polarisation term [1]. For RE ions having
less than or more than half filled 4f shell there will be an orbital
field produced. This field is in general large and may be several
megagauss. This is by far the largest field seen by the RE nucleus
and we call it H_{4f}.

There are two other fields we must consider. A magnetic ion immersed in a "sea" of conduction electrons interacts with the electrons via the exchange interaction [2]. In the case of the RE-Fe$_2$'s both the Fe and RE moments will polarise the conduction band electrons. Let us concentrate on a given RE nucleus and study the influence of the polarisation of the conduction bands. Firstly, all of the other RE ions and Fe ions will polarise the bands around them and via some form of spacially extended interaction (e.g. the RKKY [3] oscillations) this polarisation will be felt by the conduction electrons in the vicinity of the nucleus under study. Via the hyperfine interaction [4] a "transferred" field, H_{NN}, will be seen. The ion whose nucleus is under study will also polarise the conduction electrons in its near vicinity and create a self-polarisation field at the nucleus again via the hyperfine interaction. This field is termed H_{sp}. The total field is then written

$$H_T = H_{4f} + H_{sp} + H_{NN} \tag{1}$$

It is in this total field that the rare-earth nucleus will resonate if we supply electromagnetic energy of the correct polarisation and frequency. In these experiments, therefore, we do not supply a magnetic field, we use the internal field, H_T, seen by the nucleus.

In the experiments reported here we have determined the total hyperfine field seen by the RE nuclei in ^{167}ErFe$_2$, ^{163}DyFe$_2$, ^{169}TmFe$_2$, ^{159}TbFe$_2$ and we include existing results on ^{157}GdFe$_2$ and ^{165}HoFe$_2$. Measurements were made using the conventional spin echo technique at a temperature of 1.4K. The pulsed spectrometer operated in the frequency range 400 MHz to 4 GHz. Unlike conventional NMR, the high internal fields render it almost useless to vary the external field so that essentially we vary, point by point, the frequency. The necessity to do this can be easily demonstrated, suppose a resonance line has a half peak height width of 1 MHz in a central frequency of 900 MHz and this corresponds to a field H_T of 7 mega oersteds. The line width in field is then 1/900 of 7 mega oersteds, i.e. 7800 oersteds. For conventional laboratory magnetic field sweeps this represents a huge line width. "Regular" NMR line widths are usually less than several tens of oersteds. The measured hyperfine fields in our experiments are given in table 1.

Bleaney [1] has shown that:

$$H_{4f} = A<J_z> \tag{2}$$

where A is essentially constant for a given ion [5] and $<J_z>$ is the total 4f angular momentum. It has been suggested that for the RE-Fe$_2$'s the RE moments are saturated. Experiments we have performed to measure the quadrupolar splittting [7] and neutron diffraction measurements on HoFe$_2$ [8] suggest that this is not the case. In calculating the 4f hyperfine field using equation (2) we shall use the 4f moments available from the quadrupolar results and from neutrons. We will not make the assumption $<J_z> = J$. For Tb and Tm in RE-Fe$_2$ we are forced to assume $<J_z> = J$ since there are no quadrupolar or neutron results for TmFe$_2$ and for TbFe$_2$ it seems that other contributions to the quadrupolar splitting are present.

TABLE I

Comparison of the measured hyperfine fields and pure
4f fields for RE-Fe$_2$'s (all fields in kilo oersteds).

The measurement for Ho in HoFe$_2$ is from reference 13.

For H$_{4f}$ we have used equation 2 with $\langle J_z \rangle$ determined
from quadrupolar results or neutrons[8]. For Tb and Tm
we assume $\langle J_z \rangle = J$.

	Gd	Tb	Dy	Ho	Er	Tm
H$_T$	450	3584	6480	7761	8184	7014
H$_{4f}$	−332±6	3180±29	5639±106	6671±8	7500±54	6647±30

We must now deduce H$_{sp}$ and H$_{NN}$ via a process of deduction which
must explain consistently several observations. In figure 1 we show
a schematic representation of our estimation of the situation in YFe$_2$,
GdFe$_2$ and RE-Fe$_2$ (RE ≠ Gd). The first observation is that the measured
hyperfine field at the ^{89}Y site in YFe$_2$ is -222 kOe [9]. Now application of an external magnetic field to YFe$_2$ will cause the moments to
align parallel to the field (figure 1). A negative field at the Y
site means that the conduction electron polarisation at the Y site
produced by the Fe ions (a transferred polarisation) creates a field
at the nucleus opposite to the external field. This sign of field
arises either due to a positive polarisation active through a core
polarisation term [1] or a negative polarisation active through a
contact hyperfine term [4]. Now ordering of the RE and Fe moments
in RE-Fe$_2$ is opposite so that it is likely that the conduction electron polarisation near a RE site produced by the Fe ions is negative
rather than positive (the conduction electron-RE ion exchange is positive). Assuming YFe$_2$ to be symbolic of RE-Fe$_2$, a negative polarisation
coupled with the contract term seems the likely source of the
negative hyperfine field in YFe$_2$.

Ignoring momentarily the magnetism of the RE 4f electrons,
we can deduce the transferred hyperfine field at any RE site in
RE-Fe$_2$ by scaling the Y in YFe$_2$ result. One must take care of
signs, however, since in the RE-Fe$_2$ series (figure 1) an applied
magnetic field aligns primarily the RE ions and these, being
coupled antiferromagnetically to the Fe ions, cause the Fe moments
to align antiparallel to the applied field. From Figure 1 we see
that for the RE-Fe$_2$ series we can estimate the transferred hyper-

fine field from the Fe ions to be H_{NN} (Fe):

$$H_{NN}(Fe) = 222 \frac{\tilde{A}(Re)}{\tilde{A}(Y)} \qquad (4)$$

where $\tilde{A}$ are the contact hyperfine terms tabulated by Campbell [4].
One must also make allowance in equation 4 for the fact that the Fe
moment varies slightly across the RE series [10].

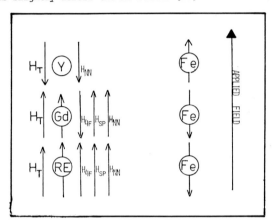

Figure 1.   Schematic representation of the ordering of magnetic moments
and fields in RE - Fe_2 systems. In $GdFe_2$ the 4f field is
opposite to the Gd moment and applied field due to core
polarisation.

The transferred field from the RE ions in any RE-Fe_2 compound
can be deduced from results for $GdFe_2$. Substitution experiments
[9,11,12] indicate that the transferred hyperfine field due to Fe
ions and Gd ions are parallel and additive. It is possible to deduce the transferred field seen by any RE ion due to its RE neighbours from the Gd result [11] using simple scaling proceedures like
equation 4. We have done this for all of the RE ions we are considering and the total transferred field (H_{NN}(Fe) + H_{NN}(RE)) values
are given in table 2. We have been able to verify the value of H_{NN}
(Er) by substitution experiments and find good agreement with the
scaled result used to deduce H_{NN} in table 2. We have not yet performed experiments on Fe substituted alloys. The general and important conclusion about the transferred hyperfine fields is that they
are positive for all of the RE ions. In table 2 we present the values
deduced for H_{sp} using equation 1 and the results of table 1 together
with the H_{NN} values. We include the table 2 the values one would have
expected for H_{sp} if the conduction band polarisation by the RE ion
locally were a "spin only" type exchange term [2], fitted to the only

pure spin ion, Gd. It is clear that such a model does not apply for
the RE-Fe$_2$ series. This leads to the interesting physical question
which was the objective of our study which is, is the spin-only model
sufficient for ferromagnetic, metallic systems.

In conclusion, we find that the self-polarisation fields in the
heavy RE-Fe$_2$'s do not follow a simple spin-only exchange model, it is
likely that orbital exchange contributions should be considered.
Furthermore, the moment for Tb in TbFe$_2$ and Tm in TmFe$_2$ are certainly
reduced below the free ion value.

TABLE II

Values of the transferred hyperfine field, H_{NN}, and
self polarisation field, H_{sp}. H_{spin} show the values
expected on a pure spin only exchange model (fitted
to Gd). The negative H_{sp} values for Tb and Tm
indicate their $<J_z>$ are less than J for these compounds.

	Gd	Tb	Dy	Ho	Er	Tm
H_{NN}	609	602	592	587	554	538
H_{sp}	173±6	−198±29	249±106	503±8	130±54	−171±30
H_{spin}	173	154	133	91	84	57

REFERENCES

[1] B. Bleaney, in "Magnetic Properties of Rare-Earth
Metals", edited by R.J. Elliot (Plenum, London, 1972).

[2] N.L. Huang-Liu, K.J. Ling, and R. Orbach, Phys. Rev.
B14 4087 (1976).

[3] See for example: K.H.J. Buschow, J.F. Fast, A.M.
Van Diepen and H.W. de Wijn Phys. Stat. Solidi 24
715 (1967).

[4] I.A. Campbell, J. Phys. C2 1338 (1969).

454

[5] Y. Berthier, R.A.B. Devine and E. Belorizky Phys. Rev.
     B17 4137 (1978).

[6] N. Koon and J.J. Rhyne in "Crystalline Electric Field
     and Structural Effects in f Electron Systems" editors
     J.E. Crow, R.P. Guertin and E.A. mihalisin (Plenum,
     New York, 1980).

[7] R.A.B. Devine and Y. Berthier, submitted for publication.

[8] H. Fuess, D. Givord, A.R. Gregory and J. Schweizer, J.
     Appl. Phys. 50 2000 (1979).

[9] A. Oppelt and K.H.J. Buschow, Phys. Rev. B13 4698 (1976).

[10] F. Hartmann-Boutron, private communication.

[11] V.A. Vasil'kovskii, N.M. Kotvum, A.K. Kuprianov, S.A.
     Nitkin and V.F. Ostrovskii, Zh. Eksp. Teor. Fiz. 65
     693 (1973).

[12] R.E. Gegenwarth, J.I. Budnick, S. Skalski and J.H.
     Wernick, Phys. Rev. Letts. 18 9 (1967).

[13] M.A.H. MacCausland and I.A. McKenzie, Advances in
     Physics 28 305 (1979).

MÖSSBAUER STUDY OF NEW LAVES PHASES RFe$_2$ COMPOUNDS (R = Pr, Nd, Yb)

C. MEYER, F. HARTMANN-BOUTRON, Y. GROS
Laboratoire de Spectrometrie Physique, B.P. 53X
38041 Grenoble-Cedex, France

ABSTRACT

A summary of the magnetic properties of three Laves Phases compounds RFe$_2$(R = Pr, Nd, Yb) synthesized under high pressure is presented. Most of the results have been obtained by Mössbauer spectroscopy. Through the ^{57}Fe resonance, the easy directions of magnetization between 4 and 300 K have been determined; the hyperfine field at Fe contains a large anisotropic contribution, only partly due to dipolar effects. From the Yb^{170} Mössbauer resonance measurements in YbFe$_2$, the exchange field of the iron and the crystalline field, acting on Yb^{3+}, have been derived.

INTRODUCTION

Among the RFe$_2$ series (R = rare earth) three compounds had never been investigated. They are: PrFe$_2$, NdFe$_2$ and YbFe$_2$. Because of size restrictions with regard to the R atom, which control the formation of the C15 structure, these three phases can only be synthesized under very high pressure [1]. For a few years we have undertaken the study of these materials after having produced them under a pressure of 80 kbars [2-7]. Recently, Shimotomai et al. have also prepared PrFe$_2$ [9] and NdFe$_2$ [10], and their results are quite similar to ours.

The aim of the present report is to gather the experimental data we have obtained on the three compounds and which have been described in detail in different papers [2-7]. We will briefly recall some of the characteristics measured by macroscopic experiments in the next section, and discuss the Mössbauer results in the following sections. In each of these sections we will also survey the results on the whole RFe$_2$ series [8].

PREPARATION, CHARACTERIZATION AND MAGNETIZATION MEASUREMENTS

Preparation and Structure

The compounds PrFe$_2$, NdFe$_2$ and YbFe$_2$ have been synthesized under 80 kbars and 1200 $^{\circ}$C. The purity of the samples was checked by X-ray diffraction. In all the preparations we observe predominantly the C15 phase. In YbFe$_2$ a small amount of non magnetic YbO could be seen. In PrFe$_2$, some iron metal remained unreacted. In addition, some other very weak diffraction lines in PrFe$_2$ and NdFe$_2$ appear, that we could not identify, but did not show up in the Mössbauer spectra. We are of the opinion that the Mössbauer spectroscopy is especially well adapted to this kind of study, because it is a local measurement and has the ability to distinguish several phases.

The refined cubic lattice parameters are in agreement with those of Cannon et al. [1]. For instance, a = 7.458 ± 0.001 Å for NdFe$_2$ and a = 7.244 ± 0.001 Å for YbFe$_2$.

Magnetic Structure and Ordering Temperatures

In YbFe$_2$ the variation of the magnetization as a function of the temperature gives an evidence of a compensation point at 31 K. This is in agreement with the current assumption of a ferrimagnetic arrangement between the iron and the rare earth moments, for the second half of the rare earth series. Following the same theory the rare earths of the first half of the series should have their moments ferromagnetically coupled to the iron atoms. This has been established for NdFe$_2$ from the measurement of the Mössbauer effect in ^{57}Fe under external field at 4.2 K.

The Curie temperatures have been measured in the three compounds from the thermal variation of the magnetization under 5000 Oe. They are presented in Table I

The Curie temperature variation among the whole series of the RFe$_2$, including the new compounds, can be described using a simple molecular field model. The T_c exhibit a regular variation as a function of the de Gennes factor $(g_J-1)^2 J(J+1)$ characteristic of the rare earth as shown in Fig. 1. Nevertheless, for a more quantitative study, a band model theory for discussing the 3d electrons should be used [11].

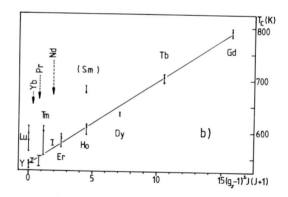

Fig. 1 Curie temperatures T_c in the RFe$_2$ versus $(g_J-1)^2 J(J+1)$

Fe^{57} MÖSSBAUER SPECTROSCOPY

Determination of the Easy Direction of Magnetization

One of the important properties of the RFe$_2$ compounds is their large magneto-crystalline anisotropy, which is basically due to the crystalline field effects on the rare earth. That gives rise to preferential magnetization directions in the cubic lattice, which are generally one of the high symmetry axes: [111], [100] or [110]. The classical macroscopic anisotropy measurements on single crystals are not possible in these new RFe$_2$ compounds, because of the difficulty in growing single crystals under high pressure. On the contrary, by Fe^{57} Mössbauer spectroscopy one can determine the easy directions even from measurements on polycrystalline materials. Indeed, the hyperfine interactions measured at the trigonal Fe site directly depend on the orientation of the Fe moment with respect to the local symmetry axis (one of the <111> directions of the cube).

TABLE I   MAGNETIC DATA IN PrFe$_2$, NdFe$_2$ AND YbFe$_2$

	Magnetic Order	T_c (K)	T_{comp} (K)	$\vec{M}$ Direction (0 K)	M Direction (300 K)	Reorientation Temperature T(K)
PrFe$_2$		543	-	[100]	[111] ?	?
NdFe$_2$	ferro	578	-	[110]	(100)	150
YbFe$_2$	ferri	543	31	[100]	(100)	50

In YbFe$_2$, NdFe$_2$ and PrFe$_2$ at 4.2 K we have found the magnetization to lie along the high symmetry axes, as it is generally the case for the RFe$_2$ at low temperature, as shown in Table I.   In YbFe$_2$ and NdFe$_2$ we have observed a rotation of the magnetization at higher temperature inside the (100) plane. At room temperature its direction lies approximately at 10° from the <100> axis.   In PrFe$_2$ this reorientation phenomenon is less clear, because of the contribution from the impurities to the spectrum. NB: when $\vec{M}$ is in an intermediate direction between the high symmetry axes, the analysis of the spectrum becomes difficult. An alternative method involving a large external field can be used. In this case the hyperfine interactions are statistically averaged and from the shape of the spectrum it is possible to derive the data necessary for the knowledge of the precise position of the magnetization at zero external field.

Study of the Hyperfine Field at Fe[57]
    1. General. The hyperfine field $\vec{H}$ measured at iron nucleus in the RFe$_2$ can be decomposed into two contributions $\vec{Hn} + \vec{Hd}$ · $\vec{Hn}$ is isotropic. It contains the usual core polarization term due to the d band splitting and proportional to the iron spin, and also a contribution from the conduction electrons polarized by the rare earth, assumed to be proportional to its spin and isotropic.

Then: $Hn = A<S_{Fe}> + B|(g_J-1) <J_{Re}>|$ where A and B are assumed to be constant in the RFe$_2$ series. Then Hn is expected to follow a linear variation as function of $|(g_J-1)J|$. As a matter of fact up to now this law seemed to be satisfied by the hyperfine fields evolution measured in the RFe$_2$ compounds [12-14].
•Hd is anisotropic. It has been usually attributed to dipolar effects arising from the neighbor moments. It could reach 10% of Hn. Generally, the measured value of Hd is much more important than the calculated one ($\approx$ 40% larger). No explanation of this discrepancy had been found until now.

    2. Experimental Results and Discussion. In Fig. 2 we have plotted the isotropic hyperfine field at iron versus $|(g_J-1)J|$, including our experiments.
In order to have more accuracy we have first subtracted from each value of Hn, the Lorentz contribution H_L^{RE} proportional to the rare earth moment, which is not negligible compared to Hn ($\approx$ 5%). Across the series this term is not con-

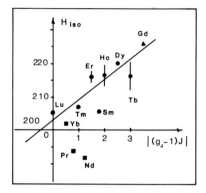

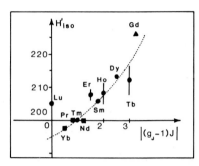

Fig. 2 Isotropic hyperfine H_{iso} at F_e versus $|(g_J-1)J|$ [Eq. (1)]

Fig. 3 Isotropic hyperfine H'_{iso} at F_e versus $|(g_J-1)J|$ [Eq. (2)]

stant and changes its sign in the middle. What is plotted is then:

$$H_{iso} = Hn - H_L^{RE}. \tag{1}$$

As one can see on the figure a rather linear variation is obtained for the RFe_2 of the second half. On the contrary the points corresponding to Pr and Nd are quite out of the mean curve.

This discrepancy can be avoided in assuming the presence of another term in the hyperfine field, a transferred contribution from the rare earth, now proportional to its orbital moment, thus to $(2-g_J)J$. Consequently, we have plotted in Fig. 3 the quantity

$$H'_{iso} = H_{iso} - C|(2-g_J)J| \text{ versus } |(g_J-1)J| \tag{2}$$

and adjusting the constant C value in the best way, we have obtained a rather good alignment of all the points along the same curve, which is not quite linear

*The measured value of the anisotropic part of the hyperfine field in the new compounds is also much larger than the calculated dipolar term. In $NdFe_2$ at 4.2 K the discrepancy is particularly dramatic. The dipolar field should be zero and one observes an enormous anisotropy of the field ($\approx 20\%$). We can thus conclude that this anisotropy does not arise only from the dipolar effects, but that another contribution interferes. In order to explain this phenomenon, we have assumed the intrinsic hyperfine structure of the iron atom to be aniso-tropic reflecting the trigonal symmetry of the iron site. As a matter of fact we can demonstrate that the corresponding additional hyperfine field would have the same form that the dipolar one.

The hyperfine anisotropy of the iron in RFe_2 seems to exist also in several other intermetallics: anomalies of the apparent dipolar field have been observed in Er_6Fe_{23} [15], $ErFe_3$ [16] and Tm_2Fe_{17} [17]. This can be related to the impor-tant hyperfine anisotropies of the cobalt detected by NMR in YCo_5 and $SmCo_5$. Therefore the hyperfine anisotropy of the 3d element in the rare earth intermetallics compounds seems to be a general feature.

Yb[170] MÖSSBAUER SPECTROSCOPY

Hyperfine Interactions on the Rare Earth

The thermal variation of the hyperfine field H_{eff} and the quadrupole interaction E_Q measured at the rare earth nucleus, reflect the electronic structure of the R^{3+} ion.

The 4f electron of the rare earth characterized by the J value is submitted to the exchange field of the iron (H_{ex}) and to the cubic crystalline field characterized by the classical parameter $A_4\langle r^4\rangle$ and $A_6\langle r^6\rangle$.

The hyperfine interactions can be calculated in the corresponding electronic levels scheme through the medium of the mean thermal values of $\langle J_z\rangle$ and $\langle 3J_z^2 - J(J+1)\rangle$ to which H_{eff} and E_Q are respectively proportional. Then by a least square fit of the experimental curves $H_{eff}(T)$ and $E_Q(T)$ in this theory, it is possible to derive the values of H_{ex}, $A_4\langle r^4\rangle$ and $A_6\langle r^6\rangle$.

Generally, in the RFe_2 one finds:

$$\frac{A_4\, a_0^4}{k_B} \simeq 40\ K, \quad \frac{A_6\, a_0^6}{k_B} \simeq - \text{ a few K} \quad \text{and} \quad \frac{\mu_B\, H_{ex}}{k_B} \simeq 100 \text{ to } 150\ K.$$

up to now these values have been assumed to be constant in the series.

Experimental Results Obtained in YbFe$_2$

Yb[170] Mössbauer spectra between 0 and 60 K have given rise to hyperfine interactions which we could fit as function of the temperature with the following parameters:

$$\left\{ \begin{array}{l} \dfrac{\mu_B\, H_{ex}}{k_B} = 111 \pm 4\ K \\[3mm] 0 < \dfrac{A_4\, a_0^4}{k_B} < 4\ K \\[3mm] 0 < \dfrac{A_6\, a_0^6}{k_B} < 1\ K \end{array} \right.$$

The magnitude of the exchange field agrees with the previous expectations, but the crystalline field is much smaller than what was assumed for the RFe_2.

Moreover this discrepancy also exists for $NdFe_2$ and $SmFe_2$ where the magnetization is parallel to $\langle 110\rangle$: indeed such an anisotropy direction requires that $|A_4\langle r^4\rangle|$ and $|A_6\langle r^6\rangle|$ should be of the same order of magnitude.

Then it seems obvious to conclude that in the RFe_2 the crystalline field parameters A_4 and A_6 are not constant across the series.

460

ACKNOWLEDGEMENTS

One of the authors (C.M.) wishes to thank B. Dunlap and G. Shenoy for helpful discussions and for their hospitality at Argonne National Laboratory.

REFERENCES

1. J. F. Cannon, D. L. Robertson and H. T. Hall, Mat. Res. Bull. $\underline{7}$, 5 (1972).

2 C. Meyer, B. Srour, Y. Gros, F. Hartmann-Boutron and J. J. Capponi, J. Physique $\underline{38}$, 1449 (1977).

3. C. Meyer, F. Hartmann-Boutron, Y. Gros, B. Srour and J. J. Capponi, J. Physique Coll. $\underline{40}$, C5-191 (1979) (Physics of Metallic Rare-Earths, Saint-Pierre-de-Chartreuse, 1978).

4. C. Meyer, Y. Gros, F. Hartmann-Boutron and J. J. Capponi, J. Physique $\underline{40}$, 403 (1979) and addendum in J. Physique $\underline{41}$, 1075 (1980).

5. C. Meyer, F. Hartmann-Boutron, Y. Gros, J. J. Capponi, J. Physique Coll. $\underline{41}$, C1-191 (1980) (Mössbauer Conference, Portoroz, 1979).

6. C. Meyer, F. Hartmann-Boutron, J. J. Capponi, J. Chappert and O. Massenet, J. Magnetism and Magnetic Mat. $\underline{15-18}$, 1229 (1980).

7. C. Meyer, Ph.D. Thesis, Grenoble, 1980.

8. C. Meyer, F. Hartmann-Boutron, Y. Gros, Y. Berthier and J. L. Buevoz, submitted to J. Physique.

9. M. Shimotomai, H. Miyake and M. Doyama, J. Phys. F Metal Phys. $\underline{10}$, 707 (1980).

10. M. Shimotomai, H. Miyake and M. Doyama, submitted for publication.

11. M. Cyrot and M. Lavagna, J. Physique $\underline{40}$, 763 (1979).

12. K. H. J. Buschow, Rep. Prog. Phys. $\underline{40}$, 1179 (1977).

13. A. P. Guimaraes and D. St. P. Bunbury, J. Phys. F $\underline{3}$, 885 (1973).

14. M. P. Dariel, U. Atzmony and D. Lebenbaum, Phys. Stat. Solidi (b) $\underline{59}$, 615 (1973).

15. P. C. M. Gubbens, Ph.D. Thesis, Delft University Press (1977).

16. A. M. van der Kraan, P. C. M. Gubbens, K. H. J. Buschow, Phys. Stat. Solidi (a) $\underline{31}$, 495 (1975).

17. P. C. M. Gubbens, K. H. J. Buschow, Physica Stat. Solidi (a) $\underline{34}$, 729 (1976)

Published 1981 by Elsevier North Holland, Inc.
Kaufmann and Shenoy, editors
Nuclear and Electron Resonance Spectroscopies Applied to Materials Science

A MÖSSBAUER STUDY OF THE AMORPHOUS SYSTEM $(Fe_xNi_{1-x})_{75}P_{16}B_6Al_3$

S. BJARMAN and R. WÄPPLING
Institute of Physics, Uppsala University, Box 530, S-751 21 Uppsala, Sweden
K.V. RAO
Department of physics, University of Illinois at Urbana-C,
Urbana. Illinois 61801, USA

ABSTRACT

In the system $(Fe_xNi_{1-x})_{75}P_{16}B_6Al_3$ an unusual magnetic behaviour is found in the composition range 0.2<x<0.3. The drastic
changes seen at low temperature in the thermomagnetic measurements is found not to affect the Mössbauer spectra. From the
width of the magnetic hyperfine field distribution as function
of temperature there seems to be a distribution in Curie temperatures. Due to the absence of spin texture it was not possible
to make a distinction between the two proposed low temperature
phases although the general results favour the spin glass alternative.

INTRODUCTION

For more than 10 years the interest in amorphous magnetism has been
continuously growing. The possibilities of producing large amounts of the metallic glasses first shown by Allied Chemical with the wellknown "Metglas" has for
the last years even made these rather exotic materials technically interesting.
On the scientific side a number of investigations using a variety of different
experimental methods have been undertaken in order to gain more knowledge on the
particular properties of these materials. Many of these studies have been
directed towards the magnetic properties of the metallic glasses [1].
In the system $(Fe_xNi_{1-x})_{75}P_{16}B_6Al_3$ the magnetic properties are essentially
determined by the iron content. Thus at low x-values, no magnetic moment is
found, while for eg. x=0.8 a Curie temperature of 600K has been determined in a
careful investigation of the magnetization of this amorphous metallic system
[2]. The magnetic structures proposed by the authors of ref. 2 involved several
different magnetic phases including, depending on composition and temperature,
spin glass, mictomagnet, ferromagnet and antiferromagnet.
The measurements quoted above were initiated by an earlier study by one of
the present authors (K.V. Rao [3]). In that work a somewhat simpler magnetic
phase diagram was proposed (Fig. 1). As can be seen there are some unusual
features at low temperatures in the composition range 0.2<x<0.4. This is further
demonstrated by the ac-susceptibility for x=0.2 as shown in Fig. 2. The drop in
the susceptibility was interpreted as due to a phase transition to a spin glass
structure as the temperature is lowered beneath a critical temperature. In the
later work [2] a similar change is seen in the static susceptibility at low
fields and is interpreted as the result of a transition to a state with competing ferro and antiferromagnetic interactions.
In the present work we have used the Mössbauer effect in order to investigate
the magnetic phases of the above mentioned system in the composition range
0.20<x<0.40. For a general introduction to the experimental method se e.g.
ref [4].

462

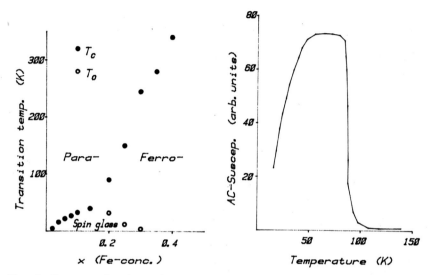

Fig. 1. The magnetic phase diagram of ref. 3.

Fig. 2. The low field susceptibility for x=0.2 as function of temperature.

EXPERIMENTAL DETAILS

The recordings of the Mössbauer spectra were all made in the conventional transmission geometry. The source was 50 mCi CoRh and the absorbers were made from ~20 microns thick ribbons formed during the quenching process [5] by placing them between two mylar foils. The absorbers were then placed in a variable temperature gas flow cryostat where the temperature stability is approx. 0.2K. The compositions investigated were x=0.20,0.25,0.35. Simultaneous calibration was made for each run by recording the spectrum of an iron metal foil using a second CoRh source attached to the other end of the vibrator of a conventional constant acceleration spectrometer.

RESULTS AND DISCUSSION

Low temperature Mössbauer spectra for the sample $(Fe_{20}Ni_{80})_{75}P_{16}B_6Al_3$ are shown in Fig. 3. The spectrum at 90K is representative for the paramagnetic state showing a well resolved quadrupole splitting. Below 90K the sample is magnetically ordered and shows a distribution of magnetic hyperfine fields characteristic for amorphous magnets. The field distribution is obtained from a least squares fit allowing for 30 discrete fields restricted to obey a Lorentian distribution at low field values and a Gaussian at high fields [6]. In Fig. 4 is shown the temperature variation of the maximum of the field distribution. From spin wave theory a $T^{3/2}$ dependence is expected and a "best fit" curve is also presented in Fig. 4 together with the corresponding curve for a linear temperature dependence. It seems that, at last at x=0.2, a marked deviation from spin wave theory occurs. The same conclusion has been drawn from a recent FMR study [7]. The temperature dependence found can be interpreted in terms of competing

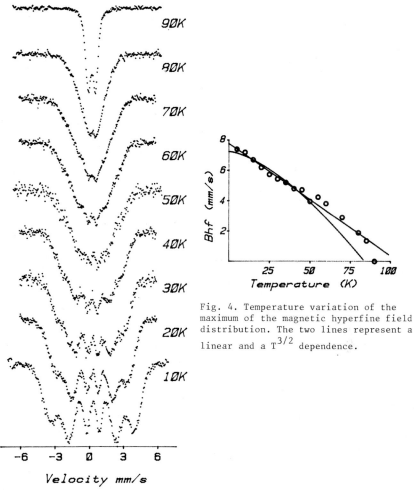

90K
80K
70K
60K
50K
40K
30K
20K
10K

-6 -3 Ø 3 6

Velocity mm/s

Fig. 3. Low temperature Mössbauer
spectra at the composition x=0.2.

Fig. 4. Temperature variation of the
maximum of the magnetic hyperfine field
distribution. The two lines represent a
linear and a $T^{3/2}$ dependence.

long range and short range interactions by comparison with the situation in the
Au-Fe system [8].

From Fig. 4, it is evident that the drastic changes seen at about 40K in the
thermomagnetic measurements have no significant effect on the Mössbauer spectra.
This is demonstrated more clearly in Fig. 5 showing spectra for x=0.25 both
above and below the low temperature transition. Since the low temperature
phase clearly is not paramagnetic (cf. the spectral difference between 90K and

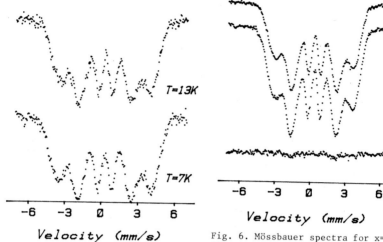

Fig. 5. The Mössbauer spectra for x=0.25 above and below the low temperature transition.

Fig. 6. Mössbauer spectra for x=0.35 at 80K recorded with the angle between the absorber plane and the gammaray direction equal to a) 90 deg. and b) 60 deg. The difference between the normalised spectra is shown in c).

10K in Fig. 3) the vanishing magnetisation is due either to a ferromagnetic/antiferromagnetic transition or a ferromagnetic/spin glass transition. A distinction between the two alternatives can often be made by a detailed inspection of the Mössbauer spectra above and below the transition as follows. Since a high cooling rate is essential in order to prevent crystallisation, at least one dimension of the material has to be small. This is the dimension in which the cooling takes place by heat conduction through the amorphous-to-be material. During the quenching stage a temperature gradient exists in this direction and as a result the as-quenched material is found to be non-isotropic [9]. In that case a texture can be detected in the Mössbauer spectra. As can be seen in Fig. 3 the electric quadrupole interaction is averaged out in the magnetically ordered state and a possible texture can only be detected through a careful inspection of line intensities. Such an inspection can be facilitated by recording spectra at different angles between the absorber and the gamma ray direction [10]. Fig. 6 displays spectra for x=0.35 recorded at 80K at two different angles together with their difference after normalization and it is clear that in the present case there is not any pronounced texture. In the ferromagnetic/antiferromagnetic transition the spin texture should in general be preserved since the anisotropy energy is not affected by the transition, whereas in a ferromagnetic/spin glass transition spin texture would disappear since the exchange energies involved would be larger than the anisotropy energy. If the competing long range and short range exchange interactions are of different signs and have different temperature dependences, both spin glass and antiferromagnetism are possible low temperature phases. In general, however, the spin glass state predominates as is found in palladium with small amounts of iron and manganese [11] and in the gold iron system [12]. The magnetic phase diagram is in the latter case very similar to the one proposed for the

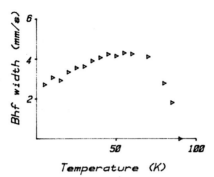

Fig. 7. The width of the hyperfine field distribution as function of temperature.

$(Fe_xNi_{1-x})_{75}P_{16}B_6Al_3$ system if the low temperature phase is interpreted as spin glass. The transition between ferromagnetism and spin glass can be explained within a random Ising model as shown by Sherrington and Kirkpatrick [13].

The paramagnetic to ferromagnetic transition is not very distinct (cf. Fig. 3). This is further demonstrated by the width of the magnetic hyperfine field distribution shown in Fig. 7. If there was only one ordering temperature, the width would show a temperature dependence similar to that of the field (Fig. 4). The increase in width with increasing temperature can be interpreted in terms of a distribution in Curie temperatures. This also gives rize to a sharper drop close to the ordering temperature as compared to a single temperature transition. In the case of a distribution in Curie temperatures, it is not possible to obtain detailed information about the system close to the tricritical point i.e. the point in the temperature-concentration plane where paramagnetism, ferromagnetism and spin glas/antiferromagnetism meet. A distribution of ordering temperatures was found also in the annealing study of a Au-Fe alloy [8].

ACKNOWLEDGMENTS

The authors would like to thank Dr H.S. Chen for supplying the samples and Dr E. Figueroa for communicating his results prior to publishing.

REFERENCES

1. See eg. J. Magn. Magn. Mat. 15-18 (1980) 1325-1425.

2. E. Figueroa, K. Gramm, O. Beckman and K.V. Rao, Uppsala University Institute of Technology Report UPTEC 8034 R, April 1980 (unpublished).

3. K.V. Rao, R. Malmhäll, S.M. Bhagat, G. Bäckström and H.S. Chen, MMM-Intermag Conf. (1980) Proc. IEEE to be published.

4. R.L. Cohen, Applications of Mössbauer Spectroscopy. Vol 1, Academic Press, New York (1976).

5. H.S. Chen, Acta Met. 24 (1976) 153.

6. C.C. Tsuei, in Amorphous Magnetism, H.O. Hooper and A.M. deGraaf, eds., Plenum, New York-London (1973), p.299.

7. S.M. Bhagat, M.L. Spano, K.V. Rao and H.S. Chen, Solid State Comm. 33 (1980) 303.

8. R.J. Borg, D.Y.F. Lai and C.E. Violet, Phys. Rev. B5 (1972) 1035.

9. P. Haasen, Phys. Blatter 344 (1978) 573.

10. T. Ericsson and R. Wäppling, J. Physique 37 (1976) C6-719.

11. B.H. Verbeek, G.J. Nieuwenhuys, H. Stocker and J.A. Mydosh, Phys. Rev. Lett. 40 (1978) 586.

12. B.R. Coles, B.V. Sarkissian and R.H. Taylor, Phil. Mag. B37 (1978) 489.

13. D. Sherrington and S. Kirkpatrick, Phys. Rev. Lett 35 (1975) 1792.

Published 1981 by Elsevier North Holland, Inc.
Kaufmann and Shenoy, editors
Nuclear and Electron Resonance Spectroscopies Applied to Materials Science

NMR AND MÖSSBAUER EFFECT STUDIES ON DILUTED HEISENBERG FERROMAGNETS AND
SPIN-GLASSES: $Eu_xSr_{1-x}S$

H. LÜTGEMEIER, CH. SAUER, AND W. ZINN
Institut für Festkörperforschung der KFA Jülich,
D-5170 Jülich, W.-Germany (F.R.G.)

ABSTRACT

The systematic variations of experimentally determined exchange and hyperfine (h.f.) interactions between Eu^{2+} ions are compared firstly within the EuX
(X=O,S,Se,Te) series of compounds and secondly in the
magnetic dilution system $Eu_xSr_{1-x}S$. Reasonably relations can be established between the individual
nearest and next nearest neighbour exchange interactions (J_1,J_2) and the transferred h.f. interactions
(ΔB_1, ΔB_2), respectively, by considering their variations with the Eu-Eu distances (R_1,R_2). Using these
results, the measured mean hyperfine field, $B_I(x)$,
and the ferromagnetic saturation h.f. field, $B^{\uparrow\uparrow}(x)$,
of the $Eu_xSr_{1-x}S$ system can be related reasonably
well to the ferro- and paramagnetic phase
boundaries, $T_c(x)$ and $\theta(x)$, respectively.

INTRODUCTION

Considerable effort has been made during the past decade to establish the
relations between the transferred magnetic hyperfine (h.f.) interactions at
the (^{151}Eu, ^{153}Eu)-nuclei and the related Eu-Eu-exchange interactions in the
Eu^{2+}-monochalcogenides EuX (X=O,S,Se,Te), which are well-known examples of
isotropic Heisenberg ferromagnets (see e.g. Refs. 1,2 and refs. therein). Due
to the mostly uncertain or unknown individual transferred h.f. fields, ΔB_r,
and related exchange energy constants, J_r, of the nearest (r=1) and next
nearest (r=2) Eu neighbours reliable relations and conclusions on their
systematic variation in the series EuX could not be established so far. In
contrast, the assumptions substituted for results often led us to even wrong
or misleading extrapolations and conclusions in the past (e.g. in Refs. 2,3,4).
Only in the last year important new experimental results on both the individual
Eu-Eu exchange interactions, between r-th Eu neighbours, J_r, and the related
individual transferred h.f. fields, ΔB_r, have been obtained on EuS [4,5], on
EuSe [6,7] and on EuO to be reported here. These results now are sufficient to
outline the above mentioned relations fairly certain for the first time. They
are shown also to provide a basis for explaining the measured h.f. fields in
the substituted system $Eu_xSr_{1-x}S$ [3,4], which has become the most extensively
studied model system for dilute Heisenberg ferromagnets revealing, in particular, spin glass behaviour for 0.13<x<0.65 (see e.g. Refs. 8,9,10 and refs.
therein).

EXCHANGE VERSUS HYPERFINE INTERACTIONS IN EuX

As discussed in detail previously [1,2] the total magnetic h.f. field, B_I,
at the Eu^{2+} nucleus is given by the sum of the intrinsic h.f. field, B^O, due

to the atom's own 4f-moment and of the transferred h.f. field, B_{THF}, due to the sum of all transferred individual contributions, ΔB_r of z_r Eu-neighbours in the r-th shells:

$$B_I = B^o + B_{THF} \tag{1}$$

B^o is related to the Eu^{++} magnetic moment ($\mu_s = 7\,\mu_B$), while the total transferred h.f. field

$$B_{THF} = (z_1 \Delta B_1 + z_2 \Delta B_2 + \ldots) \tag{2}$$

with $z_1 = 12$ and $z_2 = 6$ in undiluted EuX is expected to be related to the total exchange energy

$$k_B \theta = \frac{2S(S+1)}{3} (12J_1 + 6J_2 + \ldots) \tag{3}$$

In particular, relations between the individual ΔB_r and J_r are expected to exist.

There are two methods to determine the B_{THF} and the individual ΔB_r experimentally: i) by measuring B_I in different ordered states, e.g. $B^{\uparrow\uparrow}$ and $B^{\uparrow\downarrow}$ in the ferro- and antiferromagnetic state, and considering their differences, ΔB_I; ii) by substitution of parts of the z_r Eu-neighbours by nonmagnetic atoms like Sr. In ferromagnets, like EuO and EuS, only method ii) can be used. In Fig. 1 and Table I the experimental data on B_I and B_{THF} are summarized.

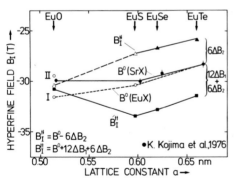

Fig.1. Eu h.f. fields in the ferromagnetic ($B_I^{\uparrow\uparrow}$) and antiferromagnetic ($B_I^{\uparrow\downarrow}$) state of EuX. The intrinsic h.f. field data B^o (SrX) for Eu^{2+} strongly diluted in SrS are taken from Ref. 11. The open circles I,II correspond to the two possible assignments for $\Delta B_{1,2}$ in EuO. The dashed B^o-line represents a reasonable extrapolation for B^o in EuX. The dashed $B_I^{\uparrow\downarrow}$-line has been deduced from the experimental B^o- and ΔB_2-data .

Included are also ESR results [11] on B^o from Eu^{2+}-ESR in SrS host-lattices. Contrary to the 4f-magnetic moment, these intrinsic h.f. fields B^o, are not constant for all EuX, but decrease systematically in magnitude with increasing lattice constant. For EuO and EuS in particular, B^o, may change between SrX and EuX as it is indicated in Fig. 1 by the systematic variations of B^o and B_I. These systematic variations of the h.f. could be established mainly by recent NMR results on the individual ΔB_1 and ΔB_2 in the Heisenberg ferromagnets EuO and EuS obtained by studies on Sr-substituted single crystal spheres. Typical spectra are shown in Fig. 2. They reveal satellite lines, which can be attributed to substitutions on either the nearest, ΔB_1, or next nearest Eu^{++}-neighbours, ΔB_2, by a Sr^{2+}-ion. These two possible ΔB_r-attributions lead to either of the two plots shown in Fig. 3b and Fig. 3c. On Fig. 3a for comparison we have plotted the results now available on the individual exchange constant J_r from the literature as completed by a recent neutron scattering study on a single crystal of ^{153}EuS [5].

TABLE I
Summary of h.f. field data (in Tesla)

	EuO		EuS	EuSe	EuTe	Ref.
B^O(Eu/SrX)		−29.9	−30.0	−29.2	−28.3	11
B^O(EuX)	−29.5	−31.5	−(30.0...31.0)	∼−29.2	∽−28.3	
ΔB_1	−0.19	+0.16	−(0.1...0.04)	−0.09	−0.06	
ΔB_2	+0.16	−0.19	−0.33	−0.41*	−0.38	*6
B_{THF}(↑↓)	−1.4	+0.8	−(3.2...2.3)	−3.6	−3.1	
B_I↑↓	−30.5	−30.4	−(28.0...29.0)	−26.7*	−26.0	*7
B_I↑↑		−30.77*	−33.44*	−32.8	−31.4	*2

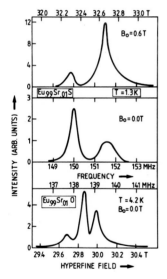

Fig.2. NMR spectra on single crystal
spheres of $Eu_{0.99}Sr_{.01}S$ in external fields
B_o=0.6 T and B_o=0T at T=1.3 K, and of
$Eu_{.99}Sr_{.01}O$ at B_o=0T and T=4.2 T. The
spectra reveal 2 satellites with EuO, but
only one with EuS (see text)

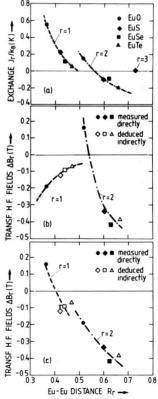

Fig. 3. (a) Exchange parameter J_r/k_B vs. the
Eu-Eu distance R_r (see Ref. 5 and refs. therein).
(b), (c) Transferred h.f. fields ΔB_r vs. R_r
with (b) and (c) representing the two possible
assignments of the NMR-satellites in $Eu_{.99}Sr_{.01}O$ to the Eu-n.n. and
n.n.n. contributions $\Delta B_{1,2}$.

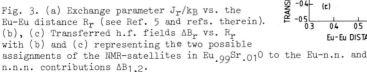

No simple relations have been found between the variations of B_I and $k_B\theta$ with lattice spacing as expected from Eqs. (2) and (3). However, the variations of the individual ΔB_r and J_r versus the related Eu-Eu distances R_r shown in Fig. 3 reveal reasonable relations, in particular, for the solution shown in Fig. 3c. There, the satellite in EuS is attributed to $\Delta B_2 = -0.33$ T (Note that this is different from the assumption in Ref. 4).The two satellites in EuO in this case have been attributed as follows: $\Delta B_1 = -0.19$ T and $\Delta B_2 = +0.16$ T which make them consistent with the other h.f. field contri-butions given in the same column of Table I. This solution seems most meaning-ful since e.g. the correlations in the signs between the J_r and B_r are always either -1 for r=1 as expected from Ref. 12, or +1 for r=2. In addition, also the magnitude of J_2 and ΔB_2 then are related as predicted theoretically for superexchange interactions in Ref. 13, i.e. $J_r \sim \Delta B_r^2$.

The completed h.f. field data on the EuX series of compounds summarized in Table I now provide a sufficient basis to consider and to discuss also the systematic variations of the h.f. interactions on dilution, e.g. $Eu_xSr_{1-x}S$ to be discussed next.

MAGNETIC HYPERFINE FIELDS AND INTERACTIONS IN DILUTED $Eu_xSr_{1-x}S$ FERROMAGNETS AND SPIN-GLASSES

By the Figs. 4 and 5 the x-dependence of the measured Eu-h.f. fields in zero applied field, $B_I(x)$, and the high-field, ferromagnetically saturated state, $B^{\uparrow\uparrow}(x)$, are compared to the phase transition energies in $Eu_xSr_{1-x}S$ as given by the measured $T_c(x)$, $T_f(x)$, and $\theta(x)$-curves.

The main findings and relations are as follows:
i) the ferromagnetic saturation h.f. field, $B^{\uparrow\uparrow}(x)$, is related to the para-magnetic Curie temperature, $\theta(x)$, as expected for a random Eu-, Sr-distribution from Eqs. (2) and (3) if one substitutes the coordination numbers z_r by $n_r = z_r x$ and takes into account the increase of the lattice spacings between EuS and SrS together with the thermal lattice expansion effect on $\theta(x)$ [10].
ii) The mean h.f. field in zero applied field, $\bar{B}_I(x)$, reproduced in Fig. 4 from Ref. 3 can be described fairly well by introducing weight functions $f(x)$ for the ferromagnetically and $1-f(x)$ for the antiferromagnetically correlated Eu-neighbour spins as follows:

$$B(x)=f(x)B^{\uparrow\uparrow}(x) + (1-f(x))B^{\uparrow\downarrow}(x) \qquad (4)$$

Here we use for $B^{\uparrow\uparrow}(x)$ our measured data discussed in i), and calculated $B^{\uparrow\downarrow}(x)$ for the antiferromagnetic ordered Eu-neighbours as e.g. in the anti-ferromagnetic NSNS-spin structure of EuTe discussed in detail in Refs. 1,2 as follows:

$$B^{\uparrow\downarrow}(x) = B^0(x)-n_2\Delta B_2 \qquad (5)$$

where ΔB_2 is the NMR result of Table I. From the maximum estimated uncertainty of the $B^0(x)$-dependence as indicated by the hatched region in Fig. 4 a corres-ponding region of uncertainty then follows for $B^{\uparrow\downarrow}(x)$ from Eq. (5) and for the weight function $f(x)$ from Eq. (4) by fitting $\bar{B}(x)$ to Eq. (4) to the measured $\bar{B}(x)$ data given by the full-drawn curve in Fig. 4.

Of course, we are well aware that Eq. (4) is only a rough description of the real and complicated spin structures in the diluted ferromagnetic and spin glass phase of $Eu_xSr_{1-x}S$ as discussed in detail in Refs. 8,9. However, we think the x-dependence of the deduced weight-function $f(x)$ corresponds fairly well to $T_c(x)$ and the magnetic phase diagram shown in Fig.5 even for this simple model. We hope, that the current more detailed and realistic Monte Carlo

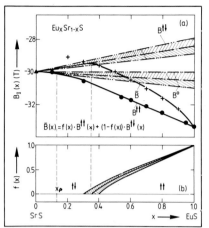

Fig. 4. (a) Eu h.f. fields in the high field ferromagnetic (●) phase, $B^{\uparrow\uparrow}$, and as measured in zero external field (+), $\bar{B}$, as function of the Eu-concentration x of the $Eu_xSr_{1-x}S$-system. The x-dependence of the intrinsic h.f. fields B^O, is still uncertain within limits indicated by the hatched region. From the B^O region and $\Delta B_2 = -0.33$ T an corresponding region for the h.f. field $B^{\uparrow\downarrow}$ of the anti-ferromagnetic spin configuration can be derived as shown.
(b) Weight factor f(x) of ferromagnetic neighbour spin correlations derived from the B^O-region and $\bar{B}(x)$ as given in the figure.

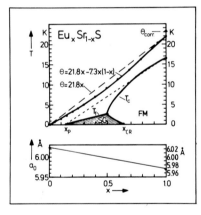

Fig. 5. (a) magnetic phase diagram of the $Eu_xSr_{1-x}S$ system after Ref. 8. (FM=ferrom. phase, SG=spin glass phase, x_p=percolation threshold, $T_{c/CR/f}$=Curie/critical/freezing temperature). The data for the paramagnetic Curie temperature, Θ, were taken from Ref. 10.
(b) Variation of the lattice constant, a_o, with x through the $Eu_xSr_{1-x}S$ system.

calculations [14] equivalent to those performed for $T_c(x)$ in Ref. 9 will provide a better agreement and correlation between the distribution of the actual Eu-neighbour spin correlations in both, the mean h.f. field, $\bar{B}_I(x)$, and the inhomogeneous magnetic h.f. line broadenings, $\Delta(2\Gamma)_B$, expected from the attributed distribution functions. For fairness we would like to mention that the binomial distribution of the n.n. and n.n.n. Eu-neighbours, $w(n_1,n_2)$, together with the full ΔB_r of Eq. (5) and Table I overestimate this line broadening considerably. In particular, our simple model can neither explain the rather small broadenings observed in the present NGR study of $B^{\uparrow\uparrow}(x)$ nor the lack of magnetic NGR-line broadening reported in the $\bar{B}_I(x)$ study previously [3].

472

CONCLUSIONS

We have shown that meaningful relations can be established between the individual transferred h.f. fields, ΔB_1, ΔB_2, and corresponding Heisenberg exchange constants, J_1, J_2, in the EuX series of model substances. Reasonable relations could be also outlined by this study between the average h.f. field in zero applied field, $\overline{B}_I(x)$, and the ferromagnetic phase boundary, $T_c(x)$. A better explanation is awaited from a more detailed theoretical treatment based on Monte Carlo calculations of the distribution of Eu-neighbour spin correlations [14] .

We acknowledge helpful discussions with K. Binder and W. Kinzel of our institute, and with A.J. Freeman of Northwestern University, Evanston (USA).

REFERENCES

1. W. Zinn, J. Magn. Magn. Materials 3, 23a (1976).

2. Ch. Sauer and W. Zinn, Physics 86-88B, 1031 (1977).

3. G. Crecelius, H. Maletta, H. Pink, and W. Zinn, J. Magn. Magn. Materials 5, 150 (1977).

4. H.G. Bohn, K.J. Fischer, U. Köbler, H. Lütgemeier, and Ch. Sauer, J. Magn. Magn. Materials, 15-18, 667 (1980).

5. H.G. Bohn, W. Zinn, B. Dorner, and A. Kollmar, J. Appl. Phys. (Proc. M&MM Conf. Nov. 1980, Dallas, to be published 1981).

6. K. Kojima, T, Hihara, T. Kamigaichi, Proc. Intern. Conf. on Ferrites, Sept. 1980, Kyoto (to be published).

7. T. Hihara, K. Kojima, T. Imai, H. Fujii, and T. Kamigaichi, J. Magn. Magn. Mater. 15-18, 665 (1980).

8. H. Maletta, J. Physique (Paris) 41, C5-115 (1980).

9. K. Binder, W. Kinzel, and D. Stauffer, Z. Physik B36, 161 (1979).

10. U. Köbler and K. Binder, J. Magn. Magn. Materials 15, 313 (1980).

11. K. Kojima, T. Komaru, T. Hihara, and Y. Koi, J. Phys. Soc. Japan 40, 1570 (1976).

12. R.E. Watson and A.J. Freeman in: Hyperfine Interactions, A.J. Freeman and R.B. Frankel eds. (Academic Press, New York, 1967) p. 53.

13. G.A. Sawatzky, W. Geertsma, and C. Haas, J. Magn. Magn. Materials 3, 37 (1976).

14. K. Binder and W. Kinzel, private communication.

Insulators and Compounds

EPR INVESTIGATIONS OF IMPURITIES IN THE LANTHANIDE ORTHOPHOSPHATES

M. M. ABRAHAM, L. A. BOATNER, and M. RAPPAZ
Solid State Division, Oak Ridge National Laboratory,[*] Oak Ridge, TN 37830

ABSTRACT

Lanthanide orthophosphates formed from elements in the
first half of the 4f transition series are analogs of the
monoclinic mineral monazite. The known geologic properties
of this mineral make the general class of lanthanide orthophosphate compounds attractive substances for long-term
containment and disposal of α-active actinide nuclear wastes.
EPR spectroscopy has been used to investigate the structural
properties and solid state chemical properties of impurities
in these materials and to compare the characteristics of
single crystals and polycrystalline bodies.

INTRODUCTION

The mineral monazite, a mixed lanthanide orthophosphate ($LnPO_4$ with Ln = La,
Ce, Nd, ...), exhibits a number of characteristics that make analogs of this
substance attractive as potential primary high-level radioactive waste
forms [1,2]. These characteristics include an established long-term stability
($\sim 10^9$ years) under different geological conditions, a known ability to contain relatively high percentages of thorium and uranium, and an apparently
high degree of resistance to metamictization due to α-particle and α-recoil
radiation damage. This potentially important application of lanthanide orthophosphates has provided the impetus for a series of investigations of the
physical and chemical properties of mixed orthophosphate-impurity systems.
In particular, the technique of electron paramagnetic resonance (EPR) has been
applied to the determination of valence states and site symmetries for various
impurities in both single crystals and powders [3] of this class of compounds.
In studying the solid state properties of mixed lanthanide orthophosphate-
impurity systems, the EPR results have been correlated with investigations
using optical techniques, Mössbauer spectroscopy, and x-ray diffraction. By
using a probe ion whose spectrum reflects the properties of the crystalline
electric field, it was also possible to obtain structural information by means
of EPR spectroscopy.

Orthophosphates of the first half of the lanthanide transition series (i.e.
La through Gd) crystallize in a monoclinic form that is the direct analog of
natural monazite while the orthophosphates of the second half of the lanthanide
series (i.e. Tb through Lu), as well as YPO_4 and $ScPO_4$, crystallize in a tetragonal form analogous to the mineral zircon. The EPR investigations were
extended to encompass both lanthanide orthophosphate structural types.

EXPERIMENTAL

The lanthanide orthophosphate single crystals employed in these investigations
were grown in a lead-based flux ($Pb_2P_2O_7$) using a variation of the technique

[*] Research sponsored by the Division of Materials Science, U.S. Department of
Energy, under contract W-7405-eng-26 with Union Carbide Corporation.

described by Feigelson [4]. Following the growth by slow cooling, the crystals were removed from the entraining flux by boiling in nitric acid for about three weeks. Orthophosphate powder samples were prepared by a metathesis reaction using the appropriate lanthanide oxide and $(NH_4)_2HPO_4$, and were precipitated in molten urea [3].

The initial EPR studies concentrated on Gd^{3+} impurities in the monoclinic monazite-type orthophosphate hosts ($LnPO_4$ with Ln = La, Ce, Pr, Nd, Sm, Eu), and in the tetragonal zircon-type hosts $LuPO_4$, YPO_4, and $ScPO_4$. Fig. 1 shows the EPR spectra obtained for Gd^{3+} in $LuPO_4$. In this figure, the magnetic field

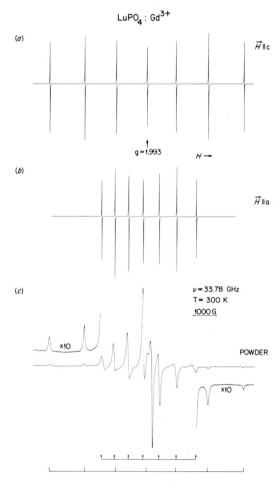

Fig. 1. EPR spectra observed for Gd^{3+} in a single crystal (a), (b) and a powder (c) of $LuPO_4$. The applied magnetic field is oriented parallel to the 4-fold tetragonal symmetry axis in (a) and perpendicular to this axis in (b). In (c) the position of the powder pattern shoulders are indicated by bars at the bottom of the figure and the divergences are marked by arrows.

is applied along the tetragonal c axis of the single crystal in the upper trace and along the a axis, i.e. perpendicular to the c axis, in the middle trace. The spectrum obtained for the corresponding powder is shown in the lower trace (i.e. Fig. 1c). It can be seen that the shoulders and divergences of the resultant powder pattern are in line with the single crystal transitions shown in the two traces above. These and similar investigations showed that the Gd^{3+} ion occupied identical substitutional sites in both the single crystal and powder specimens for either the monoclinic [5] or tetragonal systems [6]. In an actual waste, the material will undergo a series of chemical and physical processes which convert the waste oxides to the appropriate chemical form and compact this material into a high-density body. EPR spectroscopy provides a capability for verifying that following a complex processing sequence, the lanthanide or actinide impurities still occupy a known crystallographic site. The formation of potentially undesirable situations in which the α-active impurity is complexed with point defects introduced by a processing sequence (or eventually by radiation damage) or is converted to an interstitial position, can be detected. This represents an important capability for assuring predictable solid state chemical properties. It should be noted that this technique is applicable to all types of polycrystalline waste forms and is not simply limited to the lanthanide orthophosphates.

Prior to the investigation of α-active actinide-doped orthophosphate samples, a systematic study of the isoelectronic rare-earth analogs was performed. Isotopically-enriched impurities were employed and EPR spectra of the Kramers' ions, Ce^{3+}, Nd^{3+}, Dy^{3+}, Er^{3+}, and Yb^{3+} were observed in the tetragonal symmetry hosts $LuPO_4$, YPO_4, and $ScPO_4$. Unambiguous identifications of the various elements were facilitated by the use of the enriched isotopes and their corresponding characteristic hyperfine structures. The spectrum of Nd^{3+} was of particular interest since its electronic properties (i.e. $4f^3$ configuration) are analogous to those of trivalent uranium which has a $5f^3$ configuration. Therefore, an identification of the $^{238}U^{3+}$ EPR line was possible without the necessity of employing isotopically enriched uranium ^{233}U or ^{235}U. Fig. 2 shows the EPR spectrum obtained for a ^{145}Nd-doped single crystal of YPO_4 with the magnetic field along the tetragonal crystal axis. The spectrum was obtained at X-band and a temperature of 4.2 K. Additional lines due to Er^{3+} and Gd^{3+} are also present. Similar investigations of other lanthanide-actinide analog impurity systems are underway.

An unintentional Pb^{3+} impurity originating from the lead-based flux was observed in a substitutional cation site for the $LuPO_4$ and YPO_4 single crystals. A positive identification of Pb^{3+} (electronic configuration, $6s^1$) was made from the observation of the spectrum of the 20.8% naturally abundant isotope ^{207}Pb, which has a characteristically large hyperfine interaction. A five-line superhyperfine structure is evident in the EPR spectrum of even-even Pb^{3+} in YPO_4 as shown in Fig. 3. Electron nuclear double resonance (ENDOR) experiments at 4.2 K established that this structure is due to the four second nearest phosphorus neighbors surrounding the substitutional lead ion in the rare-earth site. At 4.2 K, each of the five principal superhyperfine lines in the EPR spectrum split into three components with intensities of 1:2:1, indicating an additional smaller interaction with two equivalent $I = 1/2$ neighbors. This smaller interaction arises from coupling to the two nearest ^{31}P nuclei. These results show that the s-electron wave function of Pb^{3+} is delocalized and its density is larger at the positions of the second-nearest ^{31}P neighbors than at the positions of the first-nearest neighbors.

A special set of circumstances made it possible to employ a new technique [7] in measuring the hyperfine constant of ^{207}Pb and the resulting values were the largest ever found for Pb^{3+} in a solid [> 48 GHz]. The solid state chemical restraints of the host lattice are apparently responsible for the conversion

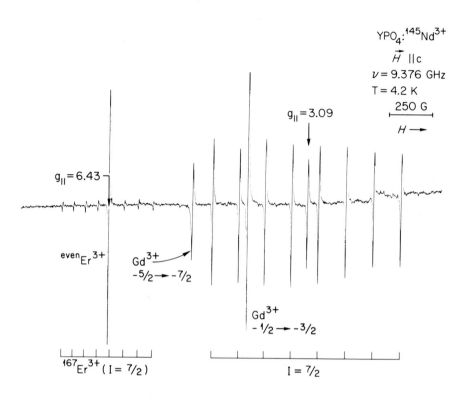

Fig. 2. EPR spectrum of $^{145}Nd^{3+}$ in YPO_4. Positions of the 8 ^{145}Nd (I = 7/2) hyperfine lines are indicated. The spectrum of naturally abundant Er^{3+} is also in evidence along with two Gd^{3+} fine structure lines.

and stabilization of Pb^{2+} to Pb^{3+}. A similar situation occurs in the stabilization of divalent rare-earth ions in the alkaline earth halides. For the case of Pb^{3+} in $LuPO_4$ and YPO_4, no irradiation or electrochemical treatments were required for the $Pb^{2+} \rightarrow Pb^{3+}$ conversion and the trivalent ion was stable at room temperature.

Although lead is normally not a component of high-level radioactive wastes produced by nuclear reactor operations, it is an end member of certain branches of the decay schemes of the actinides. Therefore, after a sufficient period of time in the nuclear waste repository, lead will be a constituent of the primary waste form.

Real high-level nuclear reactor wastes contain many impurities in addition to the lanthanides and actinides. EPR spectroscopy can also be applied to the investigation of these other components that are paramagnetic. For example, the spectra of Fe^{3+} and Mn^{2+} have been observed in single crystals and powders of $LuPO_4$, YPO_4, and $ScPO_4$.

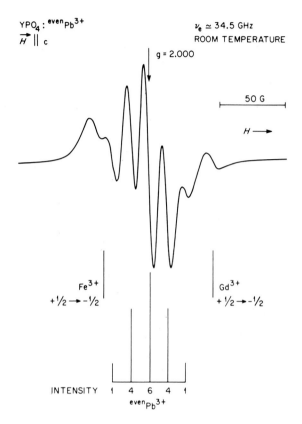

Fig. 3. EPR spectrum of the even-even isotopes of Pb^{3+} in YPO_4. The 5-line superhyperfine structure with a 1:4:6:4:1 intensity ratio is due to an inter- action with four equivalent second nearest ^{31}P neighbors. The magnetic field is oriented parallel to the four-fold crystal axis. Additional transitions due to Fe^{3+} and Gd^{3+} are also evident.

SUMMARY

The function of a crystalline substance in the role of a primary waste form is to incorporate or stabilize radioactive (and other) impurities in its crystal structure. The properties of such an impurity host ensemble can be determined, to a substantial degree, by the application of EPR spectroscopy. Here, lan- thanide orthophosphates are being evaluated as waste forms for the containment of actinide and other elements. The currently available EPR results, combined with the results of Mössbauer and optical transmission spectroscopic studies, show that it is possible to incorporate 2+, 3+, 4+, and 5+ ions in the lanthan- ide orthophosphates. The ability to determine the valence states of various impurities and to obtain information regarding the location of ions in the host orthophosphate lattice represents an important capability in the evaluation of these materials as radioactive waste forms.

ACKNOWLEDGEMENTS

Oak Ridge National Laboratory is operated by Union Carbide Corporation for the USDOE under contract W-7405-eng-26. The technical assistance of J. O. Ramey is gratefully acknowledged.

REFERENCES

1. L. A. Boatner, G. W. Beall, M. M. Abraham, C. B. Finch, P. G. Huray, and M. Rappaz, "Monazite and Other Lanthanide Orthophosphates as Alternate Actinide Waste Forms," Scientific Basis for Nuclear Waste Management, Vol. II, C. J. Northrup, ed. (Plenum Press, 1980), pp. 289-296.

2. L. A. Boatner, G. W. Beall, M. M. Abraham, C. B. Finch, R. J. Floran, P. G. Huray, and M. Rappaz, to be published in Proceedings of the International Symposium on the Management of Alpha-Contaminated Wastes, Vienna, Austria, June 1980.

3. M. M. Abraham, L. A. Boatner, T. C. Quinby, D. K. Thomas, and M. Rappaz, Radioactive Waste Management, in press.

4. R. S. Feigelson, J. Am. Ceram. Soc. 47, 257 (1964).

5. M. Rappaz, M. M. Abraham, J. O. Ramey, and L. A. Boatner, Phys. Rev. B, in press.

6. M. Rappaz, L. A. Boatner, and M. M. Abraham, J. Chem. Phys. 73, 1095 (1980).

7. M. M. Abraham, L. A. Boatner, and M. Rappaz, Phys. Rev. Lett. 45, 839 (1980).

Published 1981 by Elsevier North Holland, Inc.
Kaufmann and Shenoy, editors
Nuclear and Electron Resonance Spectroscopies Applied to Materials Science

NUCLEAR SPIN RELAXATION INVESTIGATIONS ON THE INFLUENCE OF IMPURITIES AND
TEMPERATURE ON THE MEAN FREE PATH OF MOBILE DISLOCATIONS IN NaCl

W. H. M. ALSEM, J. Th. M. De HOSSON
Dept. of Applied Physics, Materials Science Centre, University of Groningen,
Nijenborgh 18, 9747 AG Groningen, The Netherlands

H. TAMLER, H. J. HACKELÖER, O. KANERT
Institute of Physics, University of Dortmund, Postfach 50 05 00,
46 Dortmund 50, W. Germany

ABSTRACT

Dislocation motion in alkali halide single crystals is strongly impeded by the presence of impurities, apart from obstacles built by the forest dislocations. The mean free path L of
stepwise moving dislocations is measured by determination of
the spin-lattice relaxation rate $1/T_{1\rho}$ as a function of the
strain rate $\dot{\varepsilon}$, varying the content of impurities and the temperature. The latter influences the distribution of the point
defects and the activation rate of dislocations before obstacles, while the former merely shorten L, thereby raising
$1/T_{1\rho}$.

INTRODUCTION

To obtain insight in the macroscopic plasticity of crystals it is crucial to
study moving dislocations on a microscopic scale. This has been done now by the
technique of nuclear spin relaxation measurements.
The motion of dislocations under the influence of external stress is characterized as "jerky glide" (1). Dislocations are assumed to move in zero elapsed
time between obstacles. At obstacles they may be released either by thermal
fluctuations or by a change in applied stress or in the obstacle structure. In
fact one could distinguish between dislocations which are actually running and
those waiting to be released at obstacles.
Generally, the running time is small compared to the waiting time and almost
all mobile dislocations are waiting. Introducing ρ_m as the mobile dislocation
length per unit volume and L as the average distance swept out by a released
dislocation segment, one gets for the shear a produced by the dislocation
motion:

$$a = b \, \rho_m \, L \, ,
\qquad (1)$$

where $\vec{b}$ is the Burgersvector of the dislocations; the shear strain rate $\dot{a}$ is
then given by:

$$\dot{a} = b \, \rho_m \, \frac{L}{t_w} \, ,
\qquad (2)$$

where t_w is the mean waiting time before obstacles. Parameters like the mean
free path L determine to a great extent the plastic deformation behaviour of
crystals. For instance, the stress needed to continue further deformation, the
flow stress, increases with strain. This increase is well known to be correla-

ted to a decrease in L.

The main obstacles at which dislocations can be held up are points where other immobile dislocations (forest dislocations) cut the glide plane or point defects like impurities. The intersection of mobile dislocations with forest dislocations can be hard to accomplish because of formation of small nodal segments in the dislocations, which lower the local energy significantly. Therefore these are referred to as strong obstacles. In alkali halide crystals of monovalent ions divalent impurities are effective obstacles. To preserve charge neutrality in the lattice each impurity ion, e.g. a Ca^{++} ion in NaCl, is associated with a cation vacancy. Those dipoles interact with dislocations and therefore also influence the mechanical properties. At temperatures below 150°C the obstacles are fixed and can be overcome by thermal activation, whereas at higher temperatures the mobile point defects are able to follow the moving dislocations which gives rise to a drag force. Apart from single dipoles also aggregates of dipoles are present in the crystal. Their number and distribution depend strongly on the thermal treatment previous to the deformation and the deformation temperature itself (2).

The possibility of detecting moving dislocations by pulsed NMR is offered by the fact that their motion is connected with fluctuations in the local quadrupolar field. However, relative to the frequency at which the nuclei precess around the direction of an external magnetic field, their motion is too slow to influence the spin lattice relaxation time T_1. Therefore $T_{1\rho}$, the spin lattice relaxation time in the presence of a resonant magnetic r.f. field H_1 is measured on deforming crystals (3, 4, 5). The relaxation in the coordinate frame, rotating with the resonant frequency, caused by the motion of dislocations, is given by (6):

$$\frac{1}{T_{1\rho}} = f \cdot \frac{g_Q}{\tau} \, , \tag{3}$$

where f is the fraction of the effective field which fluctuates during motion of dislocations and is given by

$$f = \frac{\rho_m}{\rho_t} \frac{H_Q^2}{H_1^2 + H_D^2 + H_Q^2} \, . \tag{4}$$

H_1 is the amplitude of the r.f. magnetic field, H_D and H_Q the local dipolar and quadrupolar fields, ρ_m the mobile dislocation density and ρ_t the total dislocation density, g_Q is a quadrupolar geometrical factor almost equal to 1. τ is the correlation time for dislocation motion and as outlined above, should be identified with t_w. Eq. (3) is only valid in the so called strong collision region where $\tau > 1/\gamma H_1$. With the aid of Eq. (2) and inserting Eq. (4) one can rewrite Eq. (3) as:

$$\frac{1}{T_{1\rho}} = \frac{A}{\gamma^2 (H_1^2 + H_D^2 + H_Q^2)} \cdot \frac{1}{bL} \cdot \dot{a} \, , \tag{5}$$

where A is the mean quadrupolar energy per dislocation length and γ the gyromagnetic ratio. Considering the fact that the mean free path L is determined by the mean distance d_F between forest dislocation pinning point and by d_p, the mean distance between point defects, Eq. (5) is refined by insertion of

$$\frac{1}{L} = \frac{1}{d_F} + \frac{1}{d_p} \quad .$$

(6)

EXPERIMENTS

Relaxation measurements have been performed on NaCl (001) single crystals doped with different amounts of $CaCl_2$ ranging from 7 to 340 Mol ppm in the temperature range from 10 to 270°C. The spin lattice relaxation rate in the rotating frame is measured by means of the spin-locking technique. It consists of a $\pi/2$-pulse immediately followed by a pulse of the r.f. magnetic field H_1 aligned along the rotated magnetization and therefore called the locking pulse. The initial height of the free induction decay after the locking pulse is a measure for $1/T_{1\rho}$. The dynamic enhancement of the relaxation caused by jumping dislocations is given by

$$\left(\frac{1}{T_{1\rho}} \right)_D = \left(\frac{1}{T_{1\rho}} \right)_{TOT} - \left(\frac{1}{T_{1\rho}} \right)_{ST}$$

(7)

where $(1/T_{1\rho})_{TOT}$ is the (total) spin-lattice relaxation rate during plastic deformation of the crystals and $(1/T_{1\rho})_{ST}$ the (static) relaxation rate measured on the crystals which are not deforming. The mean free path L is obtained by measuring $1/T_{1\rho}$ as a function of the shear strain rate $\dot{a}$ and inserting the other parameters in Eq. (5). Further details and the experimental set-up have been described elsewhere (4, 5).

RESULTS AND DISCUSSION

In Fig. 1 spin lattice relaxation rates in the rotating frame $R_D^\rho \equiv (1/T_{1\rho})_D$ have been depicted as a function of the strain rate in the direction of the compres-

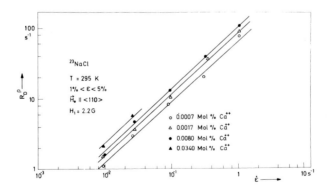

Fig. 1.   Dislocation induced part of the relaxation rate R_D^ρ as a function of the strain rate $\dot{\varepsilon}$ in ^{23}NaCl single crystals doped with different amounts of impurities.

sion axis, $\dot{\varepsilon}$, for single crystals doped with different amounts of impurities. Raising the concentration of impurities will diminish the mean distance d_p and therefore the relaxation rate will increase. The mean free path L can be determined with the aid of Eq. (5) yielding for pure crystals a value for d_F of about 1.0 µm. Evaluation of the data using Eq. (6) gives values for d_p which have been plotted in Fig. 2. The results do not correspond with the values obtained from

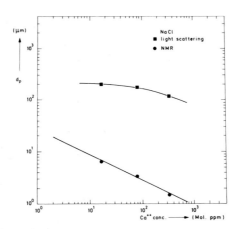

Fig. 2.   The mean distance d_p between point like obstacles as a function of the Ca^{++}-concentration in NaCl as obtained from pulsed NMR experiments and between large impurity clusters as obtained from light scattering experiments.

calculations of the mean distance between randomly distributed single point defects in the crystal with the given impurity concentration c, according to

$$d = \left(\frac{M}{c\rho N}\right)^{1/3}, \tag{8}$$

with M the molecular weight, ρ the density and N Avogadro's number. In fact, d_p turns out to be about a factor 10^2 larger than d. It is therefore concluded that impurity clusters are present in the slowly-furnace-cooled crystals. Therefore the number of local obstacles for moving dislocations diminishes. Light scattering experiments on the crystals show some of the segregations. There mean distance has been plotted in Fig. 2 too. In addition the NMR data on d_p in Fig. 2 do not exactly exhibit a concentration dependence which varies according to $c^{-1/3}$. Obviously the stronger decrease of d_p with raising Ca^{++} concentration means that the probability of cluster formation does not increase linearly with increasing Ca^{++} concentration.

In Fig. 3 spin lattice relaxation rates have been depicted as a function of temperature. These static measurements have been performed on ultra pure single crystals of NaCl (imp. conc. $\leq$ 10 ppm) and on crystals containing 40 Mol ppm Ca^{++} ions. Up to temperatures of about 150° C the lattice part of $1/T_{1\rho}$ in the undoped crystals increases with T^2 whereas an exponential increase occurs above

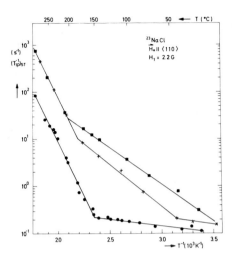

Fig. 3.   The static spin-lattice relaxation rate in the rotating frame as a
function of temperature in ^{23}NaCl ($\dot{\varepsilon} = 0$).

this temperature with an activation energy of 0.89 eV. The same energy is found
from the slope of the high temperature part of $(T_{1\rho}^{-1})_{ST}$ of doped crystals. In
this region the cation vacancy diffusion is the dominant process and the migration energy corresponds with the above mentioned value (7). In the temperature
range between 50° and 180° C $(T_{1\rho}^{-1})_{ST}$ depends on the thermal treatment of the
doped crystals. The as grown crystals show an exponential increase with a slope
of 0.36 eV while the crystals after an anneal at 300° C and a cooling to room
temperature with 50°/min have higher relaxation rates increasing with a slope
of 0.31 eV. These values are close to the dissociation energy of dimers (a configuration of two dipoles) (8, 9).
Measurements of $T_{1\rho}$ on deforming crystals as a function of temperature yielded
the data depicted in Fig. 4. They were obtained at a strain rate of $\dot{\varepsilon} = 0.4$ s^{-1}
with a deformation degree of $\varepsilon=3\%$ and 7% for the undoped crystals and about 5%
for the doped crystals. It can be seen that the room temperature value for the
mean free path for moving dislocations hardly changes in the region up to 150°C.
The very few impurities are fixed obstacles with respect to the mobile dislocations and the interaction of the moving dislocations with forest dislocations
dominate the magnitude of L. Therefore higher strained crystals show a slightly
lower value of L. Above 150° C the migration of vacancies becomes important and
the mobile dislocations are assisted in passing the obstacles. This yields the
increase in L from the same temperature where $(T_{1\rho}^{-1})_{ST}$ starts to increase (see
Fig. 3). At 260° C L is almost 4 μm for undoped crystals at $\varepsilon=3\%$.
The fraction of weaker obstacles, which can be overcome quite easily at raised
temperatures, is smaller for crystals strained to $\varepsilon=7\%$. This moderates the increase in L at higher temperatures.
Some introductory dynamic spin-lattice relaxation measurements on NaCl single
crystals doped with 40 Mol ppm Ca^{++} as a function of temperature are also given.
Using Eq. (5) the mean free path L has been obtained from these experiments and
they are depicted in Fig. 4 too. At 260° C L has decreased about a factor of 4

486

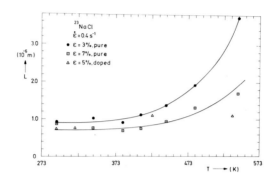

Fig. 4. The mean free path L as a function of temperature for pure and doped (40 Mol ppm Ca^{++}) NaCl <100> crystals using a strain rate $\dot{\varepsilon}$=0.4 s^{-1}.

with respect to the undoped crystals. This variation is caused by the fact that the number of obstacles, because of dissolution of dipole aggregates, and the mobility of these point defects increases with temperature; these effects are opposed by an increasing thermal activation of dislocations detained by obstacles.

ACKNOWLEDGMENT

This work is part of the research program of the Foundation for Fundamental Research on Matter (F O. M. - Utrecht) and has been made possible by financial support from the Netherlands Organization for the Advancement of Pure Research (Z. W. O. - The Hague) and the "Deutsche Forschungsgemeinschaft" (W. Germany).

REFERENCES

1. U. F. Kocks, A. S. Argon, M. F. Ashby, Thermodynamics and Kinetics of Slip, Progr. in Mat. Sci. 19, Pergamon Press (1975).

2. G. Berg, F. Fröhlich, S. Siebenhühner, Phys. Stat. Sol. (a) 31, 385 (1975).

3. W. H. M. Alsem, Faraday Symp. of the Chem. Soc. 13, London (1978) 193.

4. W. H. M. Alsem, A. W. Sleeswyk, H. J. Hackelöer, R. Münter, H. Tamler, O. Kanert, J. de Physique - Paris, C6-146 (1980).

5. W. H. M. Alsem, J. Th. M. de Hosson, H. Tamler, H. J. Hackelöer, O. Kanert, to be published in Solid State Comm. (1980).

6. D. Wolf, O. Kanert, Phys. Rev. B 16, 4776 (1977).

7. R. W. Dreyfus, A. S. Nowick, Phys. Rev. 126, 1367 (1962).

8. J. H. Crawford, Jr., J. Phys. Chem. Solids 31, 399 (1970).

9. M. Dubiel, G. Berg, F. Fröhlich, Phys. Stat. Sol. (b) 89, 595 (1978).

μ^+e^- HYPERFINE INTERACTIONS IN QUARTZ CRYSTALS

J.H. BREWER
Department of Physics, University of British Columbia, Vancouver, B.C., Canada
V6T 2A3

D.G. FLEMING AND D.P. SPENCER
Department of Chemistry, University of British Columbia, Vancouver, B.C.,
Canada, V6T 1Y6

ABSTRACT

Longitudinal muonium spin relaxation/modulation in zero
magnetic field (zf-MSR) has been used to study muonium
(μ^+e^-) atoms in single-crystal α-quartz between 5 K and room
temperature. At 6 K, three frequencies are observed,
corresponding to a triaxial hyperfine matrix whose principal
values are close to those observed for hydrogen atoms frozen
into known sites. For intermediate temperatures the Mu
atoms "hop" between sites, causing a relaxation whose rate
first increases with the hop rate and then decreases due to
motional narrowing. Finally, at room temperature, a single-
frequency oscillation is observed, corresponding to a
uniaxial motionally-averaged hyperfine interaction.

INTRODUCTION

The muonium (Mu) atom is a light isotope of the hydrogen atom, in the truest
sense ($m_{Mu} = 0.1126\ m_H$). This has led to vigorous investigations of its
chemical reactions and electronic structure in condensed matter in the past few
years, sometimes motivated by the relative resistance of H atoms to measurement
and sometimes by the promise of a precise comparison between the behavior of
two isotopes with an unprecedented mass ratio. Several reviews outline the
situation up to 1978, [1] and conference proceedings provide an update of the
extremely rapid progress since then [2]. We report here the application of a
novel technique to the study of Mu in quartz crystals, where the analogy with
H is unusually clear.

Quartz has long been the "prototype" insulator for study of muonium in
solids [1-6]; however, until several years ago only *fused* quartz had been
investigated. Apart from the propensity of muons to form Mu with high
efficiency, the behavior of Mu in fused quartz was unremarkable.

Then in 1977 the precession of triplet-state Mu atoms in weak magnetic fields
in *crystalline* quartz was found to be split by 0.79(3) MHz, reflecting a slight
uniaxial asymmetry in the hyperfine interaction of μ^+e^- in quartz, [4-6] qualitatively (though not quantitatively) similar to hyperfine matrices observed
for μ^+e^- states in silicon and germanium. [See Refs. 1 and 2]. The effective
spin hamiltonian can be written

$$H = g_e \mu_B\ \vec{S} \cdot \vec{B} - g_\mu \mu_\mu\ \vec{I} \cdot \vec{B} + \vec{S} \cdot \underset{\approx}{A} \cdot \vec{I} \tag{1}$$

where g_e and g_μ are respectively the electron and muon g-factors, $\vec{S}$ is the
electron spin, $\vec{I}$ is the muon spin, $\vec{B}$ is the magnetic field, μ_B is the Bohr

magneton, μ_μ is the muon magneton, and $\underset{\approx}{A}$ is a matrix hyperfine coupling, which can be written

$$\underset{\approx}{A} = h \langle \nu_0 \rangle \; (\underset{\approx}{1} + \underset{\approx}{\alpha}) \; , \qquad (2)$$

where $\underset{\approx}{1}$ is the unit matrix and $\underset{\approx}{\alpha}$ is a normalized "anisotropy matrix," describing the fractional distortion of the hyperfine interaction. The magnitude of the isotropic hyperfine interaction is given by $\langle \nu_0 \rangle$, which for Mu in vacuum has the value 4463 MHz. For Mu in quartz at room temperature, Brown *et al.* found $\langle \nu_0 \rangle$ = 4509 ± 3 MHz, or 1.03(7)% *larger* than the vacuum value [6]. The anisotropy matrix was found [4] to be $\underset{\approx}{\alpha}$ = -0.000175(7) $\hat{z}\hat{z}$, where $\hat{z}$ = $\hat{c}$; that is, the anisotropy is symmetric about the $\hat{c}$-axis.

A further prediction of this model is that a "modulation" of the muon polarization should be evident even in *zero* magnetic field. This has now been observed.

EXPERIMENTAL TECHNIQUE

Longitudinal muonium spin relaxation/modulation in zero magnetic field (zf-MSR) is a new experimental technique developed at TRIUMF in the last year. It is not dissimilar to transverse-field muonium spin rotation ($B_\perp$-MSR), which also reveals the low-field transition frequencies between triplet eigenstates in the form of time distributions [1], but it is considerably simpler. In quartz at 300 K, for instance, rather than two frequencies split by 0.79 MHz ($B_\perp$-MSR), with zf-MSR one sees a single frequency of 0.395 MHz. This makes a substantial saving in counting statistics. The interpretation is also simpler.

Many reviews describe the basic µSR technique [1]: one stops a beam of polarized muons in a target of interest (here, quartz shielded by mu-metal from magnetic fields), starting a high-precision time digitizer on the pulse from a counter through which the incoming muon passes, and stopping the same "clock" on a pulse from a positron detector. Due to the asymmetric decay of the muon ($\mu^+ \rightarrow e^+\nu_e\bar{\nu}_\mu$), in which the e^+ is emitted preferentially along the μ^+ spin, the time spectrum which results from many such measurements will display both the exponential decay lifetime of the muon and any time-dependence of the muon polarization. For longitudinal or zero-field measurements, the e^+ detector is positioned along the original muon polarization direction [7].

COMPARISON WITH H ATOMS IN QUARTZ

The Mu results at room temperature are quite unlike the H atom results of Perlson and Weil [8] at low temperature, which show a substantially larger distortion without axial symmetry.

It was therefore necessary to study the system at low temperature and zero magnetic field in order to make a more direct comparison with the H atom results. As soon as this was done (below about 80 K) it was obvious that the weakly perturbed uniaxial spin hamiltonian of higher temperatures was no longer in effect. Time spectra (Fig. 1) and corresponding Fourier power spectra (Fig. 2) for T = 5-6 K and B < 50 mG show *three* frequencies ν_{13} = 7.9(1) MHz, ν_{12} = 6.2(1) MHz, and ν_{23} = 1.7(1) MHz, which appear with varying *amplitudes* but do not change frequency as the crystal is rotated about the initial muon polarization direction.

These results are consistent with an effective spin hamiltonian as in Eqs. (1) and (2), with three nonequivalent principal axes to the hyperfine matrix. This splits the three F=1 energy eigenvalues (normally degenerate in zero field) into three distinct levels. Note that $\nu_{13} = \nu_{12} + \nu_{23}$, as expected for transitions between three levels; the frequencies are in fact labelled according to such a picture. This picture turns out to be quite consistent with current

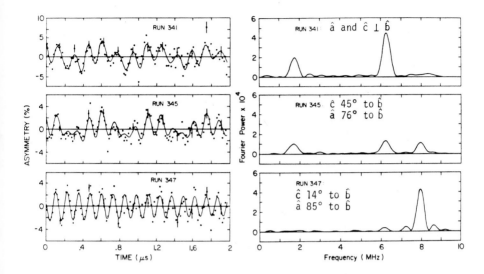

Fig. 1. Reduced longitudinal zf-MSR time spectra (constant background and muon-decay exponential removed) for Mu in a single crystal of α-quartz at 5-6 K and B < 50 mG, for several orientations of the crystal with respect to the initial muon spin direction.

Fig. 2. Fourier power spectra for data shown in Fig. 1. Orientation of ĉ (optic axis of the crystal) and â (twofold axis) with respect to b̂ (muon polarization direction) are indicated in the captions.

hydrogen atom data in quartz.

The three frequencies ν_{12}, ν_{23} and ν_{13} do not change as the crystal is rotated, but their *amplitudes* vary as the product of the direction cosines with the principal axes with which they are associated. A preliminary analysis of this variation suggests that the principal axes must make angles of 94(5)° and 115(5)° with the crystal ĉ-axis. This is in excellent agreement with the results of Perlson and Weil on H in α-quartz [8], who find that the H atom traps below ∿ 120 K at sites on a twofold axis between two Si atoms, giving principal axes at 90° and 114° to the ĉ-axis.

Moreover, when the low-temperature Mu and H results are cast in terms of the "normalized hyperfine anisotropy matrix" $\underset{\approx}{\alpha}$ defined by Eq. (2), as shown below in Table I, the similarity between Mu and H is striking.

The site of Mu and H in quartz is therefore taken to be the same. However, it is clear that Mu is slightly less distorted than H, and [if the room-temperature value of $\langle \nu_0 \rangle$(Mu) may be compared with the low-temperature value of $\langle \nu_0 \rangle$(H)] possibly only half as isotropically compressed by the trapping site. This is not surprising in view of the expected effects of the much smaller mass of Mu on the zero-point motion.

TABLE I
Hyperfine parameters of H and Mu

ATOM $(\langle\nu_0\rangle_{vac})$	TEMP	$\dfrac{\langle\nu_0\rangle}{\langle\nu_0\rangle_{vac}}$	ANISOTROPY MATRIX $\underset{\approx}{\alpha}$		(diagonal rep.)
H (1420.4 MHz)	$\lesssim$ 120 K	1.02162(3)	0.00294(8)	0	0
			0	−0.00098(4)	0
			0	0	−0.00196(4)
Mu (4463.3 MHz)	$\lesssim$ 80 K	1.0103 (?) assumed the same as at high temp.	0.00207(5)	0	0
			0	−0.00068(5)	0
			0	0	−0.00139(5)
Mu (4463.3 MHz)	$\gtrsim$ 250 K	1.0103 (7)	0	0	0
			0	0	0
			0	0	−0.000175(7)

THERMALLY ACTIVATED DIFFUSION OF MUONIUM

Below about 80 K, Mu is relaxed, presumably by the dipolar fields from ^{29}Si nuclei, at a constant rate λ of about 0.2 μs^{-1}. At higher temperatures, thermally-activated hopping between sites begins to contribute an additional relaxation mechanism. The trapping sites are all equivalent, but their principal axes are oriented differently to the muon polarization direction; thus when a Mu atom hops to a new site its spin begins to oscillate with different amplitudes for the three frequency components, resulting in an effective relaxation mechanism on the time scale of a mean oscillation quarter-period, about 50 ns. The fitted relaxation rate of these zf-MSR "quadrupolar oscillations" is plotted as a function of temperature in Fig. 3 (circular points), and its log is plotted as a function of inverse temperature in Fig. 4. Assuming that the relaxation rate is the same as the hop rate in the low temperature region, we extract an Arrhenius activation temperature for hopping T_a = 719(68) K. The preexponential factor is λ_0 = 4835 (+ 4900 or − 2400) μs^{-1}.

When the hop rate exceeds the highest local oscillation frequency (at around 140 K) the effects of the different hyperfine couplings at different (equivalent) sites begin to be motionally averaged, and the relaxation rate becomes a *decreasing* function of temperature, as can been seen from the triangular points in Figs. 3 and 4. In this region the Arrhenius fit gives an activation temperature of T_a = 1336(40) K and a preexponential factor of λ_0 = 2.3(+ 5.2 or − 1.6) × 10^3 s^{-1}. An exact model connecting this result with that at lower temperature should be feasible, but has not yet been attempted. In particular, it is not known what happens to the distortion of the hyperfine interaction at higher temperatures; in Table I we have assumed that the low temperature case is simply averaged, leaving the same isotropic $\langle\nu_0\rangle$ at high temperature, but for

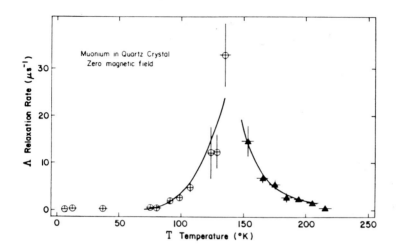

Fig. 3. Exponential relaxation rate of Mu in crystalline quartz as a function of temperature. Low-temperature points (circles) are from fits to relaxing envelope of oscillations; high-temperature points (triangles) are from fits to an exponential longitudinal relaxation function.

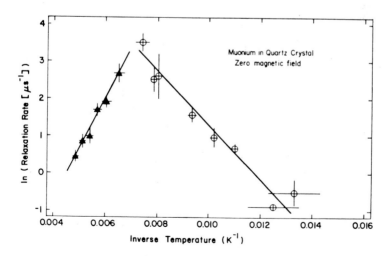

Fig. 4. Arrhenius plot of data shown in Fig. 3.

492

all we know qualitative changes may take place in the Mu atomic wave function
as the Mu atom's motion becomes very rapid.

When the temperature is high enough to "motionally narrow" the relaxation
rate to less than about 0.5 μs^{-1}, one begins to see the single zf-MSR oscillation at 0.395 MHz characteristic of the average uniaxial hyperfine interaction.
At room temperature the relaxation of this oscillation is slow (0.2 μs^{-1}); in
view of the clear indication that Mu is moving quite rapidly through the quartz
lattice at this temperature, any depolarization by nuclear dipolar fields must
be negligible, and the observed relaxation must be only the vestiges of the
quadrupolar effect at lower temperatures. Thus Mu should hardly relax at all
in quartz heated to about 400 K. However, at some temperature one must expect
Mu to reach the deeper traps provided by Al^{3+} impurities, as has been postulated
to explain the behavior of H at high temperatures [9]. This has not yet been
observed.

SUMMARY

No detailed interpretation of the differences between H and Mu in quartz has
yet been attempted, but this is the first time that such similar results have
been available for both Mu and H in the same crystal. The lessons learned from
this comparison should prove useful in the interpretation of MSR data in many
other nonmetals. Cases of immediate relevance are Si and Ge [1,2] where little
is known about the electronic structure of H but a great deal is known about Mu.
In view of the practical importance of H in determining the electrical
properties of (e.g.) amorphous Si, better understanding is crucial.

ACKNOWLEDGEMENTS

We would like to thank Doug Beder for guiding us through the first stages of
understanding how zf-MSR and $B_\perp$-MSR should appear in the case of a triaxial
hyperfine interaction. We are especially indebted to John Weil for providing
unpublished data on H in quartz for comparison, and for actively assisting in
our interpretation of the Mu results. This research was funded by the Natural
Sciences and Engineering Research Council of Canada.

REFERENCES

1. J.H. Brewer, K.M. Crowe, F.N. Gygax and A. Scienck in: *Muon Physics*,
    V.W. Hughes and C.S. Wu eds. (Academic Press, New York, 1975); J.H. Brewer
    and K.M. Crowe, Ann. Revs. Nucl. Part. Sci. 28 (1978); P.W. Percival,
    E. Roduner and H. Fischer in: "Positronium and Muonium Chemistry",
    H. Ache ed., Adv. in Chem. Series 175, 335-355 (1979); D.G. Fleming,
    D.M. Garner, L.C. Vaz, D.C. Walker, J.H. Brewer and K.M. Crowe in:
    "Positronium and Muonium Chemistry", H. Ache ed., Adv. in Chem. Series 175,
    279-334 (1979).

2. Proc. of 1st Int. Topical Meeting on Muon Spin Rotation, Rorschach,
    Switzerland, 1978, North-Holland, 1979; Proc. of 2nd Int. Topical Meeting
    on Muon Spin Rotation, Vancouver, B.C., Canada, 1980, North-Holland, to be
    published. Both are also special issues of Hyperfine Interactions.

3. G.G. Myasishcheva, Yu.V. Obukhov and V.G. Firsov, Zh. Eksp. Teor. Fiz. 53
    451 (1967) [Sov. Phys. JETP 26, 298 (1968)].

4. J.H. Brewer, D.S. Beder and D.P. Spencer, Phys. Rev. Lett. 42, 808-811
    (1979).

5. V.G. Barychevsky and S.A. Kuten, Phys. Lett. A 67, 808 (1979); D.S. Beder, Nucl. Phys. A 305, 411 (1978); D.S. Beder, Phys. Rev. 21B, 3861-3867 (1980).

6. J.A. Brown, S.A. Dodds, T.L. Estle, R.H. Heffner, M. Leon and D.A. Vanderwater, Solid State Comm. 33, 613-614 (1980).

7. R.H. Hayano, Y.J. Uemura, J.I. Imazato, N. Nishida, T. Yamazaki and R. Kubo, Phys. Rev. 20B (1979).

8. B.D. Perlson and J.A. Weil, J. Magn. Res. 15, 594 (1974); J.A. Weil, private communication.

9. A. Sosin, Radiation Effects 26, 267-271 (1975).

^{57}Fe AND ^{125}Te MÖSSBAUER STUDY OF LiFeCo$_3$TeO$_8$ AND LiFeNi$_3$TeO$_8$

ANDRE GERARD, FERNANDE GRANDJEAN AND CARLO FLEBUS
Institute of Physics, B5, University of Liège,
B-4000 Sart-Tilman, BELGIUM.

ABSTRACT

^{57}Fe and ^{125}Te Mössbauer data on the spinels LiFeNi$_3$TeO$_8$ and LiFeCo$_3$TeO$_8$ are presented. Structural information relating to site occupations by Fe, valency of Fe and Te and magnetic hyperfine fields are obtained from this study.

INTRODUCTION

The spinel compounds LiFeCo$_3$TeO$_8$ and LiFeNi$_3$TeO$_8$ may be derived from another spinel series $M_4^{2+}M^{3+}Sb^{5+}O_8$ if the couple ($M^{2+}Sb^{5+}$) is substituted by (Li^+Te^{6+}). X-ray study of the tellurates family $Li^+M^{3+}M_3^{2+}Te^{6+}O_8$ has shown that these compounds have the same orthorhombic structure that the antimonates series [1]. ^{57}Fe Mössbauer study of $M_4^{2+}Fe^{3+}Sb^{5+}O_8$ compounds where M=Mg, Co, Ni and Cu, shows that Fe^{3+} ions occupy tetrahedral (A) and octahedral (B) sites in a ratio 2/1 (2). We report here on the Mössbauer results and the magnetic behavior of LiFeCo$_3$TeO$_8$ and LiFeNi$_3$TeO$_8$.

EXPERIMENTAL

The preparation of the samples is very similar to the one used for the antimonates samples (1). ^{57}Fe and ^{125}Te Mössbauer spectrometers are of constant acceleration type. For ^{57}Fe, a 25 mC source of ^{57}Co diffused in Rh was used; for ^{125}Te, a 8 mC source of ^{125}Sb diffused in Cu was used and kept at 77 K for all experiments. The 35 keV radiation of ^{125}Sb was detected by an intrinsic Ge detector which allows a very good discrimination of the 35 keV radiation from the X-rays at 27 and 32 keV. Low temperature spectra were recorded in a Thor Cryogenics variable temperature cryostat and the temperature of the sample was controlled at $\pm$0.25 K with an Fe-Au thermocouple.

^{57}Fe MÖSSBAUER STUDY

At room temperature. Fig. 1 shows the Mössbauer spectra of LiFeCo$_3$TeO$_8$ and LiFeNi$_3$TeO$_8$ at room temperature. These spectra are very similar to those observed for the antimonates series $M_4^{2+}Fe^{3+}Sb^{5+}O_8$ and they have been fitted by the same method (2). These result from the superposition of two symmetrical doublets in the population ratio 2/1 with the hyperfine parameters given in table I. In analogy to the identification done for the antimonates, the quadrupole doublet with the smaller isomer shift, the larger quadrupole splitting and the larger intensity was attributed to the tetrahedral (A) site in the orthorhombic structure.

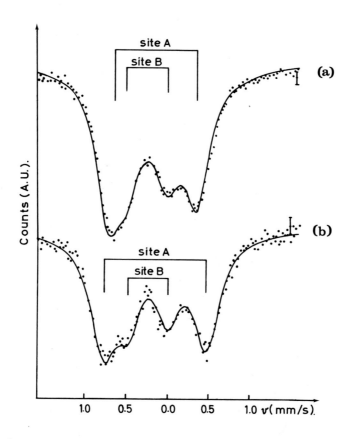

Fig. 1. Room temperature Mössbauer spectra of $LiFeCo_3TeO_8$ (a) and $LiFeNi_3TeO_8$ (b). The solid lines are the result of a least-squares fit to two symmetrical doublets as indicated by the vertical sticks.

Consequently, Fe^{3+} ions occupy both tetrahedral (A) and octahedral (B) sites in these tellurates. If we assume that Li atoms occupy A sites as usually in spinels, we may propose the following cationic distribution $(LiXFe_2)_A [X_8FeTe_3]_B$, where X=Ni or Co.

TABLE I
Hyperfine parameters of ^{57}Fe in LiFeX$_3$TeO$_8$ at room temperature

X	A site		B site	
	δ (mm/s)	Δ (mm/s)	δ (mm/s)	Δ (mm/s)
Co	0.142	1.036	0.238	0.508
Ni	0.152	1.244	0.239	0.472

δ=isomer shift referred to ^{57}Co in Rh (error ±0.02 mm/s)
Δ=quadrupole splitting (error ±0.02 mm/s)

At low temperatures. Both compounds are magnetically ordered at 77 K. Fig. 2 shows the case of LiFeNi$_3$TeO$_8$. The coexistence of two sextets with different populations in the spectrum confirms the occupation of two different crystallographic sites by Fe^{3+} ions as deduced from the paramagnetic spectra. For LiFeCo$_3$TeO$_8$, the hyperfine fields at 77 K are H$_A$=375$\pm$5 kOe and H$_B$=410$\pm$5 kOe respectively for tetrahedral and octahedral sites. For LiFeNi$_3$TeO$_8$, at 77 K, H$_A$ amounts to 370$\pm$5 kOe and H$_B$ to 420$\pm$5 kOe. The difference between the hyperfine fields of both sites is in agreement with the results on classical spinels [3] i.e. H$_A$<H$_B$.

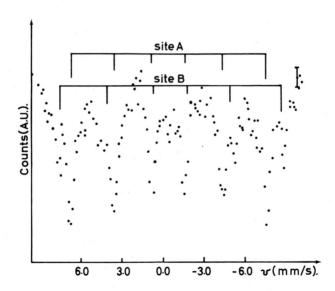

Fig. 2. Mössbauer spectrum of LiFeNi$_3$TeO$_8$ at 77 K.

For both compounds, as temperature increases, it becomes very difficult to distinguish between the two sites. Furthermore, between 122 K and 162 K for $LiFeCo_3TeO_8$, and between 142 K and 218 K for $LiFeNi_3TeO_8$, a strong paramagnetic component exists in the spectra, simultaneously with the magnetic component. At 162 K for X=Co and at 218 K for X=Ni, the magnetic transition is completed and the magnetic six line spectrum has completely disappeared. These temperatures are thus taken as Curie temperatures for both compounds. For instance, the magnetic behavior of $LiFeNi_3TeO_8$ is summarized in Fig. 3.

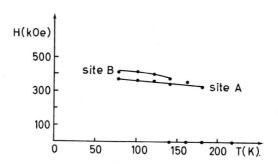

Fig. 3. Temperature dependence of [57]Fe hyperfine fields in $LiFeNi_3TeO_8$.

In conclusion, the important results are that the paramagnetic phase slowly disappears as temperature decreases, the lines are not broadened as in a relaxation process and the magnetic hyperfine fields do not appreciably change with temperature. Very similar observations have been made for $ZnCr_2O_4$ doped with Fe^{3+} [4] and have been associated with a first-order magnetic transition. By analogy, we think that we also have a first-order magnetic transition with a distribution of transition temperatures for the compounds studied.

[125]Te MÖSSBAUER STUDY

[125]Te Mössbauer spectra of $LiFeNi_3TeO_8$ and $LiFeCo_3TeO_8$ have been recorded for the samples at 77, 200, 240 K. All spectra have the shape of a broad line. The characteristics are summarized in table II. The values observed for the isomer shifts lay between the values observed in other oxides of tellurium: -1.2 to -0.4 mm/s for Te^{6+} [5]. The valency of Te is thus confirmed to be 6+. From table II, it can be seen that the observed linewidth of 6.5 mm/s at 240 K, i.e. in the paramagnetic region increases by 50% on cooling down to 77 K, i.e. in the magnetic region. At 240 K, the broadening compared to the theoretical width ($2\Gamma_n$ =5.2 mm/s) amounts to 26% and it could either be due to defects in the lattice or to a quadrupole coupling. In order to estimate quadrupole coupling-Δ and hyperfine fields-H, [125]Te Mössbauer spectra were synthesized for different values of Δ and H and compared with the experimental spectra. The observed linewidth of 6.5 mm/s at 240 K can be well reproduced with a Δ=2.5 mm/s and H=0 which give rise to a linewidth of 6.7 mm/s. Since the structure of these compounds is orthorhombic, a quadrupole splitting is very likely to exist and we conclude that it amounts to 2.50$\pm$0.15 mm/s. Such a small value does not give rise to a doublet in the spectrum since it is lower than the minimum ob-

TABLE II
^{125}Te Mössbauer characteristics of $LiFeX_3TeO_8$

X	T(K)	δ (mm/s)	Γ (mm/s)
Ni	77	-1.04	9.53
	200	-1.08	7.19
	240	-1.09	6.58
Co	77	-0.97	10.96
	200	-1.23	7.20
	240	-1.25	6.40

δ=isomer shift referred to ^{125}Sb in Cu (error +0.15 mm/s)
Γ=linewidth at half maximum (error +0.15 mm/s)

servable linewidth of $2\Gamma_n$=5.2 mm/s. With that value of the quadrupole splitting and magnetic fields of H=30 kOe, H=40 kOe, H=80 kOe and H=90 kOe, linewidths respectively of 7.00+0.15 mm/s, 7.30+0.15 mm/s, 9.00+0.15 mm/s and 9.60+0.15 mm/s were obtained on the synthesized spectrum. Thus,the spectra at 200 K would correspond to a hyperfine field amounting to 35+5 kOe and the one at 77 K, to a hyperfine field of 85+5 kOe for $LiFeNi_3TeO_8$ and 95+5 kOe for $LiFeCo_3TeO_8$.

The spectrum is not splitted by the hyperfine field because of the existence of the quadrupole splitting, it is only slightly asymmetric as can be seen by a careful analysis of the experimental data. It is clear that, because of the unsplit nature of the spectra, the analysis is not unique.

The hyperfine fields observed on the Te atoms are transferred hyperfine fields due to the presence of magnetic ions in the compounds. The nature of these fields is completely different from those observed on Fe^{3+} ions, that explains the very different values observed on both atoms: Te and Fe. If the hyperfine field on Fe^{3+} is easily related to the spin of this ion, it is difficult to account quantitatively for the observed values for Te. To our knowledge, there are very little information on magnetic hyperfine field on Te^{6+}. The only studied situation is Te^{6+} as an interstitial impurity in Fe_2O_3 and Cr_2O_3 [6] and in this case, the hyperfine field measured at 77 K amounts to 136+5 kOe. In spite of very different localizations of Te^{6+} in Cr_2O_3 or Fe_2O_3 and in our case, the Te^{6+} hyperfine fields are similar.

Most of the values of transferred hyperfine fields on Te given in the litterature are measured in intermetallic compounds [7,8] or oxides [9] where the formal valency of Te is 2-. In these cases the transferred hyperfine field is explained by the overlapping of the 3d orbitals of the magnetic atom with the 5s orbitals of the Te. In our case, Te^{6+} has the electronic structure $4d^{10}$ and an ionic radius of 0.56 Å. The Fe^{3+} ionic radius being of 0.64 Å and the average A-B and B-B distances in classical spinels being respectively 3.44 and 2.94 Å [10], the overlapping of 3d orbitals of Fe^{3+} and the 4d orbitals of Te^{6+} should be rather small and consequently cannot explain the transferred hyperfine field. A super-transfer process through the O^{2-} ions, first-neighbors is probably the most effective one in creating the hyperfine field measured on Te^{6+}.

CONCLUSIONS

^{57}Fe and ^{125}Te Mössbauer study of $LiFeNi_3TeO_8$ and $LiFeCo_3TeO_8$ was carried out. Fe^{3+} ions were found to occupy A and B sites of the spinel structure of the spinel structure of these compounds, in the population ratio 2/1. Both com-

500

pounds are magnetically ordered below 162 K for $LiFeCo_3TeO_8$ and 218 K for $LiFeNi_3TeO_8$. The Mössbauer spectra of ^{125}Te confirm the 6+ valency of Te and show a transferred magnetic hyperfine field at 77 K which is qualitatively explained by a super-transfer process. Further experiments with improved equipment on ^{125}Te will be performed in the near future in order to study, in more details, the behavior of the transferred hyperfine field with temperature.

ACKNOWLEDGMENTS

We are indebted to Prof. P. Tarte and Dr. J. Preudhomme for the preparation and the X-ray study of the samples. We thank also the Fonds National de la Recherche Scientifique for financial support.

REFERENCES

1. P. Tarte and J. Preudhomme, J. Solid St. Chem. 29, 273 (1979).
2. A. Gérard, F. Grandjean, J. Preudhomme and P. Tarte, J. de Phys., Coll. C2, 40, 339 (1979).
3. N.N. Greenwood and T.C. Gibb, Mössbauer Spectroscopy, (Chapman & Hall, London 1971) p. 259.
4. F. Varret, A. Gérard, F. Hartmann-Boutron, P. Imbert and R. Kleinberger, Proceedings of the Conference on the Application of the Mössbauer Effect (Hungarian Academy of Sciences, Budapest 1969) p. 581.
5. Ibid 3, p. 454.
6. L.P. Fefilatiev, G. Demazeau, P.B. Fabritchnyi and A.M. Babechkin, Solid St. Comm. 28, 509 (1978).
7. M. Pasternak and A.L. Spijkervet, Phys. Rev. 181, 574 (1969).
8. J. Granot and S. Bukshpan, J. Phys. C, Solid St. Phys. 8, 1435 (1975).
9. J.F. Ullrich and D.H. Vincent, Phys. Lett. 25A, 731 (1967).
10.E.W. Gorter, Phil. Res. Repts 9, 295 (1954).

Published 1981 by Elsevier North Holland, Inc.
Kaufmann and Shenoy, editors
Nuclear and Electron Resonance Spectroscopies Applied to Materials Science

THE MOTIONALLY NARROWED NMR LINE SHAPE OF AMMONIUM SELENATE*,

CHING YAO, R. HALLSWORTH and I. J. LOWE
Physics Department, University of Pittsburgh, Pittsburgh, PA 15260

ABSTRACT

The proton free induction decay $f(t)$ of ammonium selenate
is measured from temperatures of no apparent effects of
lattice motion (-185°C) upon $f(t)$, to substantial motional
effects (-145°C) upon $f(t)$. Our measured results are
consistent with a classical hindered rotational motion
of the ammonium group. The deviation of $f(t)$ from the
rigid-lattice fid $f_o(t)$ can be expanded in a Taylor series

as

$$\Delta_1 f(t) = \sum_{n=3}^{\infty} \frac{a_n(T)}{n!} \, t^n$$

From the $\Delta_1 f(t)$ data, a_3 and a_4 are derived, and they are
well fitted by the simple relations

$$a_3 = C_3 M_2 \nu = 4\alpha\omega_1^2/T_{1r}$$

$$a_4 = - C_4 M_2 \nu^2$$

where ν is the jump frequency of the rotational motion,
T_{1r} is the rotating-frame spin-lattice relaxation time,
and ω_1 the strength of the rotating magnetic field in
frequency units. α, C_3 and C_4 are calculable constants
of order unity.

INTRODUCTION

The spatial motion of atoms and molecules can affect a nuclear magnetic
resonance (nmr) line shape, and the affected line shape can be a useful tool
for studying these motions. These motions tend to average and thus reduce
the time average of the magnetic interactions among the nuclei. In a continuous wave nmr (cwnmr) experiment, the observed absorption line appears
to "narrow". For a transient experiment that produces an observed signal
called a free induction decay (fid) (the fourier transform of the cwnmr
line) the fid appears to lengthen and last for a longer time. The nmr line
shape is said to be strongly motionally narrowed when the width of the
observed line shape is much narrower than the observed width when there isn't
any motion (the so called rigid lattice). The theory of the strongly
motionally narrowed line shape has been well developed, and experimental
studies confirm its principal predictions [1]. A rigorous theory for the
moderately motionally narrowed line shape, that is the observed nmr line

*Supported by NSF Grant DMR 78-15441.

shape where lattice motion has influenced the line by only a small amount, is much more difficult and has not been developed nearly as far. For this reason,   the measurements of moderately motionally narrowed line shapes is normally used to draw qualitative and not quantitative conclusions about the motion.

In this article, we will describe the formulation of a theory of the moderately motionally narrowed line and report on the experimental verifications of some of the predictions of this theory.

In Section II, the development of the theory is sketched out, and theoretical results listed. Most of this part can be skimmed by readers not interested in details, except to learn nomenclature. Details of the nmr experiments and data processing are described in Sections III and IV respectively; conclusions are listed in Section V.

THEORY

Van Vleck, in his classic paper [2] on moments of line shapes showed that the $2n^{th}$ moment M_{2n} of the "rigid lattice" absorption line ($n = 1, 2, \ldots$) could be calculated exactly without the use of the eigenenergies and eigenstates of the spin system. From a practical point of view, only the first few moments can be measured or calculated without a heroic amount of labor. The "rigid lattice" free induction decay shape function, $f_o(t)$, can be expanded in terms of these moments as

$$f_o(t) = 1 + \sum_{n=1}^{\infty} (-1)^n \frac{M_{2n}}{(2n)!} t^{2n} . \tag{1}$$

In the region near $t = 0$, the short time region, only the first few moments are needed to accurately represent $f_o(t)$.

We have tried to extend the above ideas to the case of the moderately motionally narrowed line by obtaining a series of coefficients that are similar to moments. To do this we have described lattice motion by a stochastic model as did Anderson [3] and Kubo and Tomita [4], and also used some of their mathematical techniques. However, we have used only expansions that are valid in the short time region without making any approximations.

Very sketchily, what one does is to apply standard perturbation theory and develop the free induction decay function into a series in powers of the (secular) dipolar interaction H'. In a rigid lattice, this merely gives us the Van Vleck formula. When motion is allowed, however, one gets instead that [6]

$$f(t) = 1 + \sum_{n=1}^{\infty} I_{2n}(t), \tag{2}$$

where

$$I_{2n}(t) = \sum_{\nu_1 \cdots \nu_n} T_r \psi(\nu_1, \ldots, \nu_{2n}) \int_0^t dt_{2n} \cdots \int_0^{t_2} dt_1 \frac{1}{y_{\nu_1}(t_1) \cdots y_{\nu_{2n}}(t_{2n})} \tag{3}$$

is a summation over spin pairs $\nu_1, \nu_2, \ldots, \nu_{2n}$. The term $\psi(\nu_1, \ldots, \nu_{2n})$ is a complicated function of the spin operators of all these pairs whose trace, however, can be evaluated directly and exactly.

$$\overline{y_{\nu_1}(t_1)\cdots y_{\nu_{2n}}(t_{2n})}$$

is the stochastic average of a product of the spatial factors in the dipolar Hamiltonian. The close relation of Eq. 2 to the Van Vleck formula in Eq. 1 is revealed when $I_{2n}(t)$ is further expanded into a power series in time t (the assumed stationary and Markovian nature of the y_ν functions are very important for this step). Then it can be shown that

$$I_{2n}(t) = (-1)^n \frac{M_{2n}}{(2n)!} t^{2n} + \sum_{k=1}^{\infty} [\frac{a_{2n+k,2n}}{(2n+k)!}] t^{2n+k}$$ (4)

All the terms following the first rigid lattice term are generated by the lattice motion. There is a simple prescription for writing the coefficients $a_{2n+k,2n}$ in terms of the transition matrix W for a Markov process. We shall not give it here, but instead resum the series to make some observations. Defining $\Delta_1 f(t)$ as

$$\Delta_1 f(t) = f(t) - f_o(t)$$ (5)

and using Equations 2 and 4 yields

$$\Delta_1 f(t) = \sum_{n=1}^{\infty} \sum_{k=1}^{\infty} \frac{a_{2n+k,2n}}{(2n+k)!} t^{2n+k}$$ (6)

$$= \frac{1}{3!} a_3 t^3 + \frac{1}{4!} a_4 t^4 + \frac{1}{5!} a_5 t^5 + \cdots$$ (7)

with

$$a_3 \equiv a_{3,2}$$ (8)

$$a_4 \equiv a_{4,2}$$

$$a_5 \equiv a_{5,2} + a_{5,4}$$

The first two coefficients a_3, a_4 are of immediate interest because they are the easiest to calculate and also the most accessible to direct determination from a given set of $\Delta_1 f(t)$ data. We shall concentrate on them in the following discussion.

The earlier described prescription leads to an expression for a_3, that

$$a_3 = - [T_r[M_x^2]]^{-1} \sum_{\nu} T_r[M_x, \psi_o(\nu)][\psi_o(\nu), M_x][p \ Y_o(\nu)WY_o(\nu)].$$ (9)

a_4 has a similar expression with W replaced by W^2. $Y_o^{(\nu)}$ is a diagonal matrix whose elements are the values of the functions $y_o^{(\nu)}$ in the various configurations of the lattice. $\psi_o(\nu)$ are the spin operators such that $\sum_\nu \psi_o(\nu)y_o(\nu)=H'$ is the secular dipolar Hamiltonian.

Two general relations involving a_3 and a_4 can be established. The first connects a_3 and T_{1r}, the rotating-frame spin-lattice relaxation time. On the long correlation time side of the T_{1r} minimum, where the power spectrum of the lattice motion normally has an inverse square frequency dependence, it can be shown quite generally that[6]

$$a_3 = \alpha 4\omega_1^2/T_{1r} \qquad (10)$$

where ω_1 is the strength of the rotating-frame Zeeman field in frequency units, and α is a numerical constant which is exactly unity for a system of identical spins, and is less than unity when other species of spins also participate in the relaxation.

A further condition should be added to the list of restrictions when Eq. 10 is valid. This is that T_{1r} exist. This may not always be true because the formulas for T_{1r} are based upon the existence of a spin temperature in the rotating frame, and the assumption that first order time dependent perturbation theory can be used. Neither of these restrictions are needed to derive the formula for a_3. Unfortunately, a_3 can only be measured over the restricted temperature range where moderate motional narrowing is occuring. T_{1r} can normally be measured over a much wider temperature range.

If eq. 9 is evaluated and crudely reduced to simple forms, then $a_3 \sim \nu M_2$ where ν is some characteristic transition rate that describes the lattice motion. In a simple system where a single ν suffices, that is when $W = \nu V$ such that V is a constant matrix determined completely by the geometry of the system, the above relation can be made rigorous so that

$$a_3 = c_3 \nu M_2 \qquad (11)$$

The constant c_3 depends only upon the details of the atomic jumps.

A second relation involves a_3 and a_4. Equation (11) can be readily generalized so that in a schematic way it can be claimed that

$$a_{2n+k,2n} \sim M_{2n} W^k , \qquad (12)$$

because the formula that gives this particular a coefficient involved k factors of the matrix W. Again for a simple system, one can then write

$$a_4 = c_4 M_2 \nu^2 \qquad (13)$$

and

$$\frac{a_4}{a_3} = \frac{c_4}{c_3} \nu . \qquad (14)$$

This affords a significant test of the consistency of the theory, and also a way of measuring ν that is not complicated by considerations of M_2.

EXPERIMENTAL DETAILS

Ammonium selenate was chosen to test the previously outlined theory for the following reasons:

i.   the proton absorption line undergoes a large narrowing (second moment decreases by a factor of ~7) in a convenient temperature range centered about -160°C.

ii. the narrowing mechanism is believed to be a thermally activated and completely classical reorientation of the ammonium ion complex.

iii. T_1 and T_{1r} data are very well fitted by a single activation energy, which gives some support to a simple-minded model in which the NH_4 tetrahedron reorients itself by activated jumps about its four C_3 symmetry axes. Within this model, the calculation of c_3 and c_4 that appears in Equations 11 and 13 is very simple.

The spectrometer used in the experiment has been described fully elsewhere [5]. One novel feature in our measuring technique merits however some remark. For a study of this sort, it is imperative that we be able to calibrate the overall gain of the detection system to well within several parts in a thousand so that the fid's can be normalized and $\Delta_1 f(t)$ found.

This is achieved in the present case by inserting into our sample a small capillary of liquid propane. The shape of the liquid signal is temperature-independent, being determined by the magnetic field inhomogeneity alone. Its amplitude, however, is affected by all the same factors that will affect the solid ammonium selenate signal. We can then calibrate the measured f(t) functions by the amplitude of the propane signal. The normalization obtained in this manner is believed to be accurate to within 10^{-3}.

DATA ANALYSIS AND DISCUSSION

We found it very difficult to use brute force technique to fit a polynomial in t to our experimental free induction decay data and obtain the a_n coefficients. Instead, we used the following procedure. We first

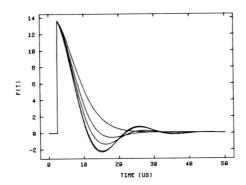

Fig. 1 A set of free induction decays of ammonium selenate that shows the effect of motional narrowing.

normalized the ammonium selenate signal against the liquid propane
signal; thus removing the signal size dependence upon the Boltzmann factor,
amplifier gain and pulse notation angle. After normalization, the rigid-
lattice decay function was subtracted from all the others. The resulting

difference function $\Delta_1 f(t)$ was plotted vs. t^3 for each temperature, and

the initial slopes of these plots measured to compute a_3 at each temperature.

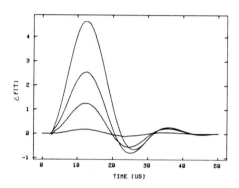

Fig. 2 A typical set of difference curves
for ammonium selenate, obtained by subtracting the rigid-lattice fid from each
of the "narrowed" fids. The temperature
ranges from about -183°C to -153°C.

Unfortunately, this step cannot be iterated to yield a_4 because the signal-
to-noise ratio deteriorates too quickly with each subtraction. We did
however analyze the data using a different scheme. Combining Equations
6 and 12, the terms in $\Delta_1 f(t)$ can be regrouped according to their power

dependence on ν. Thus

$$\Delta_1 f(t) = \nu\, f_1(t) + \nu^2\, f_2(t) + \nu^3\, f_3(t) + \ldots \qquad (15)$$

where

$$f_1(t) = \frac{1}{\nu}\left(\frac{1}{3!}\, a_3 t^3 + \frac{1}{5!}\, a_{5,4} t^5 + \frac{1}{7!}\, a_{7,6} t^7 + \ldots\right) \qquad (16)$$

$$f_2(t) = \frac{1}{\nu^2}\left(\frac{1}{4!}\, a_4 t^4 + \frac{1}{6!}\, a_{6,4} t^6 + \frac{1}{8!}\, a_{8,6} t^8 + \ldots\right) \qquad (17)$$

etc.

are independent of temperature.

For each value of t then, we can fit the values of $\Delta_1 f(t)$ at the various

temperatures as a polynomial in ν. This can be carried out for the entire
range of t, and yields the functions $f_1(t)$, $f_2(t)$, and so on. To do this,
we need to know that value of ν is to be associated with each temperature.

This information can be supplied by either a_3 we already obtained in conjunction with Equation (11), or the T_{1r} data. The two sets of data are fully consistent with one another as seen in Fig. 3. The $f_1(t)$ and $f_2(t)$ data so obtained can be plotted against t^3 and t^4 respectively, and according to Eqs. (16) and (17) the initial slopes again allow us to compute a_3 and a_4.

The set of a_3's determined direct by from $\Delta_1 f(t)$ is given in Fig. 3.

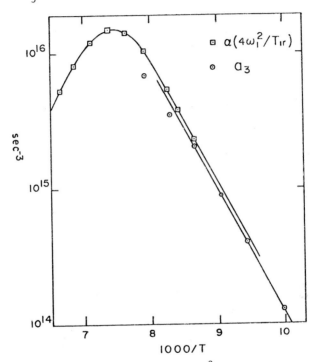

Fig. 3   A comparison of $\alpha(4\omega_1^2/T_{1r})$ and a_3. T_{1r} data is fitted by a Debye function. The line that fits a_3 data is drawn to have the same activation energy as predicted by T_{1r}. α is calculated to be 0.91 in our model.

Detailed model calculation that includes both H-H and N-H interactions within one NH_4 tetrahedron predicts that

$$a_3 = (0.91)\,\frac{4\omega_1^2}{T_{1r}}\,. \tag{18}$$

Experimentally we find that the linear dependence of a_3 upon T_{1r}^{-1}

is well obeyed, with the proportionality constant being about 0.82. The 10% difference of the measured and theoretically predicted α is well within our experimental error.

The same model calculation predicts that

$$a_3 = 4.481 \; M_2^{intra} \; \nu$$

$$a_4 = 20.81 \; M_2^{intra} \; \nu^2 \tag{19}$$

and hence

$$\frac{a_4}{a_3} = 4.644 \; \nu \tag{20}$$

The values of a_3 and a_4 derived from the $f_1(t)$ and $f_2(t)$ data are, at $\beta = 1000/T = 9.04$ (~ $-162°C$),

$$a_3 = 9.469 \times 10^{14} \; sec^{-3}$$

$$a_4 = 5.075 \times 10^{19} \; sec^{-4} \; .$$

According to Eq. 20, this yields $\nu = 11.54 \times 10^3 \; sec^{-1}$. This compared favorably with the value derived from the T_{1r} data at that temperature of $\nu = 9.101 \times 10^3 \; sec^{-1}$.

CONCLUSIONS

We have described a theory of how to use the short time free induction decay data of the moderately motionally narrowed nmr line to study lattice motion. The measured parameters for ammonium selenate are consistent with the general predictions of the theory, and also demonstrate the possible uses to which this technique might be used. With an improvement in the signal to noise ratio, and a more sophisticated data anlysis technique, one may be able to determine more motional parameters for even complicated systems, such as nuclei undergoing diffusive motion in a lattice.

REFERENCES

[1] A. Abragam, The Principles of Nuclear Magnetism (Clarendon Press, Oxford 1961).

[2] J. H. Van Vleck, Phys. Rev. 74, 1168 (1948).

[3] P. W. Anderson, J. Phys. Soc. Japan 9, 316 (1954).

[4] R. Kubo, K. Tomita, J. Phys. Soc. Japan 9, 888 (1954).

[5] R. F. Karlicek, I. J. Lowe, J. Mag. Res. 32, 199 (1979).

[6] Ching Yao, Thesis, Univ. of Pittsburgh (1980).

Published 1981 by Elsevier North Holland, Inc.
Kaufmann and Shenoy, editors
Nuclear and Electron Resonance Spectroscopies Applied to Materials Science

QUADRUPOLE HYPERFINE INTERACTION AND MAGNETIC HYPERFINE
FIELD IN FeOCl AND ITS INTERCALATES[*]

Y. MAEDA
Department of Chemistry, Faculty of Science, Kyushu University,
33 Hakozaki Higashi-Ku, Fukuoka 812 Japan

R. H. HERBER
Department of Chemistry, Rutgers University, New Brunswick, N.J. 08903 USA

ABSTRACT

FeOCl is a layered compound belonging to the orthorhombic
space group P_{mnm} (D_{2h}^{13}), with two formula units per unit cell.
Adjacent layers in the (010) crystallographic plane, in
which chlorine atoms face each other, are held together by
van der Waals forces which are readily disrupted by the insertion of appropriate Lewis base molecules. This intercalation of "guest" species causes an expansion in the b-
direction, leaving the remainder of the unit cell dimensions
unchanged. Among base molecules which are readily intercalated are NH_3, pyridine and a wide variety of substituted
pyridines, nitrogen containing heterocyclic molecules, phos-
phines and phosphites. The limiting stoichiometry ranges
from 1:3 ("guest" to Fe atom ratio) for small intercalants
to 1:6 for large "guest" molecules. Using both powder X-ray
diffraction and temperature dependent ^{57}Fe Mössbauer effect
spectroscopy, the lattice dynamics, hyperfine interactions,
magnetic ordering temperature and magnetic hyperfine field
(at 4.2 K) have been studied for the neat matrix and a number of base intercalates. The magnetic hyperfine field at
the iron atom at 4.2 K is 441∓3 KOe, independent of the magnitude of the b-axis expansion. V_{zz} is negative for neat
FeOCl and intercalates in which the "guest" to host ratio is
$\leq$ 1:3, and positive when this ratio is $\geq$ 1:4.

[*] Supported by the National Science Foundation under Grant DMR 7808615A01.

INTRODUCTION

Iron(III) oxychloride, FeOCl, is a lamellar compound belonging to the ortho-rhombic space group P_{mnm} (D_{2h}^{13}) in which two formula units, each containing a high spin iron atom, are present per unit cell [1, 2]. The crystal structure consists of a stack of double sheets of cis-$FeCl_2O_4$ octahedra, linked together with shared edges. The unit cell dimensions of the virgin material are a= 3.780, b= 7.917 and c= 3.302 Å, and are readily obtained from the appro-priate powder pattern X-ray diffraction data. The lamellar architecture of the matrix has a number of interesting chemical and solid state consequences, among which is the ability to introduce by relatively mild chemical means, a number of organic molecules into the structure [3-7]. These molecules are assumed to insert into the van der Waals layer which is formed by the facing sheets of chlorine atoms, and which is envisioned as being held together by relatively weak chemical forces. The basis of this assumption rests in part on the obser-vation that molecular intercalation into FeOCl causes a marked b-axis expan-sion, leaving the a and c dimensions of the unit cell nearly intact. The molec-ules which have been introduced into this lattice are all Lewis base type com-pounds, although the presence of a lone pair of electrons must be viewed as a necessary, rather than a sufficient condition for intercalation. The ease of the insertion process appears to depend in part on the basicity of the lone pair, and a modest correlation appears to exist between the pK_a of the base molecule and the ease with which the layered matrix can be expanded [8].

There have been numerous speculations [9, 10] concerning the orientation of the guest molecules within the lattice and the driving force for the intercalation process. Studies on a number of substituted pyridines [11] in which the length of the long molecular axis (the C_2 or C_3 axis, as the case may be) is allowed to vary by 10^{-8} cm or more, suggest that one orientation which leads to a stable intercalate has the guest molecule so positioned that the lone pair orbital can overlap with an acceptor d-orbital of the metal atom; that is, with the long molecular axis parallel or nearly parallel to the ac plane of the matrix. Such an orientation would serve to pry the FeOCl double layers apart and locate a guest molecule either in every cavity in the structure, leading to an $FeOCl(G)_{1/3}$ stoichiometry, (where G represents the guest molecular species) or in every other cavity, leading to an $FeOCl(G)_{1/6}$ stoichiometry.

Among the diverse physico-chemical techniques used to study these intercalates is ^{57}Fe Mössbauer Effect (nuclear gamma ray resonance) spectroscopy, which makes it possible to probe the effect of intercalation on the hyperfine parameters of the iron atom, and to investigate the lattice dynamical and magnetic properties of the host lattice in the presence of various intercalant molecular species. Among the parameters which can be extracted from Mössbauer effect spectra accum-ulated over a temperature range are: the isomer shift (IS), the quadrupole coupling (QS), the recoil-free fraction (f) and the magnetic hyperfine field at the iron nucleus (H_i), and the thermal coefficients of each of these parameters. In the present study, the temperature dependence of QS for FeOCl and a number of intercalates has been examined in detail over the range 78<T<300 K, and the resultant data have been interpreted in terms of a model for these unusual anisotropic solids.

EXPERIMENTAL

FeOCl was prepared by literature methods [6, 12], using an evacuated hot tube method, and characterized by its X-ray diffraction powder pattern. The various intercalates were prepared as described previously [3-5, 8, 11] and character-ized by powder patterns, elemental analysis and infra-red spectroscopy. All samples were maintained under an inert atmosphere and stored under rigorously anhydrous conditions. [57]Fe Mössbauer effect spectra were obtained as described earlier [13] and data reduction was similarly effected, using a matrix inversion least squares fitting routine [14] suitably modified to run on the Rutgers University IBM 370/168 computer.

RESULTS AND DISCUSSION

The electric quadrupole interaction removes the degeneracy of the first nuclear excited state in [57]Fe, and is due to the interaction between the nuclear electric quadrupole moment, eQ, and the principal component of the diazonalized electric field gradient at the nuclear lattice site, $\partial^2 v/\partial z^2$. In FeOCl, the major contribution to the field gradient arises from the <u>cis</u> arrangement of the FeO_4Cl_2 octahedron, and is related to the covalency of the iron atom-nearest neighbor chemical bonding interaction. As noted previously [13], the temperature dependence of the QS interaction in unintercalated FeOCl is very small, and has a value of $\sim -1.0 \times 10^{-4}$ mm sec^{-1} deg^{-1} in the temperature range $100 \leq T \leq 300$ K. The origin of this temperature dependence is primarily in the normal volume expansion, which allows a small relaxation to occur in the geometry around the metal atom. Such a thermal dependence of the quadrupole hyperfine interaction is observed in many covalent solids.

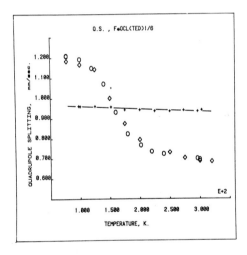

Fig. 1. Temperature dependence of the quadrupole hyperfine interaction in the triethylene diamine intercalate of FeOCl. The circles and diamonds refer to two replicate samples to show the reproducibility of the data. The data for unintercalated FeOCl (crosses) is included for comparison.

The temperature dependence of QS for the molecular intercalates of FeOCl is strikingly different. A representative example – the data for the triethylene diamine intercalate, FeOCl(TED)$_{1/6}$ – is summarized in Fig. 1 in which the data for neat FeOCl are included for comparison. The slope of the QS(T) curve for the intercalate is summarized graphically in Fig. 2 from which it is seen that the inflection point occurs at about 153 K. The full width at half maximum of the curve in Fig. 2 is approximately 60 degrees. The corresponding data for a number of other amine intercalates all of which show qualitatively similar behavior) are summarized in Table 1, which also includes pertinent data for the isomer shift parameter (and its temperature dependence in the high temperature limit), the temperature dependence of the recoil-free fraction, and a Mössbauer lattice temperature extracted from the f(T) data [13, 15].

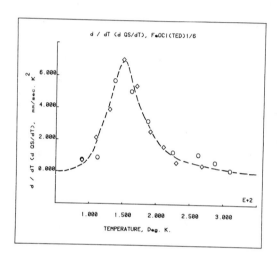

Fig. 2. Derivative with respect to temperature of the data for the triethylene diamine intercalate of FeOCl shown in Fig. 1. The ordinate units are mm sec^{-1} K^{-2} x 10^3. The maximum in this curve occurs at ~153 K and has a full width at half maximum of ~60 K.

The interpretation of the QS(T) data summarized in Figs. 1 and 2 leads to the following description of FeOCl and its amine intercalates: The neat FeOCl lattice can be viewed as a more or less classical three dimensional solid, despite the obvious anisotropy in some of its properties brought about by the presence of the van der Waals layer in the lamellar structure. The temperature dependences of the isomer shift, quadrupole splitting and recoil-free fraction parameters are similar to those observed for other covalent solids. For the intercalates, on the other hand, there appears to be a transition from behavior as a three-dimensional solid at low temperatures to that characteristic of a _quasi_ two-dimensional solid at high temperatures. This transition, which occurs over a temperature range of ~60 degrees, comes about when the mean thermal energy of the lattice is comparable to the binding energy across the van der Waals gap. From the inflection point temperature extracted from Fig. 2, this binding energy across the van der Waals gap can be estimated to be on the order of kT ~0.3 to 0.4 kcal mol^{-1}, and this is taken to be the energy characteristic of the inter layer vibrations against each other in the solid. A recent far infra-red spectroscopic study [16] has, in fact, demonstrated the existence of a lattice mode region absorption maximum at ~165 cm^{-1} which has been tentatively identified with this interlayer motion in the intercalates.

TABLE I

Summary of hyperfine parameters and related data for amine intercalates of FeOCl.

	T,K	FeOCl	FeOCl(py)^a 1/3	FeOCl(2 pic)^a 1/4	FeOCl(2,6 lut)^a 1/6	FeOCl(TED)^a 1/6	FeOCl(DMAP)^a 1/6	FeOCl(quin)^a 1/6
IS^b, mm sec^{-1}	78	0.511±0.008	0.477±0.030	0.503±0.017	0.514±0.078	0.509±0.010	0.500±0.008	0.511±0.010
	300	0.358±0.008^c	0.461±0.076	0.404±0.044	0.454±0.090	0.465±0.008	0.456±0.008	0.434±0.010
QS, mm sec^{-1}	78	0.963±0.008^c	1.252±0.022	1.146±0.015	1.158±0.020	1.010±0.008	1.125±0.015	1.118±0.010
	300	0.940±0.008	0.670±0.008	0.935±0.010	0.765±0.022	0.700±0.008	0.709±0.008	0.759±0.010
$\frac{d \ln A}{dT}$ x 10^3, K^{-1}		1.170	3.41	2.24(100<T<225) 3.20(T>250)	3.06	2.30	2.94	1.76(120<T<260)
θ_M , K		341	200	206	211	243	215	278
b , Å		7.91	13.45	13.52	14.80	13.06	13.41	13.36
Δb , Å		-	5.54	5.61	6.89	5.15	5.50	5.45
T_{max}^d , K		-	-	-	-	153±5	140±5	142±5

^{a} py = pyridine; 2-pic = 2 picoline; 2,6 lut = 2,6 lutidine; TED = diazabicyclo(2,2,2)octane; DMAD = N,N dimethyl-4 amino pyridine; quin = quinuclidine

^{b} All isomer shifts are with respect to the center of a metallic iron resonance spectrum at 296±1 K

^{c} Extrapolated from data above T_N

^{d} Maximum in d/dT (ΔQS/ΔT), see for example Fig. 2.

514

As will be noted from the data summarized in Table I, the b-axis unit cell dimension expansion is quite similar in a number of the amine intercalates so far examined. Thus, the present data are as yet insufficient to answer the question of whether the characteristic transition temperature extracted from the QS(T) data is a function of the b axis spacing or not. Due to the broadness of the derivative curve (Fig. 2) it may be necessary to examine intercalates which increase (rather than decrease) the interlayer binding energy in FeOCl intercalates. Such studies are currently underway in these laboratories.

ACKNOWLEDGEMENTS
This research was supported in part by the Division of Materials Research of the National Science Foundation under grant DMR 7808615A01, and by a grant from the Center for Computer and Information Services, Rutgers University. This support is herewith gratefully acknowledged. The authors are also indebted to T. K. McGuire for assistance with some of the computer programs used in this investigation and to R. Cassell for the preparation of FeOCl and contribution to some of the measurements.

REFERENCES AND NOTES

1. S. Goldsztaub, C. R. Acad. Sci. Paris 198, 667 (1934); ibid, Bull. Soc. Franc. Miner. Cryst. 58, 6 (1935)

2. M. D. Lind, Acta Cryst. B26, 1058 (1970)

3. F. Kanamaru, M. Shimada, M. Koizumi, M. Takano and T. Takada, J. Solid State Chem. 7, 297 (1973)

4. S. Kikkawa, F. Kanamaru and M. Koizumi, Bull. Chem. Soc. Japan 52, 963 (1979) and references therein.

5. P. Palvadeau, L. Coic, J. Rouxel and J. Portier, Mat. Res. Bull. 13, 221 (1978) and references therein.

6. T. R. Halbert and J. C. Scanlon, Mat. Res. Bull. 14, 415 (1979)

7. R. H. Herber and Y. Maeda, Inorg. Chem. 00, 000 (1980

8. R. H. Herber and Y. Maeda, submitted to Inorg. Chem.

9. F. Kanamaru and M. Koizumi, Japan J. Appl. Phys. 13, 1319 (1974)

10. R. Schöllhorn, H. D. Zagefka, T. Butz, and A. Lerf, Mat. Res. Bull. 14, 369 (1979)

11. R. H. Herber and Y. Maeda, Proc. Symp. on Practical Applications of Nuclear and Radiochemistry, Second Chemical Congress of the North American Continent, Las Vegas, Nev. 1980, to be published.

12. E. Kostiner and J. Steger, J. Solid State Chem. 3, 273 (1971)

13. R. H. Herber and Y. Maeda, Physica 99B, 352 (1980)

14. The authors are indebted to G. K. Shenoy for providing this program and to M. F. Leahy for its appropriate modification.

15. The Mössbauer lattice temperature, θ_M, is calculated making the "free-atom" mass assumption, and this parameter should be used for comparison between the structurally related intercalation compounds only.

16. R. H. Herber and A. J. Rein, unpublished data.

EPR OF Mn^{2+} IN $Ni(CH_3COO)_2 \cdot 4H_2O$ AND $K_2Ni(SO_4)_2 \cdot 6H_2O$

SUSHIL K. MISRA AND M. JALOCHOWSKI*
Physics Department, Concordia University, Montreal, Canada H3G 1M8

ABSTRACT

The technique of electron paramagnetic resonance has been applied to study the magnetic properties of nickel acetate and nickel potassium tutton salt single crystals, using Mn^{2+} ion as probe. From the values of spin Hamiltonian parameters and linewidths at room, liquid nitrogen and liquid helium temperatures it is concluded that these crystals do not become magnetically ordered as the temperature is lowered to 3.2K, and thus the transition temperature, below which the crystal would order either ferromagnetically, or antiferromagnetically, for these crystals, should be below 3.2K.

INTRODUCTION

EPR measurements at room temperature on Mn^{2+}- doped single crystals of $Ni(CH_3COO)_2 \cdot 4H_2O$ [referred to hereafter as NACE] and $K_2Ni(SO_4)_2 \cdot 6H_2O$ [referred to hereafter as PNSUL] have been previously reported by Janakiraman and Upreti [1], and by Upreti [2] and Griffiths and Owen [3] respectively. These measurements were, however, confined to room temperature only.

Since in both these samples the host lattices contain Ni^{2+} ions which are paramagnetic, it was felt that the technique of electron paramagnetic resonance using Mn^{2+} ion as probe could be used to study the low temperature magnetic ordering of the host lattices. If a magnetic ordering does indeed take place the Mn^{2+} EPR spectra would be considerably modified below the transition temperature. At all temperatures it was possible to see clearly resolved lines [1-3] due to the short spin-lattice relaxation time of the Ni^{2+} ions.

In this report the results of X-band EPR measurements on Mn^{2+} ions doping NACE and PNSUL single crystals at room, liquid nitrogen and liquid helium temperatures are described. (For details of crystal structure, experimental arrangement, etc., see ref. [4].)

DATA

(a) NACE crystal. Two sets of spectra corresponding to the two magnetically inequivalent sites for Mn^{2+} ions in the unit cell were obtained. The spectra at 62K and 3.2K have the same angular variation pattern as at room temperature with the following differences in so far as the observation of resonant line positions is concerned. While at room temperature all the 30 hyperfine lines are observed, at 62K all the 30 hyperfine lines are observed only for magnetic field orientation within an angle of $30°$ from the z axis (in the zx plane). At 3.2K the lines are harder to observe, only 24 lines are observed for magnetic field orientation along z axis while along x axis only the lowest-lying six lines are observed.

516

(b) PNSUL crystal. For this sample there are also two sets of spectra observed
corresponding to two inequivalent sites for Mn^{2+}. The spectra at lower temperature are of the same general form as at room temperature. In particular,
while all the 30 lines are observable for magnetic field orientation along the
x axis, or within an angle of 10° from it, fewer lines are observed away from
the x-axis; for the magnetic field orientation at, or about, z axis only six or
seven lines are observed.

PARAMETERS

The following spin-Hamiltonian, expressed in the usual notation
appropriately fits the particular site symmetry of Mn^{2+} ions in both hosts [1,3]

$$H_s = \beta \vec{H}.\vec{\vec{g}}.\vec{S} + \sum_{m=0,2} \frac{1}{3} b_2^m O_2^m + \sum_{m=0,2,4} \frac{1}{60} b_4^m O_4^m$$

$$+ Q'[I_z^2 - I(I+1)/3] + Q''[I_x^2 - I_y^2] + AS_z I_z + B(S_x I_x + S_y I_y)$$

In the above equation $S(=5/2)$ and $I(=5/2)$ are respectively the electronic and
nuclear spins of the Mn^{2+} ion.

The parameters are evaluated from the knowledge of resonant line
positions (fixed klystron frequency, variable magnetic field) for several orientations of the external magnetic field using a rigorous least-squares fitting
procedure [5], adapted to the electron-nuclear spin-coupled system of the
Mn^{2+} ion. The values of large parameters $g, b_2^0, b_2^2, b_4^0, A, B$ are listed in Table I
at different temperatures. The absolute signs of the parameters are determined
from the relative intentities of lines at liquid helium temperature, and the
separation of the hyperfine lines [4].

LINEWIDTHS

(a) Temperature variation. A study of the average linewidth corresponding
to the lowest and highest field hyperfine sextets reveals the following features.
For both hosts, it is seen that at low temperatures the linewidths decrease as
the temperature is decreased, while the linewidths are within experimental
errors, essentially the same from 20K and 10K respectively up to room temperatures for NACE and PNSUL hosts.

(b) Variation with magnetic field intensity. It is observed that the linewidths increase with the intensity of the external magnetic field. An attempt
was made to fit the linewidths with polynomials of type $a + bH$ (linear) and

$c + dH^2$ (quadratic). From the reulting r.m.s. values it was found that the
linewidths neither depend on H linearly, nor quadratically, in a unique manner.
This would indicate that the spin-spin interactions do not contribute to linewidth, as they exhibit quadratic dependence on H [6].

CONCLUSIONS

From the values of the linewidths and spin Hamiltonian parameters it
is clear that down to even 3.2K, for both hosts, the change in EPR spectra do
not indicate a passage through the transition temperature at which a magnetic

ordeing would take place. It is thus concluded that the magnetic orderings should take place below 3.2K. To this end, one should make EPR measurements using a He^3, or a He^3 - He^4 dilution, refrigerator.

TABLE I   SPIN HAMILTONIAN PARAMETERS.[*]

	NACE			PNSUL		
	295K	62K	3.2K	295K	60K	3.2K
g_{zz}	1.984	1.952	1.953	2.009	2.000	2.068
g_{xx}	2.000	1.966	1.989	1.994	2.000	1.997
b_2^0	1.271	1.435	1.454	-0.807	-1.000	-1.022
b_2^2	-0.254	-0.396	-0.413	0.292	0.300	0.304
b_4^0	-0.023	0.019	0.022	-0.188	-0.320	-0.322
A	-0.270	-0.295	-0.254	-0.249	-0.200	-0.202
B	-0.270	-0.205	-0.214	-0.264	-0.260	-0.256

* The b_ℓ^m and A, B are expressed in units of GHz.   The errors of g-factors are ±0.001 GHz: those of b_2^m ±0.003 GHz, while for b_4^0, A, B these are ±.001 GHz.

REFERENCES

*Present address: c/o Department of Experimental Physics, Marii-Curie University, Lublin, Poland

1. R. Jankiraman and G. C. Upreti, J. Chem. Phys. 54, 2336 (1971).
2. G. C. Upreti, J. Mag. Reson. 14, 274 (1974).
3. J. H. E. Griffiths and J. Owen, Proc. Roy. Soc. A213, 459 (1952).
4. S. K. Misra and M. Jalochowski, to be published (1980).
5. S. K. Misra, J. Mag. Reson. 23, 403 (1976).
6. T. Moriya and Y. Obata, J. Phys. Soc. 13, 1333 (1958).

Technique and Theory

Published 1981 by Elsevier North Holland, Inc.
Kaufmann and Shenoy, editors
Nuclear and Electron Resonance Spectroscopies Applied to Materials Science

521

STUDY OF INTERNAL MOTION IN MATERIALS BY MEANS OF DIRECT OBSERVATION OF
RELAXATION RESONANCES IN NMR (NMRRR)

L. VAN GERVEN, P. COPPENS[*] and P. VAN HECKE[**]
Laboratorium voor Vaste Stof - Fysika en Magnetisme, Katholieke Universiteit
Leuven, 3030 Leuven (Belgium)

ABSTRACT

A method, different from the classical ones, for studying
internal motions - in particular reorientational and tunneling
modulation of rotations - via NMR is described. Instead of
looking for matching or resonance maxima in NMR T_1^{-1} vs. T -
curves [where $\rho(T) \cong \omega_o$, with ρ: the correlation rate and ω_o:
the Larmor frequency] we look for such maxima in NMR T_1^{-1} vs.
ω_o - curves [where $\omega = \phi_o$, with ω: the (tunneling) frequency].

These method, *Nuclear Magnetic Resonance Relaxation Resonance
(NMRRR)*, has marked advantages, which are discussed. A field
cycling technique is presented. Typical results, for solid
silane, are given.

INTRODUCTION

Nuclear magnetic resonance is well known as a powerful tool for the detailed
study of materials, in particular of organic materials, like *e.g.* polymers, and
biological materials.

In most cases this research, in particular in industrial laboratories, has been
limited to the study of internal *structures*, *i.e.* of the *static* properties of
materials, for which NMR, owing to the very high resolution of its spectra, is
most adequate.

Another major advantage, however, of radiospectroscopy over optical spectroscopy
has not been widely used up to now in the study of materials. In radiospectroscopy, in contrast to higher frequencies spectroscopies, *spontaneous* (emission)
relaxation transitions do not occur, and so other relaxation transitions, *thermal*
transitions, induced by different internal relaxation mechanisms prevail. This
opens the road to the study of *dynamic* processes in materials via NMR, in
particular NMRR, *nuclear magnetic resonance relaxation*, a domain which has not
yet been fully exploited.

PHENOMENA. SOME THEORY

The modulation of the space coordinates of nuclei $r(t)$, $\theta(t)$, $\phi(t)$ in the non-
secular terms ($q > 0$) of the dipolar Hamiltonian of nuclear spin pairs

$$H_D = \sum_q F^{(q)}(r,\theta,\phi)A^{(q)} \tag{1}$$

[$F^{(q)}$ = coordinate functions; $A^{(q)}$ = spin functions]

---

[*] Research fellow of the Belgian *Interuniversitair Instituut voor Kernwetenschappen*
[**] Now with N.A.T.O., 1110 Brussels (Belgium)

gives rise to modulations of the "internal" magnetic fields $\vec{h}_D$, stimulating *spin-lattice relaxation*. In this way *internal motions*, often called *molecular motions*, motions of *molecules* or *molecular groups*, lie in many cases at the basis of spin-lattice relaxation, are a relaxation mechanism *par excellence*.

In fact the for NMR interesting modulations of $\vec{h}_D$ very often are brought about by *modulations* of internal motions. The direct cause is not the "fundamental" motions, translations or rotations - their frequencies are generally too high to induce the relatively low frequency Zeeman transitions - but merely modulations of these fundamental motions. The modulations have a *characteristic time* τ.

Which kinds of modulations there are, we will discuss now for the case that *rotation* is the "fundamental" motion (analogous things can be said about translations, including vibrations). In order to fix the idea, let us consider molecules or molecular groups (*e.g.* CH_3-, CH_4, NH_3, NH_4-, etc.).

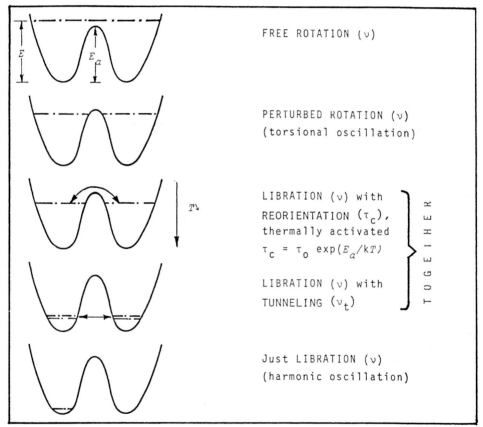

Fig. 1. Sketch of rotation and rotation modulation phenomena in a simple double potential well with increasing hindrance.

Rotations of these molecules or groups can be subdivided in

a) *Non-periodic* rotation, characterized by a *correlation time* ($\tau = \tau_c$) and a *correlation rate* $\rho = 2\pi/\tau_c$: this phenomenon has to do with rotational diffusion, but in solids it is called *reorientation*, since in most cases there is a fixed axis, around which an interrupted rotation, a hopping between equivalent equilibrium positions takes place. τ_c is the mean time between jumps.
b) *Periodic* rotations, characterized by a *period* ($\tau = T$) and a *frequency* $\omega = 2\pi/T$. Here we have to distinguish
   1. *Free rotation.*
   2. Rotational vibration in a potential well *(libration).*
   3. Rotational vibration through a potential barrier *(tunneling).*

An important parameter is the height of the barrier E_a, as compared to the energy E of the rotator. In figure 1 a sketch is given of the phenomena which would occur successively, the lower the energy of a rotating group or molecule becomes, *i.e.* the more the rotation is hindered, *i.e.* the lower the temperature is.

In figure 2 the energy levels of the tetraheder rotor with 4 protons (CH_4 or SiH_4) are given as a function of E_a/E (not on scale). $\nu (= \omega/2\pi)$ is the rotation or libration frequency; tunnel splittings correspond to the tunneling frequency $\nu_t (= \omega_t/2\pi)$. Nuclear spin isomerism occurs. A,T,E are the different isomers which exist in the corresponding levels: A has a *molecular* nuclear spin $I = 2$, T has $I = 1$, E has $I = 0$.

Reorientation and *tunneling* are the processes we are interested in. Their characteristic rates, *viz.* frequencies, often lie in the range of NMR - frequencies (10^7 - 10^8 Hz). In fact they are connected phenomena. Reorientation is a more complicated phenomenon and can be seen as a process of: phonon absorption transition to a higher librational level in the same well + "tunneling" at the higher level + phonon emission transition to the original librational level in a different well. In contrast to "simple" tunneling it is a *random, non-phase conserving, non-periodic, phonon assisted process*, therefore characterized by a *temperature* dependent *correlation* time.

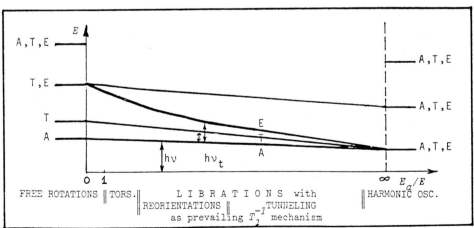

Fig. 2. Energy levels of a tetraheder rotor (not on scale) in a crystal field of high (e.g. tetrahedral) symmetry. In a lower symmetry, levels will be splitted further, e.g. T → T',T".

524

The ν's of *rotations* and *librations*, determined by the moment of inertia, are much higher: typically 10^{12} Hz. Rotations and librations *as such* do not contribute to the spin relaxation processes; their modulations - reorientation and tunneling - as said above, do.

We want to mention, from theory, just two important relations between T_1, the (Zeeman) spin-lattice relaxation time, and τ_c, *viz.* ω_t, the first for reorientation, a relaxation mechanism prevailing at higher temperatures, the second for tunneling, a relaxation mechanism prevailing at lower temperatures.

If the protons in a tetrahedric molecule can be treated as single spins (no nuclear spin isomerism), the *intra*molecular contribution to the spin-lattice relaxation rate by thermally activated rotational reorientation is given by classical relaxation theory [1]:

$$T_1^{-1} = C \frac{\gamma^4 \hbar^2}{r^6} \left(\frac{\tau_c}{1 + \omega_o^2 \tau_c^2} + \frac{4\tau_c}{1 + 4\omega_o^2 \tau_c^2} \right) \qquad (2)$$

r = distance between spins
τ_c = reorientational correlation time
ω_o = Larmor frequency determined by the external magnetic field H_o ($\omega_o \equiv \gamma H_o$)
  (at resonance = ω_s, the spectrometer frequency)
C = 0.9 for SiH_4 and CH_4
  = 0.6 for CH_3-.

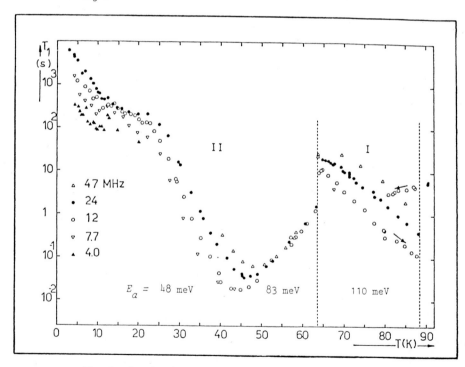

Fig. 3. T_1 in silane as a function of temperature.

There is a "matching" condition: T_1^{-1} is maximum for

$$\rho \cong \omega_o \qquad i.e. \quad \omega_o \tau_c = 0.616 \tag{3}$$

Recent work of NIJMAN and TRAPPENIERS [2] yields a relation for the *intra*molecular contribution to spin-lattice relaxation by rotational tunneling in a tetraheder:

$$T_1^{-1} \propto \sum_{n=1}^{2} \frac{\tau_c}{(\omega_t - n\omega_o)^2 \tau_c^2 + 1} \tag{4}$$

ω_t = tunneling frequency.

Here we have a "resonance" condition (ω being now ω_t), and a double one[*]: T_1^{-1} is maximum for

$$\omega_t = \omega_o \quad \text{and} \quad \omega_t = 2\omega_o \tag{5}$$

EXPERIMENTALS

Condition (3), and the related shape of the "matching" line, has been extensively used by many research workers in NMR (with $\omega_s = \omega_o$) for studying internal motions via T_1^{-1} *vs.* T - measurements, looking for T_1^{-1} - maxima at temperatures where $\rho(T) \cong \omega_o$ (see *e.g.* figure 3).
In this way ρ (and τ_c) are determined directly *at one temperature* only, the temperature of the maximum, and there are other inconveniences to this method for studying internal motions. From the slope of the T_1^{-1} *vs.* T - curve one could deduce E_a (see figure 1) and in this way calculate ρ at other temperatures. This is, however, given the fact that very often E_a/kT is quite different from unity, a questionable procedure.

A better way is to do T_1^{-1} *vs.* ω_o - measurements and to look for T_1^{-1} - maxima, where $\omega = \omega_o$ (ω can be *e.g.* ω_t, the tunneling frequency and so we come to condition (5)). This method leads to *direct* and *sharp* measurements, at *any* (reasonable) *temperature*, of ω (in casu ω_t). These are quite interesting advantages over the T_1^{-1} *vs.* T - measurements, the more that such measurements are also more selective, *i.e.* less influenced by other relaxation mechanisms.

In order to keep, for obvious reasons, ω_s high and constant, a *field-cycling* method was proposed and used.

Our field cycling and pulse sequence schemes for measuring $T_1^{-1}(\omega_o)$ are given in figure 4. In $H_o(t)$ we notice that $H_{oM}(=\omega_s/\gamma)$, the "Measuring" field, is kept constant, while the "Relaxation" field $H_{oR}(=\omega_o/\gamma)$, in which the spin-lattice relaxation occurs, is varied. The pulse sequence is the classical $\pi/2 - \pi/2$ - sequence, except that now θ, the duration of the relaxation interval, is constant and M_z^R is measured and plotted as a function of H_{oR}, not as a function of θ. M_z^R is a measure of T_1^{-1}: the faster the relaxation, the larger M_z^R.

Our actual measurements are performed on a Bruker 321 S pulse spectrometer (tunable over a frequency range from 4 to 90 MHz) at H_{oM} - values of about 1.0 and 2.2 Tesla (ca 40 and 90 MHz) and H_{oR} - values, variable between 0.01 Tesla and 2.2 Tesla. The magnet is an electromagnet. So, a strong disadvantage of our actual equipment is the long switching time - typically 35 seconds for a jump of 1 Tesla (see figure 4) - which limits our field cycling experiments to samples having a very long T_1 (at low temperatures). Indeed,

---

[*] Since in dipolar interaction, the basis of relaxation transitions (see (1)), at least 2 spins play a role, double relaxation flop-flop's, besides single relaxation flop's, are not negligible at all.

$T_1 > \theta \gg$ switching time

is an obvious requirement.

In order to investigate samples with a much shorter T_1, a fast switching magnet (something like a switching time of 10 ms/Tesla) is wanted. Two possibilities are under consideration:

- a liquid nitrogen cooled copper coil, with the heat dissipation problem as major disadvantage;
- a fast switching superconducting coil, raising, on the other hand, the problem of long term reproducibility.

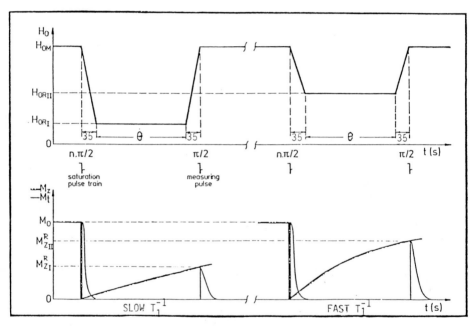

Fig. 4. Field cycling scheme and its pulse frequence scheme for measuring, via $M_z(H_{oR})$, $T_1^{-1}(\omega_o)$.

RESULTS

The method for studying internal motions, in particular their rates and frequencies, via $T_1^{-1}(\omega_o)$-measurements, which we call *Nuclear Magnetic Resonance Relaxation Resonance (NMRRR)* - and also the technique of field cycling - has yet to be fully exploited.

As a typical example we give our results for solid silane. In silane T_1 is long enough (about 10 minutes at 10 K), so that we can make θ much longer than the maximum switching time of 35 s: θ is taken around 100 s.

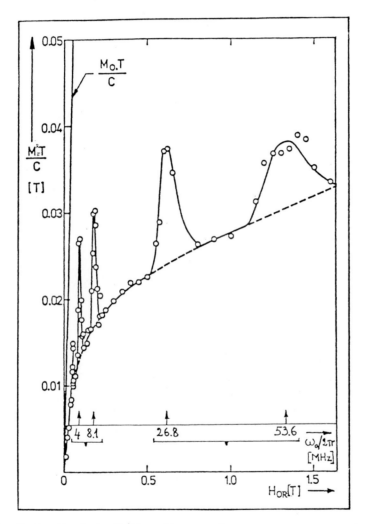

Fig. 5. Result of the field cycling experiment: the proton magnetization M_z^R (Curie corrected) in silane, recovered from saturation after a $\theta = 60$ s, as a function of H_{OR}. $T = 8.8$ K. $\omega/2\pi = 38.2$ MHz. Pairs of relaxation peaks are indicated.

A typical plot $M_z^R(H_{OR})$ is shown in figure 5. Two pairs of NMRRR-peaks appear at H_{OR}-values (hence Larmor-frequencies ω_o) where resonant relaxation according to eq. (4) and (5) occurs. They are superimposed on a continuously field dependent M_z^R, due to the "normal" field dependence of the spin-lattice relaxation time caused by non-resonant contributions from eq. (2) and (4) and by other experimental effects.

The values of the tunnel energies, deduced from our NMRRR-field cycling experiment are

(0.033 ± 0.002) μeV for the splitting T' → T" and
(0.22 ± 0.01) μeV for the splitting A → (T'T").

T' and T" are sublevels of T (see figure 2) due to a splitting of T by the crystal field in SiH_4, which therefore should be of lower than tetrahedral symmetry, since in a tetrahedral symmetry no such splitting occurs. The corresponding tunneling frequencies are 8 MHz and 53 MHz.

PROBLEMS. OUTLOOK

So many more substances and materials, containing rotating and vibrating groups, could be investigated for tunneling and other motions. For the moment we are investigating *i.a.* pentamethylbenzene, tetramethylgermanium, copperacetate (together with Professor Clough's group in Nottingham) and some other substances.

Temperature dependence of these NMRRR's should be studied.

Study of shape and line width of the NMRRR lines can give important and direct information about different connected processes, and in particular about reorientation of the molecules and groups at *different* frequencies and temperatures.

ACKNOWLEDGMENTS

We are indebted to the Belgian *Interuniversitair Instituut voor Kernwetenschappen* for financial support to the *Nuclear Magnetism* project. We are grateful to Professors S. Clough (Nottingham) and J. Hennel (Kraków) for their contribution, as Visiting Professors at the University of Leuven, to our research work and for many stimulating discussions. We also thank Mrs. B. Gabryś.

REFERENCES

1. A. Abragam, *The Principles of Nuclear Magnetism* (Oxford Press, Oxford, 1961)

2. A.J. Nijman, N.J. Trappeniers, *Proceedings XIXth Congress Ampère* (Heidelberg, 1977), 353

3. G. Janssens, P. Van Hecke, L. Van Gerven, Chem. Phys. Lett. 42, 445 (1976)

4. P. Van Hecke, G. Janssens, Phys. Rev. B, 17, 2124 (1978)

Published 1981 by Elsevier North Holland, Inc.
Kaufmann and Shenoy, editors
Nuclear and Electron Resonance Spectroscopies Applied to Materials Science

NEW METHOD FOR THE DETERMINATION OF DIFFUSION CONSTANTS FROM PARTIALLY
NARROWED NMR LINES*

D. WOLF
Materials Science Division, Argonne National Laboratory, Argonne, IL 60439

ABSTRACT

The effect of atomic and molecular motions on the NMR
free-induction decay (FID) and lineshape is investigated
theoretically in the intermediate temperature range in
which the NMR line is only partially narrowed. It is
shown that the FID may be decomposed into the weighted
sum of a rigid-lattice (background) contribution and an
exponentially decaying part containing all the information on the diffusive or reorientational motions in the
crystal in terms of the spin-spin relaxation time T_2.

INTRODUCTION

In contrast to spin-lattice relaxation processes, spin-spin relaxation is
most effective in the absence of atomic or molecular motions. The result is a
background ("rigid-lattice") free-induction decay (FID) or linewidth which
prohibits the investigation of internal motions in crystals if the mean time
between consecutive jumps of an atom or molecule, τ, is longer than the
inverse of the rigid-lattice second moment, $\Delta\omega^{RL}$.
Owing to the complexity of NMR lineshape theories in solids, it has been
difficult in the past to extract quantitative information (such as the diffusion coefficient) from the relatively simple FID or lineshape measurements
in the intermediate temperature range in which the NMR line is neither
Lorentzian (as in the motionally-narrowed region) nor practically Gaussian (as
in the rigid-lattice region).
In this article it is shown that the rigid-lattice (RL) and motional FID or
linewidth contributions are simply additive and that, therefore, the information on the atomic motions is extracted rather easily by subtracting the
experimentally measurable RL background contribution from the actual FID or
linewidth in the intermediate region. For simplicity, the following discussion is couched in terms of the FID. As is well known, the lineshape is
obtained by a simple Fourier transform of the FID.

BASIC THEORY

As discussed in some detail in Ref. 1, for NMR purposes the Hamiltonian H
of a crystal is subdivided into a "lattice" Hamiltonian, H_L, a Hamiltonian,
H_S, of the completely isolated spin system embedded in the crystal, and the
spin-lattice coupling Hamiltonian, H_{SL}, according to

$$H = H_S + H_{SL} + H_L \quad .$$

(1)

*Work supported by the U.S. Department of Energy.

By definition, $[H_S, H_L] = 0$. In general, H_S includes internal (rigid-lattice) spin-spin interactions inside the isolated spin system (such as the direct dipolar interaction Hamiltonian, H_D^{RL}), as well as Zeeman interaction Hamiltonians with externally applied time-independent (H_Z) and time-dependent $[H_1(t)]$ magnetic fields, $\vec{h}_o$ and $\vec{h}_1(t)$, respectively; hence, considering only internal dipole-dipole interactions,

$$H_S = H_D^{RL} + H_Z + H_1(t) \quad . \tag{2}$$

The "lattice"-induced fluctuations of H_{SL} are governed by the expression [1]

$$H_{SL}(t) = e^{i/\hbar \, H_L t} \, H_{SL} e^{-i/\hbar \, H_L t} = H_D(t) - H_D^{RL} \quad , \tag{3}$$

where it was observed that, by definition, $[H_D^{RL}, H_L] = 0$.

According to Eqs. (1) - (3),

$$H(t) = e^{i/\hbar \, H_L t} \, H \, e^{-i/\hbar \, H_L t} = H_o + H_L + H_1(t) + H_{SL}(t) \quad , \tag{4}$$

where $H_o = H_Z + H_D^{RL}$; i.e., $H(t)$ is explicitly time-dependent due to the time variation of the external field $\vec{h}_1(t)$ and implicitly time dependent due to the fluctuations in the crystal (such as diffusive motions).

The time evolution of the density matrix σ of the spin system (averaged over all "lattice" degrees of freedom) satisfies the Von Neumann-Liouville equation (see, e.g., Refs. 1-7)

$$\frac{d\sigma}{dt} = -\frac{i}{\hbar} \left[H_o + H_1(t) + H_{SL}(t), \, \sigma(t) \right] \quad , \tag{5}$$

with the initial condition that for $t \to -\infty$ the spin system is at thermal equilibrium with itself and with the "lattice" at temperature ϑ_L:

$$\sigma(t \to -\infty) = \sigma_{eq} = \frac{e^{-H_o/k\vartheta_L}}{Tr\left(e^{-H_o/k\vartheta_L}\right)} \quad , \tag{6}$$

where k denotes Boltzmann's constant.

Within the framework of the Bloch-Wangsness-Redfield theory [1,2,7] $H_{SL}(t)$ in Eq. (5) is considered as a perturbation while $H_1(t) \equiv 0$. For that reason the theory is restricted to the motionally-narrowed region.

In the linear-response theory of Kubo and Tomita [1,4,8] $H_1(t)$ is treated as the small perturbation, thus enabling the calculation of the FID and NMR lineshape even in a rigid lattice in which by definition, $H_{SL}(t) \equiv 0$.

In contrast to both theories, we now consider both $H_1(t)$ and $H_{SL}(t)$ simultaneously as weak perturbations on the spin system. In the usual manner Eq. (5) may thus be solved recursively by applying an iterative perturbation-type procedure. The "fast" time variation in Eq. (5) due to H_o is readily removed by defining the Heisenberg operators

$$H_{SL}^*(t) = e^{i/\hbar \, H_o t} H_{SL}(t) e^{-i/\hbar \, H_o t}; \quad \vec{\mu}(t) = e^{i/\hbar \, H_o t} \, \vec{\mu} \, e^{-i/\hbar \, H_o t} \quad , \tag{7}$$

where $\vec{\mu} = \Sigma_i \; \vec{\mu}_i = \gamma \hbar \; \Sigma_i \; \vec{I}_i$ is the magnetic-moment operator (γ = gyromagnetic ratio, $\vec{I}$ = nuclear spin vector operator), with $H_1(t) = - \vec{\mu} \; \vec{h}_1(t)$. If $H_{SL}(t)$ is assumed to be a random stationary operator with vanishing ensemble average and if we realize that the time variations of $H_{SL}(t)$ and $H_1(t)$ are not correlated, it may be shown after a straightforward calculation (compare, e.g., Ref. 3) that in the linear-response limit for $H_1(t)$ and in the second-order perturbation limit for $H_{SL}(t)$ the solution to Eq. (5) may be written as follows:

$$\sigma(t) = \sigma_{eq} + \sigma^{RL}(t) + \sigma^{SL}(t) \quad , \tag{8}$$

with

$$\sigma^{RL}(t) = \frac{i}{\hbar} \int_{-\infty}^{t} dt' \; \vec{h}_1(t') \; \overline{\left[\vec{\mu}(t'-t), \sigma_{eq} \right]} \tag{9}$$

and

$$\sigma^{SL}(t) = - \frac{1}{\hbar^2} \int_{-\infty}^{t} dt' \int_{-\infty}^{t'} dt'' \; \overline{\left[H_{SL}^*(t'-t), \; \left[H_{SL}^*(t''-t), \; \sigma_{eq} \right] \right]} \quad , \tag{10}$$

where the bar indicates an ensemble average. Note that $\sigma^{SL}(t)$ is essentially governed by $H_{SL}(t)$ while $\sigma^{RL}(t)$ contains rigid-lattice properties only.

Equations (8) - (10) may be applied to determine the time variation of the expectation value $\langle Q \rangle = \text{Tr}(\sigma(t)Q)$ of any operator Q associated with the spin system, such as the energy $\langle H_S \rangle$ or the magnetization $\vec{M} = \langle \vec{\mu} \rangle$.

FID IN THE PRESENCE OF NUCLEAR MOTIONS

From Eq. (8) the relaxation equation for the macroscopic magnetization in the x direction (perpendicular to $\vec{h}_o$, with $M_x^{eq} = 0$) becomes:

$$M_x(t) = M_x^{RL} + M_x^{SL}(t) \quad . \tag{11}$$

Completely analogous to the theory of Kubo and Tomita [8], $\vec{h}_1(t)$ in Eq. (9) may be identified with a step function perturbation associated, say, with the sudden removal of $\vec{h}_1$ at $t = 0$. Also, applying the classical dissipation-fluctuation theorem [4] and limiting ourselves to spin systems containing one sort of nuclei only, after a straightforward calculation identical to that of Abragam [9], we obtain (with $\omega_o = \gamma H_o$)

$$M_x^{RL}(t) = \frac{\cos \omega_o t}{2 \text{Tr}\left(\sigma_{eq} \mu_x^2 \right)} \; \text{Tr}\left(\sigma_{eq} \left[\mu_x^{RL}(t), \; \mu_x \right] \right) \tag{12}$$

with

$$M_x^{RL}(t) = e^{i/\hbar \; H_D^{(0)RL} t} \; \mu_x \; e^{-i/\hbar \; H_D^{(0)RL} t} \quad , \tag{13}$$

where $H_D^{(0)}$ denotes the secular part of the dipole Hamiltonian. We are thus restricting ourselves to the adiabatic lineshape (i.e., to temperatures well below the T_1 minimum due to the nuclear motions). Equation (12) is identical with the usual starting equation for the explicit determination of the rigid-lattice FID and lineshape [3,8,9]. Hence, in the rotating reference frame (index r) Eq. (12) may formally be written as follows:

$$M_{xr}^{RL}(t) = M_{xr}^{RL}(0) \ F^{RL}(t) \quad , \tag{14}$$

where $F^{RL}(t)$ is the normalized rigid-lattice FID function.

In a similar manner, starting from Eq. (10) $M_{x}^{SL}(t)$ may be determined. After a calculation similar to that of Abragam [2] it is found that in the high-field limit (in which $h_o \gg h_L$, where h_L is the dipolar local field)

$$M_{xr}^{SL}(t) = M_{xr}^{SL}(0) \ e^{-t/T_2} \quad , \tag{15}$$

with the usual expression for T_2 in the adiabatic limit (Ref. 2):

$$\frac{1}{T_2} = \frac{3}{8} \gamma^4 \hbar^2 \ I(I + 1) \ J^{(0)}(0) \quad , \tag{16}$$

where $J^{(0)}$ is the spectral density function associated with fluctuations of the secular dipole Hamiltonian. For self diffusion in crystals, $J^{(0)}(0) = A\tau$ where τ denotes the mean time between successive jumps of an atom, while the constant A depends on the diffusion mechanism [1].

Combining Eqs. (11), (14) and (15) we thus obtain:

$$M_{xr}(t) = M_{xr}(0)F(t) = M_{xr}^{RL}(0)F^{RL}(t) + M_{xr}^{SL}(0)e^{-t/T_2} \quad . \tag{17}$$

As is well known [3,8,9], the calculation of $F^{RL}(t)$ and hence of the actual FID function $F(t)$ in the presence of nuclear motions is a very difficult problem. However, to use FID's to investigate atomic motions (via T_2) it is sufficient, according to Eq. (17), to <u>measure</u> $F^{RL}(t)$ <u>rather</u> than calculate it.

INVESTIGATION OF ATOMIC MOTIONS FROM THE ONSET OF MOTIONAL NARROWING

The remaining problem in applying Eq. (17) to determine T_2 arises from the temperature-dependent prefactors in the weighted sum in Eq. (17). Since, by definition, $F^{RL}(0) = 1$, for $t = 0$ Eq. (17) yields:

$$M_{xr}(0) = M_{xr}^{RL}(0) + M_{xr}^{SL}(0) \quad . \tag{18}$$

Note that $M_{xr}(0)$ is determined experimentally by the actual FID in the intermediate temperature range. To determine $M_{xr}^{RL}(0)$ and $M_{xr}^{SL}(0)$ independently, another relationship is needed.

According to Eq. (17), two types of spins may be distinguished, namely those performing at least one jump during a time of the order of T_2^{RL} (defined as the inverse of the RL second moment, $\Delta\omega^{RL}$) and those which do not jump during that time. The probability $w_o(T_2^{RL}, \tau)$ that an atom does not perform any jumps in time T_2^{RL} if the mean time between successive jumps of a given atom is τ, is [1]

$$w_o\left(T_2^{RL}, \tau\right) = \exp\left(-T_2^{RL}/\tau\right) \quad . \tag{19}$$

Since $M_{xr}(0)$ is proportional to the total number of spins, we may write

$$M_{xr}^{RL}(0) = M_{xr}(0) \; w_o\left(T_2^{RL}, \tau\right) \quad . \tag{20}$$

Hence, with Eq. (18) we find that

$$M_{xr}^{SL}(0) = M_{xr}(0) \left[1 - w_o\left(T_2^{RL}, \tau\right)\right] \quad . \tag{21}$$

Combining Eqs. (17) - (21) we finally obtain for the FID shape function in the intermediate temperature region:

$$F(t) = e^{-T_2^{RL}/\tau} F^{RL}(t) + \left(1 - e^{-T_2^{RL}/\tau}\right) e^{-t/T_2} \quad . \tag{22}$$

For thermally activated processes $\tau = \tau_o \exp(E/kT)$, where E is the activation energy. This enables $F(t)$ in Eq. (22) to be expressed in terms of the absolute temperature T.

DISCUSSION

As pointed out above, for an assumed diffusion mechanism T_2^{-1} may be expressed in terms of τ with no adjustable parameters. Hence, Eq. (22) relates the actual FID shape function $F(t)$, for example, to the diffusion constant $D = \ell^2/6\tau$, where ℓ is the jump distance of the atoms. The normalized rigid-lattice FID function, $F^{RL}(t)$, may be determined from an FID measurement at a lower temperature. Therefore, Eq. (22) contains only one parameter, namely, the mean residence time τ. From a fit of Eq. (22) to the experimental FID, $F(t)$, in the intermediate temperature range, τ and hence D may be extracted in a rather straightforward manner.

Another, perhaps more interesting way to extract the diffusional contribution to $F(t)$ in Eq. (22) may be the use of a multiple-pulse sequence. This technique takes advantage of the idea that the part of the FID associated with $F^{RL}(t)$ is reversible (and may thus be refocussed by means of a proper pulse sequence) while the contribution arising from the nuclear motions (T_2) is irreversible.

Finally, it is pointed out that the Fourier transform of Eq. (22) yields the interesting result that the actual NMR lineshape under conditions of partial motional narrowing is a weighted average of the rigid-lattice (low-temperature) and the motionally-narrowed Lorentzian (high-temperature) lineshape.

ACKNOWLEDGMENTS

I have greatly benefited from discussions with Prof. O. Kanert.

REFERENCES

1. D. Wolf, Spin Temperature and Nuclear Spin Relaxation in Matter, Clarendon Press, Oxford, 1979.

2. A. Abragam, Principles of Nuclear Magnetism, Clarendon Press, Oxford, 1962, Chapter VIII.

3. F. Lado, J. D. Memory and G. W. Parker, Phys. Rev. B 7, 2910 (1973).

4. J. M. Deutch and I. Oppenheim, Adv. Magn. Res. 3, 43 (1968).

5. A. G. Redfield, IBM J. Res. Develop. 1, 19 (1957).

6. A. G. Redfield, Adv. Magn. Res. 1, 1 (1965).

7. C. P. Slichter, Principles of Magnetic Resonance, Harper and Row, New York, 1963, Chapter 5.

8. R. Kubo and K. Tomita, J. Phys. Soc. Japan 9, 888 (1954).

9. A. Abragam, Principles of Nuclear Magnetism, Clarendon Press, Oxford, 1962, Chapters IV and X.

Published 1981 by Elsevier North Holland, Inc.
Kaufmann and Shenoy, editors
Nuclear and Electron Resonance Spectroscopies Applied to Materials Science

535

IMPURITY DIFFUSION BY NMR*

JAMES R. BECKETT, JEAN POURQUIÉ and DAVID C. AILION
Department of Physics, University of Utah, Salt Lake City, Utah 84112

ABSTRACT

We discuss a recently developed technique for studying the
diffusion of spins with a small gyromagnetic ratio and apply
it to investigate the nature of Ag^+ diffusion in AgF. We also
investigate the potential application of this technique to
the observation of diffusing impurities. Experimental studies on Cu:3% Al and $(P_2O_5)_{0.8}(Li_2O)_{0.2}$ glass are described.

INTRODUCTION

Nuclear magnetic resonance (NMR) has long been an important technique for
studying the motion of atoms in solids. Since the sensitivity of the NMR signal
is proportional both to the gyromagnetic ratio γ and to the number of spins of a
given species [1], most NMR diffusion studies have been performed on abundant,
strongly magnetic (i.e., large γ) spin systems rather than on systems which are
either weakly magnetic (small γ) or in low abundance (impurities). Recently,
Stokes and Ailion [2] developed a technique for studying the diffusion of abundant but weakly magnetic spins (S-spins) whose gyromagnetic ratio may be too
small for direct diffusion observations by NMR. In this paper, we shall discuss
the application of this technique to two cases in which the diffusing atom is of
low abundance: 1) interstitial diffusion and 2) impurity diffusion.

DIPOLAR RELAXATION

In order to understand the underlying principles of the Stokes-Ailion technique, consider first a diatomic salt consisting of two spin species, one of
which is strongly magnetic (the I spins) and the other is weakly magnetic (the S
spins). Suppose, for the moment, that the I spins are diffusing and that the S
spins are stationary. One way [3] to observe this diffusion would be to cool the
entire dipolar spin system (e.g., by adiabatic demagnetization in the rotating
frame) [4] so that each spin is aligned preferentially parallel to the local
dipolar field due to its neighbors. A diffusion jump would then cause a local
heating of the dipolar system, since the local field at the final site will in
general have a different orientation than that at the original site, with the
result that the spin orientation will no longer be preferentially parallel to
the local field. This local heating will then be transferred to the entire
dipolar system by a series of energy conserving spin-flips between neighboring
spins. This process of equilibration of dipolar temperature among all the spins
is called spin diffusion [5] and occurs at a rate of $\sim 10^4$ s^{-1}. If the process
of spin diffusion is rapid compared to the atomic jumping, then this dipolar
relaxation time T_{1D} will be of order of the average time τ_c between diffusion
jumps of each atom. (Actually, one can show [3] that a plot of T_{1D} vs. temperature will result in T_{1D} having a minimum value at the temperature for which
$\omega_D \tau_c \sim 1$, where $\hbar \omega_D$ is the average dipolar interaction strength experienced by
the I spins.) At temperatures below the T_{1D} minimum,

$$\frac{1}{T_{1D}} \approx \frac{1}{\tau_c} \quad . \tag{1}$$

Now consider that the weak S spins are diffusing but the I spins are stationary. If we attempt to detect the diffusion as before by measuring the temperature change of the entire dipolar reservoir, we see that this temperature change will be greatly attenuated from the result for diffusing I spins. This is because only the terms in the dipolar Hamiltonian involving I-S and S-S interactions change when an S atom jumps. However, the temperature change of the dipolar reservoir is determined by the heat capacity of the entire dipolar Hamiltonian, which consists of I-I, I-S, and S-S parts. For the case that I spins are strongly magnetic and S-spins are weak, this heat capacity will normally be dominated by I-I interactions. If the jumping is by S spins, the dipolar relaxation time T_{1D} is given [2] by

$$\frac{1}{T_{1D}} \simeq \frac{1}{\tau_c} \frac{H_{DIS}^2 + H_{DSS}^2}{H_{DII}^2 + H_{DIS}^2 + H_{DSS}^2} \approx \frac{1}{\tau_c} \frac{H_{DIS}^2}{H_{DII}^2} , \tag{2}$$

where H_{DII}, H_{DIS}, and H_{DSS} are the average dipolar local field terms arising from I-I, I-S, and S-S interactions, respectively. For S weak and I strong, we will normally have that

$$H_{DII}^2 > H_{DIS}^2 > H_{DSS}^2 . \tag{3}$$

Accordingly, we see from Eq. (2) that the dipolar relaxation rate T_{1D}^{-1} arising from the diffusion of weak S spins may be difficult to observe, since it will be reduced by the heat capacity ratio $(H_{DIS}/H_{DII})^2$ in comparison to the relaxation rate in Eq. (1) for jumping I-spins.

If, in fact, only a fraction of the S-spins participate in the jumping, as in interstitial diffusion, then the terms, H_{DIS}^2 and H_{DSS}^2, in the numerator of Eq. (2) should be replaced by $H_{DIS}'^2$, and $H_{DSS}'^2$, where these terms are, respectively, the square of the strengths of the dipolar local fields between the diffusing S spins and the I spins and between the diffusing S spins and all the other S spins. Thus,

$$\frac{1}{T_{1D}} \simeq \frac{1}{\tau_c} \frac{H_{DIS}'^2 + H_{DSS}'^2}{H_{DII}^2 + H_{DIS}^2 + H_{DSS}^2} . \tag{4}$$

Note that $H_{DIS}' = H_{DIS}$ and $H_{DSS}' = H_{DSS}$ in the case of the interstitialcy or vacancy diffusion mechanisms, which involve jumping of all the S spins, but that $H_{DIS}' \ll H_{DIS}$ and $H_{DSS}' \ll H_{DSS}$ for interstitial diffusion, which involves diffusion of only a fraction of all the S spins. In this case the heat capacity ratio will be even more unfavorable than in Eq. (2).

ENHANCEMENT OF RELAXATION DUE TO DIFFUSION OF WEAKLY MAGNETIC SPINS

In order to increase the NMR relaxation rate due to diffusion of weakly magnetic spins, Stokes and Ailion effectively caused the heat capacity due to I-I interactions to vanish in the denominator of Eqs. (2) and (4). In this case, Eq. (2) will be identical to Eq. (1) and the relaxation efficiency for S diffusion will be the same as for I diffusion. If the NMR experiment is performed by irradiating the I-spins and observing changes in the I spins' resonance signal, it is appropriate to describe the dipolar relaxation in the rotating frame of the I-spins. In this case, the transformed dipolar Hamiltonian is given [6] by

$$\mathcal{H}_D^{(00)} = -\frac{1}{2} (1 - 3\cos^2 \theta_I) \mathcal{H}_{DII}^{(0)} + \cos \theta_I \, \mathcal{H}_{DIS}^{(0)} + \mathcal{H}_{DSS}^{(0)} , \tag{5}$$

where θ_I is the angle between the static external field $\vec{H}_o$ and the rotating frame field $\vec{H}_{eff} = H_1\hat{i} + (H_o - \omega/\gamma)\hat{k}$. Furthermore, a rotating frame dipolar relaxation time $T_{1D'}$ can be defined in order to characterize relaxation of $\mathcal{H}_D^{(00)}$. If we replace the various dipolar interaction terms in Eq. (2) by their rotating frame counterparts, we see that $T_{1D'}$ is given by

$$\frac{1}{T_{1D'}} = \frac{1}{\tau_c} \frac{\cos^2\theta_I \, H_{DIS}'^2}{[\frac{1}{2}(3\cos^2\theta_I - 1)]^2 H_{DII}^2 + \cos^2\theta_I H_{DIS}^2} . \tag{6}$$

(In this expression, originally derived by Stokes and Ailion [2], it is assumed that the S spins are sufficiently weak that H_{DSS}^2 and $H_{DSS}'^2$ can be neglected compared to H_{DIS}^2 and $H_{DIS}'^2$ respectively.) The significant feature of both Eqs. (5) and (6) is that θ_I can be varied easily by varying either H_o or the rf frequency ω. Moreover, by varying θ_I, the relative contributions to $\mathcal{H}_D^{(00)}$ arising from I-I and I-S can be changed. For instance, far off-resonance ($\theta_I = 0$), $\mathcal{H}_D^{(00)} = \mathcal{H}_D^{(0)}$ and Eq. (6) for $T_{1D'}$ reduces to the value in Eq. (4), so that $T_{1D'}(\theta_I = 0) \equiv T_{1D}$. However, when $3\cos^2\theta_I - 1 = 0$ ($\theta_I = 54.7°$), the I-I interaction term in Eq. (5) and in the denominator of Eq. (6) will be zero; thus the heat capacity reduction factor of Eq. (2) will not be present for strong collision [3] diffusion mechanisms like vacancy and interstitialcy for which $H_{DIS}' = H_{DIS}$. At this "magic angle" the relaxation rate $T_{1D'}^{-1}$ due to the diffusion of weak spins will be substantially enhanced and will be comparable to the relaxation rate in Eq. (1) which would result from the diffusion of strong spins. Since the heat capacity of the I-I interactions is zero at this angle, the temperature change of the dipolar reservoir due to jumping of the S-spins is greatly increased. Even for interstitial diffusion for which $H_{DIS}' \ll H_{DIS}$, there will still be considerable enhancement at the magic angle of the relaxation rate of Eq. (6). On the other hand, by going "on-resonance" ($\theta_I = 90°$), we can eliminate the I-S interaction term and our sensitivity to the motion of the S-spins is greatly decreased. Equation (6) predicts, when S-spins are diffusing, a striking dependence of $T_{1D'}$ on θ_I which has been verified experimentally for K^+ vacancy diffusion in KF:0.1% CaF_2, both in single crystals [2] and in powdered samples [7]. By varying θ_I in Eq. (6) the relative sensitivity of $T_{1D'}$ to different kinds of motion can be changed; hence, such studies promise to be fruitful in distinguishing and studying separately the motion of both I and S spins in more complex systems. (Even though $T_{1D'}$ is measured in the the presence of a large H_1 (~ 30 gauss) [2], the measured relaxation time will not include the Zeeman heat capacity since the cross relaxation time between $\mathcal{H}_Z$ and $\mathcal{H}_D^{(00)}$ will be exceedingly long [8].)

THE NATURE OF Ag^+ DIFFUSION IN AgF

We shall now proceed to discuss how this $T_{1D'}$ technique, in conjunction with ionic conductivity measurements, can be used to distinguish interstitial from interstitialcy diffusion.

Recent NMR studies by Stokes and Ailion [2] and by Raaen et al. [9] of ^{19}F relaxation in AgF showed only a contribution arising from ^{19}F diffusion above 333K but no contribution from Ag^+ diffusion down to 125K. However, Raaen et. al. did report a large electrical conductivity that is almost certainly due to the rapid motion of Ag^+ ions. From their conductivity data one can estimate that a T_{1D} minimum due to Ag^+ diffusion should occur in the temperature range 120-150K. In order to find an Ag^+ contribution to the relaxation we have performed T_{1D} measurements in the temperature region extending down to 60K. However, as can be seen in Fig. 1, no T_{1D} minimum was found. In order to enhance the relaxation contribution due to Ag^+ diffusion we also measured $T_{1D'}$ at $\theta_I = 51°$ over the same temperature region but, as with T_{1D}, did not see effects of Ag^+ diffusion. (We believe that the relaxation in both T_{1D} and $T_{1D'}$ is due to paramagnetic impurities [10].) Furthermore, we measured $T_{1D'}$ vs. θ_I and failed to see the dependence on θ_I predicted by Eq. (6) for mechanisms like vacancy or interstitialcy Ag^+ diffusion. If, on the other hand, the Ag^+ diffusion were due to interstitial jumps, then the Ag^+ contribution to the relaxation would be weaker and would not necessarily be observable. The interstitial mechanism is a much less effective relaxation mechanism, since $H_{SI}^t \ll H_{SI}$ in Eqs. (4) and (6). From an NMR point of view, the essential difference between the interstitialcy or vacancy mechanisms and the interstitial mechanism is that, with the former mechanisms, the atoms of a given species can be characterized by a correlation time τ_c which is the average time each atom sits between jumps which strongly relax the magnetization. For the interstitial mechanism, there are "two" kinds

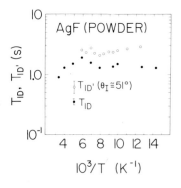

Fig. 1. $T_{1D'}(51°)$ and T_{1D} vs. $10^3/T$ in AgF (powder). Typical errors bars are indicated.

of atoms: those at the lattice site which make up the bulk of the NMR signal but are <u>not</u> jumping and the relatively low number of interstitial atoms whose jumping will relax the bulk atoms more weakly than if the bulk atoms themselves were jumping. Thus the contribution of interstitial diffusion to T_{1D} and even $T_{1D'}$ may still be masked by other mechanisms. In summary, the failure to see T_{1D} and $T_{1D'}$ minima in the presence of a large ionic conductivity indicates that the Ag^+ ions diffuse by a mechanism which gives rise to weak NMR relaxation, such as the interstitial mechanism or, possibly, by a short circuit route as with grain boundary diffusion.

IMPURITY DIFFUSION

An exciting possible application for the $T_{1D'}$ technique is to study the diffusion of impurity atoms by observing their effect on $T_{1D'}$ of the strong species. Figure 2 illustrates the kind of dependence on θ_I that would be expected for impurity diffusion for a number of different diffusion mechanisms. In this particular calculation the impurity atom is ^{27}Al and the NMR is performed on the ^{63}Cu isotope. A complication in this case is that Cu contains two isotopes ^{63}Cu and ^{65}Cu having relative abundances of 69% and 31%, respectively. Even though the I-I interaction of the ^{63}Cu spins can be eliminated by irradiation at the magic angle, the appreciable heat capacity arising from $^{65}C-^{65}Cu$ spins will remain; accordingly, the reduction of $T_{1D'}$ at 54.7° in Fig. 2 is less pronounced than in the case of only one species of I spins.

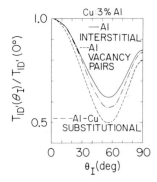

Fig. 2. $T_{1D'}(\theta_I)/T_{1D'}(0°)$ for various Al jump mechanisms in Cu:3% Al.

We attempted to see this effect in Cu:3% Al, since $T_{1\rho}$ measurements by Kanert on Al in a Cu-Al alloy indicate that the Al impurities diffuse faster than the Cu atoms [11], but unfortunately, we were unable to see any effects of Al diffusion in T_{1D} or $T_{1D'}$. The failure of this attempt is due to the fact that Cu has spin 3/2 and has therefore a quadrupole moment. This latter would not influence the validity of the $T_{1D'}$ method as long as the quadrupolar shifts of the levels are much less than the dipolar broadening of the spin levels so that rapid cross-relaxation can occur between the I and S spins. Al-impurities, however, are known to introduce quadrupolar shifts of the NMR frequencies of the nearby nearby Cu atoms [12]. These shifts exceed the dipolar broadening of the levels by several orders of magnitude and thus prevent a rapid transfer of energy from the hot S spins to the cooler I spins. Were this not the case, we would expect to see effects similar to those in Fig. 2.

$T_{1D'}$ MEASUREMENTS IN $(P_2O_5)_{0.8}(Li_2O)_{0.2}$ GLASS

In order to avoid the problems, described earlier, which arise if the I spins experience a strong quadrupole interaction, we chose next to study a $(P_2O_5)_{1-x}(Li_2O)_x$ glass, in which the I spin ^{31}P has spin 1/2 and thus no quadrupole interaction. A particularly attractive feature of this substance is that the Li concentration can be varied arbitrarily.

Göbel et al. [13] reported T_1 and $T_{1\rho}$ measurements on ^7Li in P_2O_5-Li_2O glasses of varying composition in the temperature region $20K < T < 570K$. Supported by diffusion data obtained by other techniques, they attributed their dipolar relaxation data to slow Li diffusion in glasses with similar composition to our sample at temperatures $T < 400K$.

In order to be able to make a comparison with the results of Göbel et al., we started with a glass with a Li_2O concentration that was high enough to do NMR measurements on both the Li and the P. To date we have measured the T_1 and T_{1D} of phosphorus in the temperature region $293K > T > 125K$ (see Fig. 3). The dipolar relaxation was non-exponential but could, within measuring error, be described by the sum of two exponentials having comparable amplitudes. (The T_{1D} data plotted in Fig. 3 corresponds to the initial rate of the decay. The time constant of the slow component is about one order of magnitude larger.) Such nonexponential behavior implies that the dipolar relaxation cannot be described by assuming a single dipolar temperature for the entire system. This may be due either to the fact that there is a strong spatial variation in relaxation rates in this sample or, possibly, to slowness in the spin diffusion process in a glass.

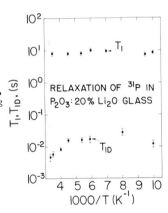

Fig. 3. T_1, T_{1D} of ^{31}P in P_2O_5:20% Li_2O glass.

Preliminary measurements of $T_{1D'}$ at a temperature of 298K showed unexpected behavior (Fig. 4). When θ_I was varied from 80° to 90°, the slower component appears to vanish and the relaxation behavior becomes approximately exponential with a time constant ~ T_{1D}, whereas at the other θ_I angles $T_{1D'}$ shows almost no dependence on θ_I. The time constants of both fast and slow components are given in Fig. 4. The weak θ_I dependence in the range 0°-80° may be understood in part from the fact that the condition given by Eq. (3) is not fulfilled, since H_{DSS}^2

540

is comparable to H_{DII}^2. Furthermore, since the dipolar coupling between the P and Li spins becomes very small at $\theta_I \simeq 90°$ [see Eq.(5)], the sudden change in relaxation behavior seems to indicate that the nonexponentiality at the other angles may be due to a dynamic coupling with an extra energy system (or constant of the motion) involving the Li nuclei. The fact that the initial rate is almost independent of θ_I indicates that the short component relaxation is probably governed by P-motions. This interpretation is further supported by the fact that these components have the same value at $\theta_I = 90°$ where the coupling between the P and the Li becomes very small.

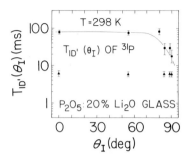

Fig. 4. Long (■) and short (▲) components of $T_{1D'}$. The curve is an approximate fit.

REFERENCES

*This work was supported by the U.S. National Science Foundation under Grant DMR 76-18966.

1.   C. P. Slichter, Principles of Magnetic Resonance, 2nd ed. (Springer-Verlag, Berlin, 1978).
2.   H. T. Stokes and D. C. Ailion, Phys. Rev. 18, 141 (1978).
3.   D. C. Ailion and C. P. Slichter, Phys. Rev. 137, A235 (1965).
4.   C. P. Slichter and W. C. Holton, Phys. Rev. 122, 1701 (1961).
5.   A. Abragam, The Principles of Nuclear Magnetism (Clarendon, Oxford, 1961) pp, 136-141.
6.   M. Goldman, Spin Temperature and Nuclear Magnetic Resonance in Solids (Clarendon, Oxford, 1970), p. 37.
7.   J. Beckett, J. Pourquié, and D. C. Ailion, in Proc. of XXIst AMPERE Congress (North Holland, Delft, 1980).
8.   H. T. Stokes and D. C. Ailion, Phys. Rev. B 16, 3056 (1977).
9.   A. M. Raaen, I. Svare and T. A. Fjeldly, Phys. Rev. B 21, 4895 (1980).
10.   J. Beckett, J. Pourquié, and D. C. Ailion, Phys. Rev. (to be published).
11.   O. Kanert, private correspondence.
12.   G. Gruner and M. Minier, Adv. Phys. 26, 231 (1977).
13.   E. Göbel, W. Müller-Warmuth, H. Olyschläger, and H. Dutz, J. Mag. Res. 36, 371 (1979).

Copyright 1981 by Elsevier North Holland, Inc.
Kaufmann and Shenoy, editors
Nuclear and Electron Resonance Spectroscopies Applied to Materials Science

A METHOD FOR DETERMINING IMPURITY-HOST FORCE-CONSTANT RATIOS

BEREND KOLK
Physics Department, Boston University, 111 Cummington Street, Boston MA 02215

ABSTRACT

One of the parameters of interest in studies on dilute
alloys is the ratio of the effective host-host to impurity-host interaction. A method is presented which
allows the evaluation of the effective impurity-host
force-constant ratio, A'/A, for substitutional, isolated, impurity atoms in a cubic lattice by combining
recoilless-fraction data (Mössbauer effect) of the impurity atoms with x-ray- or neutron-diffraction data
of the host lattice. With this method, values of A'/A
are obtained for dilute ^{57}Fe impurities in various
hosts.

INTRODUCTION

Mössbauer discovered in 1958 that there exists a non--zero probability f for
recoilless emission or absorption of a γ ray by a nucleus embedded in a solid
[1]. In these recoilless processes, no energy transfer occurs between the γ
ray and the lattice via creation or annihilation of phonons. Hence, the state
of the lattice before and after the nuclear transition is the same. The probability f is often referred to as the recoilless fraction, and its temperature
dependence is determined by

$$f(T) = exp[-2W'(T)] \qquad (1a)$$

where

$$2W'(T) = k^2 \langle x^2 \rangle_T \qquad (1b)$$

is called the Debye-Waller factor. Here $\langle x^2 \rangle_T$ represents the mean-square displacement (msd) of the γ-ray emitting or absorbing atom along the direction of
the γ-ray wave vector $\vec{k}$.

Hence, recoilless fraction measurements permit the determination of the msd
$\langle x^2 \rangle_T$. It is shown in the following section that from the temperature dependence of $\langle x^2 \rangle_T$, certain moments and corresponding anharmonic constants of the
phonon-frequency distribution can be derived. The msd of atoms in solids can
be evaluated from x-ray or neutron diffraction measurements as well. The temperature dependence of the intensity of the diffracted radiation is given by a
formula similar to eqn. (1a) with a Debye-Waller factor,

$$2W(T) = (\vec{k} - \vec{k}')^2 \langle x^2 \rangle_T \qquad (2)$$

where $\vec{k}$ and $\vec{k}'$ are the wave vectors of the incident and diffracted x-rays or
neutrons, and $\langle x^2 \rangle_T$ the msd of the atoms along the direction of $\vec{k} - \vec{k}'$. Note
that $|\vec{k} - \vec{k}'| = 4\pi sin\theta/\lambda$ where θ is the angle of incidence. The latter rela-
tion inserted in eqn. (2) yields the more familiar expression in x-ray litera-
ture,

$$2W(T) = 16\pi^2 \frac{sin^2\theta}{\lambda^2} \langle x^2 \rangle_T. \tag{3}$$

The recoilless fraction and diffraction measurements have each their specific advantages and disadvantages. Recoilless fraction measurements permit the study of the msd's of very dilute impurities in solids. However, the number of isotopes suitable for the Mössbauer effect (ME) is limited. Diffraction measurements are not very appropriate for determining msd's of very dilute impurities but are not restricted to a limited number of isotopes as is the ME.

Thus, the ME lends itself very well to the study of msd's of impurities with such dilute concentrations that the single-impurity approximation is valid, while the msd's of the host atoms can be determined from diffraction measurements. It is the purpose of this paper to show that from such measurements the force constant ratio A'/A can be derived. Here A' is the force constant between the impurity and its first nearest neighbor (nn), and A that between the host atom and its first nn.

THE MEAN-SQUARE DISPLACEMENT

In the x-ray literature the msd is expressed in terms of the Debye temperature θ_D. This quantity depends on the temperature T for two reasons. Firstly, it is assumed in the Debye model that the phonon frequency distribution is parabolic. This is a valid approximation only for the low frequency part of this distribution while appreciable deviations for the high frequency part exist. Because higher phonon frequencies become more weighted with increasing temperatures, the Debye cut-off frequency and the associated Debye temperature θ_D must be properly adjusted at each temperature. Secondly, θ_D is temperature dependent because of anharmonic effects. It is not easy to distinguish these two different contributions to the temperature dependence of θ_D. This makes it rather hard to study anharmonic effects. Moreover, each property of the lattice is associated with a different value of θ_D. Hence the Debye temperature is not a very useful quantity in lattice dynamical studies.

A better method of analyzing the msd and other lattice-dynamical properties consists of expressing these properties in terms of frequency moments.

$$\mu(n) = \int_0^\infty \omega^n g(\omega) d\omega, \tag{4}$$

or in the associated characteristic temperatures of the order $n(n \geq 3)$,

$$\theta(n) = \frac{\hbar}{k_B} \left\{ \frac{1}{3}(n + 3)\mu(n) \right\}^{1/n} \tag{5}$$

Here $g(\omega)$ is the phonon frequency distribution, and k_B the Boltzmann constant. For $n = 3$ and $n = 0$, the proper limits in eqn. (5) have to be taken. In the harmonic approximation, $\theta(n)$ is independent of temperature. In a real crystal, the phonons interact with each other because of the anharmonic terms in the vibrational Hamiltonian. This interaction leads to a temperature dependent shift in the phonon frequencies and to a finite mean life for each phonon. When the harmonic coupling between the phonons is relatively small, $\theta(n)$ can be expressed as [2]

$$\theta(n,T) = \theta_0(n)\{1 - \varepsilon(n)T\} \tag{6}$$

where $\theta_0(n)$ is the weighted temperature associated with the phonon-frequency distribution at $T = 0$, and $\varepsilon(n)$ the nth anharmonicity constant.

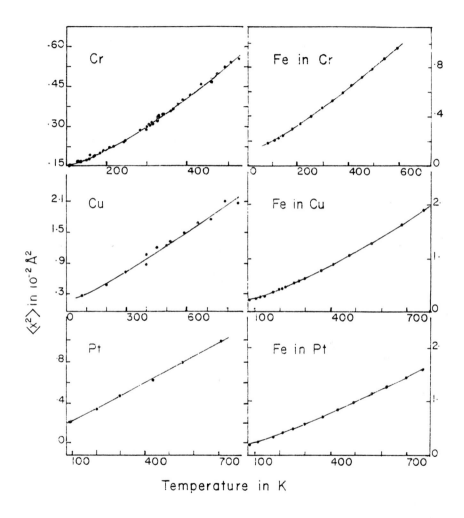

Figure 1: The curves represent least-square fits of eqn. (7) to x-ray diffraction data of the hosts and to recoilless fraction data of the ^{57}Fe impurity. The data were taken from the following references: Cr [9], Cu [10], Pt [11], FeCr [8], FeCu and FePt [4].

The msd of an atom in a monoatomic cubic lattice can be expressed for moderate and high temperatures $T > \theta_0(-2)/2\pi$ [3,4]

$$\left\langle x^2 \right\rangle_T = \frac{3\hbar^2}{mk_B} \frac{T}{[\theta_0(-2)]^2} \left[1 + 2\varepsilon(-2)T + \left\{ \frac{\theta_0(-2)}{6T} \right\}^2 \right] \qquad (7)$$

where m is the mass of the atom. X-ray diffraction data of chromium, copper, and platinum fitted with eqn. (7) are shown in fig. 1 and the corresponding values of $\theta_0(-2)$ and $\varepsilon(-2)$ are collected in table I. A more detailed discussion of the evaluation of these quantities from diffraction-line intensities is forthcoming.

The msd of an impurity atom is represented as well by eqn. (7) for temperatures $T > \theta(-2)/2\pi$, when $\theta_0(-2)$, $\varepsilon(-2)$ and m are replaced by $\theta'_0(-2)$, $\varepsilon'(-2)$ and m', respectively. Here $\theta'(-2)$ and $\varepsilon'(-2)$ are the characteristic temperature and anharmonic constant of the order -2, respectively, associated with the impurity, and m' is the mass of the impurity atom. Fits of msd's data of ^{57}Fe in chromium, copper, and platinum hosts are displayed in fig. 1 and corresponding values of $\theta'_0(-2)$ and $\varepsilon'(-2)$ are given in table I.

THE DETERMINATION OF A'/A

Various models have been developed for the msd of an isolated impurity in a solid. These models have been reviewed by Grow et al. [5]. The most sophisticated model at this moment is that of Mannheim [6], which provides an analytical expression for the msd of an isolated impurity in a cubic host for an unrestricted range of temperatures, masses and force-constant ratios. This is achieved by assuming that the lattice behaves harmonically and by considering only the first nn interactions. Grow et al. [5] derived from the Mannheim theory a useful expression:

$$\frac{A}{A'} = 1 + \left[\frac{m}{m'} \left\{ \frac{\theta_0(-2)}{\theta'_0(-2)} \right\}^2 - 1 \right] / \beta_{-2} \qquad (8)$$

where $\beta_{-2} = \mu(+2)\mu(-2)$. The values of β_{-2} lie in the range $\beta_{-2} = 0.6 \pm 0.1$. Values of β_{-2} for various metals are given in table I. Note that eqn. (8) incorporates effects due to the appearance of localized modes.

In the literature [5,7,8], Mannheim's initial expressions are evaluated using the host phonon-frequency distribution obtained from Born-von Kármán fits to dispersion relations, derived from inelastic neutron scattering data. Due to the anharmonic terms in the vibrational Hamiltonian, the phonon frequency distribution obtained in this way depends on the temperature T at which the inelastic neutron scattering measurement has been performed. This leads to ambiguities in the determination of A'/A (see [12]). Grow et al. [5] therefore define a pseudo-harmonic effective impurity-host force constant $(A'/A)_T$ which they derive from the recoilless fraction of the impurity at the same temperature at which the inelastic neutron scattering experiments of the host are performed. These pseudo-harmonic $(A'/A)_T$ ratios are represented by eqn. (8), when uncorrected (-2) characteristic temperatures, $\theta(-2,T)$ and $\theta'(-2,T)$ are used in this equation instead of $\theta_0(-2)$ and $\theta'_0(-2)$, i.e.

$$\left[\frac{A}{A'} \right]_T = 1 + \left[\frac{m}{m'} \left\{ \frac{\theta_0(-2)}{\theta'_0(-2)} \right\}^2 (1 - 2\Delta\varepsilon T) - 1 \right] (\beta_{-2})^{-1} \qquad (9)$$

where $\Delta\varepsilon = \varepsilon(-2) - \varepsilon'(-2)$. Hence, the pseudo-harmonic ratios $(A'/A)_T$ are temperature dependent unless $\Delta\varepsilon = 0$, which is not the case as an inspection of table I shows. Although the anharmonic effects on $\theta(-2,T)$ and $\theta'(-2,T)$ may be small, their impact on the value of $(A'/A)_T$ is appreciable because of the fact that the factor $(m/m')\{\theta_0(-2)/\theta'_0(-2)\}^2$ in eqn. (9) does not differ too much from one.

The advantage of using diffraction-line intensity data is that diffraction measurements are generally performed over a wide range of temperatures which permits the evaluation of $\theta_0(-2)$ and $\varepsilon(-2)$ of the host. This allows us to evaluate the harmonic impurity-host ratio A'/A which in contrast to the pseudo-harmonic ratio $(A'/A)_T$ is temperature independent. Using the values of $\theta_0(-2)$ and $\theta'_0(-2)$ given in table I, the ratios A'/A displayed in the lower part of this table are obtained. The temperature independent values of A'/A evaluated with the method presented here deviate from the pseudo-harmonic $(A'/A)_T$ values of Grow et al. [5]. In order to compare our results with those of Grow et al., the pseudo-harmonic ratios $(A'/A)_T$ were calculated by inserting the values given in the upper part of table I in eqn. (9). The ratios obtained are in agreement with those reported by Grow et al. (see table I).

Hence, by combining properly the information obtained using two different techniques, the Mössbauer effect and x-ray or neutron diffraction, it is possible to determine the temperature independent harmonic force constant ratio A'/A of a substitutional, isolated impurity in a cubic lattice.

TABLE I

The characteristic temperatures of the order -2 obtained for the host and ^{57}Fe impurity from fitting eqn. (7) to x-ray diffraction and to recoilless fraction data (see fig. 1). The values of β_{-2} were taken from Grow et al. [5]. The third column in the lower table yields the temperatures at which the pseudo-harmonic ratios (A'/A) of Grow et al. given in the fourth column were derived. In the fifth column the pseudo-harmonic ratios, obtained by inserting the values of the upper table in eqn. (9), are shown. The values of the temperature independent force-constant ratios determined by eqn. (8) are displayed in the last column.

	HOST		Fe IMPURITY	
HOST	$\theta_0(-2)\,[K]$	$2\varepsilon(-2)\,[10^{-4}K^{-1}]$	$\theta'_0(-2)\,[K]$	$2\varepsilon'(-2)\,[10^{-4}K^{-1}]$
chromium	633 ± 23	10.0 ± 2.0	449 ± 4	1.6 ± 0.03
copper	317 ± 10	1.7 ± 0.8	374 ± 1.3	2.4 ± 0.1
platinum	238 ± 8	0.1 ± 0.7	372 ± 3.0	1.07 ± 0.15

	β_{-2}	m/m'	$T\,[{}^\circ K]$	$(A'/A)_T$	$(A'/A)_T$	A'/A
chromium	0.691	0.91	300	0.70 ± 0.05	0.66 ± 0.16	0.46 ± 0.04
copper	0.559	1.12	300	1.22 ± 0.06	1.47 ± 0.22	1.54 ± 0.21
platinum	0.506	3.43	80	0.63 ± 0.06	0.55 ± 0.16	0.56 ± 0.16

546

ACKNOWLEDGMENTS

   Work supported by NSF Grant #DMR 77-19017.

REFERENCES

1. R. L. Mössbauer, Z. Phys. 151, 124 (1958); and Naturwissenschaften 45, 538 (1958).

2. L. S. Salter, Adv. in Physics 14, 1 (1965).

3. R. M. Housley and F. Hess, Phys. Rev. 146, 517 (1966).

4. R. H. Nussbaum, B. G. Howard, W. L. Nees and C. F. Steen, Phys. Rev. 173, 653 (1968).

5. J. M. Grow, D. G. Woward, R. H. Nussbaum and M. Takeo, Phys. Rev. 17, 15 (1978).

6. P. D. Mannheim, Phys. Rev. 165, 1011 (1968).

7. D. Ray and S. P. Puri, Phys. Lett. 33A, 306 (1970).

8. B. F. Brace, D. G. Howard and R. H. Nussbaum, Phys. Lett. 43A, 336 (1973).

9. R. H. Wilson, E. F. Skelton and J. L. Katz, Acta Cryst. 21, 635 (1966).

10. E. A. Owen and R. W. Williams, Proc. Roy. Soc. A 188, 509 (1947).

11. K. Alexopoulous, J. Boskovits, S. Mourikis and M. Roilos, Acta Cryst. 19, 349 (1965).

12. S. S. Cohen, R. H. Nussbaum and D. G. Howard, Phys. Rev. B12, 4095 (1975).

ISOTOPE EFFECT ON THE ELECTRONIC STRUCTURE OF HYDROGEN IN METALS

P. JENA AND C. B. SATTERTHWAITE
Physics Department, Virginia Commonwealth University, Richmond, VA 23284

ABSTRACT

   The effect of the isotopic mass on the redistribution of electron charge and
spin density around a light impurity such as hydrogen in metals has been studied.
Considered in this review are: (a) the magnetic hyperfine coupling to the hydrogen nucleus and the effect of isotopic mass, (b) the isotope effect in hydrogen impurity resistivity, (c) the reverse isotope effect in superconducting
PdH(D) and (d) the thermally induced detrapping of hydrogen from the vicinity of
solute sites in metallic hosts.

   The technological importance of the study of hydrogen in metals such as hydrogen storage, hydrogen embrittlement, and hydrogen-induced blistering has stimulated many theoretical and experimental investigations in recent years.[1] Considerable attention has been given to a fundamental understanding of the metal-
hydrogen interaction and the nature of induced electron distribution around this
light impurity. In this paper we focus our attention on a rather novel aspect
of the metal-hydrogen interaction--the quantum nature of the solute. The large
amplitude of the zero-point vibration of hydrogen and helium in their condensed
phase is known to give rise to quantum effects and the success of the quantum
theory of solids in accounting for experimental observations is well documented.[2]
Here, we emphasize the quantum effect of a light interstitial on the electronic
environment of its host. Since the positive muon, the lightest isotope of hydrogen has a mass of 1/9 proton mass and triton is three times as heavy as a
proton, it is clear that hydrogen and its isotopes are the best candidates for
studying the quantum effects. We review the experimental situation regarding
the nuclear spin-lattice relaxation rate, impurity resistivity, superconducting
transition temperature and solute trapping of hydrogen as it pertains to the
quantum-nature of hydrogen. A simple physical picture is provided to understand
the experimental trend.

   Since the electrons respond to an external perturbation almost instantaneously compared to the time it takes the impurity such as hydrogen either to execute
one complete vibration around its equilibrium site or to jump from one site to
another, the response is assumed to be adiabatic. Thus the electrons of the host
material are not supposed to be able to distinguish between the different isotopes of hydrogen. Such would be the case if the isotopes of hydrogen were vibrating in a region of space where the ambient charge distribution is homogeneous. From inelastic neutron scattering experiments[3] it can be deduced that the
root mean square amplitude of hydrogen in most metals may be as large as 0.25Å.
Using harmonic expression, the corresponding root mean square amplitudes of the
lightest (μ^+) and heaviest (triton) isotopes of hydrogen are 0.43Å and 0.19Å respectively. It was first pointed out by Miller and Satterthwaite[4] that this
large zero point vibration of hydrogen would bring it quite close to the host
metal atom and the resulting overlap between electron wave functions centered
around the hydrogen and metal atom site should have significant effect on the
electronic structure of hydrogen.

   Jena et. al.[5] have computed the electron density around a proton occupying
different sites inside a Pd-octahedron using the self-consistent molecular

cluster approach. Their result, shown in Fig. 1, clearly demonstrates that the electron density around the proton increases rapidly as the proton approaches the nearest neighbor metal atom. Thus a statistical average over all the proton configurations weighted according to its probability distribution would yield a net higher electron density at the vibrating proton site than that if the proton were static. Since the overlap effects are a direct consequence of the amplitude of vibration, it is evident that the electron density at the site of different isotopes of hydrogen would be different.

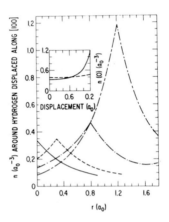

Fig. 1. Self-consistent molecular-cluster result for electron charge-density distribution along the (100) direction around a hydrogen atom located at (0,0,0) (——curve), (0.3,0,0) (---curve), (0.8,0,0) (—·—·curve), and (1.2,0,0) (– – — –curve). The inset shows a comparison between the electron charge density at the proton site in a molecular-cluster (solid curve) and pseudojellium (dashed curve) models.

The above isotope effect on the electronic structure of hydrogen can be visualized in a simple way using the local density approximation. In this approach we assume that the adiabatic response of the electron to the proton at any site is governed by the local ambient electron density at that site. Since the ambient electron density always increases as one approaches the host atom site from the interstitial location, the electron density at the proton site averaged over all configurations would be larger than that if the proton were static. This is

in agreement with the more sophisticated molecular cluster result described
earlier. With this simple picture, we now analyze various experimental
observations.

Nuclear spin-lattice relaxation rate:

In the Fermi contact interaction, the nuclear spin lattice relaxation rate
is given by

$$\frac{1}{T_1 T} = 4\pi h \gamma^2 \, k_B \, |N_s(E_F) H_{hfs}|^2 \tag{1}$$

where H_{hfs} is the hyperfine field at the probe nucleus and $N_s(E_F)$ is the density
of the s-electrons at the Fermi surface. The remaining symbols in Eq. (1) have
their usual meaning. The hyperfine field is proportional to the electron spin
density at the nucleus, $n^{\sigma}(o)$,

$$H_{hfs} \; \alpha \; \frac{n^{\uparrow}(o) - n^{\downarrow}(o)}{n^{\uparrow}_o - n^{\downarrow}_o} \; , \tag{2}$$

where n^{σ}_o is the ambient density of electrons with spin σ. NMR experiments[6] on
^{1}H and ^{2}D site in PdH and PdD yield $N_s(E_F) H_{hfs}$ to be 4.0×10^{-15} G. erg^{-1} and
4.09×10^{-15} G. erg^{-1} respectively. This difference cannot originate from band
structure effects in PdH and PdD since the lattice spacing between these two
materials differ only by 0.006Å. The quadrupolar contribution due to defects in
PdD has also been ruled out as a likely factor. Thus the experiment clearly
shows that the electron spin density at the deuteron site is significantly larger than that at the proton site.

This observation is a direct consequence of the isotope effect on the electronic structure of hydrogen. As has been noted above, as the point charge
moves closer to the metal atom, it encounters larger ambient electron density.
This leads to a more efficient screening of its charge and thus to a reduction
in the strength of the scattering potential. Since a proton has a larger zero-
point amplitude than the deuteron, its effective potential is weaker than that
of the deuteron. Consequently the perturbed spin density at the hydrogen site
is less than that at the deuterium site— in agreement with experimental result.
Jena et. al.[7] have studied this effect using both the real space analysis described above and the momentum space analysis. Their result is in quantitative
agreement with experiment in PdH and PdD.

Impurity resistivity due to hydrogen:

Since the scattering strength of the point charge decreases as its amplitude
of vibration increases, it is felt that its consequence could be observed in
the residual resistivity measurements in metal-hydrogen alloys. The impurity
resistivity due to scattering of free electrons by a very small concentration
of impurity atoms can be given by,

$$\Delta\rho = \frac{4\pi n_i h m}{k_F n_o e^2} \sum_{\ell=0}^{\infty} (\ell+1) \sin^2 |\eta_{\ell}(E_F) - \eta_{\ell+1}(E_F)| \, , \tag{3}$$

where n_i, n_o, and k_F are respectively the impurity concentration, ambient electron density at the impurity configuration and the Fermi wave vector. $\eta_{\ell}(E_F)$
is the scattering phase shift on the Fermi surface for the ℓth partial wave of
an electron with mass m and charge e. Two interesting effects can be studied
from Eq. (3). First, the isotope effect on the residual resistivity due to hydrogen and deuterium can be studied since $\eta_{\ell}(E_F)$ is directly related to the
strength of the scattering potential which in turn depends upon the amplitude of

the zero-point vibration. Second, the temperature dependence of $\Delta\rho$ can also be studied since the strength of the scattering potential is expected to decrease as the vibrational amplitude increases with temperature. Manninen and Jena[8] have studied this problem for hydrogen and deuterium in Pd and Ta matrix. Their result is summarized in Fig. 2.

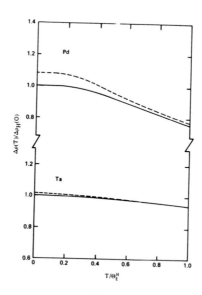

Fig. 2. Temperature dependence of the resistivity of hydrogen (solid lines) and deuterium (dashed lines) in Pd and Ta. The temperature is scaled to the Einstein temperature of hydrogen (θ_E^H) in both metals. The result for the hydrogen resistivity at $T=0K \Delta\rho(0)$, is 2.8 $\mu\Omega$-cm/at.% for Pd and 0.58$\mu\Omega$-cm/at.% for Ta.

Several interesting features should be noticed. (1) The residual resistivity due to deuterium is larger than that due to hydrogen. (2) The impurity resistivity due to both hydrogen and deutrium decreases with temperature, the difference between the two resistivities diminishing as temperature increases. (3) The magnitude of the slope of the temperature dependence is different in different metals and is largest for systems where the localized mode frequency of hydrogen is smallest. There have been many experiments where the electrical resistivity of metal-hydrogen alloys have been studied as a function of temperature.[9] However, in many instances, the quality of the data does not allow an unambiguous determination of the effect of thermal vibration of hydrogen on impurity resistivity. Some of the recent experiments agree with the prediction in Fig. 2, although there are systems where opposite isotope effect and temperature depen-

dence of the impurity resistivity have been observed. The present model ne-
glects anharmonic effects, possible trapping of hydrogen by other impurities in
the host metal and changes in the geometry and force constants due to hydrogen
in the host material. Improved theoretical calculations and precision experiments should throw considerable light on the isotope effect.

Superconducting transition temperature:
   The onset of superconductivity due to hydrogenation of metals and the change
in the superconducting transition temperature, T_c between metal-hydride and
deuteride have stirred considerable excitement in recent years. Using McMillan's
strong coupling formula,[10] it can be shown that

$$\frac{\Delta T_c}{T_c} = \frac{\Delta \Theta_D}{\Theta_D} + \frac{1.04(1+.38\mu^*)\lambda}{\lambda - \mu^*(1+0.62\lambda)^2} \frac{\Delta \lambda}{\lambda} \qquad (4)$$

where Θ_D, μ^* and λ are respectively the Debye temperature, conventional coulomb pseudopotential, and electron-phonon coupling parameter. ΔT_c is the observed change in the transition temperature. Higher value of T_c in the deuteride
phase compared to the hydride phase has been explained[11] to be predominantly due
to the force constant changes associated with the different mass of hydrogen and
deuterium. The effect due to zero-point vibration on T_c because of a redistribution of electrons around ^{1}H and ^{2}D as pointed out by Miller and Satterthwaite[4]
has yet to be considered.

Thermal induced detrapping of hydrogen from impurities:
   There are several experiments[12] that suggest that hydrogen can be trapped by
impurities in metal if the impurity and host atom sizes are different. The
strain caused by the impurity is believed to be the major factor for trapping
and may be responsible for increased solubility of hydrogen in certain binary
alloys. At some critical temperature, one expects hydrogen to be detrapped from
the impurity. This detrapping temperature may give some indication regarding
the impurity-hydrogen binding. This effect can be seen by measuring the quadrupole interaction at the impurity site in a cubic metal host as a function of
temperature. At low temperatures the presence of hydrogen in the vicinity of
the impurity will destroy the cubic symmetry giving rise to an electric field
gradient. Another interesting aspect is to study the effect when hydrogen is
replaced by deuterium. To appreciate this, imagine a simple picture where the
probability of escape is given by $\nu e^{-E/kT}$. Here ν is the attempt frequency and
E is the energy of the point charge with mass M. Assuming that the potential
well binding the hydrogen or the deuterium to the impurity is the same, it can
be easily shown that for equal probability of escape,

$$T_H/T_D \simeq E_H/E_D. \qquad (5)$$

Here T_H (T_D) is the temperature at which hydrogen (deuterium) is detrapped from
the impurity and E_H (E_D) is the corresponding energy level in the potential well
for hydrogen (deuterium). Since a heavier point charge lies deeper in the potential well, Eq. (5) suggests that deuterium will be detrapped from the impurity at higher temperature than hydrogen. The quantitative prediction of the
ratio in Eq. (5) would require an accurate knowledge of the trapping potential.
NMR and Mössbauer experiments[13] can shed considerable light on our understanding
of thermally induced detrapping of hydrogen.

There are other quantum effects of hydrogen not illustrated here. The most notable one being the diffusion of μ^+, proton and deuteron. This is a rather large field and the reader is referred to some recent reviews[14] in this field. The purpose of this paper has been to focus our attention on various electronic properties of metal-hydrogen alloys where isotope dependence can be studied. It is hoped that this concise review would stimulate further experimental and theoretical work in this field.

This work is supported in part by National Science Foundation.

REFERENCES

1. See G. Alefeld and J. Volkl, editors. "Hydrogen in Metals" (Topics in applied Physics, Springer, Berlin), vol. 28, 1978.

2. L. H. Nosanow, Phys. Rev. 146, 120 (1966) and references there in.

3. J. M. Rowe, J. J. Rush, H. G. Smith, M. Mostoller, Phys. Rev. Lett. 33, 1297 (1974).

4. R. J. Miller and C. B. Satterthwaite, Phys. Rev. Lett. 34, 144 (1975).

5. P. Jena, F. Y. Fradin, and D. E. Ellis, Phys. Rev. B20, 3543 (1979).

6. C. L. Wiley and F. Y. Fradin, Phys. Rev. B17. 3462 (1978).

7. P. Jena, C. L. Wiley, and F. Y. Fradin, Phys. Rev. Lett. 40, 578 (1978).

8. M. Manninen and P. Jena, Solid St. Commun, 34, 179 (1980).

9. See Ref. 8 for an experimental survey.

10. W. L. McMillan, Phys. Rev. 167, 331 (1968).

11. D. A. Papaconstantopoulos, B. M. Klein, E. N. Economou and L. L. Boyer, Phys. Rev. B17, 141 (1978). and references there in.

12. D. G. Westlake and J. F. Miller, J. Phys. F10, 859 (1980).

13. P. Boolchand (private communication).

14. See K. W. Kehr in Ref. 1.

AUTHOR INDEX

SUBJECT INDEX

(Page citations refer to the first page of
the article in which the subject occurs)